VOLUME FIVE HUNDRED-ONE

METHODS IN ENZYMOLOGY

Serpin Structure and Evolution

METHODS IN ENZYMOLOGY

Editors-in-Chief

JOHN N. ABELSON AND MELVIN I. SIMON

Division of Biology
California Institute of Technology
Pasadena, California

Founding Editors

SIDNEY P. COLOWICK AND NATHAN O. KAPLAN

VOLUME FIVE HUNDRED-ONE

METHODS IN ENZYMOLOGY

Serpin Structure and Evolution

EDITED BY

JAMES C. WHISSTOCK AND PHILLIP I. BIRD

Department of Biochemistry and Molecular Biology
Monash University, Clayton
Victoria, Australia

[redacted]

ELSEVIER

AMSTERDAM • BOSTON • HEIDELBERG • LONDON
NEW YORK • OXFORD • PARIS • SAN DIEGO
SAN FRANCISCO • SINGAPORE • SYDNEY • TOKYO
Academic Press is an imprint of Elsevier

Academic Press is an imprint of Elsevier
525 B Street, Suite 1900, San Diego, CA 92101-4495, USA
225 Wyman Street, Waltham, MA 02451, USA
32 Jamestown Road, London NW1 7BY, UK

First edition 2011

For information on all Academic Press publications
visit our website at elsevierdirect.com

ISBN: 978-0-12-385950-1
ISSN: 0076-6879

Printed and bound in United States of America
11 12 13 14 10 9 8 7 6 5 4 3 2 1

Contents

Contributors *xi*
Preface *xvii*
Volumes in Series *xix*

1. Intracellular Production of Recombinant Serpins in Yeast **1**

Dion Kaiserman, Corinne Hitchen, Vita Levina, Stephen P. Bottomley, and Phillip I. Bird

1. Introduction 1
2. Selection of Strain and Expression Plasmid 2
3. Growth of Yeast 3
4. Transformation of Yeast 4
5. Screening Transformants 4
6. Large-Scale Growth and Induction 5
7. Lysis 6
8. Purification 7
9. Assessing Serpin Activity and Removing Inactive Forms 9
10. Production of Polymerogenic Serpins 10

References 12

2. Production of Recombinant Serpins in *Escherichia coli* **13**

Mary C. Pearce and Lisa D. Cabrita

1. Introduction 14
2. Experimental Procedures for the Production of Serpins in *E. coli* 20
3. Some Common Variations Used for Expression of Other Serpins 21
4. Preparation of AT from Inclusion Bodies 23
5. Preparation of Soluble AT 24

References 27

3. Isolation and Characterization of the Nuclear Serpin MENT **29**

Sergei Grigoryev and Sheena McGowan

1. Introduction 30
2. Purification of the MENT Protein 31
3. Protease Inhibition/Serpin Activity 34

4. MENT Interaction with DNA and Chromatin *In Vitro* 36
5. Analysis of MENT Association with Native Chromatin *In Situ* 42
Acknowledgments 46
References 46

4. Solving Serpin Crystal Structures 49

Randy J. Read, Aiwu Zhou, and Penelope E. Stein

1. Introduction 50
2. Protein Production and Purification 50
3. Modifications to Aid Crystallization 51
4. Crystallization 51
5. Experimental Phasing 52
6. Molecular Replacement 53
7. Phase Improvement by Density Modification 57
8. Refinement and Validation 58
References 59

5. Crystallography of Serpins and Serpin Complexes 63

M. A. Dunstone and James C. Whisstock

1. Introduction 64
2. First Glimpses of Serpin Structures 64
3. The Serpin–Enzyme Complex 64
4. The Michaelis Complex 76
5. Conformational Control of Serpins—Antithrombin and Heparin 77
6. Nonconventional Serpin Complexes 77
7. Conformational Change and the Formation of the Latent Conformation 78
8. Abnormal Conformational Change—The δ-Form 79
9. Serpin Polymers 79
10. Crystallization of Serpins and Serpin Complexes 81
11. Conclusions 82
References 82

6. Serpins as Hormone Carriers: Modulation of Release 89

Robin Carrell, Xiaoqiang Qi, and Aiwu Zhou

1. Serpins and Allosteric Modulation 89
2. Hormone Carriage—TBG and CBG 90
3. Temperature Response: A Protein Thermocouple 95
4. Angiotensinogen and Its Interaction with Renin 96
Acknowledgments 100
References 101

7. Serpin–Glycosaminoglycan Interactions **105**

Chantelle M. Rein, Umesh R. Desai, and Frank C. Church

1. Quantitative Methods 106
2. Qualitative Methods 121
3. Animal Models 127

Acknowledgments 131
References 131

8. Targeting Serpins in High-Throughput and Structure-Based Drug Design **139**

Yi-Pin Chang, Ravi Mahadeva, Anathe O. M. Patschull, Irene Nobeli, Ugo I. Ekeowa, Adam R. McKay, Konstantinos Thalassinos, James A. Irving, Imran Haq, Mun Peak Nyon, John Christodoulou, Adriana Ordóñez, Elena Miranda, and Bibek Gooptu

1. Introduction 140
2. Targeting the s4A Site with Peptides in a Pathogenic Variant of α_1-antitrypsin 142
3. Computational Approaches 148
4. *In vitro* Screening of Small Molecules 155
5. Mammalian Cell Models and Beyond 166
6. Conclusion 168

References 169

9. Development of Inhibitors of Plasminogen Activator Inhibitor-1 **177**

Shih-Hon Li and Daniel A. Lawrence

1. Introduction 178
2. Serpins as Drug Targets 179
3. Development of PAI-1 Inhibitors 184
4. Concluding Remarks 200

References 200

10. Bioinformatic Approaches for the Identification of Serpin Genes with Multiple Reactive Site Loop Coding Exons **209**

Stefan Börner and Hermann Ragg

1. Introduction 210
2. Procedure for Identification of Serpin Genes with mRSL Cassette Exons 211
3. Conclusion 219

Acknowledgment 220
References 220

11. Methods to Measure the Kinetics of Protease Inhibition by Serpins **223**

Anita J. Horvath, Bernadine G. C. Lu, Robert N. Pike, and Stephen P. Bottomley

1. Introduction 223
2. Determining the Rate of Protease Inhibition (k_a) 226
3. Efficiency of the Serpin Inhibitory Reaction 230

Acknowledgments 233
References 233

12. Predicting Serpin/Protease Interactions **237**

Jiangning Song, Antony Y. Matthews, Cyril F. Reboul, Dion Kaiserman, Robert N. Pike, Phillip I. Bird, and James C. Whisstock

1. Introduction 238
2. Phage Display Methods 242
3. Sequence Analysis Methods 256
4. Concluding Remarks and Perspective 269

Acknowledgments 270
References 270

13. Amino-Terminal Oriented Mass Spectrometry of Substrates (ATOMS): N-Terminal Sequencing of Proteins and Proteolytic Cleavage Sites by Quantitative Mass Spectrometry **275**

Alain Doucet and Christopher M. Overall

1. Introduction 276
2. Overview of ATOMS 277
3. Control Experiment to Determine the Ratio Cutoff and Identify Natural N-Termini, Basal Proteolytic Products, and Outliers 281
4. Limited Proteolytic Processing of the Target Protein by the Test Protease *In Vitro* 282
5. Isotopic Labeling and Tryptic Digestion 283
6. Identification of Peptides by Liquid Chromatography-Tandem Mass Spectrometry 286
7. Mass Spectrometry Data Analysis 287
8. Discussion: Measuring the Effect of Protease Inhibitors on the Generation of Proteolytic Fragments 291

Acknowledgments 292
References 292

14. Computational Methods for Studying Serpin Conformational Change and Structural Plasticity **295**

Itamar Kass, Cyril F. Reboul, and Ashley M. Buckle

1. Introduction 296
2. Local to Global Dynamics Simulations 299
3. Pushing the Limits—Improving Conformational Sampling 304
4. I Know Where to Go—Directed Simulations 307
5. Nondynamical Methods 308
6. Software 309
7. Force Fields 311
8. Hardware 313
9. Case Study 314
10. Outlook 316

References 319

15. Probing Serpin Conformational Change Using Mass Spectrometry and Related Methods **325**

Yuko Tsutsui, Anindya Sarkar, and Patrick L. Wintrode

1. Introduction 326
2. Applications of HXMS 331
3. Determination of Thermodynamic Stability Using Hydrogen–Deuterium Exchange Combined with Mass Spectrometry 336
4. "Functional Unfolding" During the Native → Cleaved Transition 343
5. Investigating the Polymerization Pathway and Polymer Structure of α_1-AT by HXMS and Ion Mobility MS 344
6. Future Prospects 348

References 348

16. Determining Serpin Conformational Distributions with Single Molecule Fluorescence **351**

Nicole Mushero and Anne Gershenson

1. Introduction 353
2. Labeling Serpins and Proteases with Fluorophores 354
3. Overview of Single Molecule Fluorescence Techniques 355
4. Serpin Polymerization 358
5. Conformational Distributions of Protease–Serpin Complexes 368
6. Conclusions and Future Directions 373

Acknowledgments 373

References 374

17. Serpin Polymerization *In Vitro* **379**

James A. Huntington and Masayuki Yamasaki

1. Introduction 380
2. Methods of Inducing Polymerization 383
3. Kinetics of Polymerization 395
4. Effect of Mutations and "Drugs" on Polymerization 402
5. Mechanisms of Polymerization 408
6. Conclusions 415

Acknowledgments 416
References 417

18. The Serpinopathies: Studying Serpin Polymerization *In Vivo* **421**

James A. Irving, Ugo I. Ekeowa, Didier Belorgey, Imran Haq, Bibek Gooptu, Elena Miranda, Juan Pérez, Benoit D. Roussel, Adriana Ordóñez, Lucy E. Dalton, Sally E. Thomas, Stefan J. Marciniak, Helen Parfrey, Edwin R. Chilvers, Jeffrey H. Teckman, Sam Alam, Ravi Mahadeva, S. Tamir Rashid, Ludovic Vallier, and David A. Lomas

1. Introduction to Serpin Polymers and the Serpinopathies—David Lomas 423
2. Biophysical Techniques to Assess Serpin Polymers Formed *In Vivo*—James Irving, Ugo Ekeowa, Didier Belorgey, and Imran Haq 424
3. Assessment of Serpin Polymers by Electron Microscopy—Bibek Gooptu 428
4. Development of mAbs to Aberrant Conformers of α_1-Antitrypsin and Neuroserpin—Elena Miranda and Juan Pérez 436
5. Development of Cell Models to Assess the Polymerization of Antitrypsin—Adriana Ordóñez 441
6. Development of Cell Models to Assess the Polymerization of Neuroserpin—Elena Miranda, Juan Perez, and Benoit Roussel 443
7. Detection of the UPR and the OPR in the Serpinopathies—Lucy Dalton, Sally Thomas, Benoit Roussel, and Stefan Marciniak 445
8. Characterization of the Interaction Between Serpin Polymers and Neutrophils—Helen Parfrey and Edwin Chilvers 448
9. The Use of Transgenic Mice to Assess the Hepatic Consequences of Serpin Polymerization. Jeff Teckman 451
10. The Use of Transgenic Mice to Assess the Pulmonary Consequences of Serpin Polymerization—Sam Alam and Ravi Mahadeva 453
11. Characterization of Serpin Polymerization Using iPS to Generate Hepatocyte-Like Cell Lines—Tamir Rashid and Ludovic Vallier 457

Acknowledgments 461
References 461

Author Index *467*
Subject Index *493*

Contributors

Sam Alam
Respiratory Medicine Division, Department of Medicine, University of Cambridge School of Clinical Medicine, Addenbrooke's and Papworth Hospitals, Cambridge, United Kingdom

Stefan Börner
Department of Biotechnology, Faculty of Technology, Bielefeld University, Bielefeld, Germany

Didier Belorgey
Department of Medicine, Cambridge Institute for Medical Research, University of Cambridge, Cambridge, United Kingdom

Phillip I. Bird
Department of Biochemistry and Molecular Biology, Monash University, Melbourne, Victoria, Australia

Stephen P. Bottomley
Department of Biochemistry and Molecular Biology, Monash University, Melbourne, Victoria, Australia

Ashley M. Buckle
Department of Biochemistry and Molecular Biology, Monash University, Melbourne, Victoria, Australia

Lisa D. Cabrita
Department of Structural and Molecular Biology, University College London, London, United Kingdom

Robin Carrell
Cambridge Institute for Medical Research, University of Cambridge, Cambridge, United Kingdom

Yi-Pin Chang
Chemistry Research Laboratory, Department of Chemistry, University of Oxford, Oxford, United Kingdom

Edwin R. Chilvers
Respiratory Medicine Division, Department of Medicine, University of Cambridge School of Clinical Medicine, Addenbrooke's and Papworth Hospitals, Cambridge, United Kingdom

John Christodoulou
ISMB/UCL, Research Department of Structural & Molecular Biology, University College London, London, United Kingdom

Frank C. Church
Department of Pathology and Laboratory Medicine; Department of Medicine; Department of Pharmacology, and UNC McAllister Heart Institute, The University of North Carolina at Chapel Hill, Chapel Hill, North Carolina, USA

Lucy E. Dalton
Department of Medicine, Cambridge Institute for Medical Research, University of Cambridge, Cambridge, United Kingdom

Umesh R. Desai
Department of Medicinal Chemistry and Institute for Structural Biology and Drug Discovery, Virginia Commonwealth University, Richmond, Virginia, USA

Alain Doucet
Centre for Blood Research, Life Sciences Institute, University of British Columbia, Vancouver, British Columbia, Canada

M. A. Dunstone
Department of Biochemistry and Molecular Biology, Monash University, Melbourne, and Department of Microbiology, Monash University, Clayton, Victoria, Australia

Ugo I. Ekeowa
Department of Medicine, Cambridge Institute for Medical Research, University of Cambridge, Cambridge, United Kingdom

Anne Gershenson
Department of Biochemistry and Molecular Biology, University of Massachusetts, Amherst, Massachusetts, USA

Bibek Gooptu
ISMB/Birkbeck, Crystallography, Department of Biological Sciences, Birkbeck College, London, United Kingdom

Sergei Grigoryev
Department of Biochemistry & Molecular Biology, Penn State University College of Medicine, Milton S. Hershey Medical Center, University Drive, Hershey, Pennsylvania, USA

Imran Haq
Department of Medicine, Cambridge Institute for Medical Research, University of Cambridge, Cambridge, United Kingdom

Corinne Hitchen
Department of Biochemistry and Molecular Biology, Monash University, Melbourne, Victoria, Australia

Anita J. Horvath
Australian Centre for Blood Diseases, Monash University, Melbourne, Victoria, Australia

James A. Huntington
Department of Haematology, Cambridge Institute for Medical Research, University of Cambridge, Cambridge, United Kingdom

James A. Irving
Department of Medicine, Cambridge Institute for Medical Research, University of Cambridge, Cambridge, United Kingdom

Dion Kaiserman
Department of Biochemistry and Molecular Biology, Monash University, Melbourne, Victoria, Australia

Itamar Kass
Department of Biochemistry and Molecular Biology, Monash University, Melbourne, Victoria, Australia

Daniel A. Lawrence
Division of Cardiovascular Medicine, Department of Internal Medicine, University of Michigan Medical School, Ann Arbor, Michigan, USA

Vita Levina
Department of Biochemistry, La Trobe Institute for Molecular Science, Bundoora, Victoria, Australia

Shih-Hon Li
Department of Pathology, University of Michigan Medical School, Ann Arbor, Michigan, USA

David A. Lomas
Department of Medicine, Cambridge Institute for Medical Research, University of Cambridge, and Respiratory Medicine Division, Department of Medicine, University of Cambridge School of Clinical Medicine, Addenbrooke's and Papworth Hospitals, Cambridge, United Kingdom

Bernadine G. C. Lu
Australian Centre for Blood Diseases, Monash University, Melbourne, Victoria, Australia

Ravi Mahadeva
Respiratory Medicine Division, Department of Medicine, University of Cambridge School of Clinical Medicine, Addenbrooke's and Papworth Hospitals, Cambridge, United Kingdom

Stefan J. Marciniak
Department of Medicine, Cambridge Institute for Medical Research, University of Cambridge, and Respiratory Medicine Division, Department of Medicine, University of Cambridge School of Clinical Medicine, Addenbrooke's and Papworth Hospitals, Cambridge, United Kingdom

Antony Y. Matthews
Department of Biochemistry and Molecular Biology, Monash University, Melbourne, Victoria, Australia

Sheena McGowan
Department of Biochemistry and Molecular Biology and ARC Centre of Excellence for Structural and Functional Microbial Genomics, Monash University, Clayton, Victoria, Australia

Adam R. McKay
ISMB/UCL, Research Department of Structural & Molecular Biology, University College London, London, United Kingdom

Elena Miranda
Dipartimento di Biologia e Biotecnologie 'Charles Darwin', Università di Roma La Sapienza, Piazzale Aldo Moro 5, Roma, Italy

Nicole Mushero
University of Massachusetts, School of Medicine, Worcester, Massachusetts, USA

Irene Nobeli
ISMB/Birkbeck, Crystallography, Department of Biological Sciences, Birkbeck College, London, United Kingdom

Mun Peak Nyon
ISMB/Birkbeck, Crystallography, Department of Biological Sciences, Birkbeck College, London, United Kingdom

Adriana Ordóñez
Department of Medicine, Cambridge Institute for Medical Research, University of Cambridge, Cambridge, United Kingdom

Christopher M. Overall
Centre for Blood Research, Life Sciences Institute, University of British Columbia, Vancouver, British Columbia, Canada

Juan Pérez
Departamento de Biología Celular, Genética y Fisiología, Universidad de Málaga, Facultad de Ciencias, Campus de Teatinos, Malaga, Spain

Helen Parfrey
Respiratory Medicine Division, Department of Medicine, University of Cambridge School of Clinical Medicine, Addenbrooke's and Papworth Hospitals, Cambridge, United Kingdom

Anathe O.M. Patschull
ISMB/Birkbeck, Crystallography, Department of Biological Sciences, Birkbeck College, London, United Kingdom

Mary C. Pearce
Department of Biochemistry, Monash University, Clayton, Australia

Robert N. Pike
Department of Biochemistry and Molecular Biology, Monash University, Melbourne, Victoria, Australia

Xiaoqiang Qi
Cambridge Institute for Medical Research, University of Cambridge, Cambridge, United Kingdom

Hermann Ragg
Department of Biotechnology, Faculty of Technology, Bielefeld University, Bielefeld, Germany

S. Tamir Rashid
Department of Medicine, Cambridge Institute for Medical Research, and Laboratory for Regenerative Medicine, University of Cambridge, Cambridge, United Kingdom

Randy J. Read
Department of Haematology, Cambridge Institute for Medical Research, University of Cambridge, Cambridge, United Kingdom

Cyril F. Reboul
Department of Biochemistry and Molecular Biology, Monash University, Melbourne, Victoria, Australia

Chantelle M. Rein
Department of Pathology and Laboratory Medicine, and Department of Medicine, The University of North Carolina at Chapel Hill, Chapel Hill, North Carolina, USA

Benoit D. Roussel
Department of Medicine, Cambridge Institute for Medical Research, University of Cambridge, Cambridge, United Kingdom

Anindya Sarkar
Department of Physiology & Biophysics, Case Western Reserve University, Cleveland, Ohio, USA

Jiangning Song
Department of Biochemistry and Molecular Biology, Monash University, Melbourne, Victoria, Australia, and State Engineering Laboratory for Industrial

Enzymes and Key Laboratory of Systems Microbial Biotechnology, Tianjin Institute of Industrial Biotechnology, Chinese Academy of Sciences, Tianjin, China

Penelope E. Stein
Department of Medicine, University of Cambridge, Addenbrooke's Hospital, Cambridge, United Kingdom

Jeffrey H. Teckman
St. Louis University School of Medicine, Cardinal Glennon Children's Hospital, St. Louis, Missouri, USA

Konstantinos Thalassinos
ISMB/UCL, Research Department of Structural & Molecular Biology, University College London, London, United Kingdom

Sally E. Thomas
Department of Medicine, Cambridge Institute for Medical Research, University of Cambridge, Cambridge, United Kingdom

Yuko Tsutsui
Division of Chemistry and Chemical Engineering, California Institute of Technology, Pasadena, California, USA

Ludovic Vallier
Laboratory for Regenerative Medicine, University of Cambridge, Cambridge, United Kingdom

James C. Whisstock
Department of Biochemistry and Molecular Biology, Monash University, Melbourne, and ARC Centre of Excellence in Structural and Functional Microbial Genomics, Monash University, Clayton, Victoria, Australia

Patrick L. Wintrode
Department of Physiology & Biophysics, Case Western Reserve University, Cleveland, Ohio, USA

Masayuki Yamasaki
Department of Haematology, Cambridge Institute for Medical Research, University of Cambridge, Cambridge, United Kingdom

Aiwu Zhou
Department of Haematology, and Cambridge Institute for Medical Research, University of Cambridge, Cambridge, United Kingdom

PREFACE

Serpins comprise a family of proteins that has fascinated investigators for over 60 years. Besides highlighting the importance of the homeostatic regulation of proteolysis, serpin research has yielded precious insights into protein–protein interactions and protein conformational change, and the relationship between protein misfolding and disease. Further, the family contains striking examples of the use of a common protein fold for different biological purposes: besides protease inhibition, serpins are used as hormone precursors, hormone carriers, and protein-folding chaperones. Insights into the flexibility of the serpin fold are pointing the way toward development of therapeutics that will ameliorate some common human afflictions.

At the time of writing, there are more than 1000 serpins, mostly appearing as predicted proteins derived from genome sequencing projects. The majority of these are anticipated to be protease inhibitors, based upon the presence of conserved motifs in the crucial reactive center loop region. It is, however, evident that serpins occur in all kingdoms of life and perform a wide array of evolutionary and biological roles (including important functions outside the inhibition of proteases). This diversity is reflected in the broad scope and multiple systems covered by the chapters in these volumes. The challenge for the field is to elucidate the role of each particular serpin in its own niche, and this will typically require examination of structure, tissue distribution and cellular localization, identification of partners or targets, and assessment of physiological significance via forward or reverse genetics. Examples of these approaches will be found within these pages.

What then are the "big questions" or achievements remaining in serpin research? We would nominate (in no particular order) the precise physiological mechanism of serpin polymerization and associated cytotoxicity, the roles of noninhibitory serpins and the contribution of conformational change to their function, and the development of additional targeted anti-serpin therapeutics. Finally, as we move into the era of the 1000 genomes project and the routine sequencing of human genomes for medical purposes, we expect that a greater number of serpin mutations will be associated with a wide range of human diseases. The next generation of serpin researchers will make great inroads into all these problems.

We acknowledge and thank all the authors for contributing chapters that neatly illustrate the state-of-play in our understanding of serpins and the cutting-edge approaches from a variety of disciplines that are being used to expand the frontiers of our knowledge. We hope that these chapters in these two volumes will prove a very useful resource for both seasoned serpinologists and new hands.

In closing, we dedicate these volumes to our teachers, mentors, colleagues, and students, who continue to inspire us in our personal journeys of discovery.

March 2011

James C. Whisstock and Phillip I. Bird

Methods in Enzymology

Volume I. Preparation and Assay of Enzymes
Edited by Sidney P. Colowick and Nathan O. Kaplan

Volume II. Preparation and Assay of Enzymes
Edited by Sidney P. Colowick and Nathan O. Kaplan

Volume III. Preparation and Assay of Substrates
Edited by Sidney P. Colowick and Nathan O. Kaplan

Volume IV. Special Techniques for the Enzymologist
Edited by Sidney P. Colowick and Nathan O. Kaplan

Volume V. Preparation and Assay of Enzymes
Edited by Sidney P. Colowick and Nathan O. Kaplan

Volume VI. Preparation and Assay of Enzymes *(Continued)*
Preparation and Assay of Substrates
Special Techniques
Edited by Sidney P. Colowick and Nathan O. Kaplan

Volume VII. Cumulative Subject Index
Edited by Sidney P. Colowick and Nathan O. Kaplan

Volume VIII. Complex Carbohydrates
Edited by Elizabeth F. Neufeld and Victor Ginsburg

Volume IX. Carbohydrate Metabolism
Edited by Willis A. Wood

Volume X. Oxidation and Phosphorylation
Edited by Ronald W. Estabrook and Maynard E. Pullman

Volume XI. Enzyme Structure
Edited by C. H. W. Hirs

Volume XII. Nucleic Acids (Parts A and B)
Edited by Lawrence Grossman and Kivie Moldave

Volume XIII. Citric Acid Cycle
Edited by J. M. Lowenstein

Volume XIV. Lipids
Edited by J. M. Lowenstein

Volume XV. Steroids and Terpenoids
Edited by Raymond B. Clayton

VOLUME XVI. Fast Reactions
Edited by KENNETH KUSTIN

VOLUME XVII. Metabolism of Amino Acids and Amines (Parts A and B)
Edited by HERBERT TABOR AND CELIA WHITE TABOR

VOLUME XVIII. Vitamins and Coenzymes (Parts A, B, and C)
Edited by DONALD B. MCCORMICK AND LEMUEL D. WRIGHT

VOLUME XIX. Proteolytic Enzymes
Edited by GERTRUDE E. PERLMANN AND LASZLO LORAND

VOLUME XX. Nucleic Acids and Protein Synthesis (Part C)
Edited by KIVIE MOLDAVE AND LAWRENCE GROSSMAN

VOLUME XXI. Nucleic Acids (Part D)
Edited by LAWRENCE GROSSMAN AND KIVIE MOLDAVE

VOLUME XXII. Enzyme Purification and Related Techniques
Edited by WILLIAM B. JAKOBY

VOLUME XXIII. Photosynthesis (Part A)
Edited by ANTHONY SAN PIETRO

VOLUME XXIV. Photosynthesis and Nitrogen Fixation (Part B)
Edited by ANTHONY SAN PIETRO

VOLUME XXV. Enzyme Structure (Part B)
Edited by C. H. W. HIRS AND SERGE N. TIMASHEFF

VOLUME XXVI. Enzyme Structure (Part C)
Edited by C. H. W. HIRS AND SERGE N. TIMASHEFF

VOLUME XXVII. Enzyme Structure (Part D)
Edited by C. H. W. HIRS AND SERGE N. TIMASHEFF

VOLUME XXVIII. Complex Carbohydrates (Part B)
Edited by VICTOR GINSBURG

VOLUME XXIX. Nucleic Acids and Protein Synthesis (Part E)
Edited by LAWRENCE GROSSMAN AND KIVIE MOLDAVE

VOLUME XXX. Nucleic Acids and Protein Synthesis (Part F)
Edited by KIVIE MOLDAVE AND LAWRENCE GROSSMAN

VOLUME XXXI. Biomembranes (Part A)
Edited by SIDNEY FLEISCHER AND LESTER PACKER

VOLUME XXXII. Biomembranes (Part B)
Edited by SIDNEY FLEISCHER AND LESTER PACKER

VOLUME XXXIII. Cumulative Subject Index Volumes I-XXX
Edited by MARTHA G. DENNIS AND EDWARD A. DENNIS

VOLUME XXXIV. Affinity Techniques (Enzyme Purification: Part B)
Edited by WILLIAM B. JAKOBY AND MEIR WILCHEK

VOLUME LIII. Biomembranes (Part D: Biological Oxidations)
Edited by SIDNEY FLEISCHER AND LESTER PACKER

VOLUME LIV. Biomembranes (Part E: Biological Oxidations)
Edited by SIDNEY FLEISCHER AND LESTER PACKER

VOLUME LV. Biomembranes (Part F: Bioenergetics)
Edited by SIDNEY FLEISCHER AND LESTER PACKER

VOLUME LVI. Biomembranes (Part G: Bioenergetics)
Edited by SIDNEY FLEISCHER AND LESTER PACKER

VOLUME LVII. Bioluminescence and Chemiluminescence
Edited by MARLENE A. DELUCA

VOLUME LVIII. Cell Culture
Edited by WILLIAM B. JAKOBY AND IRA PASTAN

VOLUME LIX. Nucleic Acids and Protein Synthesis (Part G)
Edited by KIVIE MOLDAVE AND LAWRENCE GROSSMAN

VOLUME LX. Nucleic Acids and Protein Synthesis (Part H)
Edited by KIVIE MOLDAVE AND LAWRENCE GROSSMAN

VOLUME 61. Enzyme Structure (Part H)
Edited by C. H. W. HIRS AND SERGE N. TIMASHEFF

VOLUME 62. Vitamins and Coenzymes (Part D)
Edited by DONALD B. MCCORMICK AND LEMUEL D. WRIGHT

VOLUME 63. Enzyme Kinetics and Mechanism (Part A: Initial Rate and Inhibitor Methods)
Edited by DANIEL L. PURICH

VOLUME 64. Enzyme Kinetics and Mechanism (Part B: Isotopic Probes and Complex Enzyme Systems)
Edited by DANIEL L. PURICH

VOLUME 65. Nucleic Acids (Part I)
Edited by LAWRENCE GROSSMAN AND KIVIE MOLDAVE

VOLUME 66. Vitamins and Coenzymes (Part E)
Edited by DONALD B. MCCORMICK AND LEMUEL D. WRIGHT

VOLUME 67. Vitamins and Coenzymes (Part F)
Edited by DONALD B. MCCORMICK AND LEMUEL D. WRIGHT

VOLUME 68. Recombinant DNA
Edited by RAY WU

VOLUME 69. Photosynthesis and Nitrogen Fixation (Part C)
Edited by ANTHONY SAN PIETRO

VOLUME 70. Immunochemical Techniques (Part A)
Edited by HELEN VAN VUNAKIS AND JOHN J. LANGONE

Volume 71. Lipids (Part C)
Edited by John M. Lowenstein

Volume 72. Lipids (Part D)
Edited by John M. Lowenstein

Volume 73. Immunochemical Techniques (Part B)
Edited by John J. Langone and Helen Van Vunakis

Volume 74. Immunochemical Techniques (Part C)
Edited by John J. Langone and Helen Van Vunakis

Volume 75. Cumulative Subject Index Volumes XXXI, XXXII, XXXIV–LX
Edited by Edward A. Dennis and Martha G. Dennis

Volume 76. Hemoglobins
Edited by Eraldo Antonini, Luigi Rossi-Bernardi, and Emilia Chiancone

Volume 77. Detoxication and Drug Metabolism
Edited by William B. Jakoby

Volume 78. Interferons (Part A)
Edited by Sidney Pestka

Volume 79. Interferons (Part B)
Edited by Sidney Pestka

Volume 80. Proteolytic Enzymes (Part C)
Edited by Laszlo Lorand

Volume 81. Biomembranes (Part H: Visual Pigments and Purple Membranes, I)
Edited by Lester Packer

Volume 82. Structural and Contractile Proteins (Part A: Extracellular Matrix)
Edited by Leon W. Cunningham and Dixie W. Frederiksen

Volume 83. Complex Carbohydrates (Part D)
Edited by Victor Ginsburg

Volume 84. Immunochemical Techniques (Part D: Selected Immunoassays)
Edited by John J. Langone and Helen Van Vunakis

Volume 85. Structural and Contractile Proteins (Part B: The Contractile Apparatus and the Cytoskeleton)
Edited by Dixie W. Frederiksen and Leon W. Cunningham

Volume 86. Prostaglandins and Arachidonate Metabolites
Edited by William E. M. Lands and William L. Smith

Volume 87. Enzyme Kinetics and Mechanism (Part C: Intermediates, Stereo-chemistry, and Rate Studies)
Edited by Daniel L. Purich

Volume 88. Biomembranes (Part I: Visual Pigments and Purple Membranes, II)
Edited by Lester Packer

Volume 105. Oxygen Radicals in Biological Systems
Edited by Lester Packer

Volume 106. Posttranslational Modifications (Part A)
Edited by Finn Wold and Kivie Moldave

Volume 107. Posttranslational Modifications (Part B)
Edited by Finn Wold and Kivie Moldave

Volume 108. Immunochemical Techniques (Part G: Separation and Characterization of Lymphoid Cells)
Edited by Giovanni Di Sabato, John J. Langone, and Helen Van Vunakis

Volume 109. Hormone Action (Part I: Peptide Hormones)
Edited by Lutz Birnbaumer and Bert W. O'Malley

Volume 110. Steroids and Isoprenoids (Part A)
Edited by John H. Law and Hans C. Rilling

Volume 111. Steroids and Isoprenoids (Part B)
Edited by John H. Law and Hans C. Rilling

Volume 112. Drug and Enzyme Targeting (Part A)
Edited by Kenneth J. Widder and Ralph Green

Volume 113. Glutamate, Glutamine, Glutathione, and Related Compounds
Edited by Alton Meister

Volume 114. Diffraction Methods for Biological Macromolecules (Part A)
Edited by Harold W. Wyckoff, C. H. W. Hirs, and Serge N. Timasheff

Volume 115. Diffraction Methods for Biological Macromolecules (Part B)
Edited by Harold W. Wyckoff, C. H. W. Hirs, and Serge N. Timasheff

Volume 116. Immunochemical Techniques
(Part H: Effectors and Mediators of Lymphoid Cell Functions)
Edited by Giovanni Di Sabato, John J. Langone, and Helen Van Vunakis

Volume 117. Enzyme Structure (Part J)
Edited by C. H. W. Hirs and Serge N. Timasheff

Volume 118. Plant Molecular Biology
Edited by Arthur Weissbach and Herbert Weissbach

Volume 119. Interferons (Part C)
Edited by Sidney Pestka

Volume 120. Cumulative Subject Index Volumes 81–94, 96–101

Volume 121. Immunochemical Techniques (Part I: Hybridoma Technology and Monoclonal Antibodies)
Edited by John J. Langone and Helen Van Vunakis

Volume 122. Vitamins and Coenzymes (Part G)
Edited by Frank Chytil and Donald B. McCormick

VOLUME 123. Vitamins and Coenzymes (Part H)
Edited by FRANK CHYTIL AND DONALD B. MCCORMICK

VOLUME 124. Hormone Action (Part J: Neuroendocrine Peptides)
Edited by P. MICHAEL CONN

VOLUME 125. Biomembranes (Part M: Transport in Bacteria, Mitochondria, and Chloroplasts: General Approaches and Transport Systems)
Edited by SIDNEY FLEISCHER AND BECCA FLEISCHER

VOLUME 126. Biomembranes (Part N: Transport in Bacteria, Mitochondria, and Chloroplasts: Protonmotive Force)
Edited by SIDNEY FLEISCHER AND BECCA FLEISCHER

VOLUME 127. Biomembranes (Part O: Protons and Water: Structure and Translocation)
Edited by LESTER PACKER

VOLUME 128. Plasma Lipoproteins (Part A: Preparation, Structure, and Molecular Biology)
Edited by JERE P. SEGREST AND JOHN J. ALBERS

VOLUME 129. Plasma Lipoproteins (Part B: Characterization, Cell Biology, and Metabolism)
Edited by JOHN J. ALBERS AND JERE P. SEGREST

VOLUME 130. Enzyme Structure (Part K)
Edited by C. H. W. HIRS AND SERGE N. TIMASHEFF

VOLUME 131. Enzyme Structure (Part L)
Edited by C. H. W. HIRS AND SERGE N. TIMASHEFF

VOLUME 132. Immunochemical Techniques (Part J: Phagocytosis and Cell-Mediated Cytotoxicity)
Edited by GIOVANNI DI SABATO AND JOHANNES EVERSE

VOLUME 133. Bioluminescence and Chemiluminescence (Part B)
Edited by MARLENE DELUCA AND WILLIAM D. MCELROY

VOLUME 134. Structural and Contractile Proteins (Part C: The Contractile Apparatus and the Cytoskeleton)
Edited by RICHARD B. VALLEE

VOLUME 135. Immobilized Enzymes and Cells (Part B)
Edited by KLAUS MOSBACH

VOLUME 136. Immobilized Enzymes and Cells (Part C)
Edited by KLAUS MOSBACH

VOLUME 137. Immobilized Enzymes and Cells (Part D)
Edited by KLAUS MOSBACH

VOLUME 138. Complex Carbohydrates (Part E)
Edited by VICTOR GINSBURG

Volume 139. Cellular Regulators (Part A: Calcium- and Calmodulin-Binding Proteins)
Edited by Anthony R. Means and P. Michael Conn

Volume 140. Cumulative Subject Index Volumes 102–119, 121–134

Volume 141. Cellular Regulators (Part B: Calcium and Lipids)
Edited by P. Michael Conn and Anthony R. Means

Volume 142. Metabolism of Aromatic Amino Acids and Amines
Edited by Seymour Kaufman

Volume 143. Sulfur and Sulfur Amino Acids
Edited by William B. Jakoby and Owen Griffith

Volume 144. Structural and Contractile Proteins (Part D: Extracellular Matrix)
Edited by Leon W. Cunningham

Volume 145. Structural and Contractile Proteins (Part E: Extracellular Matrix)
Edited by Leon W. Cunningham

Volume 146. Peptide Growth Factors (Part A)
Edited by David Barnes and David A. Sirbasku

Volume 147. Peptide Growth Factors (Part B)
Edited by David Barnes and David A. Sirbasku

Volume 148. Plant Cell Membranes
Edited by Lester Packer and Roland Douce

Volume 149. Drug and Enzyme Targeting (Part B)
Edited by Ralph Green and Kenneth J. Widder

Volume 150. Immunochemical Techniques (Part K: *In Vitro* Models of B and T Cell Functions and Lymphoid Cell Receptors)
Edited by Giovanni Di Sabato

Volume 151. Molecular Genetics of Mammalian Cells
Edited by Michael M. Gottesman

Volume 152. Guide to Molecular Cloning Techniques
Edited by Shelby L. Berger and Alan R. Kimmel

Volume 153. Recombinant DNA (Part D)
Edited by Ray Wu and Lawrence Grossman

Volume 154. Recombinant DNA (Part E)
Edited by Ray Wu and Lawrence Grossman

Volume 155. Recombinant DNA (Part F)
Edited by Ray Wu

Volume 156. Biomembranes (Part P: ATP-Driven Pumps and Related Transport: The Na, K-Pump)
Edited by Sidney Fleischer and Becca Fleischer

VOLUME 157. Biomembranes (Part Q: ATP-Driven Pumps and Related Transport: Calcium, Proton, and Potassium Pumps)
Edited by SIDNEY FLEISCHER AND BECCA FLEISCHER

VOLUME 158. Metalloproteins (Part A)
Edited by JAMES F. RIORDAN AND BERT L. VALLEE

VOLUME 159. Initiation and Termination of Cyclic Nucleotide Action
Edited by JACKIE D. CORBIN AND ROGER A. JOHNSON

VOLUME 160. Biomass (Part A: Cellulose and Hemicellulose)
Edited by WILLIS A. WOOD AND SCOTT T. KELLOGG

VOLUME 161. Biomass (Part B: Lignin, Pectin, and Chitin)
Edited by WILLIS A. WOOD AND SCOTT T. KELLOGG

VOLUME 162. Immunochemical Techniques (Part L: Chemotaxis and Inflammation)
Edited by GIOVANNI DI SABATO

VOLUME 163. Immunochemical Techniques (Part M: Chemotaxis and Inflammation)
Edited by GIOVANNI DI SABATO

VOLUME 164. Ribosomes
Edited by HARRY F. NOLLER, JR., AND KIVIE MOLDAVE

VOLUME 165. Microbial Toxins: Tools for Enzymology
Edited by SIDNEY HARSHMAN

VOLUME 166. Branched-Chain Amino Acids
Edited by ROBERT HARRIS AND JOHN R. SOKATCH

VOLUME 167. Cyanobacteria
Edited by LESTER PACKER AND ALEXANDER N. GLAZER

VOLUME 168. Hormone Action (Part K: Neuroendocrine Peptides)
Edited by P. MICHAEL CONN

VOLUME 169. Platelets: Receptors, Adhesion, Secretion (Part A)
Edited by JACEK HAWIGER

VOLUME 170. Nucleosomes
Edited by PAUL M. WASSARMAN AND ROGER D. KORNBERG

VOLUME 171. Biomembranes (Part R: Transport Theory: Cells and Model Membranes)
Edited by SIDNEY FLEISCHER AND BECCA FLEISCHER

VOLUME 172. Biomembranes (Part S: Transport: Membrane Isolation and Characterization)
Edited by SIDNEY FLEISCHER AND BECCA FLEISCHER

Volume 173. Biomembranes [Part T: Cellular and Subcellular Transport: Eukaryotic (Nonepithelial) Cells]
Edited by Sidney Fleischer and Becca Fleischer

Volume 174. Biomembranes [Part U: Cellular and Subcellular Transport: Eukaryotic (Nonepithelial) Cells]
Edited by Sidney Fleischer and Becca Fleischer

Volume 175. Cumulative Subject Index Volumes 135–139, 141–167

Volume 176. Nuclear Magnetic Resonance (Part A: Spectral Techniques and Dynamics)
Edited by Norman J. Oppenheimer and Thomas L. James

Volume 177. Nuclear Magnetic Resonance (Part B: Structure and Mechanism)
Edited by Norman J. Oppenheimer and Thomas L. James

Volume 178. Antibodies, Antigens, and Molecular Mimicry
Edited by John J. Langone

Volume 179. Complex Carbohydrates (Part F)
Edited by Victor Ginsburg

Volume 180. RNA Processing (Part A: General Methods)
Edited by James E. Dahlberg and John N. Abelson

Volume 181. RNA Processing (Part B: Specific Methods)
Edited by James E. Dahlberg and John N. Abelson

Volume 182. Guide to Protein Purification
Edited by Murray P. Deutscher

Volume 183. Molecular Evolution: Computer Analysis of Protein and Nucleic Acid Sequences
Edited by Russell F. Doolittle

Volume 184. Avidin-Biotin Technology
Edited by Meir Wilchek and Edward A. Bayer

Volume 185. Gene Expression Technology
Edited by David V. Goeddel

Volume 186. Oxygen Radicals in Biological Systems (Part B: Oxygen Radicals and Antioxidants)
Edited by Lester Packer and Alexander N. Glazer

Volume 187. Arachidonate Related Lipid Mediators
Edited by Robert C. Murphy and Frank A. Fitzpatrick

Volume 188. Hydrocarbons and Methylotrophy
Edited by Mary E. Lidstrom

Volume 189. Retinoids (Part A: Molecular and Metabolic Aspects)
Edited by Lester Packer

Volume 206. Cytochrome P450
Edited by Michael R. Waterman and Eric F. Johnson

Volume 207. Ion Channels
Edited by Bernardo Rudy and Linda E. Iverson

Volume 208. Protein–DNA Interactions
Edited by Robert T. Sauer

Volume 209. Phospholipid Biosynthesis
Edited by Edward A. Dennis and Dennis E. Vance

Volume 210. Numerical Computer Methods
Edited by Ludwig Brand and Michael L. Johnson

Volume 211. DNA Structures (Part A: Synthesis and Physical Analysis of DNA)
Edited by David M. J. Lilley and James E. Dahlberg

Volume 212. DNA Structures (Part B: Chemical and Electrophoretic Analysis of DNA)
Edited by David M. J. Lilley and James E. Dahlberg

Volume 213. Carotenoids (Part A: Chemistry, Separation, Quantitation, and Antioxidation)
Edited by Lester Packer

Volume 214. Carotenoids (Part B: Metabolism, Genetics, and Biosynthesis)
Edited by Lester Packer

Volume 215. Platelets: Receptors, Adhesion, Secretion (Part B)
Edited by Jacek J. Hawiger

Volume 216. Recombinant DNA (Part G)
Edited by Ray Wu

Volume 217. Recombinant DNA (Part H)
Edited by Ray Wu

Volume 218. Recombinant DNA (Part I)
Edited by Ray Wu

Volume 219. Reconstitution of Intracellular Transport
Edited by James E. Rothman

Volume 220. Membrane Fusion Techniques (Part A)
Edited by Nejat Düzgüneş

Volume 221. Membrane Fusion Techniques (Part B)
Edited by Nejat Düzgüneş

Volume 222. Proteolytic Enzymes in Coagulation, Fibrinolysis, and Complement Activation (Part A: Mammalian Blood Coagulation Factors and Inhibitors)
Edited by Laszlo Lorand and Kenneth G. Mann

VOLUME 223. Proteolytic Enzymes in Coagulation, Fibrinolysis, and Complement Activation (Part B: Complement Activation, Fibrinolysis, and Nonmammalian Blood Coagulation Factors)
Edited by LASZLO LORAND AND KENNETH G. MANN

VOLUME 224. Molecular Evolution: Producing the Biochemical Data
Edited by ELIZABETH ANNE ZIMMER, THOMAS J. WHITE, REBECCA L. CANN, AND ALLAN C. WILSON

VOLUME 225. Guide to Techniques in Mouse Development
Edited by PAUL M. WASSARMAN AND MELVIN L. DEPAMPHILIS

VOLUME 226. Metallobiochemistry (Part C: Spectroscopic and Physical Methods for Probing Metal Ion Environments in Metalloenzymes and Metalloproteins)
Edited by JAMES F. RIORDAN AND BERT L. VALLEE

VOLUME 227. Metallobiochemistry (Part D: Physical and Spectroscopic Methods for Probing Metal Ion Environments in Metalloproteins)
Edited by JAMES F. RIORDAN AND BERT L. VALLEE

VOLUME 228. Aqueous Two-Phase Systems
Edited by HARRY WALTER AND GÖTE JOHANSSON

VOLUME 229. Cumulative Subject Index Volumes 195–198, 200–227

VOLUME 230. Guide to Techniques in Glycobiology
Edited by WILLIAM J. LENNARZ AND GERALD W. HART

VOLUME 231. Hemoglobins (Part B: Biochemical and Analytical Methods)
Edited by JOHANNES EVERSE, KIM D. VANDEGRIFF, AND ROBERT M. WINSLOW

VOLUME 232. Hemoglobins (Part C: Biophysical Methods)
Edited by JOHANNES EVERSE, KIM D. VANDEGRIFF, AND ROBERT M. WINSLOW

VOLUME 233. Oxygen Radicals in Biological Systems (Part C)
Edited by LESTER PACKER

VOLUME 234. Oxygen Radicals in Biological Systems (Part D)
Edited by LESTER PACKER

VOLUME 235. Bacterial Pathogenesis (Part A: Identification and Regulation of Virulence Factors)
Edited by VIRGINIA L. CLARK AND PATRIK M. BAVOIL

VOLUME 236. Bacterial Pathogenesis (Part B: Integration of Pathogenic Bacteria with Host Cells)
Edited by VIRGINIA L. CLARK AND PATRIK M. BAVOIL

VOLUME 237. Heterotrimeric G Proteins
Edited by RAVI IYENGAR

VOLUME 238. Heterotrimeric G-Protein Effectors
Edited by RAVI IYENGAR

Volume 239. Nuclear Magnetic Resonance (Part C)
Edited by Thomas L. James and Norman J. Oppenheimer

Volume 240. Numerical Computer Methods (Part B)
Edited by Michael L. Johnson and Ludwig Brand

Volume 241. Retroviral Proteases
Edited by Lawrence C. Kuo and Jules A. Shafer

Volume 242. Neoglycoconjugates (Part A)
Edited by Y. C. Lee and Reiko T. Lee

Volume 243. Inorganic Microbial Sulfur Metabolism
Edited by Harry D. Peck, Jr., and Jean LeGall

Volume 244. Proteolytic Enzymes: Serine and Cysteine Peptidases
Edited by Alan J. Barrett

Volume 245. Extracellular Matrix Components
Edited by E. Ruoslahti and E. Engvall

Volume 246. Biochemical Spectroscopy
Edited by Kenneth Sauer

Volume 247. Neoglycoconjugates (Part B: Biomedical Applications)
Edited by Y. C. Lee and Reiko T. Lee

Volume 248. Proteolytic Enzymes: Aspartic and Metallo Peptidases
Edited by Alan J. Barrett

Volume 249. Enzyme Kinetics and Mechanism (Part D: Developments in Enzyme Dynamics)
Edited by Daniel L. Purich

Volume 250. Lipid Modifications of Proteins
Edited by Patrick J. Casey and Janice E. Buss

Volume 251. Biothiols (Part A: Monothiols and Dithiols, Protein Thiols, and Thiyl Radicals)
Edited by Lester Packer

Volume 252. Biothiols (Part B: Glutathione and Thioredoxin; Thiols in Signal Transduction and Gene Regulation)
Edited by Lester Packer

Volume 253. Adhesion of Microbial Pathogens
Edited by Ron J. Doyle and Itzhak Ofek

Volume 254. Oncogene Techniques
Edited by Peter K. Vogt and Inder M. Verma

Volume 255. Small GTPases and Their Regulators (Part A: Ras Family)
Edited by W. E. Balch, Channing J. Der, and Alan Hall

Volume 256. Small GTPases and Their Regulators (Part B: Rho Family)
Edited by W. E. Balch, Channing J. Der, and Alan Hall

Volume 257. Small GTPases and Their Regulators (Part C: Proteins Involved in Transport)
Edited by W. E. Balch, Channing J. Der, and Alan Hall

Volume 258. Redox-Active Amino Acids in Biology
Edited by Judith P. Klinman

Volume 259. Energetics of Biological Macromolecules
Edited by Michael L. Johnson and Gary K. Ackers

Volume 260. Mitochondrial Biogenesis and Genetics (Part A)
Edited by Giuseppe M. Attardi and Anne Chomyn

Volume 261. Nuclear Magnetic Resonance and Nucleic Acids
Edited by Thomas L. James

Volume 262. DNA Replication
Edited by Judith L. Campbell

Volume 263. Plasma Lipoproteins (Part C: Quantitation)
Edited by William A. Bradley, Sandra H. Gianturco, and Jere P. Segrest

Volume 264. Mitochondrial Biogenesis and Genetics (Part B)
Edited by Giuseppe M. Attardi and Anne Chomyn

Volume 265. Cumulative Subject Index Volumes 228, 230–262

Volume 266. Computer Methods for Macromolecular Sequence Analysis
Edited by Russell F. Doolittle

Volume 267. Combinatorial Chemistry
Edited by John N. Abelson

Volume 268. Nitric Oxide (Part A: Sources and Detection of NO; NO Synthase)
Edited by Lester Packer

Volume 269. Nitric Oxide (Part B: Physiological and Pathological Processes)
Edited by Lester Packer

Volume 270. High Resolution Separation and Analysis of Biological Macromolecules (Part A: Fundamentals)
Edited by Barry L. Karger and William S. Hancock

Volume 271. High Resolution Separation and Analysis of Biological Macromolecules (Part B: Applications)
Edited by Barry L. Karger and William S. Hancock

Volume 272. Cytochrome P450 (Part B)
Edited by Eric F. Johnson and Michael R. Waterman

Volume 273. RNA Polymerase and Associated Factors (Part A)
Edited by Sankar Adhya

VOLUME 274. RNA Polymerase and Associated Factors (Part B)
Edited by SANKAR ADHYA

VOLUME 275. Viral Polymerases and Related Proteins
Edited by LAWRENCE C. KUO, DAVID B. OLSEN, AND STEVEN S. CARROLL

VOLUME 276. Macromolecular Crystallography (Part A)
Edited by CHARLES W. CARTER, JR., AND ROBERT M. SWEET

VOLUME 277. Macromolecular Crystallography (Part B)
Edited by CHARLES W. CARTER, JR., AND ROBERT M. SWEET

VOLUME 278. Fluorescence Spectroscopy
Edited by LUDWIG BRAND AND MICHAEL L. JOHNSON

VOLUME 279. Vitamins and Coenzymes (Part I)
Edited by DONALD B. MCCORMICK, JOHN W. SUTTIE, AND CONRAD WAGNER

VOLUME 280. Vitamins and Coenzymes (Part J)
Edited by DONALD B. MCCORMICK, JOHN W. SUTTIE, AND CONRAD WAGNER

VOLUME 281. Vitamins and Coenzymes (Part K)
Edited by DONALD B. MCCORMICK, JOHN W. SUTTIE, AND CONRAD WAGNER

VOLUME 282. Vitamins and Coenzymes (Part L)
Edited by DONALD B. MCCORMICK, JOHN W. SUTTIE, AND CONRAD WAGNER

VOLUME 283. Cell Cycle Control
Edited by WILLIAM G. DUNPHY

VOLUME 284. Lipases (Part A: Biotechnology)
Edited by BYRON RUBIN AND EDWARD A. DENNIS

VOLUME 285. Cumulative Subject Index Volumes 263, 264, 266–284, 286–289

VOLUME 286. Lipases (Part B: Enzyme Characterization and Utilization)
Edited by BYRON RUBIN AND EDWARD A. DENNIS

VOLUME 287. Chemokines
Edited by RICHARD HORUK

VOLUME 288. Chemokine Receptors
Edited by RICHARD HORUK

VOLUME 289. Solid Phase Peptide Synthesis
Edited by GREGG B. FIELDS

VOLUME 290. Molecular Chaperones
Edited by GEORGE H. LORIMER AND THOMAS BALDWIN

VOLUME 291. Caged Compounds
Edited by GERARD MARRIOTT

VOLUME 292. ABC Transporters: Biochemical, Cellular, and Molecular Aspects
Edited by SURESH V. AMBUDKAR AND MICHAEL M. GOTTESMAN

Volume 293. Ion Channels (Part B)
Edited by P. Michael Conn

Volume 294. Ion Channels (Part C)
Edited by P. Michael Conn

Volume 295. Energetics of Biological Macromolecules (Part B)
Edited by Gary K. Ackers and Michael L. Johnson

Volume 296. Neurotransmitter Transporters
Edited by Susan G. Amara

Volume 297. Photosynthesis: Molecular Biology of Energy Capture
Edited by Lee McIntosh

Volume 298. Molecular Motors and the Cytoskeleton (Part B)
Edited by Richard B. Vallee

Volume 299. Oxidants and Antioxidants (Part A)
Edited by Lester Packer

Volume 300. Oxidants and Antioxidants (Part B)
Edited by Lester Packer

Volume 301. Nitric Oxide: Biological and Antioxidant Activities (Part C)
Edited by Lester Packer

Volume 302. Green Fluorescent Protein
Edited by P. Michael Conn

Volume 303. cDNA Preparation and Display
Edited by Sherman M. Weissman

Volume 304. Chromatin
Edited by Paul M. Wassarman and Alan P. Wolffe

Volume 305. Bioluminescence and Chemiluminescence (Part C)
Edited by Thomas O. Baldwin and Miriam M. Ziegler

Volume 306. Expression of Recombinant Genes in Eukaryotic Systems
Edited by Joseph C. Glorioso and Martin C. Schmidt

Volume 307. Confocal Microscopy
Edited by P. Michael Conn

Volume 308. Enzyme Kinetics and Mechanism (Part E: Energetics of Enzyme Catalysis)
Edited by Daniel L. Purich and Vern L. Schramm

Volume 309. Amyloid, Prions, and Other Protein Aggregates
Edited by Ronald Wetzel

Volume 310. Biofilms
Edited by Ron J. Doyle

VOLUME 311. Sphingolipid Metabolism and Cell Signaling (Part A)
Edited by ALFRED H. MERRILL, JR., AND YUSUF A. HANNUN

VOLUME 312. Sphingolipid Metabolism and Cell Signaling (Part B)
Edited by ALFRED H. MERRILL, JR., AND YUSUF A. HANNUN

VOLUME 313. Antisense Technology
(Part A: General Methods, Methods of Delivery, and RNA Studies)
Edited by M. IAN PHILLIPS

VOLUME 314. Antisense Technology (Part B: Applications)
Edited by M. IAN PHILLIPS

VOLUME 315. Vertebrate Phototransduction and the Visual Cycle (Part A)
Edited by KRZYSZTOF PALCZEWSKI

VOLUME 316. Vertebrate Phototransduction and the Visual Cycle (Part B)
Edited by KRZYSZTOF PALCZEWSKI

VOLUME 317. RNA–Ligand Interactions (Part A: Structural Biology Methods)
Edited by DANIEL W. CELANDER AND JOHN N. ABELSON

VOLUME 318. RNA–Ligand Interactions (Part B: Molecular Biology Methods)
Edited by DANIEL W. CELANDER AND JOHN N. ABELSON

VOLUME 319. Singlet Oxygen, UV-A, and Ozone
Edited by LESTER PACKER AND HELMUT SIES

VOLUME 320. Cumulative Subject Index Volumes 290–319

VOLUME 321. Numerical Computer Methods (Part C)
Edited by MICHAEL L. JOHNSON AND LUDWIG BRAND

VOLUME 322. Apoptosis
Edited by JOHN C. REED

VOLUME 323. Energetics of Biological Macromolecules (Part C)
Edited by MICHAEL L. JOHNSON AND GARY K. ACKERS

VOLUME 324. Branched-Chain Amino Acids (Part B)
Edited by ROBERT A. HARRIS AND JOHN R. SOKATCH

VOLUME 325. Regulators and Effectors of Small GTPases
(Part D: Rho Family)
Edited by W. E. BALCH, CHANNING J. DER, AND ALAN HALL

VOLUME 326. Applications of Chimeric Genes and Hybrid Proteins
(Part A: Gene Expression and Protein Purification)
Edited by JEREMY THORNER, SCOTT D. EMR, AND JOHN N. ABELSON

VOLUME 327. Applications of Chimeric Genes and Hybrid Proteins
(Part B: Cell Biology and Physiology)
Edited by JEREMY THORNER, SCOTT D. EMR, AND JOHN N. ABELSON

VOLUME 328. Applications of Chimeric Genes and Hybrid Proteins (Part C: Protein–Protein Interactions and Genomics)
Edited by JEREMY THORNER, SCOTT D. EMR, AND JOHN N. ABELSON

VOLUME 329. Regulators and Effectors of Small GTPases (Part E: GTPases Involved in Vesicular Traffic)
Edited by W. E. BALCH, CHANNING J. DER, AND ALAN HALL

VOLUME 330. Hyperthermophilic Enzymes (Part A)
Edited by MICHAEL W. W. ADAMS AND ROBERT M. KELLY

VOLUME 331. Hyperthermophilic Enzymes (Part B)
Edited by MICHAEL W. W. ADAMS AND ROBERT M. KELLY

VOLUME 332. Regulators and Effectors of Small GTPases (Part F: Ras Family I)
Edited by W. E. BALCH, CHANNING J. DER, AND ALAN HALL

VOLUME 333. Regulators and Effectors of Small GTPases (Part G: Ras Family II)
Edited by W. E. BALCH, CHANNING J. DER, AND ALAN HALL

VOLUME 334. Hyperthermophilic Enzymes (Part C)
Edited by MICHAEL W. W. ADAMS AND ROBERT M. KELLY

VOLUME 335. Flavonoids and Other Polyphenols
Edited by LESTER PACKER

VOLUME 336. Microbial Growth in Biofilms (Part A: Developmental and Molecular Biological Aspects)
Edited by RON J. DOYLE

VOLUME 337. Microbial Growth in Biofilms (Part B: Special Environments and Physicochemical Aspects)
Edited by RON J. DOYLE

VOLUME 338. Nuclear Magnetic Resonance of Biological Macromolecules (Part A)
Edited by THOMAS L. JAMES, VOLKER DÖTSCH, AND ULI SCHMITZ

VOLUME 339. Nuclear Magnetic Resonance of Biological Macromolecules (Part B)
Edited by THOMAS L. JAMES, VOLKER DÖTSCH, AND ULI SCHMITZ

VOLUME 340. Drug–Nucleic Acid Interactions
Edited by JONATHAN B. CHAIRES AND MICHAEL J. WARING

VOLUME 341. Ribonucleases (Part A)
Edited by ALLEN W. NICHOLSON

VOLUME 342. Ribonucleases (Part B)
Edited by ALLEN W. NICHOLSON

VOLUME 343. G Protein Pathways (Part A: Receptors)
Edited by RAVI IYENGAR AND JOHN D. HILDEBRANDT

VOLUME 344. G Protein Pathways (Part B: G Proteins and Their Regulators)
Edited by RAVI IYENGAR AND JOHN D. HILDEBRANDT

VOLUME 362. Recognition of Carbohydrates in Biological Systems (Part A)
Edited by YUAN C. LEE AND REIKO T. LEE

VOLUME 363. Recognition of Carbohydrates in Biological Systems (Part B)
Edited by YUAN C. LEE AND REIKO T. LEE

VOLUME 364. Nuclear Receptors
Edited by DAVID W. RUSSELL AND DAVID J. MANGELSDORF

VOLUME 365. Differentiation of Embryonic Stem Cells
Edited by PAUL M. WASSAUMAN AND GORDON M. KELLER

VOLUME 366. Protein Phosphatases
Edited by SUSANNE KLUMPP AND JOSEF KRIEGLSTEIN

VOLUME 367. Liposomes (Part A)
Edited by NEJAT DÜZGÜNEŞ

VOLUME 368. Macromolecular Crystallography (Part C)
Edited by CHARLES W. CARTER, JR., AND ROBERT M. SWEET

VOLUME 369. Combinational Chemistry (Part B)
Edited by GUILLERMO A. MORALES AND BARRY A. BUNIN

VOLUME 370. RNA Polymerases and Associated Factors (Part C)
Edited by SANKAR L. ADHYA AND SUSAN GARGES

VOLUME 371. RNA Polymerases and Associated Factors (Part D)
Edited by SANKAR L. ADHYA AND SUSAN GARGES

VOLUME 372. Liposomes (Part B)
Edited by NEJAT DÜZGÜNEŞ

VOLUME 373. Liposomes (Part C)
Edited by NEJAT DÜZGÜNEŞ

VOLUME 374. Macromolecular Crystallography (Part D)
Edited by CHARLES W. CARTER, JR., AND ROBERT W. SWEET

VOLUME 375. Chromatin and Chromatin Remodeling Enzymes (Part A)
Edited by C. DAVID ALLIS AND CARL WU

VOLUME 376. Chromatin and Chromatin Remodeling Enzymes (Part B)
Edited by C. DAVID ALLIS AND CARL WU

VOLUME 377. Chromatin and Chromatin Remodeling Enzymes (Part C)
Edited by C. DAVID ALLIS AND CARL WU

VOLUME 378. Quinones and Quinone Enzymes (Part A)
Edited by HELMUT SIES AND LESTER PACKER

VOLUME 379. Energetics of Biological Macromolecules (Part D)
Edited by JO M. HOLT, MICHAEL L. JOHNSON, AND GARY K. ACKERS

VOLUME 380. Energetics of Biological Macromolecules (Part E)
Edited by JO M. HOLT, MICHAEL L. JOHNSON, AND GARY K. ACKERS

VOLUME 381. Oxygen Sensing
Edited by CHANDAN K. SEN AND GREGG L. SEMENZA

VOLUME 382. Quinones and Quinone Enzymes (Part B)
Edited by HELMUT SIES AND LESTER PACKER

VOLUME 383. Numerical Computer Methods (Part D)
Edited by LUDWIG BRAND AND MICHAEL L. JOHNSON

VOLUME 384. Numerical Computer Methods (Part E)
Edited by LUDWIG BRAND AND MICHAEL L. JOHNSON

VOLUME 385. Imaging in Biological Research (Part A)
Edited by P. MICHAEL CONN

VOLUME 386. Imaging in Biological Research (Part B)
Edited by P. MICHAEL CONN

VOLUME 387. Liposomes (Part D)
Edited by NEJAT DÜZGÜNEŞ

VOLUME 388. Protein Engineering
Edited by DAN E. ROBERTSON AND JOSEPH P. NOEL

VOLUME 389. Regulators of G-Protein Signaling (Part A)
Edited by DAVID P. SIDEROVSKI

VOLUME 390. Regulators of G-Protein Signaling (Part B)
Edited by DAVID P. SIDEROVSKI

VOLUME 391. Liposomes (Part E)
Edited by NEJAT DÜZGÜNEŞ

VOLUME 392. RNA Interference
Edited by ENGELKE ROSSI

VOLUME 393. Circadian Rhythms
Edited by MICHAEL W. YOUNG

VOLUME 394. Nuclear Magnetic Resonance of Biological Macromolecules (Part C)
Edited by THOMAS L. JAMES

VOLUME 395. Producing the Biochemical Data (Part B)
Edited by ELIZABETH A. ZIMMER AND ERIC H. ROALSON

VOLUME 396. Nitric Oxide (Part E)
Edited by LESTER PACKER AND ENRIQUE CADENAS

VOLUME 397. Environmental Microbiology
Edited by JARED R. LEADBETTER

VOLUME 398. Ubiquitin and Protein Degradation (Part A)
Edited by RAYMOND J. DESHAIES

Volume 399. Ubiquitin and Protein Degradation (Part B)
Edited by Raymond J. Deshaies

Volume 400. Phase II Conjugation Enzymes and Transport Systems
Edited by Helmut Sies and Lester Packer

Volume 401. Glutathione Transferases and Gamma Glutamyl Transpeptidases
Edited by Helmut Sies and Lester Packer

Volume 402. Biological Mass Spectrometry
Edited by A. L. Burlingame

Volume 403. GTPases Regulating Membrane Targeting and Fusion
Edited by William E. Balch, Channing J. Der, and Alan Hall

Volume 404. GTPases Regulating Membrane Dynamics
Edited by William E. Balch, Channing J. Der, and Alan Hall

Volume 405. Mass Spectrometry: Modified Proteins and Glycoconjugates
Edited by A. L. Burlingame

Volume 406. Regulators and Effectors of Small GTPases: Rho Family
Edited by William E. Balch, Channing J. Der, and Alan Hall

Volume 407. Regulators and Effectors of Small GTPases: Ras Family
Edited by William E. Balch, Channing J. Der, and Alan Hall

Volume 408. DNA Repair (Part A)
Edited by Judith L. Campbell and Paul Modrich

Volume 409. DNA Repair (Part B)
Edited by Judith L. Campbell and Paul Modrich

Volume 410. DNA Microarrays (Part A: Array Platforms and Web-Bench Protocols)
Edited by Alan Kimmel and Brian Oliver

Volume 411. DNA Microarrays (Part B: Databases and Statistics)
Edited by Alan Kimmel and Brian Oliver

Volume 412. Amyloid, Prions, and Other Protein Aggregates (Part B)
Edited by Indu Kheterpal and Ronald Wetzel

Volume 413. Amyloid, Prions, and Other Protein Aggregates (Part C)
Edited by Indu Kheterpal and Ronald Wetzel

Volume 414. Measuring Biological Responses with Automated Microscopy
Edited by James Inglese

Volume 415. Glycobiology
Edited by Minoru Fukuda

Volume 416. Glycomics
Edited by Minoru Fukuda

Volume 417. Functional Glycomics
Edited by Minoru Fukuda

Volume 418. Embryonic Stem Cells
Edited by Irina Klimanskaya and Robert Lanza

Volume 419. Adult Stem Cells
Edited by Irina Klimanskaya and Robert Lanza

Volume 420. Stem Cell Tools and Other Experimental Protocols
Edited by Irina Klimanskaya and Robert Lanza

Volume 421. Advanced Bacterial Genetics: Use of Transposons and Phage for Genomic Engineering
Edited by Kelly T. Hughes

Volume 422. Two-Component Signaling Systems, Part A
Edited by Melvin I. Simon, Brian R. Crane, and Alexandrine Crane

Volume 423. Two-Component Signaling Systems, Part B
Edited by Melvin I. Simon, Brian R. Crane, and Alexandrine Crane

Volume 424. RNA Editing
Edited by Jonatha M. Gott

Volume 425. RNA Modification
Edited by Jonatha M. Gott

Volume 426. Integrins
Edited by David Cheresh

Volume 427. MicroRNA Methods
Edited by John J. Rossi

Volume 428. Osmosensing and Osmosignaling
Edited by Helmut Sies and Dieter Haussinger

Volume 429. Translation Initiation: Extract Systems and Molecular Genetics
Edited by Jon Lorsch

Volume 430. Translation Initiation: Reconstituted Systems and Biophysical Methods
Edited by Jon Lorsch

Volume 431. Translation Initiation: Cell Biology, High-Throughput and Chemical-Based Approaches
Edited by Jon Lorsch

Volume 432. Lipidomics and Bioactive Lipids: Mass-Spectrometry–Based Lipid Analysis
Edited by H. Alex Brown

Volume 448. RNA Turnover in Eukaryotes: Nucleases, Pathways and Analysis of mRNA Decay
Edited by Lynne E. Maquat and Megerditch Kiledjian

Volume 449. RNA Turnover in Eukaryotes: Analysis of Specialized and Quality Control RNA Decay Pathways
Edited by Lynne E. Maquat and Megerditch Kiledjian

Volume 450. Fluorescence Spectroscopy
Edited by Ludwig Brand and Michael L. Johnson

Volume 451. Autophagy: Lower Eukaryotes and Non-Mammalian Systems (Part A)
Edited by Daniel J. Klionsky

Volume 452. Autophagy in Mammalian Systems (Part B)
Edited by Daniel J. Klionsky

Volume 453. Autophagy in Disease and Clinical Applications (Part C)
Edited by Daniel J. Klionsky

Volume 454. Computer Methods (Part A)
Edited by Michael L. Johnson and Ludwig Brand

Volume 455. Biothermodynamics (Part A)
Edited by Michael L. Johnson, Jo M. Holt, and Gary K. Ackers (Retired)

Volume 456. Mitochondrial Function, Part A: Mitochondrial Electron Transport Complexes and Reactive Oxygen Species
Edited by William S. Allison and Immo E. Scheffler

Volume 457. Mitochondrial Function, Part B: Mitochondrial Protein Kinases, Protein Phosphatases and Mitochondrial Diseases
Edited by William S. Allison and Anne N. Murphy

Volume 458. Complex Enzymes in Microbial Natural Product Biosynthesis, Part A: Overview Articles and Peptides
Edited by David A. Hopwood

Volume 459. Complex Enzymes in Microbial Natural Product Biosynthesis, Part B: Polyketides, Aminocoumarins and Carbohydrates
Edited by David A. Hopwood

Volume 460. Chemokines, Part A
Edited by Tracy M. Handel and Damon J. Hamel

Volume 461. Chemokines, Part B
Edited by Tracy M. Handel and Damon J. Hamel

Volume 462. Non-Natural Amino Acids
Edited by Tom W. Muir and John N. Abelson

Volume 463. Guide to Protein Purification, 2nd Edition
Edited by Richard R. Burgess and Murray P. Deutscher

Volume 464. Liposomes, Part F
Edited by Nejat Düzgüneş

Volume 465. Liposomes, Part G
Edited by Nejat Düzgüneş

Volume 466. Biothermodynamics, Part B
Edited by Michael L. Johnson, Gary K. Ackers, and Jo M. Holt

Volume 467. Computer Methods Part B
Edited by Michael L. Johnson and Ludwig Brand

Volume 468. Biophysical, Chemical, and Functional Probes of RNA Structure, Interactions and Folding: Part A
Edited by Daniel Herschlag

Volume 469. Biophysical, Chemical, and Functional Probes of RNA Structure, Interactions and Folding: Part B
Edited by Daniel Herschlag

Volume 470. Guide to Yeast Genetics: Functional Genomics, Proteomics, and Other Systems Analysis, 2nd Edition
Edited by Gerald Fink, Jonathan Weissman, and Christine Guthrie

Volume 471. Two-Component Signaling Systems, Part C
Edited by Melvin I. Simon, Brian R. Crane, and Alexandrine Crane

Volume 472. Single Molecule Tools, Part A: Fluorescence Based Approaches
Edited by Nils G. Walter

Volume 473. Thiol Redox Transitions in Cell Signaling, Part A Chemistry and Biochemistry of Low Molecular Weight and Protein Thiols
Edited by Enrique Cadenas and Lester Packer

Volume 474. Thiol Redox Transitions in Cell Signaling, Part B Cellular Localization and Signaling
Edited by Enrique Cadenas and Lester Packer

Volume 475. Single Molecule Tools, Part B: Super-Resolution, Particle Tracking, Multiparameter, and Force Based Methods
Edited by Nils G. Walter

Volume 476. Guide to Techniques in Mouse Development, Part A Mice, Embryos, and Cells, 2nd Edition
Edited by Paul M. Wassarman and Philippe M. Soriano

Volume 477. Guide to Techniques in Mouse Development, Part B Mouse Molecular Genetics, 2nd Edition
Edited by Paul M. Wassarman and Philippe M. Soriano

Volume 478. Glycomics
Edited by Minoru Fukuda

VOLUME 479. Functional Glycomics
Edited by MINORU FUKUDA

VOLUME 480. Glycobiology
Edited by MINORU FUKUDA

VOLUME 481. Cryo-EM, Part A: Sample Preparation and Data Collection
Edited by GRANT J. JENSEN

VOLUME 482. Cryo-EM, Part B: 3-D Reconstruction
Edited by GRANT J. JENSEN

VOLUME 483. Cryo-EM, Part C: Analyses, Interpretation, and Case Studies
Edited by GRANT J. JENSEN

VOLUME 484. Constitutive Activity in Receptors and Other Proteins, Part A
Edited by P. MICHAEL CONN

VOLUME 485. Constitutive Activity in Receptors and Other Proteins, Part B
Edited by P. MICHAEL CONN

VOLUME 486. Research on Nitrification and Related Processes, Part A
Edited by MARTIN G. KLOTZ

VOLUME 487. Computer Methods, Part C
Edited by MICHAEL L. JOHNSON AND LUDWIG BRAND

VOLUME 488. Biothermodynamics, Part C
Edited by MICHAEL L. JOHNSON, JO M. HOLT, AND GARY K. ACKERS

VOLUME 489. The Unfolded Protein Response and Cellular Stress, Part A
Edited by P. MICHAEL CONN

VOLUME 490. The Unfolded Protein Response and Cellular Stress, Part B
Edited by P. MICHAEL CONN

VOLUME 491. The Unfolded Protein Response and Cellular Stress, Part C
Edited by P. MICHAEL CONN

VOLUME 492. Biothermodynamics, Part D
Edited by MICHAEL L. JOHNSON, JO M. HOLT, AND GARY K. ACKERS

VOLUME 493. Fragment-Based Drug Design
Tools, Practical Approaches, and Examples
Edited by LAWRENCE C. KUO

VOLUME 494. Methods in Methane Metabolism, Part A
Methanogenesis
Edited by AMY C. ROSENZWEIG AND STEPHEN W. RAGSDALE

VOLUME 495. Methods in Methane Metabolism, Part B
Methanotrophy
Edited by AMY C. ROSENZWEIG AND STEPHEN W. RAGSDALE

VOLUME 496. Research on Nitrification and Related Processes, Part B
Edited by MARTIN G. KLOTZ AND LISA Y. STEIN

VOLUME 497. Synthetic Biology, Part A
Methods for Part/Device Characterization and Chassis Engineering
Edited by CHRISTOPHER VOIGT

VOLUME 498. Synthetic Biology, Part B
Computer Aided Design and DNA Assembly
Edited by CHRISTOPHER VOIGT

VOLUME 499. Biology of Serpins
Edited by JAMES C. WHISSTOCK AND PHILLIP I. BIRD

VOLUME 500. Methods in Systems Biology
Edited by DANIEL JAMESON, MALKHEY VERMA, AND HANS V. WESTERHOFF

VOLUME 501. Serpin Structure and Evolution
Edited by JAMES C. WHISSTOCK AND PHILLIP I. BIRD

CHAPTER ONE

Intracellular Production of Recombinant Serpins in Yeast

Dion Kaiserman,* Corinne Hitchen,* Vita Levina,† Stephen P. Bottomley,* *and* Phillip I. Bird*

Contents

1. Introduction 1
2. Selection of Strain and Expression Plasmid 2
3. Growth of Yeast 3
4. Transformation of Yeast 4
5. Screening Transformants 4
6. Large-Scale Growth and Induction 5
7. Lysis 6
8. Purification 7
9. Assessing Serpin Activity and Removing Inactive Forms 9
10. Production of Polymerogenic Serpins 10
References 12

Abstract

Yeast are a valuable system for recombinant serpin production due to their ability to synthesize large amounts of heterologous gene products as well as their expression of folding chaperones and lack of endogenous serpin genes. In this chapter, we describe a method for intracellular expression of cytoplasmic serpins in the yeast *Pichia pastoris*. We also give details on how this system can be exploited to produce polymer-forming mutants of secretory serpins.

1. Introduction

The production of high-quality recombinant protein is a critical factor in many biochemical endeavors. As such, a plethora of avenues are available to investigators ranging from purification of natively produced material from the source, to the use of heterologous expression systems in

* Department of Biochemistry and Molecular Biology, Monash University, Melbourne, Victoria, Australia
† Department of Biochemistry, La Trobe Institute for Molecular Science, Bundoora, Victoria, Australia

Methods in Enzymology, Volume 501
ISSN 0076-6879, DOI: 10.1016/B978-0-12-385950-1.00001-8

prokaryotes as well as lower eukaryotes and mammalian cell culture. Each system brings its own set of advantages and disadvantages. While purification of naturally synthesized protein ensures that the end product will contain all of the necessary posttranslational modifications, it may be difficult to obtain in sufficient quantities, and there are ethical and personal risks associated with the purification of human proteins. Although the use of mammalian expression systems typically ensures correct modifications, it is expensive and often results in poor yields. Conversely, bacterial systems usually guarantee high yield, but bacteria may not be able to correctly synthesize eukaryotic proteins that follow complicated folding pathways or contain certain posttranslational modifications, such as disulfide bonds.

The use of lower eukaryotes such as yeast represents an important alternative. The availability of practicable, high-level expression systems allows useful yields without the need for cumbersome bioreactors or complex culture media. Importantly, yeast possess a range of conserved cytoplasmic and secretory pathway chaperones that can assist in protein folding and have the ability to produce mammalian-like posttranslational modifications. In the production of recombinant serpins, yeast have an additional advantage over higher eukaryotic systems as they do not produce endogenous serpins, thereby simplifying purification. While the expression of clade B (intracellular) serpins within the yeast cytoplasm is a natural extension of their *in vivo* localization, this system can also be extended to the production of medically relevant mutants of normally secreted serpins that have a propensity to polymerize.

In this chapter, we describe methods for the production of recombinant serpins in the cytoplasm of yeast. We note that secreted serpins can also be successfully produced in yeast systems using suitable vectors and the approaches described below.

2. Selection of Strain and Expression Plasmid

There are many different species of yeast available for gene expression. We have made use of the methylotrophic yeast *Pichia pastoris* with its ability to grow to high densities and highly inducible systems of heterologous gene expression. Although a number of *P. pastoris* strains are available, we utilize the auxotrophic histidine-dependent strains GS115 (genotype: *his4*) and SMD1163 (genotype: *his4 pep4 prb1*). While the former is essentially wild type except for its dependence on histidine, the latter is deficient in the vacuolar proteinases: Proteinase A and Proteinase B. This strain exhibits significantly reduced activity of a number of vacuolar proteases that can cleave serpin reactive center loops (RCLs) during the lysis and purification stages, leading to decreased protein recovery.

A range of yeast expression vectors are now commercially available. We make use of the Invitrogen^{TM} vectors pHIL-D2, pPIC9, and pPIC9K. The pHIL-D2 vector allows intracellular expression, while pPIC9 encodes a yeast secretion signal from the *Saccharomyces cerevisiae* α-factor to direct the gene of interest through the secretory pathway. pPIC9K contains the α-factor secretion signal as well as the kanamycin resistance gene that confers copy number-dependent resistance to G418 for detection of multiple integration events (see Section 5).

3. Growth of Yeast

Untransformed yeast are grown in rich YPD medium utilizing glucose as a carbon source. It is important to note that glucose will crystallize if autoclaved for 20 min and therefore the following procedure should be followed:

3.1. Prepare 1× YP by dissolving 10 g yeast extract and 20 g bacteriological peptone in 900 ml water and autoclaving at 121 °C for 20 min.
3.2. Dissolve 200 g D-glucose in 1 l water and autoclave at 121 °C for 10 min.
3.3. Add 100 ml glucose solution to YP.
3.4. Solid agar plates can be produced by adding 15 g bacteriological agar to the YP. When adding glucose to YP-Agar, the components should be preequilibrated to 60 °C before mixing.
3.5. Yeast grow optimally at 30 °C with constant shaking at 200 rpm to enhance oxygenation of the culture, while plates should be incubated inverted in a 30 °C humidified incubator for 3–4 days.

Since the selection of transformants is positive pressure (i.e., ability to synthesize histidine) rather than negative pressure (i.e., death caused by antibiotics), transformed yeast are maintained under constant selection pressure on minimal medium RDB plates:

3.6. Dissolve 186 g D-sorbitol and 20 g bacteriological agar in 790 ml water and autoclave at 121 °C for 20 min.
3.7. Dissolve 134 g yeast nitrogen base (with ammonium sulfate, without amino acids) in 1 l water. Heat slightly to dissolve and filter sterilize.
3.8. Melt sorbitol-agar and cool to 60 °C, then add the following solutions prewarmed to 45–50 °C: 100 ml 20% (w/v) D-glucose, 100 ml yeast nitrogen base, 2 ml 0.2 g/l biotin (filter sterilized), and 10 ml amino acid solution (500 mg/l each of L-glutamic acid, L-methionine, L-lysine, L-leucine, and L-isoleucine; filter sterilized).

4. Transformation of Yeast

It was recently shown that the transformation efficiency of *P. pastoris* can be enhanced by preincubation with lithium acetate and dithiothreitol (DTT). We have found this method to be easier and more reproducible than preparation of spheroplasts, particularly for those with little experience. This method is essentially as described by Wu and Letchworth (2004).

4.1. Pick a single colony of yeast from a YPD plate and grow overnight at 30 °C with shaking in YPD.
4.2. Inoculate 100 ml of YPD with 1 ml of the overnight culture and grow to $OD_{600\ nm}$ of 1.5.
4.3. Calculate the concentration of yeast by 1 $OD_{600} = 5 \times 10^7$ cells/ml.
4.4. Collect 8×10^8 cells by centrifugation at 1500×*g* for 5 min.
4.5. Resuspend cells in 8 ml pretreatment solution (0.1 *M* lithium acetate, 0.6 *M* sorbitol, 10 mM DTT, 10 mM Tris–HCl, pH 7.5) and incubate at room temperature for 30 min.
4.6. Collect cells and wash twice with 30 ml ice cold 1 *M* sorbitol.
4.7. Resuspend cells in ice cold 1 *M* sorbitol at 1×10^{10} cells/ml.
4.8. Incubate 80 μl cells with 1 μl linearized DNA in a cold, 2 mm gap electroporation cuvette (Molecular BioProducts catalog #5520) and incubate on ice for 5 min.
4.9. Electroporate yeast at 1.5 kV, 25 μF, 186 Ω. This should generate a time constant of approximately 5 ms.
4.10. Immediately remove yeast into 1 ml of ice cold 1 *M* sorbitol and spread aliquots onto RDB plates.
4.11. Incubate plates inverted at 30 °C for 3–4 days to allow yeast colonies to grow.
4.12. Pick single, well-isolated colonies and streak purify them onto RDB agar plates.

5. Screening Transformants

While most yeast will incorporate a single copy of the expression cassette into their genome, a small proportion will naturally integrate multiple copies. There are many methods available to select and detect transformed yeast bearing multiple insertions including the introduction of G418 resistance and Southern hybridization of yeast genomic DNA. In general, expression of cytoplasmic serpins is high enough that it is not necessary to select clones with multiple integration events; however, it may assist when expression is found to be consistently low.

5.1. Select single colonies of transformed yeast from RDB plates as well as an untransformed control colony from a YPD plate and grow 2 ml starter cultures in YPD for 48 h at 30 °C with shaking.
5.2. Inoculate 5 ml YPD with 50 μl of starter culture and grow, shaking, at 30 °C for 40 h.
5.3. Pellet the yeast by centrifugation at 1500×*g* for 10 min and pour off the supernatant.
5.4. Resuspend the pellet in 5 ml YPM (1% (w/v) yeast extract, 2% (w/v) peptone, 3% (v/v) methanol) and induce expression at 30 °C in a shaking incubator. We have found that maximal induction is achieved after approximately 48 h and remains constant until ~72 h, after which yields begin to drop, however, a time course of expression (typically 12 h time points) should be performed to determine optimum culture conditions.
5.5. Take 1 ml samples at the desired time points, collect cells by centrifugation, wash once with 1 ml Tris-buffered saline (TBS; 150 m*M* NaCl, 20 m*M* Tris–HCl, pH 7.4) and store them at −80 °C until use.
5.6. Lyse yeast by resuspending the pellet in 200 μl sodium dodecyl sulfate solution (10%, w/v) and boiling them for 20 min. The cell wall makes yeast highly resistant to chemical lysis, and efficiency can be increased by pretreatment with cell wall degrading enzymes such as Zymolase. However, we have found that the lysis achieved by boiling in SDS is sufficient for small scale screening.
5.7. To 30 μl aliquots of lysates add 10 μl 1 *M* DTT and 10 μl loading buffer (50% (v/v) glycerol, 0.1% bromophenol blue, 100 m*M* Tris–HCl, pH 6.8) and boil for 5 min.
5.8. Induction can be assessed by resolving lysates samples by SDS-PAGE and immunoblotting for the gene of interest.

6. Large-Scale Growth and Induction

During large-scale growth of yeast, aeration is a critical issue. To ensure adequate aeration, yeast should be grown in conical flasks in which the culture volume does not exceed 25% of total flask volume. The use of baffled flasks can increase aeration but often leads to excessive foam which can lead to contamination if it erupts from the mouth of the flask. To avoid this, foam-retardant chemicals such as Antifoam 204 (Sigma Aldrich catalog #A6426) can be included without affecting growth of the yeast or subsequent yields. This method assumes smooth-walled flasks. The use of baffled flasks can lead to faster growth, and therefore, growth and induction times may need to be optimized.

Typically, we purify serpins from 1 l of yeast culture by inoculating two 500 ml batches of YPD in 2 l conical flasks. This batch-based culturing allows easy scalability in the growth and induction phases.

6.1. Pick a single well-isolated colony of yeast from a RDB plate into 5 ml YPD and grow overnight at 30 °C in a shaking incubator.
6.2. Set up two 2 l conical flasks with 500 ml YPD and inoculate each with 1 ml of starter culture.
6.3. Grow yeast for 48 h in a 30 °C incubator with vigorous (200–250 rpm) shaking.
6.4. Remove flasks from the incubator and place on a flat surface overnight (~16 h) at room temperature. This allows the yeast to settle to the bottom of the flask and form a pellet.
6.5. Pour off the YPD and replace with 500 ml YPM. While most of the yeast mass will form a pellet, some may be lost when pouring out the YPD. Also, the use of glycerol as an alternative carbon source will inhibit the settling. In this case, or when losses from pouring are unacceptably high, yeast can be collected by centrifugation at 6000×*g* for 10 min and resuspended in YPM then returned to the conical flasks.
6.6. Place flasks back in shaking incubator and induce for 72 h (or previously determined optimal induction time).

7. Lysis

Chemical lysis of yeast typically results in the release of digestive proteases from the vacuole that can cleave the serpin RCL and decrease yields. Therefore, we prefer to mechanically lyse yeast by vortexing in the presence of glass beads. This process generates a lot of heat which can lead to polymerization. Therefore, it is important to cool the sample as much as possible during lysis.

7.1. Collect induced yeast by centrifugation at 6000×*g* for 10 min at 4 °C.
7.2. Wash once with TBS and resuspend in 60 ml/l culture lysis buffer (1 *M* NaCl, 10 m*M* β-mercaptoethanol, 20 m*M* Tris–HCl, pH 7.4, 2 μg/ml leupeptin, 2 μg/ml aprotinin, 2 μg/ml pepstatin, 1 μ*M* AEBSF).
7.3. Split the sample into 10 ml aliquots and add ~5 ml glass beads (~0.5 mm diameter; Sigma catalog #G9268) on ice.
7.4. Vortex samples at maximum speed in 30 s bursts eight times, with at least 90 s on ice in between. If desired, lysis can be monitored by light microscopy.

7.5. Centrifuge lysates at 3000×*g* for 5 min to separate the glass beads.
7.6. Pool supernatants and clarify by centrifugation at 27,000×*g* for 30 min.

8. Purification

We typically engineer N-terminal hexahistidine tags to assist in the purification of serpins. This can be followed by protease cleavage sites such as TEV to remove the tag, so long as the protease does not also cleave the serpin RCL. Intracellular serpins tend to have low isoelectric points, allowing further purification by anion exchange. Speed and temperature are crucial factors in the purification to avoid cleavage by contaminating proteases. Robotic chromatography apparatus can increase the speed of purification as well as allowing gradient elutions and continuous monitoring of eluted proteins via UV light absorbance.

It should be noted that some factors such as elution points vary between proteins and must be empirically determined and optimized. As such, the following protocol does not include any stringent washing. Once elution points have been identified, washes containing appropriate concentrations of elution buffer can be added at steps 8.2 and 8.11 to shorten the gradient and decrease the elution time.

8.1. Load the clarified lysate directly onto a 5 ml HisTrap column (GE Lifesciences).
8.2. Wash the column with His Wash Buffer (1 *M* NaCl, 10 m*M* β-mercaptoethanol, 20 m*M* Tris, pH 7.4) until the UV absorbance reading reaches baseline.
8.3. Wash the column with 5% His Elution Buffer (1 *M* NaCl, 10 m*M* β-mercaptoethanol, 0.5 *M* imidazole, 20 m*M* Tris, pH 7.4) until the UV absorbance reading reaches baseline.
8.4. Elute with a continuous gradient of 5–100% His Elution Buffer (Fig. 1.1A and B)
8.5. Pool serpin-containing fractions and dialyze against 100 volumes TEV buffer (150 m*M* NaCl, 10 m*M* β-mercaptoethanol, 5% (v/v) glycerol, 20 m*M* Tris–HCl, pH 8.0) for 2 h at 4 °C.
8.6. Estimate protein concentration and add TEV protease to a final molar ratio of 20 serpin: 1 TEV protease.
8.7. Incubate overnight at 4 °C.
8.8. Pass the sample over a 1 ml HisTrap column. Tagged serpin and protease will bind to the column, while the tagless protein will flow

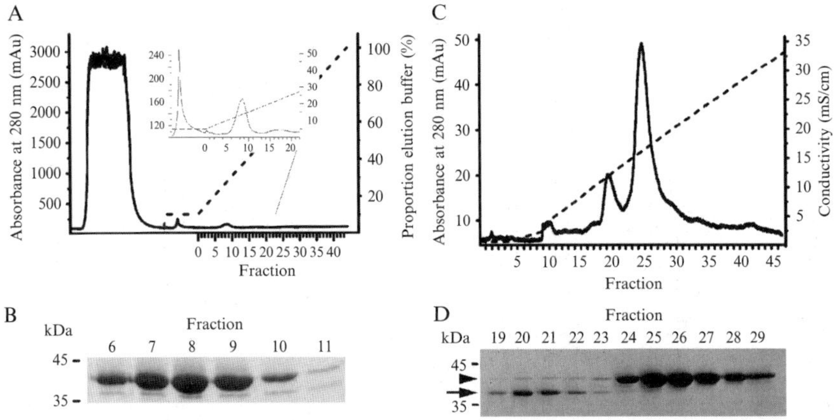

Figure 1.1 *Serpin purification from yeast lysates.* (A) Serpin was purified from clarified yeast lysates over a 1 ml HisTrap column. Inset shows a zoom of the serpin peak eluting over fractions 5–12. Solid line shows UV absorption (left *y*-axis) and dotted line shows proportion of elution buffer (right *y*-axis). (B) Samples of each fraction were resolved by 12.5% SDS-PAGE and stained with Coomassie Brilliant blue to confirm the presence of the serpin. (C) Serpin-containing fractions were pooled and further purification carried out via anion exchange on 1 ml HiTrap Q column. The two peaks represent cleaved and intact serpin. Solid line shows UV absorption (left *y*-axis), and dotted line shows proportion of elution buffer (right *y*-axis). (D) Samples of each fraction were resolved by 12.5% SDS-PAGE and stained with Coomassie Brilliant blue to show the separation of cleaved (arrow) and intact (arrow head) serpin.

through. If desired, the remaining tagged serpin can be recovered by elution with imidazole as described above.

8.9. Collect the flow through (tagless serpin) or pooled serpin-containing fractions (if the tag was not removed) and dialyze overnight at 4 °C against 100 volumes IEX Wash Buffer (20 m*M* HEPES, 10 m*M* β-mercaptoethanol, pH 7.4). Although HEPES is not recommended for anion exchange, we have found that it allows us to separate natively folded serpins from the cleaved, inactivated form (Fig. 1.1C and D).

8.10. Load the semi-purified serpin onto a 1-ml HiTrap Q column (GE Lifesciences). The volume of sample will have increased during the dialysis, and so concentration may be required before loading the anion exchange column.

8.11. Wash the column with IEX Wash Buffer until the UV absorbance reading reaches baseline.

8.12. Elute with a continuous gradient of 0–100% IEX Elution Buffer (1 *M* NaCl, 10 m*M* β-mercaptoethanol, 20 m*M* HEPES, pH 7.4) and pool the serpin-containing fractions (Fig. 1.1C and D).

8.13. If necessary, serpins can be concentrated; however, high concentration leads to polymerization of most serpins, although the concentration at which the serpin will polymerize varies depending on the protein and buffer conditions such as pH.

9. Assessing Serpin Activity and Removing Inactive Forms

Inactive serpin may be recovered as either full-length or cleaved species. Cleaved serpin is normally due to protease activity within the exposed RCL and can be easily identified by resolving the purified protein by reducing SDS-PAGE, since cleavage liberates a C-terminal fragment of approximately 4 kDa. Full-length inactive material may be due to either polymerization or latency. Serpin polymers are noncovalent and break down to monomers in the presence of SDS, while latent serpin adopts a subtly different, but totally inactive, fold. The simplest way to assay serpin activity is to take advantage of the susceptibility of natively folded serpin to proteolysis within the RCL.

9.1. Incubate serpin with and without a molar excess of protease at 37 °C for 30 min.
9.2. Resolve products by 12.5% SDS-PAGE and visualize by Coomassie Brilliant blue (Fig. 1.2).
9.3. Serpin incubated in the absence of protease should appear as a single product of approximately 42 kDa. Incubation with protease should lead to loss of the native serpin, while new species should appear corresponding to cleaved serpin and the serpin–protease complex.

Importantly, this method does not require a protease to be specifically inhibited by the serpin, only that it is able to cleave the RCL sequence. The presence of cleaved serpin in the absence of protease indicates cleavage occurred during the purification process, while retention of the full-length serpin indicates the presence of misfolded (polymeric or latent) material. Most serpins can be purified at close to 100% activity, although introduction of destabilizing mutations can affect this. Higher activity can also be ensured by increasing the concentration of protease inhibitors and decreasing temperature during the purification.

Should optimization fail to increase the active yield, serpin polymers can be easily removed through the use of size exclusion chromatography, while it has been shown that the extracellular serpin PAI1 can be converted from the latent to active fold by partial unfolding induced by heat or chemical denaturants such as guanidine (Hekman and Loskutoff, 1985; Katagiri *et al.*, 1988; Lomas *et al.*, 1995). Further, all three forms of inactive serpin can

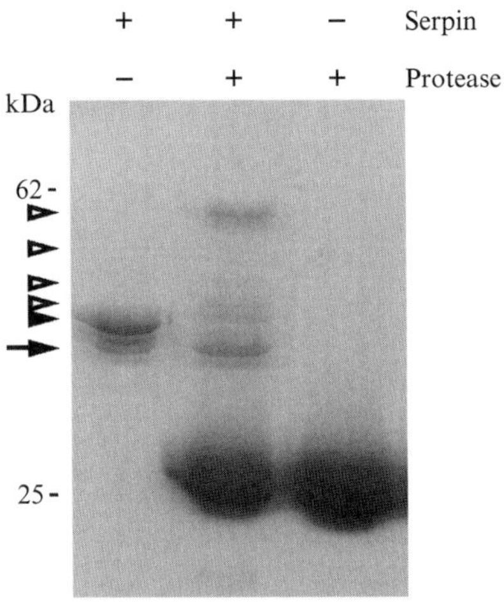

Figure 1.2 *Identifying active serpin.* Serpin was incubated with and without protease for 20 min at 37 °C then resolved by 12.5% SDS-PAGE and stained with Coomassie Brilliant blue. Native serpin (black arrow head) is either cleaved by the protease (black arrow) or forms an inhibitory complex that is subsequently degraded by free protease (white arrow heads).

exhibit subtly different surface properties allowing separation from native serpin by ion exchange, heparin affinity, or hydrophobic interaction chromatography (Fig. 1.1C and D) (Heger *et al.*, 2002; Karlsson and Winge, 2001; Lomas *et al.*, 1995).

10. Production of Polymerogenic Serpins

The metastable nature of the serpin fold is required for its inhibitory function but can make the production of recombinant protein difficult due to its tendency to polymerize. This issue is exacerbated in the case of the serpinopathies, a group of diseases characterized by the polymerization of serpins *in vivo*. *In vitro* investigations of these serpin mutants are hindered by their intrinsically unstable nature that can activate the unfolded protein response in the ER or decrease the efficiency of refolding procedures. Intracellular expression in yeast offers a new pathway to the production of these medically relevant proteins. By simply removing the signal peptide from the N-terminus, the serpin is directed into the cytoplasm to fold, allowing it to avoid stressing the ER, while still folding in a eukaryotic, chaperoned system. Recently, this system was used to produce recombinant Z α_1-antitrypsin (Z AAT) for the first time (Levina *et al.*, 2009).

The Z mutation (E342K) of α_1-antitrypsin destabilizes the folding pathway, resulting in the accumulation of toxic polymers within the liver and a correlating drop in circulating protein that ultimately results in emphysema (Perlmutter, 2006; Stoller and Aboussouan, 2005). Although recombinant protein would allow more in-depth studies into the mechanisms of Z AAT disease, its propensity to polymerize has long frustrated attempts at recombinant production and it is usually purified from plasma of affected patients.

10.1. Z AAT containing an N-terminal hexahistidine tag was cloned into pHILD2 for intracellular yeast expression.
10.2. The plasmid was linearized with *Sac*I and transformation of the SMD1163 strain and consequent screening performed as described in Sections 4 and *5*.
10.3. Large-scale cultures (4 × 500 ml) in YPD were grown, shaking, at 30 °C for 30 h and yeast allowed to settle overnight.
10.4. The medium was replaced with YPM and cells were induced for 72 h.
10.5. Cells were collected by centrifugation at 6000×*g* for 10 min, resuspended in 100 ml lysis buffer (500 m*M* NaCl, 25 m*M* imidazole, 1 m*M* β-mercaptoethanol, 2.5 m*M* EDTA, one complete protease inhibitor tablet, 2 m*M* PMSF, 10 m*M* Tris–HCl, pH 8.0) and lysed by the glass bead method.
10.6. The glass beads were removed by centrifugation at 3500×*g* for 5 min after which 100 μl Triton X-100 (final concentration 0.1%, v/v) and 1 mg DNAse (final concentration 1 μg/ml) were added to the lysates.
10.7. Lysates were clarified by centrifugation at 48,000×*g* for 60 min at 15 °C with a slow deceleration so as to not disrupt the lipid layer.
10.8. The lipid layer (top layer) was removed by pipetting and discarded, while the underlying protein supernatant was collected and filtered through a 0.45-μm membrane.
10.9. The clarified lysate was diluted fourfold with 500 m*M* NaCl, 25 m*M* imidazole, 1 m*M* β-mercaptoethanol, 0.1% (v/v) Triton X-100, 20 m*M* Tris–HCl, pH 8.0 to lower the concentration of EDTA and then loaded onto a 1 ml HisTrap column.
10.10. The column was washed with 500 m*M* NaCl, 25 m*M* imidazole, 0.5% (v/v) Triton X-100, 1 m*M* β-mercaptoethanol, 20 m*M* Tris–HCl, pH 8.0 and then a low salt wash in 50 m*M* NaCl, 25 m*M* imidazole, 0.5% (v/v) Triton X-100, 1 m*M* β-mercaptoethanol, 20 m*M* Tris–HCl, pH 8.0.
10.11. Protein was eluted in 50 m*M* NaCl, 1 m*M* β-mercaptoethanol, 0.5% (v/v) Triton X-100, 20 m*M* Tris–HCl, pH 8.0 along a 0–500 m*M* imidazole gradient.
10.12. Protein-containing fractions were pooled and diluted 1:3 with 75 m*M* NaCl, 50 m*M* Tris–HCl, pH 8.0 before loading onto a 1-ml HiTrap Q column.

10.13. The column was washed with 75 m*M* NaCl, 50 m*M* Tris, pH 8.0 and Z AAT was eluted in 50 m*M* Tris, pH 8.0 over a 75–300 m*M* NaCl gradient. The native Z AAT typically elutes below 200 m*M* NaCl, while polymers elute at higher concentrations.

This method recovers approximately one-third of the protein as intact, functional monomers with the remaining protein present primarily as polymers and a small amount of cleaved protein.

REFERENCES

Heger, A., Grunert, T., Schulz, P., Josic, D., and Buchacher, A. (2002). Separation of active and inactive forms of human antithrombin by heparin affinity chromatography. *Thromb. Res.* **106,** 157–164.

Hekman, C. M., and Loskutoff, D. J. (1985). Endothelial cells produce a latent inhibitor of plasminogen activators that can be activated by denaturants. *J. Biol. Chem.* **260,** 11581–11587.

Karlsson, G., and Winge, S. (2001). Separation of native and latent forms of human antithrombin by hydrophobic interaction high-performance liquid chromatography. *Protein Expr. Purif.* **21,** 149–155.

Katagiri, K., Okada, K., Hattori, H., and Yano, M. (1988). Bovine endothelial cell plasminogen activator inhibitor. Purification and heat activation. *Eur. J. Biochem.* **176,** 81–87.

Levina, V., Dai, W., Knaupp, A. S., Kaiserman, D., Pearce, M. C., Cabrita, L. D., Bird, P. I., and Bottomley, S. P. (2009). Expression, purification and characterization of recombinant Z alpha(1)-antitrypsin—The most common cause of alpha(1)-antitrypsin deficiency. *Protein Expr. Purif.* **68,** 226–232.

Lomas, D. A., Elliott, P. R., Chang, W. S., Wardell, M. R., and Carrell, R. W. (1995). Preparation and characterization of latent alpha 1-antitrypsin. *J. Biol. Chem.* **270,** 5282–5288.

Perlmutter, D. H. (2006). Pathogenesis of chronic liver injury and hepatocellular carcinoma in alpha-1-antitrypsin deficiency. *Pediatr. Res.* **60,** 233–238.

Stoller, J. K., and Aboussouan, L. S. (2005). Alpha1-antitrypsin deficiency. *Lancet* **365,** 2225–2236.

Wu, S., and Letchworth, G. J. (2004). High efficiency transformation by electroporation of *Pichia pastoris* pretreated with lithium acetate and dithiothreitol. *Biotechniques* **36,** 152–154.

CHAPTER TWO

Production of Recombinant Serpins in *Escherichia coli*

Mary C. Pearce* *and* Lisa D. Cabrita†

Contents

1. Introduction 14
   1.1. Recombinant expression of serpins in *Escherichia coli* 14
   1.2. Choice of vector and gene construct 14
   1.3. Host strain for serpin expression 16
   1.4. Expression conditions to improve soluble expression 16
   1.5. Purification strategies for isolation of recombinant serpins 17
2. Experimental Procedures for the Production of Serpins in *E. coli* 20
   2.1. Insoluble expression of AT in *E. coli* 20
   2.2. Expression of soluble AT in *E. coli* 21
3. Some Common Variations Used for Expression of Other Serpins 21
   3.1. Uniform isotopic labeling 21
   3.2. Nonnatural amino acid incorporation 22
4. Preparation of AT from Inclusion Bodies 23
5. Preparation of Soluble AT 24
   5.1. Optional purification strategies 26
References 27

Abstract

Serpins represent a diverse family of proteins that are found in a wide range of organisms and cellular locations. In order to study them, most need to be produced recombinantly, as isolation from their source is not always possible. Due to their relatively uncomplicated structure (single domain, few posttranslational modifications), the serpins are usually amenable to expression in *Escherichia coli*, which offers a fast and cost-effective solution for the generation of large amounts of protein. This chapter outlines the general procedures used in the expression and subsequent purification of serpins in *E. coli*, with a particular focus on the methods used for antitrypsin, the archetypal member of the family.

* Department of Biochemistry, Monash University, Clayton, Australia
† Department of Structural and Molecular Biology, University College London, London, United Kingdom

Methods in Enzymology, Volume 501
ISSN 0076-6879, DOI: 10.1016/B978-0-12-385950-1.00002-X

1. Introduction

1.1. Recombinant expression of serpins in *Escherichia coli*

The production of recombinant proteins is critical for many avenues of research that rely on isolating significant quantities of sample that are otherwise difficult to obtain from their natural source. To this end, much work has focused on developing prokaryotic and eukaryotic expression systems that produce specific recombinant proteins of interest, with variants of *Escherichia coli* representing the most popular prokaryotic organism used. There are several distinct advantages in using *E. coli*: the genetics are well established and the organism can be manipulated easily and cultured quickly, where in most cases, the protein can be overexpressed and the cells ready for protein purification within 24 hours. It is also an affordable system, which can easily be scaled up to produce large quantities, both within the laboratory and in industry alike.

Despite the diversity found within the amino acid sequence and the organisms from which they are derived, many serpins have been produced successfully in *E. coli* in suitable quantities for structural, biophysical, and functional analyses (Bird *et al.*, 2004). This is in part owing to the advancements made in the development of *E. coli* expression systems, which allow conditions to be modulated that can improve the overexpression of a given protein.

Serpins are single domain proteins that generally possess very few basic posttranslational modifications which typically makes them generally amenable to expression in prokaryotic systems. For example, antitrypsin (AT) is a 45-kDa serpin that is glycosylated within plasma, although its glycosylation has not been found to be critical for either its structure or function. Some serpins contain disulfide bonds, which can be easily formed in some instances, such as in aeropin (using a suitable strain of *E.coli*), however this is not straightforward in all cases, for example the expression in *E.coli* of the disulfide-containing and glycoslated antithrombin (ATIII) is not well tolerated.

As many serpins have been produced recombinantly for a range of applications, to present the methods for the expression and purification of each serpin is well outside the scope of this article. We will therefore present the general features associated with the recombinant production of serpins, with a focus on the most well-characterized member of the family: AT.

1.2. Choice of vector and gene construct

Within *E. coli*, the stability of mRNA is in of the order of minutes, where stabilizing motifs at the 5′- and 3′-ends can greatly enhance the lifetime of mRNA transcripts and where plasmid copy number can often influence plasmid stability (Coburn and Mackie, 1999). For the expression of serpins, however, the choice of promoter within a given vector has the greatest influence on the distribution of soluble and insoluble protein during overexpression. Several different systems have been used for the production of recombinant AT in order to improve the yield of soluble, monomeric protein following protein purification. Early production of AT was driven by a strong T7 promoter, where expression in the presence of the bacterial RNA polymerase inhibitor rifampicin helped drive the protein into inclusion bodies (Hopkins *et al.*, 1993), and the recovery of monomeric material was possible through a renaturation procedure. While the early inclusion body preparation method was adequate for wild-type AT, or for variants with conservative mutations, variants that possessed a detrimental change in sequence would result in little to no protein following purification (S.P. Bottomley, unpublished). More recently, the use of a softer T5 promoter (pQE vector; Qiagen) has been the most successful approach in the isolation of significant quantities of soluble material (3–5 mg/L; Zhou *et al.*, 2001). Other promoters that have been used successfully for serpins include the IPTG-inducible *tac* promoter for squamous cell carcinoma antigen 1 (SCCA1; Katagiri *et al.*, 2006) and SCCA2 (Barnes *et al.*, 2000), antichymotrypsin (ACT; Rubin *et al.*, 1990), the tryptophan-inducible *trp* promoter plasminogen activator inhibitor-1 (PAI-1; Sancho *et al.*, 1994), and the temperature-sensitive promoter λpl which has been used for pigment endothelium-derived factor (PEDF; Becerra *et al.*, 1993), AT (Johansen *et al.*, 1987), and ACT (Rubin *et al.*, 1990). In addition to the strength of promoter, it has been found that truncation of the N-termini of serpins, by 20–40 amino acids, can improve the yield of soluble expression as observed for AT (Hopkins *et al.*, 1993), antiplasmin (Law *et al.*, 2008), and maspin (Blacque and Worrall, 2002).

At present, most serpins are constructed with an N-terminal hexahistidine tag, which is the most efficient means of isolating protein. The production of serpin variants possessing detrimental mutations is often limited due to the propensity for these proteins to aggregate, even in the soluble pQE expression system; therefore, there have been several attempts at improving yield through the addition of solubility tags, which have had mixed success. Fusing MBP to AT has previously resulted in the production of soluble AT, a similar approach, however, showed no significant improvement in the recovery of the destabilized ZAT variant (Glu342Lys; Levina *et al.*, 2009). Although more recently, yields of 7–9 mg/L have since been reported (Agarwal *et al.*, 2010), which represents a significant improvement in the recovery of soluble material. The use of GST fusions have also been successful in the production of SCCA1 and 2 (Bartuski *et al.*, 1998) and PEDF (Zhang *et al.*, 2005).

Research groups differ in their preference for the soluble versus insoluble expression systems for AT and the expression of most serpins follows a strategy very similar to the one presented in this chapter.

1.3. Host strain for serpin expression

The *E. coli* strain BL21(DE3) and its derivatives are the typical workhorses for the overexpression of serpins under a T7 promoter and offer the advantage of circumventing general recombinant expression issues relating to toxicity, codon bias, and protease sensitivity.

No toxicity in *E. coli* has been reported for any of the dozens of serpins expressed in *E. coli*; however, the risk of proteolysis is an issue due to the protease-sensitive reactive center loop. Proteolysis is typically an undesirable effect observed during expression (and purification), which results in monomeric, but inactive material. Although the cleaved material can be removed during purification (see Section 5.1), the use of *E. coli* strains deficient in proteases of the cytoplasm, Lon, and ClpYQ, such as BL21 (DE3), also serves to reduce such effects.

As most serpins are largely of eukaryotic origin (e.g., humans), the risk of a codon usage bias may exist during expression within *E. coli*. Certain codons for arginine (AGG/AGA/CGA), isoleucine (ATA), proline (CCC), and leucine (CTA) are associated with low tRNA levels within *E. coli* and can result in low expression. The use of *E. coli* strains carrying plasmids coding for these rare tRNAs, such as Rosetta (Novagen), has been reported to improve the expression of MBP-fused AT (Agarwal *et al.*, 2010) and protein C inhibitor (Li *et al.*, 2007), bovine endopin (Hwang *et al.*, 2005), as well as serpins from microbial organisms such as *Pyrobaculum aerophilum* (Cabrita *et al.*, 2007) and *Eimeria acervulina* (Fetterer *et al.*, 2008).

Posttranslational modifications are not commonly performed by *E. coli*; however, several strains exist that carry deletions within the thioredoxin reductase genes, for example, in Rosetta-gami(DE3) (Novagen) and AD494(DE3) (Novagen), which have been used to express the disulfide-bond-containing serpins aeropin (Cabrita *et al.*, 2007) and, most recently, bomapin (Przygodzka *et al.*, 2010), respectively.

1.4. Expression conditions to improve soluble expression

Conditions used for the overexpression of serpins range considerably and are often determined empirically. The production of serpins for various applications has also been developed, including procedures for selective isotopic labeling for NMR studies (see Section 3.1) and nonnatural amino acid incorporation for biophysics (see Section 3.2).

Some typical expression conditions used for AT are described in Section 2 and include the induction of expression at OD_{600} 0.5–0.8 for a period of 3–5 h at a temperature of 37 °C. These conditions can be easily applied to other serpins, where variations in temperature (16–37 °C) and induction time (up to 20 h) can be used to modulate the extent to which soluble material is produced. Improvements in the yield of soluble material for a destabilized C-terminal truncation variant of thermopin were achieved by a combination of lower induction temperature (21 °C compared to wild-type 37 °C) and the addition of 5% (v/v) ethanol (Irving *et al.*, 2003).

1.5. Purification strategies for isolation of recombinant serpins

The choice of purification strategy of a serpin depends largely on the level of expression and the properties of the serpin itself and can require optimization at several levels. Moreover, following the isolation of soluble serpins or as insoluble inclusions, procedures are required to ensure that the protein recovered is homogeneous and has inhibitory activity (for an inhibitory serpin). Figure 2.1 depicts a typical approach for purifying a serpin that is to be attempted in *E. coli* for the first time. The first step in a typical purification procedure for a serpin (or indeed any protein produced recombinantly in *E. coli*) requires lysis of the cellular membrane which is then followed by a series of chromatographic steps to isolate the protein of interest.

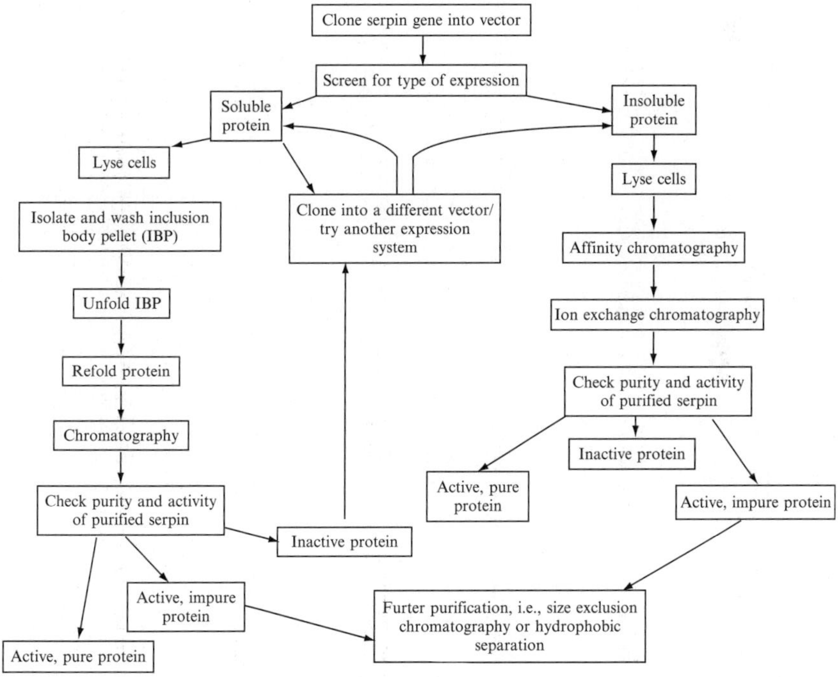

Figure 2.1 Workflow to follow when attempting to express a serpin in *E. coli.*

1.5.1. Lysis

For the purification of serpins, a buffer containing lysozyme and protease inhibitors is often used to resuspend the cellular pellet (Sections 4 and 5), which produces spheroplasts of the *E. coli*. There are several options available for lysis that are judged empirically, with the most common means including freeze-thawing and sonication. A French press and a cell homogenizer are alternatives that offer equivalent results in lysis; these are gentler methods that minimize heat generation often associated with sonication, and therefore reduce the possibility of polymerization as well as reduce the possibility of proteolytic attack.

1.5.2. Recovering overexpressed protein—Soluble versus insoluble purification

Samples are often assessed for the distribution of soluble and insoluble expression of the protein following lysis. SDS-PAGE is used to compare an uninduced sample with lysate containing soluble protein and the pelleted material, which contains the inclusion bodies of aggregated protein. The cell pellet is often a rich source of protein, which is typically free from proteolysis; however, it requires more extensive processing compared to purification of the soluble fraction. A series of centrifugation steps using a mild detergent to wash the pellet removes cellular components (e.g., membranes, nucleic acids), and the pellet is then dissolved in a denaturant (e.g., GdnHCl or urea) together with a reducing agent (e.g., DTT), which then requires renaturation (refolding). There are many strategies for refolding proteins, which are judged empirically. For more information, there are some extensive reviews on this subject (Baneyx, 1999; Cabrita and Bottomley, 2004). More recently, the REFOLD database (http://refold.med.monash.edu.au; Buckle *et al.*, 2005; Chow *et al.*, 2006) was developed which compiles known refolding protocols. One strategy favored for the refolding of serpins involves the dropwise dilution of the denatured material (Section 2.2) into a refolding buffer, which has been used successfully for AT (Hopkins *et al.*, 1993) and, more recently, neuroserpin (Belorgey *et al.*, 2002). Following refolding, the protein is purified further, typically with ion exchange and/or size exclusion chromatography.

In contrast to the procedures required to isolate material from the insoluble fraction, purification of the soluble material can be relatively straightforward, where affinity tags (His, MBP, GST) are often used to selectively purify the protein. The most common affinity tag, hexa-histidine, is often the first choice for a serpin as it typically does not perturb the structural or functional characteristics of the protein, and the purification process is both highly selective and involves relatively mild conditions (Section 5).

The dynamic nature of the serpin structure is an important consideration when attempting to isolate monomeric and active material (where it is appropriate) for structural or functional studies. Serpins are metastable and under appropriate conditions can adopt more stabilized conformations, such as the cleaved, latent, and polymeric states. The extent to which a serpin may adopt an alternative conformation relates to both the type of serpin and the impact of mutation. For example, wild-type AT can typically be found as both monomeric and inactive polymers during purification (see Fig. 2.2 depicting separation on a Q Sepharose column), while ACT has been shown recently to have the propensity for forming the inactive, delta conformation (Pearce *et al.*, 2010).

Of the serpins that have been studied, typically a combination of purification steps are required; following lysis, this often begins with an affinity step (e.g., hexa-histidine tag purification) which efficiently recovers the serpin from the lysate, but can contain varying levels of impurities and does not distinguish between the various native and nonnative conformations. The sample is often further purified with a second chromatographic step, often ion exchange, which takes advantage of the intrinsic charge of the serpin. This can be used to further purify the sample, and in the case of anion exchange (e.g., use of a HiTrap Q Sepharose column; GE Healthcare), this can separate monomer from polymer species, as observed for AT (Fig. 2.2). The size exclusion chromatography (e.g., Superdex 200 or Superdex 75; GE Healthcare) can also be used to separate monomer from the polymer species and is often used as a final "polishing" step. In some instances, the monomeric species is a mixed population of both native and nonnative species, such as latent or cleaved. A hydrophobic interaction column (e.g., Phenyl Sepharose; GE Healthcare) can be used to separate these species (Fig. 2.3, Section 5.1).

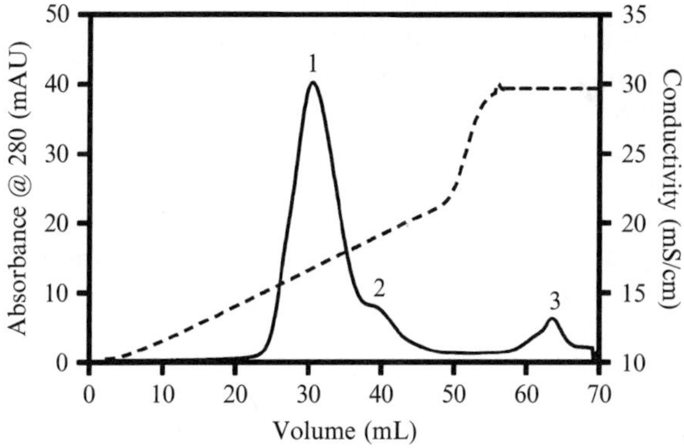

Figure 2.2 Typical elution profile for AT from a Q Sepharose column. The solid line represents absorbance, and the dashed line represents conductivity. The peaks are numbered and correspond to the following species during purification: (1) native AT, (2) cleaved AT, and (3) polymeric AT.

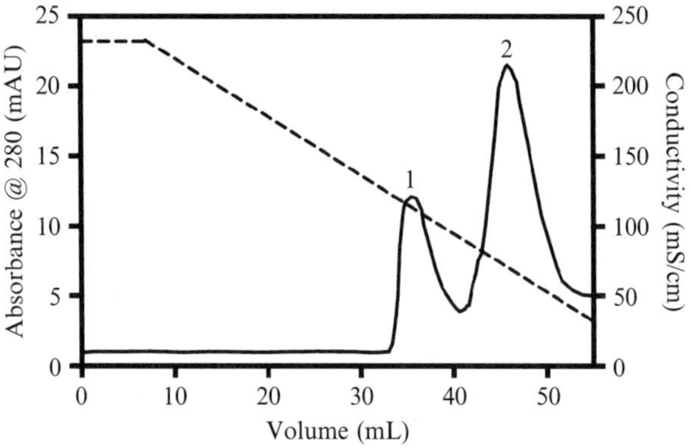

Figure 2.3 Typical elution profile for ACT from a Phenyl Sepharose column. The solid line represents absorbance, and the dashed line represents conductivity. The peaks are numbered and correspond to the following species during purification: (1) delta ACT and (2) native ACT. Polymeric ACT will elute at lower salt concentrations.

2. Experimental Procedures for the Production of Serpins in *E. coli*

Here, we present detailed methods for expression of AT for the production of both soluble and insoluble protein. In addition, we will present common variations to this method that may have been used for either AT or other members of the serpin family. Following this, the purification strategies most commonly used for AT are presented.

2.1. Insoluble expression of AT in *E. coli*

The first method of producing AT in *E. coli* utilized the pTermat expression vector, under the control of the T7 promoter, and resulted in a significant proportion of the recombinant protein being found in inclusion bodies. This method has been adapted from several sources (Hopkins *et al.*, 1993; Kwon *et al.*, 1995) and, despite the current strategies for producing soluble material, remains a common procedure for AT.

1. Pick a single colony of BL21(DE3) cells containing the pTermat AT expression vector from a fresh 2xYT agar plate and inoculate a 50-mL starter culture of 2xYT medium containing ampicillin 100 μg/mL.
2. Grow the starter culture overnight at 37 °C, shaking at 200 rpm for 12–16 h.
3. Use 10 mL of starter culture to inoculate 1 L of 2xYT medium containing 100 μg/mL ampicillin in a 2-L flask.

4. Incubate the flask at 37 °C, shaking at 200 rpm for approximately 2–3 h until the OD_{600} is 0.5–0.7, at which point add IPTG to a final concentration of 0.5–1 m*M*.
5. After 30 min, add 500 μL of 100 mg/mL rifampicin to each 1 L of culture.
6. Following a further 4 h of shaking at 200 rpm at 37 °C, harvest the cells by centrifugation at 5500 rpm for 10 min in a rotor precooled to 4 °C.
7. Decant the supernatant and resuspend the pellet in a small volume of 50 m*M* Tris, 100 m*M* NaCl, 10 m*M* EDTA, pH 8.0, so that the final total volume of 20 mL is used per 1 L of culture. Freeze the cell pellets and store at − 80 °C until purification.

2.2. Expression of soluble AT in *E. coli*

Placing AT expression under the control of the T5 promoter in the pQE30 vector (Qiagen) results in high levels of soluble material for purification, which is facilitated by an N-terminal hexa-histidine tag fused to the N-terminus to aid in purification.

1. Pick a single colony of SG13009 cells containing both the pREP4 and pQE30 AT expression vectors from a fresh 2YT agar plate and inoculate a 50-mL starter culture of 2xYT medium containing 100 μg/mL ampicillin and 30 μg/mL kanamycin.
2. Grow the starter culture at 37 °C, shaking at 200 rpm for 12–16 h.
3. Use 10 mL of starter culture to inoculate 1 L of 2xYT medium containing 100 μg/mL ampicillin and 30 μg/mL kanamycin in a 2-L flask.
4. Incubate the flask at 200 rpm, 37 °C for approximately 2–3 h until the OD_{600} is 0.5–0.7, at which point add IPTG to a final concentration of 0.5–1 m*M*.
5. Following 4 h of shaking at 200 rpm at 37 °C, harvest the cells by centrifugation at 5500 rpm for 10 min in a rotor precooled to 4 °C.
6. Remove the supernatant and resuspend the pellet in a small volume of 50 m*M* Tris, 100 m*M* NaCl, 10 m*M* EDTA, pH 8.0, so that a final total volume of 20 mL is used per 1 L of culture. Freeze the cell pellets and store at − 80 °C until purification.

3. Some Common Variations Used for Expression of Other Serpins

3.1. Uniform isotopic labeling

For the production of uniformly isotopically labeled AT, M9 minimal medium is supplemented with 2 m*M* $MgSO_4$, 0.1 m*M* $CaCl_2$, and where the carbon and nitrogen sources are replaced, ^{13}C glucose (2 g/L) and/or ^{15}N

ammonium chloride (1 g/L), respectively, where appropriate. The pQE soluble expression system has been used previously for this method; however, the insoluble fraction has been used for purification following a combination of the two methods presented in Section 2.1 (Peterson *et al.*, 2000).

1. Select a single colony of SG13009 cells containing AT in the pQE30 vector and inoculate a 50-mL starter culture of 2xYT medium containing 100 μg/mL ampicillin and 30 μg/mL kanamycin.
2. Grow the starter culture at 37 °C, shaking at 200 rpm for 12–16 h.
3. Use 10 mL of starter culture to inoculate 1 L of M9 minimal medium supplemented with 1 g/L ^{15}N ammonium chloride or 2 g/L of ^{13}C glucose containing 100 μg/mL ampicillin and 30 μg/mL kanamycin in a 2-L flask.
4. Shake the flask at 200 rpm, 37 °C, for approximately 2–3 h until the OD_{600} is 0.5, at which point add IPTG to a final concentration of 1 m*M*.
5. Following 4.5 h of shaking at 200 rpm, 37 °C, harvest the cells by centrifugation at 5500 rpm for 10 min in a rotor precooled to 4 °C.
6. Decant the supernatant and resuspend the pellet in a small volume of 50 m*M* sodium phosphate buffer, pH 7.4, containing 300 m*M* sodium chloride.

3.2. Nonnatural amino acid incorporation

The production of tryptophan-substituted PAI-1 variants has been achieved with the use of an auxotrophic strain of *E. coli*, W3110TrpA88(DE3)pLysS together with the T7 promoter containing pET-24d vector, which has been successful for the generation of selectively labeled protein for both NMR spectroscopy (Abbott *et al.*, 2004) and in fluorescence studies (Blouse *et al.*, 2002).

1. Select a single colony of W3110TrpA88(DE3)pLysS cells containing PAI-1 in the pET-24d vector and inoculate a 50-mL starter culture of 2xYT medium containing ampicillin 100 μg/mL.
2. Grow the starter culture at 37 °C, shaking at 200 rpm for 12–16 h.
3. Use 10 mL of starter culture to inoculate 1 L of M9 minimal medium supplemented with 2% (w/v) casamino acids, 0.01% (w/v) thiamine, 2 m*M* $MgSO_4$, 0.4% (v/v) glucose, 0.4% (v/v) glycerol, 0.1 m*M* $CaCl_2$, and 0.02 m*M* L-tryptophan.
4. Grow the cells at 37 °C, shaking at 200 rpm for 16 h to an OD_{600} of 0.8–1.0, at which point all the L-tryptophan will be depleted.
5. Add the tryptophan analogue (e.g., DL-5F-Trp for NMR or DL-7A-Trp for fluorescence) to the minimal medium to a final concentration of 0.5 m*M* and incubate at 37 °C, shaking at 200 rpm for 10 min.
6. Induce expression by the addition of IPTG (final concentration of 0.5–1 m*M*) and incubate at 37 °C, shaking at 200 rpm for 2–3 h.

7. Following expression, harvest the cells by centrifugation at 5500 rpm for 10 min in a rotor precooled to 4 °C.

4. Preparation of AT from Inclusion Bodies

Materials

Buffer P: 50 m*M* Tris–HCl, 100 m*M* NaCl, 10 m*M* EDTA, pH 8.0.
Buffer P wash: 50 m*M* Tris–HCl, 100 m*M* NaCl, 10 m*M* EDTA, 0.5% (v/v) Triton X-100.
Unfolding buffer: 50 m*M* Tris–HCl, 50 m*M* NaCl, 100 m*M* DTT, 6 *M* guanidine–HCl, pH 8.0.
Refolding buffer: 50 m*M* Tris–HCl, 50 m*M* NaCl, 5 m*M* DTT, pH 8.0.
Buffer A: 50 m*M* Tris–HCl, 50 m*M* NaCl, pH 8.0.
Buffer B: 50 m*M* Tris–HCl, 350 m*M* NaCl, pH 8.0.

1. Resuspend the cells from a 2-L expression of pTermat AT in 50 mL of buffer P together with the addition of 100 mg of lysozyme (in 1 mL of buffer P) and 1 m*M* PMSF (dissolved in isopropanol).
2. Incubate the cells on ice for 20 min, and then sonicate 4 × 30 s on/off at 10 kHz, also on ice.
3. Separate the inclusion bodies from other cellular debris by centrifugation at 8000 rpm in an appropriate high-speed, fixed angle rotor for 20 min at 4 °C.
4. Resuspend the inclusion bodies in 30 mL buffer P by placing a small magnetic stir bar in the tube and agitate a vortex until the pellet is dissolved, and then remove the stir bar.
5. Add 500 μL of 1 mg/mL DNaseI to the tube, then sonicate again 4 × 30 s on/off at 10 kHz on ice.
6. Centrifuge the sample at 6000 rpm in an appropriate high-speed, fixed angle rotor for 20 min at 4 °C.
7. Wash the pellet three times with buffer P wash, resuspending the sample each time as described in (4) of this protocol, followed by centrifugation at 6000 rpm at 4 °C for 20 min.
8. Resuspend the pellet in 30 mL buffer P as described in (4) above, followed by centrifugation at 6000 rpm at 4 °C for 20 min.
9. Add 1 mL of buffer P to the pellet and resuspend to a thick slurry using a magnetic stir bar as described above (4) to loosen the pellet.
10. Add 10 mL of unfolding buffer to the slurry and mix until no visible lumps remain. The solution should be clear and slightly brown.
11. Use a peristaltic pump, such as the P1 pump by GE Healthcare, to add the unfolded protein dropwise to 2 L of refolding buffer. This step

should be performed at 4 °C over 5–12 h with constant stirring to ensure the local protein concentration does not get too high and to enable the slow process of refolding to take place.

12. Remove the precipitated material by centrifugation of the refolded solution at 10,000 rpm at 4 °C for 30 min. Carefully decant the supernatant.
13. Equilibrate a 20-mL Q Sepharose column (loose Q Sepharose resin obtained from GE Healthcare) with buffer A at 2 mL/min.[1]
14. Load refolded, clarified protein solution onto Q Sepharose column at 2 mL/min.
15. Wash out any unbound protein by equilibrating the column with buffer A until Abs_{280} returns to baseline.
16. Elute AT using a 0–350 m*M* linear NaCl gradient using buffer B over 20 column volumes at 2 mL/min, collecting 5-mL fractions.
17. Analyze fractions for purity by running samples of fractions containing protein on 10% (w/v) SDS and native PAGE gels to look for contamination by aggregates or other proteins.
18. Perform an inhibitory activity assay (described elsewhere in this book) to determine which peak contains active, monomeric protein. The peak at approximately 100 m*M* NaCl will contain monomeric AT.
19. Pool all active, monomeric fractions and concentrate[2] to the desired level, then aliquot and store at − 80 °C.

5. Preparation of Soluble AT

The most common form of AT to be produced is with an N-terminal hexa-histidine tag. Purification is therefore facilitated by an immobilized metal affinity resin such as Ni–NTA, followed by separation of the various conformers present on an anion exchange column.

Buffers

Ni–NTA

Equilibration buffer: 25 m*M* $NaPO_4$, 500 m*M* NaCl, pH 8.0.

Wash buffer 1: 25 m*M* $NaPO_4$, 500 m*M* NaCl, 25 m*M* imidazole, pH 8.0.

Wash buffer 2: 25 m*M* $NaPO_4$, 50 m*M* NaCl, 25 m*M* imidazole, pH 8.0.

Elution: 25 m*M* $NaPO_4$, 50 m*M* NaCl, 250 m*M* imidazole, pH 8.0.

[1] Loose Q Sepharose resin can be substituted for a HiTrap Q Sepharose column (GE Healthcare). Select flow rates according to company recommendations, maintain gradient at 20 column volumes, and reduce fraction size.

[2] AT can be concentrated to approximately 10–15 mg/mL and remain monomeric but must be frozen.

Q Sepharose
Buffer A: 50 m*M* Tris, 50 m*M* NaCl, 5 m*M* EDTA, pH 8.0.
Buffer B: 50 m*M* Tris, 1 *M* NaCl, 5 m*M* EDTA, pH 8.0.

Columns
5 mL HisTrap HP (GE Healthcare) [or 5 mL Ni–NTA agarose resin, e.g., Qiagen].
5 mL Q Sepharose HiTrap HP (GE Healthcare).

1. Resuspend the cells from 4 L of culture medium in equilibration buffer (50 mL) containing 100 mg of lysozyme and 1 m*M* PMSF (dissolved in isopropanol).
2. Incubate the cells on ice or on a rotary shaker at 4 °C for 20 min.
3. Sonicate the cells 6 × 30 s on/off on ice at 10 kHz.
4. Clear the cell lysate by centrifugation at 20,000 rpm at 4 °C for 35 min in an appropriate high-speed, fixed angle rotor.
5. Equilibrate the HisTrap column with 10 column volumes of equilibration buffer at 5 mL/min[3] on an AKTA FPLC (GE Healthcare).
6. Dilute the cell lysate with an equal volume of equilibration buffer, and then load onto the HisTrap column at 3 mL/min.
7. Wash out any unbound protein with wash buffer 1 at 3 mL/min until Abs_{280} reaches baseline (typically 10–15 column volumes).
8. Wash the column with 5 column volumes of wash buffer 2 to reduce the NaCl concentration on the column.
9. Elute bound protein with a 0–250 m*M* imidazole gradient using elution buffer over 20 column volumes at 3 mL/min, collecting 5–10-mL fractions.[4]
10. Assess protein purity by running samples of fractions using 10% (w/v) SDS-PAGE. Pool fractions that contain AT and mix with buffer A in equal volumes.
11. Load protein solution onto a pre-equilibrated (with buffer A) 5 mL Q Sepharose HiTrap HP column at 3 mL/min.
12. Wash out any unbound protein with buffer A until A_{280} reaches baseline.
13. Elute AT by applying a 0.05–1 *M* NaCl to the column over 20 column volumes, collecting 1-mL fractions.

[3] As an alternative, the lysate can also be bound batchwise to 5 mL of Ni–NTA resin for a period of 2 h (up to overnight) at 4 °C using a rotary shaker.
[4] The protein can also be eluted using a step gradient of imidazole (250 m*M*), and fractions of 5–10 mL are collected. This is a common procedure used during a batch-binding purification strategy.

14. Analyze fractions for purity by running samples of fractions containing protein on 10% (w/v) SDS and native PAGE gels to look for contamination by aggregates or other proteins.
15. Perform an inhibitory activity assay (described elsewhere in this book) to determine which peak contains active, monomeric protein. The peak at approximately 100 m*M* NaCl will contain monomeric AT.
16. Pool all active, monomeric fractions and concentrate[2] to the desired level, and then aliquot and store at −80 °C.

5.1. Optional purification strategies

Most serpins are particularly prone to adopting nonnative conformations, which can be initiated by cleavage or by misfolding. In order to separate out these conformations, further purification is often required. The most popular methods used utilize either size exclusion or hydrophobic chromatography. A Superdex 200 column (GE Healthcare)[5] generally offers excellent separation of monomeric protein from aggregates, which elute in the void volume.

Hydrophobic separation of serpin conformations relies on the differences in exposure of the hydrophobic core between each of the structures. In this way, it is possible to separate native, latent, cleaved, and polymeric forms of the same protein with good resolution. A method for separation of native ACT from the delta form (Pearce *et al.*, 2010) is presented here and depicted graphically in Fig. 2.3.

Materials

5 mL HiTrap Phenyl Sepharose column (GE Healthcare).

Buffers

Low salt buffer: 50 m*M* Tris, pH 8.0.
High salt buffer: 50 m*M* Tris, 4 *M* NaCl, pH 8.0.

Method

1. Dilute the protein with high salt buffer so the final NaCl concentration is 3.5 *M*.[6]
2. Equilibrate the column with high salt buffer at 5 mL/min on an AKTA FPLC.

[5] Suggested buffer: 50 m*M* Tris, 150 m*M* NaCl, 5 m*M* EDTA, pH 8.0.
[6] If the protein of interest is not stable in such high salt concentrations, ammonium sulfate can be used as an alternative. In this case, the high salt buffer must contain 2 *M* ammonium sulfate, and protein must be diluted into this buffer to at least 1.7 *M* before being loaded onto the column.

3. Load protein onto column at 3 mL/min and wash out any unbound protein with high salt buffer.
4. Elute protein with a 4–0 *M* NaCl gradient over 10–15 column volumes at 3 mL/min, collecting 3-mL fractions.
5. Check fractions for protein by running samples on 10% (w/v) SDS-PAGE gels, and monitor fractions for inhibitory activity against chymotrypsin according to protocols listed elsewhere in this book.

REFERENCES

Abbott, G. L., Blouse, G. E., Perron, M. J., Shore, J. D., Luck, L. A., and Szabo, A. G. (2004). 19F NMR studies of plasminogen activator inhibitor-1. *Biochemistry* **43,** 1507–1519.

Agarwal, S., Jha, S., Sanyal, I., and Amla, D. V. (2010). Expression and purification of recombinant human alpha1-proteinase inhibitor and its single amino acid substituted variants in Escherichia coli for enhanced stability and biological activity. *J. Biotechnol.* **147,** 64–72.

Baneyx, F. (1999). Recombinant protein expression in Escherichia coli. *Curr. Opin. Biotechnol.* **10,** 411–421.

Barnes, R. C., Coulter, J., and Worrall, D. M. (2000). Immunoreactivity of recombinant squamous cell carcinoma antigen and leupin/SCCA-2: Implications for tumor marker detection. *Gynecol. Oncol.* **78,** 62–66.

Bartuski, A. J., Kamachi, Y., Schick, C., Massa, H., Trask, B. J., and Silverman, G. A. (1998). A murine ortholog of the human serpin SCCA2 maps to chromosome 1 and inhibits chymotrypsin-like serine proteinases. *Genomics* **54,** 297–306.

Becerra, S. P., Palmer, I., Kumar, A., Steele, F., Shiloach, J., Notario, V., and Chader, G. J. (1993). Overexpression of fetal human pigment epithelium-derived factor in Escherichia coli. A functionally active neurotrophic factor. *J. Biol. Chem.* **268,** 23148–23156.

Belorgey, D., Crowther, D. C., Mahadeva, R., and Lomas, D. A. (2002). Mutant Neuroserpin (S49P) that causes familial encephalopathy with neuroserpin inclusion bodies is a poor proteinase inhibitor and readily forms polymers in vitro. *J. Biol. Chem.* **277,** 17367–17373.

Bird, P. I., Pak, S. C., Worrall, D. M., and Bottomley, S. P. (2004). Production of recombinant serpins in Escherichia coli. *Methods* **32,** 169–176.

Blacque, O. E., and Worrall, D. M. (2002). Evidence for a direct interaction between the tumor suppressor serpin, maspin, and types I and III collagen. *J. Biol. Chem.* **277,** 10783–10788.

Blouse, G. E., Perron, M. J., Thompson, J. H., Day, D. E., Link, C. A., and Shore, J. D. (2002). A concerted structural transition in the plasminogen activator inhibitor-1 mechanism of inhibition. *Biochemistry* **41,** 11997–12009.

Buckle, A. M., Devlin, G. L., Jodun, R. A., Fulton, K. F., Faux, N., Whisstock, J. C., and Bottomley, S. P. (2005). The matrix refolded. *Nat. Methods.* **2,** 3.

Cabrita, L. D., and Bottomley, S. P. (2004). Protein expression and refolding—A practical guide to getting the most out of inclusion bodies. *Biotechnol. Annu. Rev.* **10,** 31–50.

Cabrita, L. D., Irving, J. A., Pearce, M. C., Whisstock, J. C., and Bottomley, S. P. (2007). Aeropin from the extremophile Pyrobaculum aerophilum bypasses the serpin misfolding trap. *J. Biol. Chem.* **282,** 26802–26809.

Chow, M. K., Amin, A. A., Fulton, K. F., Whisstock, J. C., Buckle, A. M., and Bottomley, S. P. (2006). REFOLD: an analytical database of protein refolding methods. *Protein Expr. Purif.* **46,** 166–171.

Coburn, G. A., and Mackie, G. A. (1999). Degradation of mRNA in Escherichia coli: An old problem with some new twists. *Prog. Nucleic Acid Res. Mol. Biol.* **62,** 55–108.
Fetterer, R. H., Miska, K. B., Jenkins, M. C., Barfield, R. C., and Lillehoj, H. (2008). Identification and characterization of a serpin from Eimeria acervulina. *J. Parasitol.* **94,** 1269–1274.
Hopkins, P. C., Carrell, R. W., and Stone, S. R. (1993). Effects of mutations in the hinge region of serpins. *Biochemistry* **32,** 7650–7657.
Hwang, S. R., Stoka, V., Turk, V., and Hook, V. Y. (2005). The novel bovine serpin endopin 2C demonstrates selective inhibition of the cysteine protease cathepsin L compared to the serine protease elastase, in cross-class inhibition. *Biochemistry* **44,** 7757–7767.
Irving, J. A., Cabrita, L. D., Rossjohn, J., Pike, R. N., Bottomley, S. P., and Whisstock, J. C. (2003). The 1.5 Å crystal structure of a prokaryote serpin: Controlling conformational change in a heated environment. *Structure* **11,** 387–397.
Johansen, H., Sutiphong, J., Sathe, G., Jacobs, P., Cravador, A., Bollen, A., Rosenberg, M., and Shatzman, A. (1987). High-level production of fully active human alpha 1-antitrypsin in Escherichia coli. *Mol. Biol. Med.* **4,** 291–305.
Katagiri, C., Nakanishi, J., Kadoya, K., and Hibino, T. (2006). Serpin squamous cell carcinoma antigen inhibits UV-induced apoptosis via suppression of c-JUN NH2-terminal kinase. *J. Cell Biol.* **172,** 983–990.
Kwon, K. S., Lee, S., and Yu, M. H. (1995). Refolding of alpha 1-antitrypsin expressed as inclusion bodies in Escherichia coli: Characterization of aggregation. *Biochim. Biophys. Acta* **1247,** 179–184.
Law, R. H., Sofian, T., Kan, W. T., Horvath, A. J., Hitchen, C. R., Langendorf, C. G., Buckle, A. M., Whisstock, J. C., and Coughlin, P. B. (2008). X-ray crystal structure of the fibrinolysis inhibitor alpha2-antiplasmin. *Blood* **111,** 2049–2052.
Levina, V., Dai, W., Knaupp, A. S., Kaiserman, D., Pearce, M. C., Cabrita, L. D., Bird, P. I., and Bottomley, S. P. (2009). Expression, purification and characterization of recombinant Z alpha(1)-antitrypsin—The most common cause of alpha(1)-antitrypsin deficiency. *Protein Expr. Purif.* **68,** 226–232.
Li, W., Adams, T. E., Kjellberg, M., Stenflo, J., and Huntington, J. A. (2007). Structure of native protein C inhibitor provides insight into its multiple functions. *J. Biol. Chem.* **282,** 13759–13768.
Pearce, M. C., Powers, G. A., Feil, S. C., Hansen, G., Parker, M. W., and Bottomley, S. P. (2010). Identification and characterization of a misfolded monomeric serpin formed at physiological temperature. *J. Mol. Biol.* **403,** 459–467.
Peterson, F. C., Gordon, N. C., and Gettins, P. G. (2000). Formation of a noncovalent serpin-proteinase complex involves no conformational change in the serpin. Use of 1H-15N HSQC NMR as a sensitive nonperturbing monitor of conformation. *Biochemistry* **39,** 11884–11892.
Przygodzka, P., Ramstedt, B., Tengel, T., Larsson, G., and Wilczynska, M. (2010). Bomapin is a redox-sensitive nuclear serpin that affects responsiveness of myeloid progenitor cells to growth environment. *BMC Cell Biol.* **11,** 30.
Rubin, H., Wang, Z. M., Nickbarg, E. B., McLarney, S., Naidoo, N., Schoenberger, O. L., Johnson, J. L., and Cooperman, B. S. (1990). Cloning, expression, purification, and biological activity of recombinant native and variant human alpha 1-antichymotrypsins. *J. Biol. Chem.* **265,** 1199–1207.
Sancho, E., Tonge, D. W., Hockney, R. C., and Booth, N. A. (1994). Purification and characterization of active and stable recombinant plasminogen-activator inhibitor accumulated at high levels in Escherichia coli. *Eur. J. Biochem.* **224,** 125–134.
Zhang, T., Guan, M., and Lu, Y. (2005). Production of active pigment epithelium-derived factor in E. coli. *Biotechnol. Lett.* **27,** 403–407.
Zhou, A., Carrell, R. W., and Huntington, J. A. (2001). The serpin inhibitory mechanism is critically dependent on the length of the reactive center loop. *J. Biol. Chem.* **276,** 27541–27547.

CHAPTER THREE

Isolation and Characterization of the Nuclear Serpin MENT

Sergei Grigoryev* *and* Sheena McGowan†

Contents

1. Introduction 30
2. Purification of the MENT Protein 31
   2.1. Collection and processing of chicken blood 31
   2.2. Isolation of cell nuclei from white blood cells 32
   2.3. Isolation of native MENT protein from white blood cell nuclei 33
   2.4. Expression and purification of recombinant MENT 33
3. Protease Inhibition/Serpin Activity 34
   3.1. Determination of kinetic parameters 34
   3.2. Native acid PAGE 35
   3.3. Intrinsic tryptophan fluorescence to assess conformational change 35
4. MENT Interaction with DNA and Chromatin *In Vitro* 36
   4.1. Electrophoretic mobility shift assays with double-stranded DNA (dsDNA) 36
   4.2. Chromatin association assays 37
   4.3. Deoxynucleoprotein electrophoresis of reconstituted nucleosomes 40
5. Analysis of MENT Association with Native Chromatin *In Situ* 42
   5.1. Isolation and fractionation of chicken blood cells 42
   5.2. ChIP analysis of MENT association with nuclear DNA *in situ* 43
Acknowledgments 46
References 46

Abstract

A balance between proteolytic activity and protease inhibition is required to maintain the appropriate function of biological systems in which proteases play a role. The Myeloid and Erythroid Nuclear Termination protein, MENT, is a nonhistone heterochromatin-associated serpin that is an effective inhibitor of

* Department of Biochemistry & Molecular Biology, Penn State University College of Medicine, Milton S. Hershey Medical Center, University Drive, Hershey, Pennsylvania, USA
† Department of Biochemistry and Molecular Biology and ARC Centre of Excellence for Structural and Functional Microbial Genomics, Monash University, Clayton, Victoria, Australia

Methods in Enzymology, Volume 501
ISSN 0076-6879, DOI: 10.1016/B978-0-12-385950-1.00003-1

the papain-like cysteine proteases. Our laboratories have extensively investigated the dual functions of this protein, namely, chromatin condensation and protease inhibition. Unlike other serpins to date, MENT contains a unique insertion between the C- and D-helices known as the "M-loop." This loop contains two critical functional motifs that allow the nuclear function of MENT, namely, nuclear localization and DNA binding. However, the nuclear function of MENT is not restricted to the activities of the M-loop alone. *In vitro*, MENT brings about the dramatic remodeling of chromatin into higher-order structures by forming protein bridges via its reactive center loop. Further, we have determined that in a protease-mediated effect, DNA can act as a cofactor to accelerate the rate at which MENT can inhibit its target proteases. In this chapter, we discuss the isolation of MENT from native chicken blood as well as recombinant protein produced in *Escherichia coli*. Various techniques including *in vitro* functional assays and biophysical characterization are explained that can be used to elucidate the ability of the protein to interact with DNA and other deoxynucleoprotein complexes. *In situ* chromatin precipitation using natively purified MENT is also detailed.

1. Introduction

In eukaryotes, DNA is repeatedly coiled around histone octamers to form nucleosomes (Richmond and Davey, 2003). The nucleosome zigzag arrays fold into 30 nm higher-order fibers (Schalch *et al.*, 2005). The arrays and fibers are referred to as primary and secondary levels of chromatin folding, respectively. In the nucleus, the chromatin fibers are further folded by histone and nonhistone architectural proteins to form more compact structures that are associated with tertiary, quaternary, and higher levels of compaction (Luger and Hansen, 2005; Woodcock and Dimitrov, 2001). In the nucleus, two major types of chromatin can be distinguished by their packing density: open, less compact euchromatin and highly condensed heterochromatin (Grigoryev *et al.*, 2006). Heterochromatin contains predominantly repressed genes and spreads to the bulk of nuclear chromatin during terminal cell differentiation when most of the genome becomes inactive (Grigoryev *et al.*, 2006).

The first identified and one of the best-characterized nonhistone chromatin-condensing proteins is the Myeloid and Erythroid Nuclear Termination stage-specific protein, MENT. This abundant nuclear protein accumulates in terminally differentiating chicken blood cells, binds to heterochromatin, and promotes its condensation during terminal cell differentiation (Grigoryev and Woodcock, 1998; Grigoryev *et al.*, 1992). *In vitro*, MENT brings about a dramatic remodeling and condensation of chromatin higher-order structure by promoting nucleosome array folding into the 30 nm fibers and by forming protein "bridges" connecting separate nucleosome arrays and forming tertiary chromatin structures (Springhetti *et al.*, 2003).

Remarkably, this chromatin architectural protein MENT belongs to the intracellular branch of the serpin superfamily (Silverman *et al.*, 2001). Alongside its ability to condense chromatin, MENT is an effective inhibitor of cathepsin L (Irving *et al.*, 2002), a cysteine protease residing in lysosomes and the nucleus (Goulet *et al.*, 2004). *In vitro*, MENT is an effective inhibitor of the papain-like cysteine proteinase cathepsins K, L, and V making MENT the first known chromatin-associated cysteine proteinase inhibitor.

Multiple sequence alignments and analysis of the X-ray crystal structures of MENT revealed that MENT contains a large insertion, the "M-loop," between the C- and D-helices. This loop contains two critical functional motifs: a classical nuclear localization signal (NLS) that is required for nuclear import and an AT-hook motif that is involved in chromatin and DNA binding (Grigoryev *et al.*, 1999). Like other serpins, MENT possesses a reactive center loop (RCL) through which interaction with its cognate proteinase occurs. *In vivo*, the RCL sequence of MENT has been demonstrated to be essential for proper MENT-mediated chromatin compaction and controlling cell proliferation (Irving *et al.*, 2002). It is also able to mediate formation of a loop-sheet oligomer, suggesting how the structural plasticity of serpins has adapted to mediate physiological, rather than pathogenic, loop-sheet linkages.

2. Purification of the MENT Protein

MENT is present in three main avian blood cell types (erythrocytes, lymphocytes, and granulocytes). In granulocytes, it is the most abundant nonhistone protein associated with compact heterochromatin (Grigoryev and Woodcock, 1998) making granulocytes or the crude white cell fraction (buffy coat) from peripheral chicken blood the optimal source for isolation of native wild-type MENT. Recombinant wild-type and mutant MENT proteins have been successfully produced and used extensively for *in vitro* functional and structural studies.

2.1. Collection and processing of chicken blood

In order to obtain native MENT, fresh chicken blood must be available for purchase or collection and used within 24 h of exsanguination. Chicken blood preserved from coagulation by sodium citrate or EDTA is available from Pel-Freez (Rogers, AR) or collected from freshly killed animals as described below. The entire procedure for fractionating chicken blood cells is done on ice.

1. Collect chicken blood (from freshly killed animals) by filtering through two layers of cheesecloth into PBS (0.14 *M* NaCl, 2.7 m*M* KCl, 10 m*M* Na_2HPO_4, 1.7 mm KH_2PO_4, pH 7.5) containing 3% sodium citrate. We recommend mixing 333 mL of PBS–3% citrate with fresh blood in a 1 L bottle and adding 10% citrate to final concentration of 3% to prevent coagulation.
2. Blood cells are separated from serum by centrifugation at 800×*g*/10 min/4 °C in 50 mL culture tubes (not 250 mL bottles).
3. Remove white cells (layer between red cells and serum) and wash cells three times with PBS medium plus 1% sodium citrate (80 mL per wash).
4. The upper whitish layers (buffy coats enriched with leukocytes) were removed after each centrifugation and used for isolation of white blood cell nuclei (see below) or blood cell fractionation (Section 5.1). The bottom layer containing pure erythrocytes (red cells) was used for isolation of chicken erythrocyte nuclei and chromatin (see Section 4.2).

2.2. Isolation of cell nuclei from white blood cells

For isolation of MENT, cell nuclei may be isolated from crude leukocyte preparations (buffy coats).

1. Resuspend white cells from 200 mL blood in 40 mL PBS–1% sodium citrate and centrifuge for 3 min/1000×*g*.
2. Resuspend the cell pellet in 0.5% Nonidet P-40 in ~25 mL RSB solution (3 m*M* $MgCl_2$, 10 m*M* NaCl, 10 m*M* Tris–Cl, pH 7.6) with 1 m*M* phenylmethylsulfonyl fluoride (PMSF) at 4 °C. Take an aliquot of the nuclear suspension, dissolve in solution of 1% SDS, and measure DNA concentration spectrophotometrically (1 mg/mL of nuclear DNA has an optical density $A_{260} = 20$). Adjust the DNA concentration in the main nuclear suspension to $A_{260} = 200$ (10 mg/mL of DNA).
3. Transfer the cell suspension to large (40 mL) prechilled Dounce homogenizer.
4. Homogenize the cell suspensions by 20–30 strokes of pestle B in the Dounce homogenizer over 30 min on ice.
5. Centrifuge the nuclei for 10 min/7600×*g* and resuspend nuclear pellets in ~50 mL RSB with 0.5% Nonidet P-40, 1 m*M* PMSF.
6. Pellet nuclei by spinning 7600×*g*/5 min/4 °C and discard supernatant.
7. Repeat wash until majority of red color is gone from pellet and supernatant (~3 times).
8. Resuspend the nuclei in RSB with 1 m*M* PMSF and 50% glycerol. Store at −20 °C for no more than 1 week.

2.3. Isolation of native MENT protein from white blood cell nuclei

1. Resuspend the nuclei in 30 mL RSB with 0.5 m*M* PMSF until A_{260} ~ 200 (dissolve 5 μL in 1 mL 1% SDS to read A_{260}).
2. Add an equal volume of RSB–HEPES (3 m*M* $MgCl_2$, 10 m*M* NaCl, 10 m*M* HEPES, pH 7.6) with 0.7 *M* NaCl.
3. Incubate 30 min/ice with shaking and occasional inverting.
4. Centrifuge 10 min/12,000 rpm/4 °C/SS-34 rotor (Sorvall).
5. Dilute the supernatant (contains MENT) with four volumes of 10 m*M* HEPES; 2 m*M* EDTA.
6. Centrifuge 10 min/12,000 rpm/4 °C/SS-34 rotor.
7. Load supernatant onto a 5-mL HiTrap SP-Sepharose FPLC column equilibrated with 0.2 *M* NaCl; 20 m*M* HEPES, pH 7.6. Elute proteins with 20 m*M* HEPES, pH 7.6, in a 0.2–1.0 *M* NaCl gradient and analyze by 15% SDS-PAGE. MENT protein (~47 kDa) is eluted as a single peak at NaCl concentration of ca. 0.5 *M*.
8. Combine MENT-containing fractions and dilute to 50 mL in 20 m*M* HEPES, pH 7.5.
9. Load onto 1 mL HiTrap SP-Sepharose FPLC column. Repeat elution with the same gradient and collect the purified MENT peak.
10. Dialyze MENT-containing fractions into 20 m*M* HEPES, pH 7.5, 50 m*M* NaCl.

2.4. Expression and purification of recombinant MENT

The wild-type MENT gene was cloned (*Xho*I/*Bam*HI) into the commercial expression vector pET-15b (Novagen). The pET-15b constructs were used to express MENT and variants with an N-terminal hexahistidine tag. To produce a recombinant wild-type MENT protein without an N-terminal His-tag, the wild-type MENT gene was cloned (*Nde*I/*Bam*HI) into the commercial expression vector, pET-3A (Novagen). These constructs can be used to overexpress and purify MENT (and mutant) proteins as described below, with the exception that metal affinity chromatography of the pET-3A lysates is omitted and cation exchange is repeated to enrich MENT protein.

Purification buffers

Lysis buffer: 25 m*M* $NaPO_4$, 0.5 *M* NaCl, 10 m*M* imidazole, pH 8.0
Wash buffer: 50 m*M* HEPES, pH 7.8, 50 m*M* NaCl
Elution buffer: 50 m*M* HEPES, pH 7.8, 50 m*M* NaCl, 250 m*M* imidazole
Low salt buffer: 50 m*M* HEPES, pH 7.8, 50 m*M* NaCl
High salt buffer: 50 m*M* HEPES, pH 7.8, 1.0 *M* NaCl
Gel filtration buffer (storage buffer): 50 m*M* HEPES, pH 7.8, 50 m*M* NaCl

Protein expression and purification

1. Transform plasmid of interest into *Escherichia coli* BL21(DE3)(pLysS).
2. Inoculate 3 mL (per liter) starter culture and grow 2 h/37 °C/250 rpm.
3. Inoculate 1 L (6×) 2× YT and grow 37 °C/250 rpm until $OD_{600} = 1.0$.
4. Induce bacterial cultures with 0.5 m*M* IPTG (final concentration) and grow for >6 h/28 °C/250 rpm.
5. Harvest cells by centrifugation at 4000 rpm/4 °C.
6. Pellet can be stored at −80 °C until required.
7. Resuspend cells in 50 mL of lysis buffer (per 6 L of culture).
8. Sonicate 6 × 30 s (10 KHz) with 2 min rests on ice in between.
9. Centrifuge at 12,000 rpm/SS-34 rotor/4 °C/1 h.
10. Filter supernatant through 0.45 μ*M* filter.
11. Add filtered supernatant to pre-equilibrated metal affinity resin (5 mL bed volume) and bind for 1 h/4 °C/rotating.
12. Return batched resin to gravity-flow column and wash 150 mL lysis buffer.
13. Wash resin with further 150 mL of wash buffer.
14. Elute from metal affinity resin with 50 mL elution buffer.
15. Equilibrate 5 mL HiTrap SP-Sepharose with low salt buffer (five column volumes (CV)).
16. Load 50 mL protein sample from metal affinity elution on HiTrap column.
17. Elute proteins 20 m*M* HEPES, pH 7.6, in a 0.2–1.0 *M* NaCl gradient over 20 CV.
18. Analyze by 12% SDS-PAGE for fractions containing MENT protein (~50 kDa).
19. Pool fractions containing purified protein and dialyze in gel filtration buffer.
20. Concentrate sample to reduce volume to 500 μL and load Superdex S200 10/30 column and elute proteins at a flow rate of 0.5 mL/min.
21. Confirm monomeric nature of purified MENT protein.
22. Purity of protein (single band) can be verified using SDS-PAGE.
23. The concentration of the protein can be determined using an A_{280} measurement.
24. Recombinant MENT can then be concentrated to 1 mg/mL and stored at −80 °C until use.

3. Protease Inhibition/Serpin Activity

3.1. Determination of kinetic parameters

In vitro, MENT is an effective inhibitor of the papain-like cysteine proteinase cathepsins K, L, and V. The stoichiometry of inhibition (SI) and second-order association rate constant (k_a) can be assessed using the fluorescent substrate N-Cbz-Phe-Arg-methylcoumarin as described elsewhere (Irving *et al.*, 2002; McGowan *et al.*, 2006; Ong *et al.*, 2007) and is further outlined in Chapter 11. Enzyme assays between MENT and target cathepsins are undertaken at 30 °C in assay buffer (0.1 *M* acetate, pH 5.5, 1 m*M* EDTA, 0.1% ,w/v Brij-35, 10 m*M* cysteine) and all cathepsins are preactivated by incubation in cathepsin buffer (0.1 *M* acetate, 1 m*M* EDTA, 0.1% ,w/v Brij-35, 0.02% ,w/v sodium azide, 10 m*M* cysteine, pH 5.5) for at least 20 min at room temperature before use. The active enzyme concentration was determined with *trans*-epoxysuccinyl-L-leucylamido-(4-guanidino)-butane (E-64) titration.

3.2. Native acid PAGE

The oligomeric state of MENT (and also complexes between MENT and target proteases) can be visualized using a 12% native acid gel protocol (Zhou *et al.*, 2001) where methyl green is used as a tracking dye and proteins migrate toward the negative electrode. The electrode buffer contained 40 m*M* β-alanine adjusted to pH 4.0 with acetic acid. Electrophoresis is generally carried out at room temperature. Resolution of the MENT protein and protein complexes was successful when gels were run at 80 V for ~4–5 h. Native acid PAGE gels were made fresh on the day of use. Proteins are visualized by staining with Coomassie Brilliant Blue R-250.

12% native acid PAGE recipe	Stacking gel	Resolving gel
40% acrylamide (29:1) (mL)	3.0	0.5
0.52 *M* acetic acid, pH 5.0 (mL)	5.0	–
0.18 *M* acetic acid, pH 6.0 (mL)	–	2.5
dH_20 (mL)	1.95	2.0
TEMED (μL)	12.5	6.5
10% APS (μL)	50.0	25.0
Riboflavin (1 mg/mL) (μL)	50.0	25.0

4× loading buffer: 1 mL glycerol; 4 mL 0.225 *M* acetic acid, pH 6.0; methyl green.

3.3. Intrinsic tryptophan fluorescence to assess conformational change

Potential cofactor-induced conformational change in MENT proteins can be studied by monitoring the intrinsic tryptophan fluorescence changes in the presence and absence of cofactors. MENT proteins (0.2 μ*M*) were incubated with increasing amounts of the cofactor in 10 m*M* HEPES, pH 7.0, containing 0.5 mM EDTA, 40 m*M* NaCl, and 0.1% (w/v) Brij-35 in a final volume of 2 mL. Experiments are repeated in triplicate. The emission was scanned $\geq$ five times over the range of 300–400 nm at 60 nm/min, using excitation and emission slit widths of between 5 and 8 nm. The emission spectrum of buffer in the presence and absence of cofactor, which was used as a control, was subtracted from the protein emission spectra. Acrylic cuvettes were used throughout to avoid adsorption to the cuvette walls.

When using DNA-containing cofactors, an excitation wavelength of 295 nm was used to minimize inner filter effects of measured fluorescence and DNA absorbance at excitation wavelengths was monitored and did not exceed 0.1 for any sample. Fluorescence quenching experiments were performed in the presence and absence of DNA cofactors using increasing concentrations of acrylamide (0–0.5 *M*). From the recorded titration spectra, the extent of quenching and the accessibility of tryptophan residues can be calculated from Stern–Volmer plots as previously described by Lehrer (1971).

4. MENT Interaction with DNA and Chromatin *In Vitro*

4.1. Electrophoretic mobility shift assays with double-stranded DNA (dsDNA)

The ability of MENT to interact with naked (nonchromatin) DNA was investigated by assessing the protein's ability to retard DNA from migrating through agarose gels. Varying the length and structure of double-stranded DNA can give information on the requirements of the target DNA for the protein.

4.1.1. Preparation of dsDNA for EMSAs

HPLC-purified synthetic oligonucleotides of variable lengths were purchased from commercial companies and annealed to form target dsDNA for electrophoretic mobility shift assays (EMSAs). Prior to annealing, the oligonucleotides were diluted to 100 μM and 5′-end-labeled (phosphorylated) using T4 Polynucleotide Kinase according to manufacturers' protocol. Complementary phosphorylated oligonucleotides are then added to a reaction tube and boiled (100 °C) for 10 min. The reaction is cooled to room temperature over a period of 4 h. Annealed oligonucleotides are collected in tube by gentle centrifugation.

Prior to use, dsDNA was purified to remove any single-stranded DNA that had failed to anneal. The dsDNA sample was loaded onto a 5 mL MonoQ HR FPLC column equilibrated with 1× TE. DNA was eluted with 1× TE buffer in a 0.0–1.0 *M* NaCl gradient and analyzed by agarose gel (% agarose varying dependent on desired size of dsDNA). DNA was quantitated by A_{260} before use.

4.1.2. Electrophoretic mobility shift assays

For EMSAs, 20 μL reactions contained 10 m*M* HEPES, pH 7.0, 0.5 m*M* EDTA, 40 m*M* NaCl, 0.5% (w/v) Brij-35, <0.3 m*M* of annealed oligonucleotides and between 1 and 10 μ*M* MENT protein. Reactions are incubated at room temperature for 15 min. Samples are then loaded onto 2% agarose gel after the addition of 5 μL of 50% glycerol. Electrophoresis was performed for 30 min at 100 V. Gels are visualized using ethidium bromide and imaged on an AlphaImager (Alpha Innotech). DNA that remained unbound in each reaction was quantified (ImageJ) from digital images. Affinities can be calculated by plotting the log of the protein concentration versus the log of ($b/1-b$), where b is the fraction of bound DNA (Allain *et al.*, 1999). When the value of log ($b/1-b$) is zero, 50% of DNA is bound. Apparent equilibrium dissociation constants (K_D) can be estimated from an average of at least three independent experiments. Any smearing seen at high protein concentration is indicative of the presence of multiple nucleoprotein species (Carruthers *et al.*, 1998; Georgel *et al.*, 2001, 2003). Incubation with >1 μ*M* $MENT_{WT}$ produced nucleoprotein complexes that were too large to migrate through an agarose gel (0.5–2.0% agarose).

4.2. Chromatin association assays

4.2.1. Fractionation and isolation of chicken erythrocyte nuclei

1. Erythrocytes are obtained from the chicken blood (Section 2.1) by centrifugation at 3500 rpm/10 min/4 °C in 50 mL culture tubes (not 250 mL bottles).
2. Remove white cells (layer between red cells and serum) and wash red cells from the bottom layer containing pure erythrocytes three times with PBS medium with 1% sodium citrate (80 mL per wash).
3. Resuspend fractionated erythrocytes in ~25 mL RSB with 0.5% Nonidet P-40, 1 m*M* PMSF. The final volume should be less than or equal to 40 mL.
4. Transfer suspension to large prechilled Dounce homogenizer.
5. Homogenize cells with 3× 20 strokes of Dounce pestle B over the span of 30 min on ice.
6. Pellet nuclei by spinning 4000 rpm/5 min/4 °C and discard supernatant.
7. Wash nuclei with ~50 mL RSB with 0.5% Nonidet P-40, 1 m*M* PMSF.
8. Centrifuge in a fixed angle rotor 4000 rpm/5 min/4 °C and discard supernatant.
9. Repeat wash until majority of red color is gone from pellet and supernatant (~3 times).
10. Store nuclei in RSB with 50% glycerol at −20 °C.

4.2.2. Preparation of soluble chromatin from chicken erythrocytes

Preparation of soluble chromatin uses micrococcal nuclease (MNase) digestion of isolated (frozen) erythrocyte nuclei (see Section 4.2.1). MNase is added at 3–30 units/mL depending on the nature of the nuclei and the desired extent of digestion. A small-scale preparation to optimize enzyme digestion is generally required prior to large-scale preparation of soluble chromatin.

Micrococcal nuclease digestion of nuclear suspension

1. (After storage at −20 °C) Dilute the nuclear suspension approximately threefold in RSB (3 m*M* $MgCl_2$, 10 m*M* NaCl, 10 m*M* Tris–HCl, pH 7.6) and spin down 5 min/1000×*g*. Remove the supernatant.
2. Dilute nuclear suspension in RSB until solution has an $A_{260} = 20$; add 0.5 m*M* PMSF.
3. Dilute stock solution of MNase (Roche Nuclease S7, cat# 10107921001, 15,000 units/mL in water) 10-fold in water immediately prior to use and keep on ice.
4. Place four Eppendorf tubes containing 2 μL of 50 m*M* EDTA (5 m*M* final concentration) on ice.
5. Pipette 100 μL of nuclear suspension to an empty Eppendorf tube and add 2 μL of 50 m*M* $CaCl_2$ (1 m*M* final concentration).
6. Incubate 1 min/37 °C.
7. Transfer a 20 μL aliquot (zero time point) of the nuclear suspension to one Eppendorf tube containing EDTA.
8. Initiate nuclease digestion by adding 1 μL of diluted MNase (step 3) to nuclear suspension tube and continue incubation at 37 °C.
9. At various time points (1, 5, 10 min), transfer a 20 μL aliquot from reaction tube to tubes containing EDTA on ice.
10. Terminate digestion reaction at 15 min by adding EDTA (final concentration of 10 m*M*) to the reaction, and transfer tube on ice.
11. To each sample, add 2 μL of 10% SDS (1% final concentration) and 2 μL of 0.1 mg/mL proteinase K (0.01 mg/mL final concentration) and incubate 1 h/55 °C.
12. Add 2.5 μL of 50% glycerol in 5× TAE to each tube.
13. Load an equal amount from each sample on 1% Sigma Type I Agarose gel (in TAE buffer) together with DNA markers and run a gel to determine efficiency of digestion reaction.

(Optional) In parallel, repeat the digestion reaction (steps 1–11) using 0.5 μL of MNase.

(Optional) If digestion is not complete after 15 min of incubation, repeat steps 1–12 and use 45–60 min incubation time.

4.2.2.1. Large-scale MNase digestion and preparation of soluble chromatin For preparation of soluble chromatin, the "Micrococcal Nuclease Digestion of Nuclear Suspension" protocol above should provide preliminary information about the optimal time and concentration of MNase for the large-scale digestion reaction to obtain DNA fragments with mean size between 400 and 6000 bp.

1. (After storage at −20 °C) Dilute the nuclear suspension approximately threefold in RSB (3 m*M* $MgCl_2$, 10 m*M* NaCl, 10 m*M* Tris–HCl, pH 7.6) and spin down 5 min/1000×*g*. Remove the supernatant.
2. Prepare an 80 mL nuclear suspension with RSB until solution has an $A_{260} = 20$; add 0.5 m*M* PMSF.
3. Add 1.6 mL of 50 m*M* $CaCl_2$ (1 m*M* final concentration), mix, and incubate 10 min/37 °C.
4. Start MNase digestion by adding appropriate amount of MNase (estimated and scaled up from prior protocol). Incubate at 37 °C for optimal period of time with occasional agitation.
5. (Optional) For large-scale preparation, it is worth to conduct the MNase digestion on a magnetic stirrer plate with small stirring bar (avoid bubbles). In this case, also perform the test digestion by "Micrococcal Nuclease Digestion of Nuclear Suspension" protocol on stirrer plate.
6. Terminate digestion by adding 0.5 *M* EDTA to 2 m*M* final concentration, mix, and rapidly transfer the tube on ice.
7. Centrifuge the digestion tube for 5 min/10,000 rpm/4 °C in a fixed-angle rotor. Harvest supernatant and store on ice. This supernatant constitutes soluble chromatin "S1."
8. Resuspend pellet in 20 mL of TE buffer (10 m*M* Tris–HCl, pH 7.5, 1 m*M* EDTA).
9. Incubate suspension for 20 min/ice. Centrifuge the digestion tube for 5 min/10,000 rpm/4 °C in a fixed-angle rotor. Harvest supernatant and store on ice. This supernatant constitutes soluble chromatin "S2."
10. Store S1, S2, and the final nuclear pellet at 4 °C.
11. Determine the concentration of DNA in S1, S2, and the pellet (by adding SDS to 1% to the pellet) at OD_{260} nm. If the yield of DNA in S2 is low (less than 50% of starting material), repeat step 9 to yield soluble chromatin "S3."

Supernatants S2 and S3 are the soluble chromatin and may be purified on sucrose gradient by the "Sucrose Gradient Purification of Soluble Chromatin" protocol.

4.2.3. Sucrose gradient purification of soluble chromatin

1. Prepare 5–25% linear sucrose gradients in TE buffer (10 m*M* Tris–HCl, pH 7.5, 1 m*M* EDTA) in ultracentrifuge tubes (12 or 38 mL for SW41 or SW28 rotors, respectively, depending on amount of chromatin to be purified). Form gradient at room temperature and then keep the tubes at 4 °C for less than 15 h, taking care not to disturb the gradients.

2. Load samples on gently onto the surface of gradient, taking care to avoid disturbing the gradients. Volume of sample should not exceed 1.5 mL for 38 mL tubes and 0.75 mL for 12 mL tubes.
3. Place samples in an ultracentrifuge. Ensure that centrifuge brakes are not used. For sedimentation of centrifuge samples in an ultracentrifuge native chromatin fragments having on average 12 nucleosomes (ca. 2500 bp) requires 8 h in SW41 rotor at 35,000 rpm or 14 h in SW28 rotor at 26,000 rpm.
4. Aliquot gradients into 1 mL (12 mL gradient) or 2 mL (38 mL gradient) fractions and store at 4 °C.
5. Measure DNA concentration in each fraction at A_{260}.
6. Take a 20-μL sample of each fraction and add 2 μL 10% SDS and 2.5 μL of 50% glycerol in 5× TAE. Load each sample on 1% agarose gel (Sigma Type I in TAE buffer) together with DNA markers and run a gel to determine efficiency of purification and separation of chromatin with different size.
7. Combine fractions with appropriate desired size (e.g., trimer fraction(s) contain predominantly 630 bp DNA fragment and 12-mer fractions contain 2500 bp DNA), add 0.5 m*M* PMSF, and dialyze against TE buffer for 36–40 h to remove sucrose.
8. Determine concentration of DNA (A_{260}) in dialyzed samples and concentrate, if appropriate, on Microcentrifuge devices.
9. Check samples from the dialyzed fraction on 15% SDS-PAGE to determine integrity of histones.

4.2.4. Chromatin association assays

Chromatin association reactions were made up in a final volume of 20 μL containing 10 m*M* HEPES, pH 7.0, 0.5 m*M* EDTA, 40 m*M* NaCl, 0.5% (w/v) Brij-35, S2 chromatin (A_{260} = 1.6 units of DNA), and 1–4 μ*M* MENT protein. Samples were incubated on ice for 30 min before being centrifuged at 12,000 rpm for 10 min. The supernatant is added to 5 μL of 50% glycerol before electrophoresis using 1% (w/v) Agarose Type IV (Amresco) gels for 1 h at 50 V. In the chromatin association assay, $MENT_{WT}$ reacted with soluble oligonucleosomes causing either their gel retardation or self-association in agarose gels. It is suggested that the self-associated chromatin complexes represent oligomeric suprastructures formed by the bridging of nucleosome arrays (Georgel *et al.*, 2003).

4.3. Deoxynucleoprotein electrophoresis of reconstituted nucleosomes

Nucleosomes can be reconstituted from Lowary and Widom "601" positioning DNA and chicken erythrocyte core histones. Deoxynucleoprotein (DNP) electrophoresis can then be visualized by ethidium bromide or if further sensitivity is required, [^{32}P]-ATP end-labeled DNA can be incorporated during reconstitution. For the native agarose gel electrophoresis of protein–DNA complexes, we regularly employed the $MENT_{OV}$ mutant, in which the sequence of the RCL has been replaced with that of the noninhibitory serpin, ovalbumin. The $MENT_{OV}$ mutant is a useful reagent since it binds DNA and chromatin with the same affinity as $MENT_{WT}$ but does not cause self-association (Springhetti *et al.*, 2003), thus minimizing nucleosome smearing during electrophoresis.

4.3.1. Nucleosome reconstitution

Nucleosome monomers are reconstituted from 213-bp-long DNA (monomer) templates: atcagtgatatcggaccctatacgcgGCCGCCCTGGAGAATCCC GGTGCCGAGGCCGCTCAATTGGTCGTAGCAAGCTCTAGCA CCGCTTAAACGCACGTACGCGCTGTCCCCCGCGTTTTAACC GCCAAGGGGATTACTCCCTAGTCTCCAGGCACGTGTCAGAT ATATACatcctgtgcatgtggatccgaattcatgattaattaattaatctagtgat containing a 146 bp nucleosome positioning sequence from clone 601 (Lowary and Widom, 1998) marked by capital letters. DNA fragments are 5′-end-labeled with T4 Kinase (Invitrogen) and the 3′ recessed ends filled in with Klenow DNA polymerase (Roche) according to manufacturer's instructions. Only one 5′-end on each DNA was labeled, while the other end remained protected during the labeling and was later opened by a restriction enzyme cutting. Core histones were isolated from chicken erythrocyte nuclei as per Meersseman *et al.* (1991).

Nucleosome trimers were prepared from a 639-bp-long DNA (trimer) templates containing three repeats of the above 213-bp-long mononucleosome positioning sequence. Trinucleosomes were reconstituted from pure 639-bp-long DNA fragment and chicken erythrocyte core histones using gradient NaCl dialysis based on the method described by Luger *et al.* (1999). Purified DNA and core histones were mixed at a molar ratio of 1:1.05 in 10 m*M* HEPES, pH 7.4, 2 *M* NaCl, 0.2 m*M* EDTA, 1 m*M* PMSF, 0.1% (v/v) Nonidet P-40. After 30 min on ice, the reaction was subjected to dialysis using a 3500 MWCO. Salt gradient dialysis decreased the salt concentration from 2 to 0.4 *M* over a 72-h time frame at 4 °C. The reaction was then dialyzed against 10 m*M* HEPES, pH 7.4, 5 m*M* NaCl, 0.2 m*M* EDTA at 4 °C. Assembled nucleosome particles were concentrated using a 10 MWCO concentrator and assembly confirmed using agarose gel electrophoresis. Reconstituted nucleosomes were stored on ice and used within 1 week of assembly.

4.3.2. DNP electrophoresis of nucleosome monomers

Reconstituted mononucleosomes were employed in a derivative of an EMSA to determine if the MENT protein could bind to the linker DNA of the mononucleosome. The MENT protein is mixed with nucleosome monomers (at varying molecular ratios of between 0.2 and 3.2 MENT:

nucleosome) and incubated on ice in 10 μL reactions containing 0.6 A_{260} units of DNA, 10 m*M* HEPES, pH 7.5, 37 m*M* NaCl, 0.4 m*M* EDTA, and 5% (v/v) glycerol for 30 min. Samples were centrifuged at 16,000×*g* at 4 °C for 10 min. Supernatants were loaded directly onto 1% Type IV (Sigma) Agarose gels in 20 m*M* HEPES, pH 8.0, 0.2 m*M* EDTA, and were resolved at 80 V for 90 min. Gels were fixed for 30 min in 7% (w/v) trichloroacetic acid, then dried on a glass plate under a stack of paper towels, and radio-autographed on Kodak BioMax MR film.

4.3.3. DNase I protection experiments of nucleosome trimers

Reconstituted ^{32}P-labeled nucleosome trimers (intact or mixed with MENT at a ratio of 2 MENT per nucleosome) were placed in reaction mixtures containing 0.6 A_{260} units of nucleosome DNA, 20 m*M* Tris–HCl, pH 8.0, 20 m*M* NaCl, 1 m*M* $CaCl_2$, 0.84 m*M* EDTA, and 5% glycerol. DNase I was added to 20 mg/mL and incubated for 2–40 min at 37 °C. Twenty microliters of aliquots were collected, deproteinized with proteinase K digestion, and phenol/chlorophorm, purified, and analyzed on 40-cm-long 6% polyacrylamide/urea "sequencing" gels as described (Ausubel *et al.*, 2005). Gels were dried and exposed to a Molecular Dynamics Phosphorimager for autoradiography.

5. Analysis of MENT Association with Native Chromatin *In Situ*

MENT is nonrandomly localized at specific sites in the nuclei of avian erythrocytes, lymphocytes, and granulocytes. In granulocytes, it is mostly associated with abundant compact heterochromatin (Grigoryev and Woodcock, 1998) and in erythrocytes, it has been shown to be selectively associated with heterochromatin fraction marked by histone H3K9 dimethylation (Istomina *et al.*, 2003). *In vitro*, MENT does not show any distinct DNA sequence-specific binding and its association with specific sites in nuclear chromatin is likely determined by chromatin structure and/or histone modifications. To identify native MENT-binding sites in the nuclear DNA, one can use *in situ* chromatin cross-linking and immunoprecipitation (ChIP) with nuclei of fractionated chicken blood cells (Istomina *et al.*, 2003).

5.1. Isolation and fractionation of chicken blood cells

1. Prepare a 60% Percoll containing PBS gradient 2 h prior to experiment by mixing 24 mL of stock Percoll suspension (density 1.13 g/mL, GE Healthcare) with 16 mL of 2.5× PBS in a 50 mL centrifuge tube and by centrifugation of the resulting mixture at 20 °C/30 min/23,400×*g*/SS-34 rotor.

2. Load the leukocyte-enriched material (buffy coat) from 200 mL of fresh blood onto two preformed 40 mL gradients and centrifuge at 1000×*g*/20 min/25 °C.
3. Five distinct zones are obtained postcentrifugation (Fig. 3.1, *left*). The cell fractions corresponding to each zone can be taken from the gradient and resuspended in 40 mL of PBS.
4. Centrifuge each collected fraction at 1000×*g*/5 min and resuspend in 2 mL of PBS. Prepare smears of the cell fractions, stain using Hema 3 Stat Pack staining kit (Fisher Scientific) as described in the product manual, and examine under a light microscope. Zone I contains thrombocytes and many cell aggregates, zone II contains predominantly lymphocytes, zone III is a mixture of different mononuclear white blood cells, zone IV contained highly enriched granulocytes easily identifiable by their multilobed nuclear morphology (Fig. 3.1, *right*), and zone V contains the residual erythrocytes.

5.2. ChIP analysis of MENT association with nuclear DNA *in situ*

1. Take a suspension of nuclei in RSB–HEPES buffer in 50% glycerol stored at −20 °C.

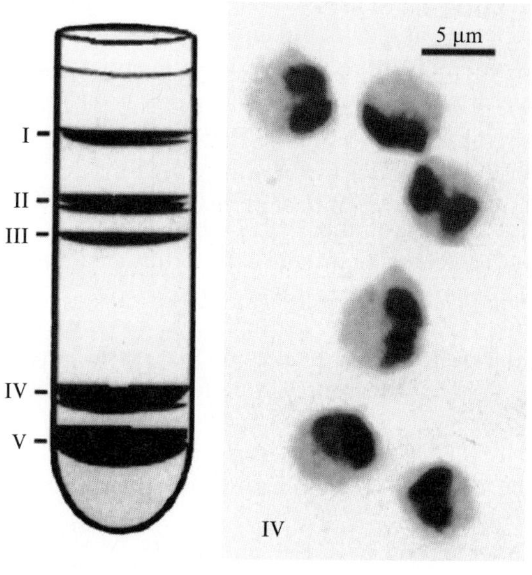

Figure 3.1 Fractionation of chicken blood and isolation of granulocytes. *Left*, scheme of chicken blood fractionation on 60% Percoll. *Right*, zone IV contains polymorphonuclear granulocytes. Scale bar: 5 μm.

2. Pellet the nuclei at 1575×*g*/5 min, 4 °C. Usually for one probe 1.5 mL of suspension ($\sim A_{260} = 8$) is sufficient.
3. Wash the pellet with RSB–HEPES buffer for 5 min/1575×*g*.
4. Resuspend the pellet in 1350 μL of RSB–HEPES. *Note*: Do not use Tris buffers during the cross-linking procedure. The concentration of nuclei during cross-linking should not exceed 10 A_{260} units.
5. Add 43 μL of 37% formaldehyde and incubate for 5 min at room temperature. Place the control (not crosslinked) probe to 4 °C.
6. Stop the reaction by the addition of 200 μL of 1 *M* glycine and incubate for 5 min at room temperature. For the control probe mix, premix 200 μL of 1 *M* glycine and 43 μL of 37% formaldehyde. Add the mixture to nuclei suspension and incubate for 5 min at room temperature.
7. Spin at 1575×*g*/5 min. Wash the pellet with RSB–HEPES (× 2).
8. Resuspend in 200 μL of L-ChIP buffer. Add protease inhibitors (protease inhibitor cocktail. Sigma Cat# P 2714) to this buffer and incubate 10 min on ice. If 200 μL is too small a volume for sonication, dissolve the pellet in 300 μL of L-ChIP buffer.
9. Sonicate using Branson Sonifier 250 set at 30–40% output. Perform a 90% duty cycle with 5 s (6×) with 20 s intervals for cooling on ice. There is little benefit to perform sonication in ice. Place the tubes in ice during the breaks. The conditions of sonication can depend on the type of sonicator.
10. Spin at 18,500×*g*/15 min/4 °C. Do not discard pellet as it can be used as a control for gene specific precipitation of DNA.
11. Dilute the supernatant with 10× D-ChIP buffer (with protease inhibitors) to get 2 mL of solution.
12. Spin at 18,500×*g*/15 min/4 °C. Do not throw away the pellet as it can be used as a control (see step 9).
13. Check size of DNA fragments using 1% agarose electrophoresis. The fragments of DNA should be <1000 bp.
14. Add antibodies (ABs) or preimmune serum at desired concentration. Incubate 1 h/RT/or overnight 4 °C.
15. Add 20 μL of 50% Protein A-Sepharose suspension. Pre-equilibrate beads with W-ChIP buffer before use (3× wash). *Note*: RIPA buffer is commonly used at this step. We prefer W-ChIP; however, RIPA is another option.
16. Incubate 1 h/RT on a rocking platform.
17. Pellet the beads at 14,000 rpm/5 s. Retain supernatant as control for the unbound fraction.
18. Wash with 1 mL of LS-ChIP buffer, HS-ChIP buffer, LiCl–ChIP buffer, TE two times.
19. Incubate beads in 250 μL of E-ChIP buffer for 15 min at room temperature before eluting protein and protein–DNA complexes.

20. Repeat the procedure.
21. Take 500 μL of final solution and add 20 μL of 5 *M* NaCl. To reverse the cross-linking incubate at 65 °C 5 h or overnight. (To reverse the cross-linking in the unbound fractions and pellets, simply resuspend the pellets in 500 μL of E-ChIP buffer and add NaCl to a final concentration 0.2 *M*).
22. Add 1 mL of 96% ethanol and incubate at −20 °C/2 h.
23. Pellet by centrifugation at 14,000 rpm/4 °C/15 min.
24. Wash the pellet with 70% ethanol and air-dry.
25. Suspend in 100 μL of TE, pH 8.0.
26. Add 20 μg of DNAse-free RNAse and incubate at 37 °C/1 h.
27. Add 11 μL of 10× proteinase K buffer and 1 μL of 20 mg/mL of proteinase K. Incubate at 50 °C/30 min.
28. Extract with phenol/chloroform/isoamyl-alcohol. Re-extract the first organic phase once with same volume of TE, pH 8.0. Pool the two aqueous solutions and repeat phenol extraction once.
29. Precipitate DNA with ethanol and resuspend final product in 200 μL of TE, pH 8.0.
30. Mix the DNA samples each containing 100 ng DNA with 1 μg of carrier herring sperm DNA (Gibco-BRL) and denature by heating to 95 °C before chilling on ice.
31. Apply DNA samples to Hybond N + membranes (GE Healthcare) with a slot blot apparatus, hybridize with 30 ng of probe DNA ^{32}P labeled with a Random Primer Extension Labeling System (Perkin-Elmer), and wash the membranes as described in the Hybond-N + vendor's instruction.
32. After hybridization, expose the membranes to a PhosphorImager (Molecular Dynamics), and quantify the radioactive signal with ImageQuant software.
33. Relative enrichment (RE) at each probed DNA site can be calculated for each series of experiments (obtained with DNA from one ChIP probed with different hybridization probes) according to the formula $RE = [(HN - HN_{cont})/HN_{tot}]/[(H_{max} - H_{maxcont})/H_{maxtot}]$. In this formula, HN is the hybridization signal after immunoprecipitation with MENT antibody probed with DNA probe *N*, HN_{cont} is the hybridization signal after immunoprecipitation of the same material with control nonimmune serum probed with *N*, HN_{tot} is the hybridization signal from total crosslinked DNA of the same material hybridized with probe *N*, H_{max} is the hybridization signal obtained with MENT antibody and probed with DNA that provides the maximal hybridization signal in the given experimental series (*max*), $H_{maxcont}$ is the hybridization signal after immunoprecipitation with control nonimmune serum probed with *max*, and H_{maxtot} is the hybridization signal from total crosslinked DNA probed with *max*.

Buffers

L-ChIP (lysis): 50 m*M* Tris–Cl, pH 8.0; 10 m*M* EDTA, 1% SDS.
D-ChIP (Dilution): 16.7 m*M* Tris–Cl, pH 8.0, 1.2 m*M* EDTA, 167 m*M* NaCl, 0.01% SDS, 1.1% Triton X-100.
LiCl–ChIP: 10 m*M* Tris–Cl, pH 8.0, 1 m*M* EDTA, 0.25 *M* LiCl, 1% Nonidet P-40, 10% deoxycholate.
LS-ChIP (low salt): 20 m*M* Tris–Cl, pH 8.0, 2 m*M* EDTA, 150 m*M* NaCl, 0.1% SDS, 1% Triton X-100.
HS-ChIP (high salt): 20 m*M* Tris–Cl, pH 8.0, 2 m*M* EDTA, 500 m*M* NaCl, 0.1% SDS, 1% Triton X-100.
E-ChIP (elution): 0.1 *M* $NaHCO_3$, 1% SDS.
RIPA: 50 m*M* Tris–Cl, pH 7.5, 1 m*M* EDTA, 150 m*M* NaCl, 0.25% SDS, 1% Nonidet P-40.
W-ChIP (wash): one part of L-ChIP buffer + nine parts of D-ChIP buffer.
10× proteinase K: 0.1 *M* Tris–Cl, pH 7.8, 50 m*M* EDTA, 5% SDS.
RSB–HEPES: 10 m*M* HEPES, pH 7.5, 10 m*M* NaCl, 3 m*M* $MgCl_2$.

ACKNOWLEDGMENTS

We thank the National Health and Medical Research Council (Australia), the Australian Research Council, and the USA National Science Foundation (grant MCB-1021681) for funding support.

REFERENCES

Allain, F. H., *et al.* (1999). Solution structure of the HMG protein NHP6A and its interaction with DNA reveals the structural determinants for non-sequence-specific binding. *EMBO J.* **18,** 2563–2579.
Ausubel, F. M., *et al.* (2005). *Current Protocols in Molecular Biology*. John Wiley and Sons, Inc., Hoboken, New York.
Carruthers, L. M., *et al.* (1998). Linker histones stabilize the intrinsic salt-dependent folding of nucleosomal arrays: Mechanistic ramifications for higher-order chromatin folding. *Biochemistry* **37,** 14776–14787.
Georgel, P. T., *et al.* (2003). Chromatin Compaction by Human MeCP2: Assembly of novel secondary chromatin structures in the absence of DNA methylation. *J. Biol. Chem.* **278,** 32181–32188.
Georgel, P. T., *et al.* (2001). Sir3-dependent assembly of supramolecular chromatin structures *in vitro*. *Proc. Natl. Acad. Sci. USA* **98,** 8584–8589.
Goulet, B., *et al.* (2004). A cathepsin L isoform that is devoid of a signal peptide localizes to the nucleus in S phase and processes the CDP/Cux transcription factor. *Mol. Cell* **14,** 207–219.
Grigoryev, S. A., *et al.* (1999). MENT, a heterochromatin protein that mediates higher order chromatin folding, is a new serpin family member. *J. Biol. Chem.* **274,** 5626–5636.
Grigoryev, S. A., *et al.* (2006). The end adjusts the means: heterochromatin remodelling during terminal cell differentiation. *Chromosome Res.* **14,** 53–69.

Grigoryev, S. A., *et al.* (1992). A novel nonhistone protein (MENT) promotes nuclear collapse at the terminal stage of avian erythropoiesis. *Exp. Cell Res.* **198,** 268–275.

Grigoryev, S. A., and Woodcock, C. L. (1998). Chromatin structure in granulocytes. A link between tight compaction and accumulation of a heterochromatin-associated protein (MENT). *J. Biol. Chem.* **273,** 3082–3089.

Irving, J. A., *et al.* (2002). Inhibitory activity of a heterochromatin-associated serpin (MENT) against papain-like cysteine proteinases affects chromatin structure and blocks cell proliferation. *J. Biol. Chem.* **277,** 13192–13201.

Istomina, N. E., *et al.* (2003). Insulation of the chicken β-globin chromosomal domain from a chromatin-condensing protein, MENT. *Mol. Cell. Biol.* **23,** 6455–6468.

Lehrer, S. S. (1971). Solute perturbation of protein fluorescence. The quenching of the tryptophyl fluorescence of model compounds and of lysozyme by iodide ion. *Biochemistry* **10,** 3254–3263.

Lowary, P. T., and Widom, J. (1998). New DNA sequence rules for high affinity binding to histone octamer and sequence-directed nucleosome positioning. *J. Mol. Biol.* **276,** 19–42.

Luger, K., and Hansen, J. C. (2005). Nucleosome and chromatin fiber dynamics. *Curr. Opin. Struct. Biol.* **15,** 188–196.

Luger, K., *et al.* (1999). Preparation of nucleosome core particle from recombinant histones. *Methods Enzymol.* **304,** 3–19.

McGowan, S., *et al.* (2006). X-ray crystal structure of MENT: evidence for functional loop-sheet polymers in chromatin condensation. *EMBO J.* **25,** 3144–3155.

Meersseman, G., *et al.* (1991). Chromatosome positioning on assembled long chromatin. Linker histones affect nucleosome placement on 5 S rDNA. *J. Mol. Biol.* **220,** 89–100.

Ong, P. C., *et al.* (2007). DNA accelerates the inhibition of human cathepsin V by serpins. *J. Biol. Chem.* **282,** 36980–36986.

Richmond, T. J., and Davey, C. A. (2003). The structure of DNA in the nucleosome core. *Nature* **423,** 145–150.

Schalch, T., *et al.* (2005). X-ray structure of a tetranucleosome and its implications for the chromatin fibre. *Nature* **436,** 138–141.

Silverman, G. A., *et al.* (2001). The serpins are an expanding superfamily of structurally similar but functionally diverse proteins. Evolution, mechanism of inhibition, novel functions, and a revised nomenclature. *J. Biol. Chem.* **276,** 33293–33296.

Springhetti, E. M., *et al.* (2003). Role of the M-loop and reactive center loop domains in the folding and bridging of nucleosome arrays by MENT. *J. Biol. Chem.* **278,** 43384–43393.

Woodcock, C. L., and Dimitrov, S. (2001). Higher-order structure of chromatin and chromosomes. *Curr. Opin. Genet. Dev.* **11,** 130–135.

Zhou, A., *et al.* (2001). Polymerization of plasminogen activator inhibitor-1. *J. Biol. Chem.* **276,** 9115–9122.

CHAPTER FOUR

Solving Serpin Crystal Structures

Randy J. Read,* Aiwu Zhou,* *and* Penelope E. Stein[†]

Contents

1. Introduction 50
2. Protein Production and Purification 50
3. Modifications to Aid Crystallization 51
4. Crystallization 51
5. Experimental Phasing 52
6. Molecular Replacement 53
   6.1. Serpin conformational states 54
   6.2. Rigid-body movements 54
   6.3. Using ensembles of multiple models 56
   6.4. Using electron density as a model 56
   6.5. Using homology models 57
7. Phase Improvement by Density Modification 57
8. Refinement and Validation 58
References 59

Abstract

Essentially the same steps are required to solve the crystal structure of a serpin as for any other protein: produce and purify protein, grow crystals, collect diffraction data, find estimates of the phase angles, and then refine and validate the structure. For the phasing step, experimental phasing methods involving heavy atom soaks were required for the first few structures, but with the large number of serpin structures now available, molecular replacement has become the method of choice. Two things are special about serpins. First, because of the central role of conformational change in serpin mechanism, it is advisable to consider a variety of molecular replacement models in different conformations and then to allow for rigid-body motions in the initial refinement steps. Second, probably owing to the flexibility of serpins, the average serpin crystal is significantly less well ordered than the average crystal of another protein, which increases the difficulty of solving and refining their structures.

* Department of Haematology, Cambridge Institute for Medical Research, University of Cambridge, Cambridge, United Kingdom
† Department of Medicine, University of Cambridge, Addenbrooke's Hospital, Cambridge, United Kingdom

Methods in Enzymology, Volume 501
ISSN 0076-6879, DOI: 10.1016/B978-0-12-385950-1.00004-3

1. Introduction

For the most part, the process of solving the crystal structure of a serpin is the same as for most other proteins: it is necessary to produce sufficient quantities (usually milligrams) of highly purified protein, induce the protein to crystallize, measure diffraction data, solve the phase problem by molecular replacement or experimental phasing methods, and then refine and validate the structure. This review only briefly discusses the general methods, which are covered well elsewhere, and concentrates instead on special features and complications of serpin structural biology. Much more detail on all aspects of protein crystallography can be found in recent textbooks (e.g., Drenth, 1999; Rupp, 2009).

2. Protein Production and Purification

Some serpins are abundant in their natural sources and can therefore be obtained by purification. For example, both α_1-antitrypsin (Laurell *et al.*, 1975) and antithrombin (McKay, 1981) can be purified from plasma, and ovalbumin can be purified from hen egg whites (adding an ion exchange step to reduce heterogeneity in phosphoserine modifications; Stein *et al.*, 1990). Other serpins are less abundant (e.g., neuroserpin; Briand *et al.*, 2001) or are too unstable to be purified from plasma (e.g., plasminogen activator inhibitor-1, PAI-1; Sharp *et al.*, 1999), so it is necessary to produce them by protein expression. This is also the case, of course, for site-directed mutants.

The majority of serpins studied structurally are eukaryotic proteins, which might be expected to require expression in eukaryotic cell-expression systems containing appropriate chaperones and other folding mediators. Fortunately, it has generally been found that serpins can readily be expressed in bacterial expression systems, perhaps because they possess few disulphide bridges, if any. The vast majority of serpins from structures deposited in the Protein Data Bank (Berman *et al.*, 2003) have been expressed in *Escherichia coli*. Nonetheless, there are cases where eukaryotic expression is necessary or at least desirable. For example, site-directed mutants of antithrombin have been expressed in BHK cells for crystallization (Huntington *et al.*, 2000a); attempts to express antithrombin in *E. coli* failed (James Huntington, personal communication). In addition, expression in mammalian cells allows the role of glycosylation and the structure of glycosylation sites to be studied, as in the case of PAI-1 expressed in CHO cells (Xue *et al.*, 1998).

Protein purification can be aided greatly by the addition of His-tags, often at the N-terminus but occasionally at the C-terminus. Even though His-tags are rarely ordered in crystal structures, their presence does not necessarily interfere with crystallization. A substantial number of serpins have been crystallized with the His-tags left on, even though these are not visible in the electron density and are presumably disordered (e.g., maspin; Al-Ayyoubi *et al.*, 2004). In the case of thyroxine-binding globulin, it was necessary to remove the His-tag before crystals could be grown (Zhou *et al.*, 2006).

Expression of poorly soluble or unstable proteins can be enhanced by fusing with a protein tag such as glutathione-*S*-transferase, mannose-binding protein, or SUMO, which is almost always removed prior to crystallization. For example, corticosteroid-binding globulin (CBG) could not be prepared in sufficient quantities without using a SUMO fusion construct (Zhou *et al.*, 2008).

3. Modifications to Aid Crystallization

PAI-1 in the active state converts spontaneously to the latent state with a half-life too short for crystallization, so the first PAI-1 structure to be determined was in the latent state (Mottonen *et al.*, 1992). The active state could only be crystallized by expressing a quadruple mutant selected to increase stability (Sharp *et al.*, 1999).

Initial attempts to crystallize maspin failed because of poor protein solubility, which was improved when all eight cysteine residues were mutated to either serine or alanine (Al-Ayyoubi *et al.*, 2004), leading to successful crystallization. Interestingly, maspin could also be crystallized without mutating the cysteine residues, but under conditions where added β-mercaptoethanol derivatized a number of surface cysteine residues (Law *et al.*, 2005).

Glycosylation sites on proteins are chemically and conformationally heterogeneous, which can be a detriment to crystallization. Glycosylated proteins can be crystallized successfully if the carbohydrate chains project into a solvent channel or if crystal-packing contacts involve only the common core carbohydrate residues; one example is β-antithrombin (McCoy *et al.*, 2003), for which ordered carbohydrate density can be seen at a number of glycosylation sites. On the other hand, it was not possible to crystallize C1-inhibitor without deglycosylation (Beinrohr *et al.*, 2007).

4. Crystallization

Proteins are generally crystallized by slowly reducing their solubility; under the right conditions, they come out of solution packed regularly in a crystal lattice instead of as an irregular aggregate. Serpin crystallization, on the whole, requires the same process of testing a variety of conditions that have been found to work with other proteins, for example, adding precipitants such as polyethylene glycol or ammonium sulfate and varying pH, salts, and other additives. A detailed summary of the theory and practice of protein crystallization can be found in a recent textbook (Bergfors, 2009).

The most typical crystallization method involves mixing the protein with some precipitant solution and then allowing it to equilibrate by vapor diffusion with the precipitant solution, either in a hanging drop or in a sitting drop. With the recent development of robotic technology for crystallization, the size of drops that can be used has decreased from the microliter range to the nanoliter range so that a small quantity of protein goes much further. This was essential, for instance, in crystallizing the complex between renin and angiotensinogen (Zhou *et al.*, 2010), as only small quantities of inactivated, deglycosylated renin could be purified.

Although most serpin crystallizations are unremarkable compared to other proteins, antithrombin provided a special case. Crystals of antithrombin were difficult to grow, taking a long time and being difficult to reproduce. When the structure was determined, it emerged that the crystals were formed from heterodimers of active and latent antithrombin, so they could only grow when the conditions allowed the formation of latent antithrombin and only after enough time had passed for sufficient quantities of the latent form to be present in the drop (Carrell *et al.*, 1994; Schreuder *et al.*, 1994). Once this was understood, crystals could be grown quickly and reproducibly by mixing equal quantities of active and latent antithrombin (Jin *et al.*, 1997). Even more usefully, it was possible to mix site-directed mutants of active antithrombin with wild-type latent antithrombin and grow crystals reliably (e.g., Huntington *et al.*, 2000a); this process was termed cassette crystallization (Carrell and Huntington, 2003).

5. Experimental Phasing

In the diffraction experiment, the intensities of the measured spots give information about the amplitudes of waves used to build up a picture of the electron density in the crystal, but not the phases that determine where the peaks in the features appear. If an atomic model of the crystal is known, the phase angles can be computed from the model, but if the structure is unknown, the phases must be deduced by experimental phasing methods. In experimental phasing, the diffraction pattern is perturbed either by binding a small number of heavy atoms to the protein molecules in the crystal (isomorphous replacement method) or by changing the wavelength, which changes the diffraction from atoms possessing an absorption edge near the wavelength (anomalous diffraction method, either with multiple wavelengths or with a single wavelength).

In the early days of serpin structural biology, when the serpin fold was not known and the conformational changes were not understood, experimental phasing was the only available option. For instance, the first serpin structure, that of cleaved α_1-antitrypsin, was solved by multiple isomorphous replacement of one crystal form, using six heavy atom derivatives, and single isomorphous replacement (one derivative) for another crystal form (Loebermann *et al.*, 1984). The first representative of the native serpin fold (lacking the insertion of the reactive center loop into the A-sheet) was plakalbumin (cleaved ovalbumin), which was determined by multiple isomorphous replacement in a single crystal form (Wright *et al.*, 1990).

Since those early days, it has been possible to use known structures to determine phases for almost all new serpin structures by the molecular replacement method (discussed below). Nonetheless, it can be useful to supplement model phase information with experimental phases, particularly when the resolution of the data is limited. The structure of cleaved antithrombin could be solved by molecular replacement using cleaved α_1-antitrypsin as a model, but with diffraction data limited to 3.2 Å resolution, structure refinement did not proceed well; when phases were obtained by multiple isomorphous replacement, a much better atomic model could be constructed (Mourey *et al.*, 1993). A crystal of uncleaved α_1-antitrypsin diffracted to the much better resolution of 2.1 Å, but density for the mobile reactive center loop was very difficult to interpret until multiple isomorphous replacement was used to supplement the phase information (Kim *et al.*, 2001).

6. Molecular Replacement

The technique of molecular replacement is used when there is a reasonable atomic model of the protein in the crystal, either from a different crystal form of the same protein or from the structure of a sufficiently close homologue. With many serpin structures now known (125 deposits in the Protein Data Bank, as of December 2010), this has become the method of choice.

In fact, molecular replacement was applied for some of the earliest serpin structures, including intact ovalbumin (Stein *et al.*, 1990), which used plakalbumin as a model, and cleaved α_1-antichymotrypsin (Baumann *et al.*, 1991), which used cleaved α_1-antitrypsin as a model.

To apply molecular replacement, it is necessary to find the orientation and position of the model (or models) in the unit cell of the unknown structure. It is possible, and computationally efficient, to carry this out in two steps: a rotation search to find the orientation and a translation search to find the position of the oriented model; these steps can be repeated if there is more than one copy or more than one model to find. Most molecular replacement programs, including those based on Patterson search (AMoRe: Navaza, 2001; Molrep: Vagin and Teplyakov, 1997) or on likelihood scoring (Phaser: McCoy *et al.*, 2007), use such a subdivision of the problem. However, it is also possible to carry out $6n$-dimensional searches for n molecules simultaneously with stochastic search methods such as genetic algorithms (EPMR; Kissinger *et al.*, 1999) or Monte Carlo (Queen of Spades; Glykos and Kokkinidis, 2000).

Compared to the traditional Patterson search methods, the likelihood-based methods implemented in *Phaser* tend to be more sensitive in finding solutions using distant relatives or in assembling complexes with multiple components, but serpin structures have been solved with a variety of molecular replacement approaches.

6.1. Serpin conformational states

Serpins are of great interest to structural biologists because of their remarkable conformational changes. The most dramatic changes occur between the native conformation, with an exposed reactive center loop and a five-stranded A-sheet (e.g., uncleaved ovalbumin; Stein *et al.*, 1990), and the various forms in which the reactive center loop has inserted to contribute a sixth strand to the A-sheet: cleaved (e.g., α_1-antitrypsin; Loebermann *et al.*, 1984), latent (e.g., PAI-1; Mottonen *et al.*, 1992), and protease-complexed (e.g., α_1-antitrypsin: trypsin; Huntington *et al.*, 2000b). In addition, smaller conformational changes mediate allosteric effects. For example, the binding of heparin to uncleaved antithrombin triggers the expulsion of the N-terminal portion of the reactive center loop from the top of the A-sheet (Jin *et al.*, 1997).

Because of this conformational variability, it is always a good idea to try several alternative molecular replacement models when solving a new serpin structure, to see which gives the strongest signal and thus provides the best starting model. If serpin structures in the expected conformational state do not give a clear solution, then it is worthwhile to try structures in other conformational states. For instance, the structure of an α_1-antitrypsin polymer was determined from crystals that were intended to contain a complex of α_1-antitrypsin with inactivated S195A thrombin (Huntington *et al.*, 1999). Presumably there was a contaminant protease that, over the course of the year required to grow crystals, cleaved α_1-antitrypsin at the P7–P6 site of its reactive center loop.

6.2. Rigid-body movements

As soon as structures were known for representatives of cleaved (α_1-antitrypsin) and uncleaved (ovalbumin) serpins, it was possible to describe the conformational change in terms of rigid-body movements of domains and subdomains (Stein and Chothia, 1991). This showed that, to a good approximation, most of the serpin forms a rigid core, and the insertion of the reactive center loop causes a rigid-body shift of a subdomain comprising strands 1–3 of the A-sheet and helix F; helices D and E act as flexible joints (see Fig. 4.1).

This analysis was complicated by the sequence differences between the two serpins. As the structural database has expanded, so that individual serpins have been seen in different conformational states, greater sophistication in the analysis of serpin conformational change has become possible (Langdown *et al.*, 2009; Whisstock *et al.*, 2000a,b). In the most recent analysis, much of antithrombin provides a constant core (group 1) against which three other rigid bodies move: group 2 is formed by helices E and F plus strand 1A and the lower portions of strands 2A and 3A, group 3 comprises helix D, and group 4 is formed by the top halves of strands 2A and 3A.

These rigid subdomains can be used either to assemble a model by molecular replacement from its constituents (starting with the constant core domain and then adding the other fragments) or to improve an initial

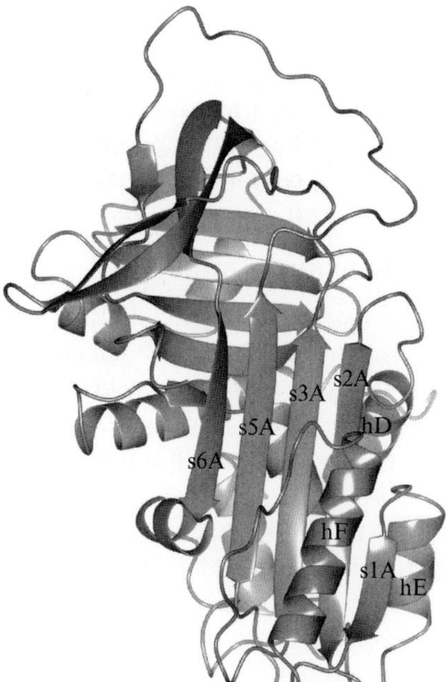

Figure 4.1 Cartoon diagram of native α_1-antitrypsin (Kim *et al.*, 2001), with labels showing the secondary structural elements involved in rigid-body movements during serpin conformational changes. Strands in the major A-sheet are labeled s1A through s6A, while helices D, E, and F are labeled hD through hF.

model by rigid-body refinement of the separate groups. For example, the structure of active PAI-1 was solved using the latent structure (Mottonen *et al.*, 1992) as a model and then carrying out rigid-body refinement of two major fragments (Sharp *et al.*, 1999).

6.3. Using ensembles of multiple models

Instead of testing alternative models individually for molecular replacement, it can be helpful to use them together as an ensemble average or composite model. Conserved features in the core of the different models, which are likely to be maintained in the new structure, will be preserved in the average, whereas features that differ among the models will be down-weighted. A simple average can be useful, but the program *Phaser* (McCoy *et al.*, 2007) computes a statistically weighted average that takes account of the correlations among the models.

For example, the structure of maspin was solved using an ensemble of four native serpin models (Law *et al.*, 2005). The structure of C1 inhibitor could not be solved with individual serpin models but was successfully solved using a composite obtained by superimposing 10 alternative models from which structurally nonconserved regions were deleted (Beinrohr *et al.*, 2007). Similarly, the structure of human angiotensinogen could not be solved with any individual model (all of which had sequence identities around 20%) but could be solved in *Phaser* by using the ensemble-weighted average of three models from which the regions that vary significantly among the models had been trimmed (Zhou *et al.*, 2010).

6.4. Using electron density as a model

To be precise, the molecular replacement problem for human angiotensinogen could be solved using an ensemble model, but the structure was not really solved at this point. The phases computed from the distant homologues were too poor to reveal the details of electron density required to improve the starting model, especially given the modest resolution of 3.2 Å. Even when the phase information was supplemented with a poor $GdCl_3$ derivative, the map was not sufficiently clear (Zhou *et al.*, 2010). The structure was ultimately solved by a rather circuitous procedure (discussed below under density modification), the first step of which involved solving the structure of rat angiotensinogen with electron density from the human angiotensinogen map. The molecular replacement models that worked for human angiotensinogen did not succeed with rat or mouse angiotensinogen (two crystal forms each). Even though the electron density map for human angiotensinogen, computed with phases from a combination of molecular replacement and isomorphous replacement, was not interpretable, the introduction of the structure factor amplitudes from the diffraction experiment would be expected to add some correct features to the density. Indeed, when the electron density from this map was used as a molecular replacement model, a clear solution was obtained for one of the two crystal forms of rat angiotensinogen. This was a key step toward the ultimate structure solution. In fact, subsequent calculations have shown that the experimental phases from the $GdCl_3$ derivative were not required; the electron density from a human angiotensinogen map phased only with the molecular replacement ensemble is sufficient to solve the rat angiotensinogen structure (R. J. Read, unpublished results).

6.5. Using homology models

Until recently, experiences with using homology models for molecular replacement were rather poor. In most cases, better results were obtained with the unmodified templates than with the models derived from them. Nonetheless, there have been isolated examples of success with this approach. One such example is the structure of neuroserpin (Briand *et al.*, 2001), which was solved using a homology model constructed with the program *Modeller* (Sali and Blundell, 1993) based on a collection of 10 templates. Perhaps, this succeeded because a series of 500 alternative models were constructed with *Modeller*; the best one was chosen by how well it satisfied the Ramachandran plot, and regions that did not agree among the models were removed (Christophe Briand, personal communication).

Methods for homology modeling continue to improve substantially, so such models are becoming more and more useful for molecular replacement. Using *Rosetta*, one of the most advanced molecular modeling packages, it was shown that a variety of homology models (and NMR structures) could be improved substantially for molecular replacement (Qian *et al.*, 2007). Remarkably, in one case, even an *ab initio* folding model succeeded in a molecular replacement calculation.

7. Phase Improvement by Density Modification

The phase information derived by either experimental phasing or molecular replacement can be improved by a method termed density modification, in which the map is changed to reduce errors in the electron density and then the phases are back-calculated from the modified map (reviewed by Kleywegt and Read, 1997). For instance, we know that the regions containing disordered solvent (typically about half of the crystal volume) should have constant density, so solvent flattening can be applied to reduce phase errors. When there is a view of the electron density for more than one copy of the molecule, the even more powerful constraint of electron density averaging can be applied. Both techniques, solvent flattening and averaging of the copies seen in two crystal forms, were applied from the start in solving the structure of α_1-antitrypsin (Loebermann *et al.*, 1984).

As noted above, density modification played a central role in determining the structure of angiotensinogen (Zhou *et al.*, 2010). One crystal form of rat angiotensinogen could be solved by placing two copies of the noisy electron density from the human angiotensinogen crystal, though this did not succeed for the other crystal form of rat angiotensinogen or for either crystal form of mouse angiotensinogen. However, after the density for the first rat angiotensinogen crystal form was improved by carrying out twofold electron density averaging, this density could be used to find two copies in the second crystal form of rat angiotensinogen. At this point, cross-crystal averaging was carried out, averaging the four copies of density in the two rat crystal forms, yielding an excellent map that could be interpreted easily. The model of rat angiotensinogen could then be used to solve two crystal forms of mouse angiotensinogen (one of which diffracted to 2.1 Å resolution), and finally, the refined model of mouse angiotensinogen could be used to re-solve the structure of human angiotensinogen, this time with an excellent model.

8. Refinement and Validation

Once a crystal structure has been determined, it is essential to carry out careful refinement and to validate it against what is known about protein structures. In casual conversations with serpin structural biologists, one gets the impression that this part of the process is generally more difficult for serpins than for typical proteins. To begin with, the reactive center loop is flexible and, in most cases, is probably only well ordered if it is involved in a close crystal-packing contact. Conformational change is at the heart of the serpin mechanism, so the crystals formed by serpins might be expected to behave worse than those of more rigid proteins.

In fact, there is some objective evidence for the notion that serpin crystals are more poorly behaved than average. An examination of the entire PDB showed that the median resolution for all protein crystal structures is about 2.1 Å (Read and Kleywegt, 2009). For the 125 structures containing proteins with the serpin fold that were present in the PDB at the end of 2010, the median resolution is significantly worse, at 2.6 Å. Nonetheless, these structures provide the essential underpinning of our detailed understanding of serpin mechanism and biology.

REFERENCES

Al-Ayyoubi, M., Gettins, P. G. W., and Volz, K. (2004). Crystal structure of human maspin, a serpin with antitumor properties. *J. Biol. Chem.* **279,** 55540–55544.

Baumann, U., Huber, R., Bode, W., Grosse, D., Lesjak, M., and Laurell, C. B. (1991). Crystal structure of cleaved human α_1-antichymotrypsin at 2.7 Å resolution and its comparison with other serpins. *J. Mol. Biol.* **218,** 595–606.

Beinrohr, L., Harmat, V., Dobó, J., Lörincz, Z., Gál, P., and Závodszky, P. (2007). C1 inhibitor serpin domain structure reveals the likely mechanism of heparin potentiation and conformational disease. *J. Biol. Chem.* **282,** 21100–21109.

Bergfors, T. M. (2009). Protein Crystallization. 2nd edn. International University Line, La Jolla.

Berman, H. M., Henrick, K., and Nakamura, H. (2003). Announcing the worldwide Protein Data Bank. *Nat. Struct. Biol.* **10,** 98.

Briand, C., Kozlov, S. V., Sonderegger, P., and Grütter, M. G. (2001). Crystal structure of neuroserpin: A neuronal serpin involved in a conformational disease. *FEBS Lett.* **505,** 18–22.

Carrell, R. W., and Huntington, J. A. (2003). How serpins change their fold for better and for worse. *Biochem. Soc. Symp.* **70,** 163–178.

Carrell, R. W., Stein, P. E., Fermi, G., and Wardell, M. R. (1994). Biological implications of a 3 Å structure of dimeric antithrombin. *Structure* **2,** 257–270.

Drenth, J. (1999). Principles of Protein X-ray Crystallography. 2nd edn. Springer-Verlag, New York.

Glykos, N. M., and Kokkinidis, M. (2000). A stochastic approach to molecular replacement. *Acta Cryst.* **D56,** 169–174.

Huntington, J. A., Pannu, N. S., Hazes, B., Read, R. J., Lomas, D. A., and Carrell, R. W. (1999). A 2.6 Å structure of a serpin polymer and implications for conformational disease. *J. Mol. Biol.* **293,** 449–455.

Huntington, J. A., McCoy, A., Belzar, K. J., Pei, X. Y., Gettins, P. G., and Carrell, R. W. (2000a). The conformational activation of antithrombin. A 2.85 Å structure of a fluoroscein derivative reveals an electrostatic link between the hinge and heparin binding regions. *J. Biol. Chem.* **275,** 15377–15383.

Huntington, J. A., Read, R. J., and Carrell, R. W. (2000b). Structure of a serpin-protease complex shows inhibition by deformation. *Nature* **407,** 923–926.

Jin, L., Abrahams, J. P., Skinner, R., Petitou, M., Pike, R. N., and Carrell, R. W. (1997). The anticoagulant activation of antithrombin by heparin. *Proc. Natl. Acad. Sci. USA* **94,** 14683–14688.

Kim, S.-J., Woo, J.-R., Seo, E. J., Yu, M.-H., and Ryu, S.-E. (2001). A 2.1 Å resolution structure of an uncleaved α_1-antitrypsin shows variability of the reactive center and other loops. *J. Mol. Biol.* **306,** 109–119.

Kissinger, C. R., Gehlhaar, D. K., and Fogel, D. B. (1999). Rapid automated molecular replacement by evolutionary search. *Acta Cryst.* **D55,** 484–491.

Kleywegt, G. J., and Read, R. J. (1997). Ways & Means: Not your average density. *Structure* **5,** 1557–1569.

Langdown, J., Belzar, K. J., Savory, W. J., Baglin, T. P., and Huntington, J. A. (2009). The critical role of hinge-region expulsion in the induced-fit heparin binding mechanism of antithrombin. *J. Mol. Biol.* **386,** 1278–1289.

Laurell, C.-B., Pierce, J., Persson, U., and Thulin, E. (1975). Purification of α_1-antitrypsin from plasma through thiol-disulfide interchange. *Eur. J. Biochem.* **57,** 107–113.

Law, R. H. P., Irving, J. A., Buckle, A. M., Ruzyla, K., Buzza, M., Bashtannyk-Puhalovich, T. A., Beddoe, T. C., Nguyen, K., Worrall, D. M., Bottomley, S. P., Bird, P. I., Rossjohn, J., *et al.* (2005). The high resolution crystal structure of the human tumor suppressor maspin reveals a novel conformational switch in the G-helix. *J. Biol. Chem.* **280,** 22356–22364.

Loebermann, H., Tokuoka, R., Deisenhofer, J., and Huber, R. (1984). Human α_1-proteinase inhibitor: Crystal structure analysis of two crystal modifications, molecular model and preliminary analysis of the implications for function. *J. Mol. Biol.* **177,** 531–556.

McCoy, A. J., Pei, X. Y., Skinner, R., Abrahams, J.-P., and Carrell, R. W. (2003). Structure of β-antithrombin and the effect of glycosylation on antithrombin's heparin affinity and activity. *J. Mol. Biol.* **326,** 823–833.

McCoy, A. J., Grosse-Kunstleve, R. W., Adams, P. D., Winn, M. D., Storoni, L. C., and Read, R. J. (2007). Phaser crystallographic software. *J. Appl. Cryst.* **40,** 658–674.

McKay, E. J. (1981). A simple two-step procedure for the isolation of antithrombin III from biological fluids. *Thromb. Res.* **21,** 375–382.

Mottonen, J., Strand, A., Symersky, J., Sweet, R. M., Danley, D. E., Geoghegan, K. F., Gerard, R. D., and Goldsmith, E. J. (1992). Structural basis of latency in plasminogen activator inhibitor-1. *Nature* **355,** 270–273.

Mourey, L., Samama, J.-P., Delarue, M., Petitou, M., Choay, J., and Moras, D. (1993). Crystal structure of cleaved bovine antithrombin III at 3.2 Å resolution. *J. Mol. Biol.* **232,** 223–241.

Navaza, J. (2001). Implementation of molecular replacement in AMoRe. *Acta Cryst.* **D57,** 1367–1372.

Qian, B., Raman, S., Das, R., Bradley, P., McCoy, A. J., Read, R. J., and Baker, D. (2007). High-resolution structure prediction and the crystallographic phase problem. *Nature* **450,** 259–264.

Read, R. J., and Kleywegt, G. J. (2009). Case-controlled structure validation. *Acta Cryst.* **D65,** 140–147.

Rupp, B. (2009). *Biomolecular Crystallography: Principles, Practice, and Application to Structural Biology.* Garland Science, New York.

Sali, A., and Blundell, T. L. (1993). Comparative protein modelling by satisfaction of spatial restraints. *J. Mol. Biol.* **234,** 779–815.

Schreuder, H. A., de Boer, B., Dijkema, R., Mulders, J., Theunissen, H. J. M., Grootenhuis, P. D. J., and Hol, W. G. J. (1994). The intact and cleaved human antithrombin III complex as a model for serpin-proteinase interactions. *Nat. Struct. Biol.* **1,** 48–54.

Sharp, A. M., Stein, P. E., Pannu, N. S., Carrell, R. W., Berkenpas, M. B., Ginsburg, D., Lawrence, D. A., and Read, R. J. (1999). The active conformation of plasminogen activator inhibitor-1, a target for drugs to control fibrinolysis and cell adhesion. *Structure* **7,** 111–118.

Stein, P. E., and Chothia, C. (1991). Serpin tertiary structure transformation. *J. Mol. Biol.* **221,** 615–621.

Stein, P. E., Leslie, A. G. W., Finch, J. T., Turnell, W. G., McLaughlin, P. J., and Carrell, R. W. (1990). Crystal structure of ovalbumin as a model for the reactive centre of serpins. *Nature* **347,** 99–102.

Vagin, A., and Teplyakov, A. (1997). MOLREP: An automated program for molecular replacement. *J. Appl. Cryst.* **30,** 1022–1025.

Whisstock, J. C., Skinner, R., Carrell, R. W., and Lesk, A. M. (2000a). Conformational changes in serpins: I. The native and cleaved conformations of α_1-antitrypsin. *J. Mol. Biol.* **296,** 685–699.

Whisstock, J. C., Pike, R. N., Jin, L., Skinner, R., Pei, X. Y., Carrell, R. W., and Lesk, A. M. (2000b). Conformational changes in serpins: II. The mechanism of activation of antithrombin by heparin. *J. Mol. Biol.* **301,** 1287–1305.

Wright, H. T., Qian, H. X., and Huber, R. (1990). Crystal structure of plakalbumin, a proteolytically nicked form of ovalbumin. *J. Mol. Biol.* **213,** 513–528.

Xue, Y., Björquist, P., Inghardt, T., Linschoten, M., Musil, D., Sjölin, L., and Deinum, J. (1998). Interfering with the inhibitory mechanism of serpins: Crystal structure of a complex formed between cleaved plasminogen activator inhibitor type 1 and a reactive-centre loop peptide. *Structure* **6,** 627–636.

Zhou, A., Wei, Z., Read, R. J., and Carrell, R. W. (2006). Structural mechanism for the carriage and release of thyroxine in the blood. *Proc. Natl. Acad. Sci. USA* **103,** 13321–13326.

Zhou, A., Wei, Z., Stanley, P. L., Read, R. J., Stein, P. E., and Carrell, R. W. (2008). The S-to-R transition of corticosteroid-binding globulin and the mechanism of hormone release. *J. Mol. Biol.* **380,** 244–251.

Zhou, A., Carrell, R. W., Murphy, M. P., Wei, Z., Yan, Y., Stanley, P. L., Stein, P. E., Broughton Pipkin, F., and Read, R. J. (2010). A redox switch in angiotensinogen modulates angiotensin release. *Nature* **468,** 108–111.

CHAPTER FIVE

Crystallography of Serpins and Serpin Complexes

M. A. Dunstone*,†,1 *and* James C. Whisstock*,‡,1

Contents

Abstract

The serpin superfamily of protease inhibitors undergoes a remarkable conformational change to inhibit target proteases. To date, over 80 different serpin crystal structures have been determined. These data reveal that the serpin monomer can adopt five different conformations (native, partially inserted native, δ-form, latent, and cleaved). Further, recent studies have also revealed that serpins can domain swap; biochemical data suggest such an event underlies serpin polymerization in diseases such as antitrypsin deficiency. Here, we provide a comprehensive analysis on crystallization of serpins in context of the structural landscape of the serpin superfamily.

* Department of Biochemistry and Molecular Biology, Monash University, Melbourne, Victoria, Australia
† Department of Microbiology, Monash University, Clayton, Victoria, Australia
‡ ARC Centre of Excellence in Structural and Functional Microbial Genomics, Monash University, Clayton, Victoria, Australia
1 These authors contributed equally to this work.

Methods in Enzymology, Volume 501
ISSN 0076-6879, DOI: 10.1016/B978-0-12-385950-1.00005-5

1. Introduction

Serpins comprise the largest family of protease inhibitors identified to date, with family members identifiable in all kingdoms of life (Irving *et al.*, 2000, 2002). These proteins are characterized by an unusual conformational rearrangement that is central to their function (Whisstock and Bottomley, 2006). In this review, we characterize the history and current status of serpin structural biology. Table 5.1 lists the current structures of serpins that have been published to date.

2. First Glimpses of Serpin Structures

The first serpin structure, that of cleaved antitrypsin, was determined in 1984 by Huber and colleagues (Fig. 5.1A; PDB ID: 7API; Loebermann *et al.*, 1984). These data revealed the surprising finding that the region proposed to interact with target proteases (the reactive center loop, RCL) lay buried as a β-strand in the center of a large central β-sheet (the A-sheet). Thus it was suggested that in order to bind to target proteases, native serpins must adopt a conformationally distinct form to the cleaved state. Specifically, it was proposed that the RCL must originally be positioned as an exposed loop that, upon being cleaved, could insert into the A-sheet as a fourth β-strand. These events would result in the A-sheet opening and switching from a five to a six β-stranded state. This hypothesis was finally supported through structural studies with the determination of the structure of the noninhibitory serpin, ovalbumin, as well as other native inhibitory serpins (Fig. 5.1B; PDB ID: 1OVA; Stein *et al.*, 1990). Taken together with biophysical studies, these data further revealed that native serpins were *metastable* since the transition to the cleaved form was accompanied by a dramatic increase in stability. The conformational change was thus termed the S (stressed) to R (relaxed) transition.

3. The Serpin–Enzyme Complex

Following the determination of the structures of cleaved antitrypsin and ovalbumin, there was considerable debate in the field about the precise role of the remarkable serpin conformational change (Whisstock *et al.*, 1996; Wright and Scarsdale, 1995) and, specifically, how serpins could inhibit target proteases. Finally, in 2000, the first structure of a serpin in complex with a target protease (antitrypsin in complex with trypsin) was

Table 5.1 Summary of all published serpin structures (up to December, 2010) including the PDB identifier (PDB) and conformational state (State). Details of the Source of the serpin (Compound; source organism;expression host/tissue source) and any genetic modification (Mutation, added tags) are provided

PDB	State	Compound; source organism; expression host/tissue source	Mutation, added tags	References
7API	Cleaved, inserted	ALPHA-1 ANTITRYPSIN; Homo sapiens; human plasma	No ("M variant")	Loebermann *et al.* (1984) and Engh *et al.* (1989)
8API	Cleaved, inserted	ALPHA-1 ANTITRYPSIN (CHAIN A); Homo sapiens; human plasma	No ("M variant")	Loebermann *et al.* (1984) and Engh *et al.* (1989)
9API	Cleaved, inserted	ALPHA-1 ANTITRYPSIN; Homo sapiens; human plasma	No ("M variant")	Engh *et al.* (1989)
No PDB entry	Cleaved, inserted	ALPHA-1 ANTITRYPSIN; Homo sapiens; human plasma (of S variant carrier)	S variant E364V	Engh *et al.* (1989)
No PDB entry	Cleaved, not inserted	Plakalbumin (subtilisin cleaved ovalbumin); *Gallus gallus*; egg white	No	Wright *et al.* (1990)
1OVA	Uncleaved, not inserted	OVALBUMIN; *Gallus gallus*; egg white	No	Stein *et al.* (1990) and Stein *et al.* (1991)
2ACH	Cleaved, inserted	ALPHA-1 ANTICHYMOTRYPSIN; Homo sapiens; plasma	No	Baumann *et al.* (1991)
1HLE	Cleaved, not inserted	HORSE LEUKOCYTE ELASTASE INHIBITOR; *Equus caballus*; leukocytes	No	Baumann *et al.* (1992)
1ATT	Cleaved, inserted	ANTITHROMBIN III; *Bos taurus*; plasma	No	Mourey *et al.* (1993)
1ANT	Dimer of a inhibitory confirmation with a latent (joined at C-sheet)	ANTITHROMBIN; Homo sapiens; plasma	No	Carrell *et al.* (1994)
1ATH	Dimer of a cleaved inserted and inhibitory confirmation	ANTITHROMBIN III; Homo sapiens; plasma	No	Schreuder *et al.* (1994)
No PDB entry	Uncleaved inhibitory serpin structure	ALPHA-1 ANTICHYMOTRYPSIN; Homo sapiens; *Escherichia coli*	T356I, L357P, L358M, A360I, L361P, (antitrypsin-like RCL) A349GG, A350T, V368T	Wei *et al.* (1994)
1C5G	Latent	PLASMINOGEN ACTIVATOR INHIBITOR-1; Homo sapiens; *Escherichia coli*	Start codon, no tags	Mottonen *et al.* (1992), Tucker *et al.* (1995)
1KCT	Uncleaved	ALPHA-1 ANTITRYPSIN; Homo sapiens; *Escherichia coli* BL21(DE3)	No tags, refolded	Song *et al.* (1995)

(*Continued*)

Table 5.1 (*Continued*)

PDB	State	Compound; source organism; expression host/tissue source	Mutation, added tags	References
9PAI	Cleaved, inserted	Chain A = PROTEIN (PLASMINOGEN ACTIVATOR INHIBITOR-1) residues 19-364, Chain B = PROTEIN (PLASMINOGEN ACTIVATOR INHIBITOR-1) residues 365–397; Chain A = B = Homo sapiens; Chain A = B = *Escherichia coli*	A335P	Aertgeerts *et al.* (1995)
1ATU	Uncleaved	ALPHA-1 ANTITRYPSIN; Homo sapiens; *Escherichia coli*	F51L, T59A, T68A, A70G, M374I, S381A, K387R, no tags, refolded	Ryu *et al.* (1996)
1PSI	Uncleaved canonical conformation	ALPHA-1 ANTITRYPSIN; Homo sapiens; *Escherichia coli*	F51L	Elliott *et al.* (1996)
No PDB entry	Cleaved, inserted	ANTICHYMOTRYPSIN; Homo sapiens; *Escherichia coli*	A347R	Lukacs *et al.* (1996)
No PDB entry	Cleaved, inserted	ANTICHYMOTRYPSIN; Homo sapiens; *Escherichia coli*	T345R	Lukacs *et al.* (1996)
1AZX	Dimer of a inhibitory confirmation with a latent (joined at C-sheet) plus heparin bound	ANTITHROMBIN; Homo sapiens; plasma	No	Jin *et al.* (1997)
2ANT	Dimer of a inhibitory confirmation with a latent (joined at C-sheet)	ANTITHROMBIN; Homo sapiens; plasma	No	Skinner *et al.* (1997)
1A7C	Uncleaved, two peptides inserted into A-sheet	Chain A = PLASMINOGEN ACTIVATOR INHIBITOR TYPE-1, Chain B = C = pentapeptide; Chain A = B = C = Homo sapiens; Chain A = *Cricetulus griseus* (Chinese hamster ovary cells), Chain B = C = synthetic	A335E	Xue *et al.* (1998)
1AS4	Cleaved, inserted	ANTICHYMOTRYPSIN; Homo sapiens; *Escherichia coli*	A349R	Lukacs *et al.* (1998)

1BR8	Dimer of a inhibitory confirmation with a latent (joined at C-sheet)	PROTEIN (ANTITHROMBIN-III); Homo sapiens; plasma	No	Skinner *et al.* (1998)
2PSI	Canonical inhibitory conformation	ALPHA-1 ANTITRYPSIN; HOMO SAPIENS; *Escherichia coli*	E1M	Elliott *et al.* (1998)
3CAA	Cleaved, inserted	ANTICHYMOTRYPSIN; Homo sapiens; *Escherichia coli*	A347R	Lukacs *et al.* (1998)
4CAA	Cleaved, inserted	ANTICHYMOTRYPSIN; Homo sapiens; *Escherichia coli*	T345R	Lukacs *et al.* (1998)
1B3K	Active uncleaved	PLASMINOGEN ACTIVATOR INHIBITOR-1; Homo sapiens; *Escherichia coli*	N15H, K154T, Q319L, M354I	Sharp *et al.* (1999)
1BY7	Uncleaved	PROTEIN (PLASMINOGEN ACTIVATOR INHIBITOR-2); Homo sapiens; *Escherichia coli*	Loop deletion mutant: residues 66–98	Harrop *et al.* (1999)
1QMB	Cleaved polymer	ALPHA-1 ANTITRYPSIN; HOMO SAPIENS; *ESCHERICHIA COLI*	M358R	Huntington *et al.* (1999)
1SEK	Uncleaved canonical conformation	SERPIN K; *Manduca sexta*; *Escherichia coli*	N-terminal His tag	Li *et al.* (1999)
1C8O	Cleaved, inserted	ICE INHIBITOR; Cowpox virus; *Escherichia coli*	All nine cysteines mutated to serine	Simonovic *et al.* (2000)
1D5S	Cleaved polymer	P1-ARG ANTITRYPSIN; Homo sapiens; *Escherichia coli*	M349R	Dunstone *et al.* (2000)
1DB2	Uncleaved	PLASMINOGEN ACTIVATOR INHIBITOR-1; Homo sapiens; *Escherichia coli*	N150H, K154T, Q301P, Q319L, M354I	Nar *et al.* (2000)
1DVM	Uncleaved, active form	PLASMINOGEN ACTIVATOR INHIBITOR-1; Homo sapiens; *Escherichia coli*	N150H, K154T, Q319L, M354I	Stout *et al.* (2000)
1DVN	Latent	PLASMINOGEN ACTIVATOR INHIBITOR-1; Homo sapiens; *Escherichia coli*	No	Stout *et al.* (2000)
1DZG	Uncleaved, active form	ANTITHROMBIN-III; HOMO SAPIENS; CRICETIDAE SP.	N135Q S380C	Huntington *et al.* (2000b)
1DZH	Uncleaved, active form	ANTITHROMBIN-III; HOMO SAPIENS; CRICETIDAE SP.	N135Q S380C-fluorescein	Huntington *et al.* (2000b)
1EZX	Complex of cleaved and inserted serpin with protease	Chain A and B = ALPHA-1 ANTITRYPSIN, Chain C = trypsin; Chain A and B = Homo sapiens, Chain C = bovine; Chain A and B = *Escherichia coli*, Chain C = bovine plasma	N-terminal His tag on alpha-1 antitrypsin	Huntington *et al.* (2000b)

(*Continued*)

Table 5.1 *(Continued)*

PDB	State	Compound; source organism; expression host/tissue source	Mutation, added tags	References
1F0C	Cleaved, inserted	ICE INHIBITOR; Cowpox virus; *Escherichia coli* BL21 (DE3)	No	Renatus *et al.* (2000)
1M93	Cleaved, inserted	Serine proteinase inhibitor-2; Cowpox virus; *Escherichia coli*	All 9 cysteines mutated to serine	Simonovic *et al.* (2000)
1QLP	Canonical inhibitory conformation	ALPHA-1 ANTITRYPSIN; HOMO SAPIENS; *ESCHERICHIA COLI*	E1M	Elliott *et al.* (2000)
1QMN	Uncleaved, delta	ALPHA-1 ANTICHYMOTRYPSIN; HOMO SAPIENS; *ESCHERICHIA COLI*	L55P	Gooptu *et al.* (2000)
1HP7	Uncleaved	ALPHA-1 ANTITRYPSIN; Homo sapiens; *Escherichia coli*	M1E, A70G, polymorphism 101H and 376D	Kim *et al.* (2001)
1IMV	Uncleaved	PIGMENT EPITHELIUM-DERIVED FACTOR; Homo sapiens; *Cricetulus griseus* (BHK cells)	No	Simonovic *et al.* (2001)
1JJO	Cleaved, inserted	NEUROSERPIN; *Mus musculus*; *Escherichia coli*	Last 13 residues deleted, plus C-term His tag	Briand (2001)
1JRR	Uncleaved with peptide in s4A position	Chain A = PLASMINOGEN ACTIVATOR INHIBITOR-2, Chain P = peptide; Chain A = Homo sapiens, Chain P = human RCL; Chain A = *Escherichia coli*, Chain P = synthetic	Loop deletion mutant: 66–98	Jankova *et al.* (2001)
1K9O	Michaelis serpin–enzyme complex	Chain E = TRYPSIN II ANIONIC, Chain I = ALASERPIN; Chain E = *Rattus norvegicus*, Chain I = *Manduca sexta*; Chain E = *Saccharomyces cerevisiae*, Chain I = *Escherichia coli*	Serpin 1K: A353K and N-term His tag, rat trypsinogen S195A	Ye *et al.* (2001)
1IZ2	Latent (referred to as kinetic trap by authors)	alpha-1 antitrypsin; Homo sapiens; *Escherichia coli*	Heptamutant (1ATU) plus V364A mutation	Im *et al.* (2002)
1JMJ	Uncleaved, partially inserted (like AT)	HEPARIN COFACTOR II; Homo sapiens; plasma	No	Baglin *et al.* (2002)
1JMO	Michaelis serpin–enzyme complex	Chain A = HEPARIN COFACTOR II, Chain H = Thrombin (heavy chain), Chain L = Thrombin (light chain); Chain A = Homo sapiens, Chain H = L = Homo sapiens; Chain A = plasma, Chain H = L = *Cricetulus griseus* (BHK cells)	Chains H and L: S195A HPC4 antibody epitope	Baglin *et al.* (2002)

1JTI	Cleaved, inserted	Ovalbumin; *Gallus gallus*; *Escherichia coli* BL21(DE3)	R339T	Yamasaki *et al.* (2002)
1E03	Dimer of a inhibitory confirmation with a latent (joined at C-sheet) plus heparin bound	ANTITHROMBIN-III; HOMO SAPIENS; plasma	No	McCoy *et al.* (2003)
1E04	Dimer of a inhibitory confirmation with a latent (joined at C-sheet)	ANTITHROMBIN-III; HOMO SAPIENS; plasma	No	McCoy *et al.* (2003)
1E05	Dimer of a inhibitory confirmation with a latent (joined at C-sheet)	ANTITHROMBIN-III; HOMO SAPIENS; plasma	No	McCoy *et al.* (2003)
1LK6	Uncleaved, two peptides inserted into A-sheet	Chain I = L = antithrombin, Chain C = hexapeptide (P14-9), Chain D = tri-peptide (P6-4); Chain I = L = C = D = Homo sapiens; Chain I and L = plasma Chain, C = D = synthetic	No	Zhou *et al.* (2003b)
1LQ8	Cleaved, inserted	Plasma serine protease inhibitor; Homo sapiens; plasma	No	Huntington *et al.* (2003)
1MTP	Cleaved, inserted	Serine proteinase Inhibitor (SERPIN), Chain A; *Thermobifida fusca*; *Escherichia coli*	N-term His tag	Irving *et al.* (2003)
1NQ9	Dimer of a INTERMEDIATE STATE with a latent (joined at C-sheet) plus heparin bound	Antithrombin-III; Homo sapiens; plasma	No	Johnson and Huntington (2003)
1OC0	Uncleaved bound to vitronectin	Chain A = PLASMINOGEN ACTIVATOR INHIBITOR-1, Chain B VITRONECTIN; Chain A = B = Homo sapiens; Chain A = B = *ESCHERICHIA COLI*	Chain A: K154T, Q319L, M354I, N150H, Chain B: vitronectin residues 1-110, N-terminal His tag, C-term GST tag	Zhou *et al.* (2003b)
1OO8	Uncleaved canonical conformation	Alpha-1 antitrypsin precursor; Homo sapiens; *Escherichia coli*	Chain A: MetP1Arg (357), C232S heptamutant (F51L, T59A, T68A, A70G, M374I, S381A, K387R)	Dementiev *et al.* (2003)

(*Continued*)

Table 5.1 (*Continued*)

PDB	State	Compound; source organism; expression host/tissue source	Mutation, added tags	References
1OPH	Michaelis serpin–enzyme complex	Chain A = Alpha-1 antitrypsin precursor, Chain B = trypsinogen; Chain A = Homo sapiens, Chain B = bovine; Chain A = B = *Escherichia coli*	Chain A: MetP1Arg (357), C232S heptamutant (F51L, T59A, T68A, A70G, M374I, S381A, K387R), Chain B = S196A	Dementiev *et al.* (2003)
1JVQ	Dimer of a inhibitory confirmation with a latent (joined at C-sheet) with peptides inserted in A sheet	Chain C = D = P14–P8 reactive loop peptide Chain I = antithrombin uncleaved L = latent antithrombin; Chain C = D = I = L = Homo sapiens; Chain C = D = synthetic peptide Chain I = L = plasma	No	Zhou *et al.* (2004)
1OYH	Dimer of a cleaved inserted and inhibitory confirmation	Antithrombin-III; Homo sapiens; *Cricetulus griseus*	E381A	Johnson and Huntington (2004)
1SR5	Complex of antithrombin–anhydrothrombin and heparin	Chain A = B = thrombin, Chain C = antithrombin-III; Chain A = B = C = Homo sapiens; Chain A = B = C = plasma	No	Dementiev *et al.* (2004)
1TB6	Complex of antithrombin–anhydrothrombin and heparin	Chain H and L = thrombin, Chain I = antithrombin; Chain H and L = Homo sapiens, Chain I = Homo sapiens; Chain H and L = *Cricetulus griseus*, Chain I = *Cricetulus griseus*	Thrombin (Chain H and L) = S195A, antithrombin Chain I = S137A, V317C, T401C	Li *et al.* (2004)
1XQG	Uncleaved	Maspin; Homo sapiens; *Escherichia coli*	N-terminal His tag and refolded, all cysteines mutated to alanine	Al-Ayyoubi *et al.* (2004)
1XQJ	Uncleaved	Maspin; Homo sapiens; *Escherichia coli*	N-terminal His tag and refolded, all cysteines mutated to alanine	Al-Ayyoubi *et al.* (2004)
1SNG	Unlceaved	COG4826: Serine protease inhibitor; *Thermobifida fusca*; *Escherichia coli*	No	Fulton *et al.* (2005)
1WZ9	Uncleaved	Maspin precursor; Homo sapiens; *Escherichia coli* BL21	N-terminal His tag removed with TEV protease	Law *et al.* (2005)

1XU8	Uncleaved	Maspin; Homo sapiens; *Escherichia coli* BL21	N-terminal His tag removed with TEV protease	Law *et al.* (2005)
1YXA	Uncleaved, partially inserted	Serine (or cysteine) proteinase inhibitor, clade A, member 3N (antichymotrypsin); *Mus musculus*; *Escherichia coli*	Deletion of signal peptide addition of N-terminal His tag	Horvath *et al.* (2005)
2ARQ	Uncleaved with peptide in s4A position	Chain A = plasminogen activator inhibitor-2, Chain P = 14-mer from plasminogen activator inhibitor-2; Chain A = P = Homo sapiens; Chain A = *ESCHERICHIA COLI*, Chain P = synthetic	Loop deletion mutant: residues 66–98	Di Giusto *et al.* (2005)
2ARR	Uncleaved with peptide in s4A position	Chain A = plasminogen activator inhibitor-2, Chain P = 14-mer from plasminogen activator inhibitor-2; Chain A = P = Homo sapiens; Chain A = *ESCHERICHIA COLI*, Chain P = synthetic	Loop deletion mutant: residues 66–98	Di Giusto *et al.* (2005)
1T1F	Uncleaved, partially inserted	Antithrombin-III; Homo sapiens; *Cricetulus griseus*	S137A and V317C/T401C (engineered disulphide bond)	Johnson *et al.* (2006b)
2B5T	Complex where thrombin homodimer is linked to ATIII via a long chain heparin	Chain A = B = C = D = thrombin, Chain E = antithrombin; Chain A, B, C, D, E = Homo sapiens; Chain A, B, C, D, E = *Mesocricetus auratus* (hamster cells)	Chain A, B, C, D = S195A, Chain I = S137A and V317C/T401C (engineered disulphide bond)	Johnson *et al.* (2006b)
2BEH	Dimer of a cleaved inserted and inhibitory confirmation	Chain I and L = antithrombin-III; Chain I and L = Homo sapiens; Chain I = *Mesocricetus auratus*, chain L = plasma	Chain I = S137A and V317C/T401C (engineered disulphide bond), Chain L = latent antithrombin from human plasma	Johnson *et al.* (2006b)
2CEO	Uncleaved, partially inserted (like AT)	THYROXINE-BINDING GLOBULIN; HOMO SAPIENS; *ESCHERICHIA COLI*	Residues 19–415, N-terminal His tag	Zhou *et al.* (2006)
2D26	Complex of cleaved and inserted serpin with protease	Chain A = B = Alpha-1 antitrypsin, Chain C = porcine pancreatic elastase; Chain A = B = Homo sapiens, Chain C = *Sus scrufa*; Chain A = B = *Escherichia coli*, Chain C = Not stated (purchased from Calbiochem)	Chain A = B C232S, R101H, heptamutant (F51L, T59A, T68A, A70G, M374I, S381A, K387R), plus His tag, Chain C = not documented	Dementiev *et al.* (2006)

(*Continued*)

Table 5.1 (*Continued*)

PDB	State	Compound; source organism; expression host/tissue source	Mutation, added tags	References
2DUT	Uncleaved, partially inserted (like AT)	Heterochromatin-associated protein MENT; *Gallus gallus*; *Escherichia coli*	Deletion of M-loop	McGowan *et al.* (2006)
2GD4	Michaelis serpin–enzyme complex	Chain A = L = factor Xa light chain, Chain B = H = factor Xa heavy chain, Chain C = I = antithrombin; Chain A = L = B = H = C = I = HOMO SAPIENS; A = L = B = H = *E. coli* C = I =CRICETIDAE SP.	Chain C = I = E347A, K348A, K350, S137A beta-glycoform, Chain B = H = S195A factor Xa, refolded	Johnson *et al.* (2006b)
2H4P	Cleaved, inserted	Heterochromatin-associated protein MENT; *Gallus gallus*; *Escherichia coli* BL21	No	McGowan *et al.* (2006)
2H4Q	Cleaved, inserted	Heterochromatin-associated protein MENT; *Gallus gallus*; *Escherichia coli*	Deletion of M-loop	McGowan *et al.* (2006)
2H4R	Uncleaved, partially inserted (like AT)	Heterochromatin-associated protein MENT; *Gallus gallus*; *Escherichia coli* BL21	No	McGowan *et al.* (2006)
2HI9	Uncleaved, active form	Plasma serine protease inhibitor; Homo sapiens; *Escherichia coli*	Deletion of first 16 residues, N-terminal His tag	Li *et al.* (2007)
2OAY	Latent	Plasma protease C1 inhibitor; Homo sapiens; *Pichia pastoris*	V458M, first 96 residues deleted, N-term His tag	Beinrohr *et al.* (2007)
2OL2	Uncleaved, active form	Plasma serine protease inhibitor; Homo sapiens; *Escherichia coli*	Deletion of first 16 residues, N-terminal His tag	Li *et al.* (2007)
2PEE	Uncleaved, partially inserted	Serine protease inhibitor; *Thermoanaerobacter tengcongensis* MB4; *Escherichia coli*	N-terminal 31 residues deleted	Zhang *et al.* (2007)
2PEF	Latent	Serine protease inhibitor; *Thermoanaerobacter tengcongensis* MB4; *Escherichia coli*	N-terminal 51 residues deleted	Zhang *et al.* (2007)
2V95	Uncleaved, active form	CORTICOSTEROID-BINDING GLOBULIN; *Rattus norvegicus*; *ESCHERICHIA COLI*	Residues 5–374, GST tag	Klieber *et al.* (2007)
2QUG	Uncleaved canonical conformation	Alpha-1 antitrypsin; Homo sapiens; *Escherichia coli*	Refold	Pearce *et al.* (2008)
3CWL	Uncleaved canonical conformation	Alpha-1 antitrypsin; Homo sapiens; *Escherichia coli*	Refold	Pearce *et al.* (2008)

3CWM	Uncleaved canonical conformation	Alpha-1 antitrypsin; Homo sapiens; *Escherichia coli*	Refold	Pearce *et al.* (2008)
2R9Y	Uncleaved, not inserted	Alpha-2 antiplasmin; *Mus musculus*; *Escherichia coli* BL21(DE3)	N-terminal 43 residues deleted	Law *et al.* (2008)
2VDX	Cleaved, inserted	CORTICOSTEROID-BINDING GLOBULIN; HOMO SAPIENS; *ESCHERICHIA COLI*	Deletion of N-terminus 11 residues, chimera of the RCL (P14–P1) replaced with antitrypsin Pittsburgh	Zhou *et al.* (2008)
2VDY	Cleaved, inserted	CORTICOSTEROID-BINDING GLOBULIN; HOMO SAPIENS; *ESCHERICHIA COLI*	Deletion of N-terminus 11 residues, chimera of the RCL (P14–P1) replaced with antitrypsin Pittsburgh	Zhou *et al.* (2008)
2VH4	Latent C-sheet polymer	TENGPIN; *Thermoanaerobacter tengcongensis*; *ESCHERICHIA COLI*	Deletion of first 42 residues	Zhang *et al.* (2008)
2ZNH	Uncleaved polymer	Antithrombin-III; Homo sapiens; plasma	No	Yamasaki *et al.* (2008)
3B9F	Michaelis serpin–enzyme complex	Chain H = L = prothrombin, Chain I = protein C inhibitor; Chain H = L = I = Homo sapiens; Chain H = L = *Cricetulus griseus* (BHK cells), I = *Escherichia coli*	Chain H and L = S195A, Chain I = deletion of 16 N-terminal residues, N-terminal His tag	Li *et al.* (2008)
3CVM	Cleaved, inserted	Plasminogen activator inhibitor 1; Homo sapiens; *Escherichia coli*	N150H-K154T-Q319L-M354I, N-terminal His tag	Jensen and Gettins (2008)
3DY0	Cleaved, inserted	Chain A = N-terminus Plasma serine protease inhibitor, Chain B = C-terminus plasma serine protease inhibitor; Chain A = B = Homo sapiens; Chain A = B = *Escherichia coli*	N-terminal 17 residues deleted	Li and Huntington (2008)
2ZV6	Uncleaved, not inserted	Serpin B3; Homo sapiens; *Escherichia coli*	N-terminal His tag	Zheng *et al.* (2009)
3DRM	Uncleaved, not inserted	Alpha-1 antitrypsin; Homo sapiens; *Escherichia coli*	T114F, His tag	Gooptu *et al.* (2009)
3DRU	Uncleaved, not inserted	Alpha-1 antitrypsin; Homo sapiens; *Escherichia coli*	G117F, His tag	Gooptu *et al.* (2009)
3EOX	Cleaved, inserted	Plasminogen activator inhibitor 1; Homo sapiens; *Escherichia coli*	N150H, K154T, Q301P, Q319L, M354I	Dewilde *et al.* (2009)

(*Continued*)

Table 5.1 (*Continued*)

PDB	State	Compound; source organism; expression host/tissue source	Mutation, added tags	References
3F02	Cleaved, inserted	Neuroserpin; Homo sapiens; *Escherichia coli*	N-terminal His tag	Ricagno *et al.* (2009)
3F5N	Uncleaved, not inserted	Neuroserpin; Homo sapiens; *Escherichia coli*	N-terminal His tag	Ricagno *et al.* (2009)
3F1S	Complex of uncleaved serpin with cofactor	Chain A = Protein Z-dependent protease inhibitor chain, Chain B = Vitamin K-dependent protein Z; Chain A = B = Homo sapiens; Chain A = *Escherichia coli*, Chain B = Homo sapiens (HEK293.EBNA cells)	A = full length, B = residues 84–360, C-terminal His tag	Wei *et al.* (2009)
3FGQ	Uncleaved, not inserted	Neuroserpin; Homo sapiens; *Escherichia coli*	C-terminal truncation delta10	Takehara *et al.* (2009)
3EVJ	Dimer of a INTERMEDIATE STATE with a latent (joined at C-sheet) plus heparin bound	Antithrombin-III; Homo sapiens; plasma	No	Langdown *et al.* (2009)
3DLW	Uncleaved, delta	Alpha-1 antichymotrypsin; Homo sapiens; *Escherichia coli*	No	Pearce *et al.* (2010)
3H5C	Complex of uncleaved serpin with cofactor	Chain A = Protein Z-dependent protease inhibitor, Chain B = Vitamin K-dependent protein Z; Chain A = B = Homo sapiens; Chain A = SPODOPTERA FRUGIPERDA, Chain B = plasma (thrombin cleaved)	423	Huang *et al.* (2010)
3KCG	Michaelis serpin–enzyme complex	Chain H and L = Coagulation factor IXa heavy chain, Chain I = antithrombin; Chain i = L = H = Homo sapiens; Chain H = L = *Escherichia coli*, Chain I = BHK cells (*Cricetulus griseus*)	Chains H and L = 103–431 residues, S195A mutation, Chain I = S137A	Johnson *et al.* (2010)
3NDD	Cleaved, inserted	Alpha-1 antitrypsin; Homo sapiens; *Escherichia coli*	P10 Pro/P9–P6 Asp mutant on background of C232A and M358R, N-terminal His tag	Yamasaki *et al.* (2010)
3NDF	Cleaved, inserted	Alpha-1 antitrypsin; Homo sapiens; *Escherichia coli*	P8–P6 Asp mutant on background of C232A and M358R, N-terminal His tag	Yamasaki *et al.* (2010)

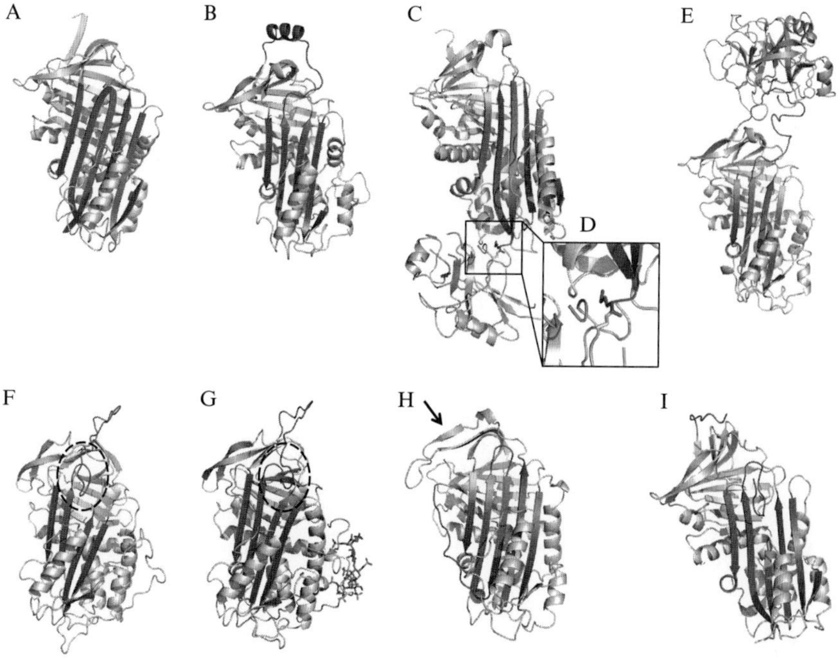

Figure 5.1 The different conformations that can be adopted by serpin molecules. In all structures the A β-sheet is colored as blue, the reactive center loop (RCL) is colored red, and complexed proteases are colored green. (A) Cleaved and inserted structure of antitrypsin (PDB ID: 7API). (B) The uncleaved structure of ovalbumin (PDB ID: 1OVA). (C) The structure of antitrypsin in complex with trypsin trapped in the acyl-enzyme intermediate state showing distortion of the catalytic triad (PDB ID: 1EZX). (D) Inset of the antitrypsin–trypsin complex showing the bond between the catalytic triad Ser195 side chain (yellow sticks) and the P1-Arginine of the serpin (red sticks). (E) The Michaelis serpin–protease complex where the protease is noncovalently bound to the RCL (PDB ID: 1K9O). (F) The structure of native antithrombin shows partial insertion of the RCL into the A-sheet (dashed circle) (PDB ID: 2ANT). (G) In the presence of bound heparin (purple sticks), the RCL of antithrombin is expelled out of the top of the A-sheet (dashed circle) (PDB ID: 1AZX). (H) The latent conformation of PAI-1 where the RCL is fully inserted into the A-sheet. This is achieved by peeling away the outside β-strand of the C-sheet (C-sheet shown with arrow) (PDB ID: 1C5G). (I) The δ-conformation of antichymotrypsin where there is partial insertion of the RCL into the top of the A-sheet and insertion of the region from the last turn of the F-helix and subsequent loop into the bottom of the A-sheet (colored orange) (PDB ID: 1QMN). (See Color Insert.)

determined (Fig. 5.1C; PDB ID: 1EZX; Huntington et al., 2000a). This structure revealed that, as in the cleaved form, the serpin RCL is fully inserted. Crucially, however, the protease remains covalently attached to the RCL through a covalent bond between the active site serine (Ser195) and the P1 residue of the serpin (Fig. 5.1D). Remarkably, the active site of the protease is significantly disrupted and part of the enzyme is not visible in electron density. These data together suggest that the following docking and cleavage of the RCL by a target protease, the serpin conformational change is so rapid that it distorts the catalytic triad before the reaction can be fully completed. The protease is thus trapped at the acyl-enyzme intermediate part of the reaction cycle. The additional disruption of the enzyme structure most likely arises through disruption of key intramolecular bonds that are made when the protease is activated from the zymogen precursor to the mature form. Essentially, the serpin thus functions to drive the protease back from the mature form to a conformation more reminiscent of the zymogen-like state.

Taken together, these data shed comprehensive insight into a wide body of serpin biochemistry. In particular, the partial unraveling of the protease domain was consistent with the observation that cryptic "buried" cleavage sites are exposed in certain proteases following complex formation with serpins (Kaslik *et al.*, 1995; Stavridi *et al.*, 1996). These data also reveal why serpins are "one use only" suicide inhibitors. Finally, these results provided a structural explanation for the observation that mutations that disrupted RCL insertion into the A-sheet resulted in substrate-like rather than inhibitory behavior. The serpin/protease interaction could be thought of as a "race" with rapid RCL insertion (and concomitant protease distortion) competing with the attempts of the catalytic machinery to release the substrate.

Since 2000, a second serpin–protease complex has been determined (antitrypsin in complex with porcine elastase; PDB ID: 2D26; Dementiev *et al.*, 2006). These data similarly reveal full translocation of the target protease together with distortion of the catalytic triad. However, little distortion is seen outside the immediate vicinity of the active site, suggesting that "unfolding" of the target protease may not be seen in all protease/serpin complexes.

4. The Michaelis Complex

Shortly after the determination of the serpin–enyzme complex, the first encounter complex between a serpin and a protease was reported (between an insect serpin and a catalytically inert [Ser195Ala mutation] form of trypsin; Fig. 5.1E; Ye *et al.*, 2001). The structure revealed that the serpin RCL docked in a similar (canonical) fashion to a target substrate. Interestingly, this structure, as well as other encounter complex structures, reveals that the serpin RCL is able to form more extensive interfaces with target proteases in comparison to small "lock and key"-type protein inhibitors (e.g., Kunitz-type inhibitors such as bovine pancreatic trypsin inhibitor). Further, the encounter complex structures of several serpins with cognate–protease partners reveal that the serpin body can also provide additional specificity through the provision of significant exosite interactions. This is particularly the case for serpins involved in controlling coagulation proteases (Baglin *et al.*, 2002; Dementiev *et al.*, 2004; Johnson *et al.*, 2006a, 2010; Li *et al.*, 2004, 2008).

5. Conformational Control of Serpins—Antithrombin and Heparin

A key function of the serpin fold is its ability for its conformation to be modulated through the influence of cofactors. Nowhere is this more apparent than in the serpin, antithrombin. The structure of uncleaved antithrombin, which was determined in 1994, revealed the surprising finding that the RCL was partially inserted into the top of the A-sheet (Carrell *et al.*, 1994). These data, together with later findings from the Huntington group, revealed that the key P1 arginine residue of antithrombin is mostly occluded from target proteases in the native "resting" state of antithrombin (Fig. 5.1F; Carrell *et al.*, 1994; Johnson *et al.*, 2006a). Remarkably, biochemical and structural studies revealed that binding of a high-affinity saccharide sequence present in heparin promoted RCL expulsion from the A-sheet together with concomitant exposure of the P1 residue (Fig. 5.1G; Jin *et al.*, 1997). The exposure of the P1 residue, together with the templating activity of heparin, dramatically increases the activity of antithrombin against target proteases such as factor Xa and thrombin.

The antithrombin binding site for heparin is formed by the D-helix, together with a cup-shaped structure formed by residues from the N-terminus. Interestingly, the structure of plasminogen activator inhibitor-1 (PAI-1) in complex with the vitronectin somatomedin B domain (Zhou et al., 2003a; see Section 7), as well as other serpins such as the bacterial serpin tengpin (Zhang *et al.*, 2007), suggests a general role for the region around the D-helix in the control of serpin conformation.

6. Nonconventional Serpin Complexes

The structure the serpin PZ-dependent inhibitor (ZPI) (a factor Xa inhibitor) in complex with the cofactor Protein Z (PZ) revealed that ZPI had evolved an unusual solution to the problem of achieving specificity against a target protease. *In vivo,* PZ binds to ZPI to accelerate ZPI inhibition of the protease factor Xa. PZ is a vitamin K-dependent plasma protein that is homologous to the chymotrypsin-like serine proteases. The catalytic site of PZ is mutated, however, and it lacks proteolytic activity. The complex structure revealed that PZ forms interactions with the C-sheet region of ZPI, placing it in a position such that it would be able to form a platform for the recruitment of factor Xa (PDB ID: 3F1S; Wei *et al.*, 2009; PDB ID: 3H5C; Huang *et al.*, 2010). Both structures, together with extensive mutagenesis studies, suggest that PZ functions to enhance ZPI inhibition activity via a template mechanism.

A second interesting set of serpin–cofactor complex structures include two noninhibitory serpins that function as hormone carriers—thyroxine binding globulin (TBG) (PDB ID: 2CEO; Zhou *et al.*, 2006) and corticosteroid binding globulin (CBG) (PDB ID: 2V95; Klieber *et al.*, 2007). The structure of TBG shows that thyroxine binds at the back of the serpin fold in a pocket formed by the B-sheet (strands 3–5) and helices H and A. It is suggested that partial entry of the RCL into the A-sheet may trigger release of thyroxine from the thyroxine binding pocket. Conversely, expulsion of the RCL may confer the ability to bind thyroxine in a high-affinity state.

7. Conformational Change and the Formation of the Latent Conformation

Another remarkable utility of the serpin fold is the ability of certain serpins to autoinactivate, as illustrated by PAI-1. Initially secreted as an active inhibitory serpin, PAI-1 rapidly forms highly stable "latent" conformation that is no longer able to interact with target proteases. The structure of latent PAI-1, reported by Mottonen *et al.* (1992), revealed that PAI-1 could spontaneously undergo the S-to-R transition *without* RCL cleavage. As with cleaved serpins, the RCL in latent PAI-1 is fully inserted; however, in order to do so, the first β-strand of the C β-sheet must peel away (Fig. 5.1H). Thus, PAI-1 intrinsically possesses the ability to rapidly autoinactivate. The presence of cofactors, such as the vitronectin somatomedin B domain, prevents this spontaneous rearrangement. The structure of PAI-1 in complex with the somatomedin B domain reveals that this domain functions by binding to s1A and the E-helix (Zhou et al., 2003a). Accordingly, by locking to s1A, somatomedin B prevents movement throughout the A-sheet, as well as concomitant RCL insertion. The spontaneous transition to the latent form has also been observed in antithrombin, although the physiological role of this change remains to be fully understood.

Recently, the structure of tengpin revealed that this serpin required an extended structure in the N-terminus to maintain the native, metastable state (Zhang *et al.*, 2007). Interestingly, like in PAI-1/somatomedin B, the

N-terminus of tengpin also forms similar interactions with s1A and the E-helix, and mutational studies reveal a short (13 amino acid) stretch of residues function to trap the molecule in the native, metastable state.

8. Abnormal Conformational Change—The δ-Form

The requirement for conformational mobility in serpins renders these molecules susceptible to mutations that promote inappropriate or unwanted conformational change. Numerous mutations have been identified that result in spontaneous formation of the latent conformation (discussed above)—these changes are suggested to destabilize the native state and promote spontaneous RCL insertion. Further, a substantial number of serpin mutations have been identified that result in serpin misfolding and the formation of polymers (discussed below). Interestingly, recent studies on antichymotrpysin have revealed that a mutation in the shutter region results in a conformation that can be described as a "halfway house" on the transition to latency (Fig. 5.1I; Gooptu *et al.*, 2000). In this structure, the RCL is inserted and fills the top half of the A-sheet. The presence of an intact C-sheet presumably prevents further insertion. The bottom half of the A-sheet is filled via the F-helix partially unwinding and inserting as a short β-strand. If nothing else, these data serve to highlight the remarkable flexibility of the serpin fold. It has variously been speculated that this structure may represent an intermediate in the transition from the native to the cleaved state, or indeed the polymeric state—although these ideas remain to be further supported.

9. Serpin Polymers

In 1992, Lomas and colleagues made the crucial discovery that serpin polymerization in the endoplasmic reticulum of liver cells most likely formed the basis of antitrypsin deficiency and the disease now known as alpha-1. The most common allele linked to Alpha-1—the Z-variant—causes disease via at least two distinct mechanisms. First, simple deficiency of antitrypsin renders an individual susceptible to emphysema through a failure to control inflammatory proteases (and particularly elastase) in the lung following infection. Second, the accumulation of antitrypsin polymers in liver cells (the site of synthesis) can eventually result in severe tissue damage (cirrhosis). A wide body of studies (Lomas *et al.*, 1992, #151) suggests that antitrypsin is a misfolding disease, and that pathogenic polymers form through stabilization of a polymeric folding intermediate.

Much current debate is ongoing in regard to the precise structure of a serpin polymer.

Serpin polymers are highly stable and can be induced through heating or mild chemical insult. Indeed, the stability of the polymeric form suggests strongly that the serpin is in the relaxed "R" conformation within the polymer. Initially, it was proposed that serpins most likely polymerized through an *in trans* RCL insertion event, whereby the RCL of one molecule inserted into the A-sheet of another (Fig. 5.2A). Supporting this idea, two structures of cleaved antitrypsin polymers (Dunstone *et al.*, 2000; Huntington *et al.*, 1999; Fig. 5.2A) demonstrated that it is possible for such RCL linkages to form *in trans*, albeit through cleavage in a nonstandard position in the RCL. The nature of these polymers, however, did not precisely reflect the properties of polymers formed *in vivo*. In particular, biochemical analysis of the material crystallized in one of these studies (Dunstone *et al.*, 2000) revealed that the polymers were not stable and could dissociate. In contrast, studies on antitrypsin polymers formed *in vivo* or through heating native material reveal that these polymers are extremely stable.

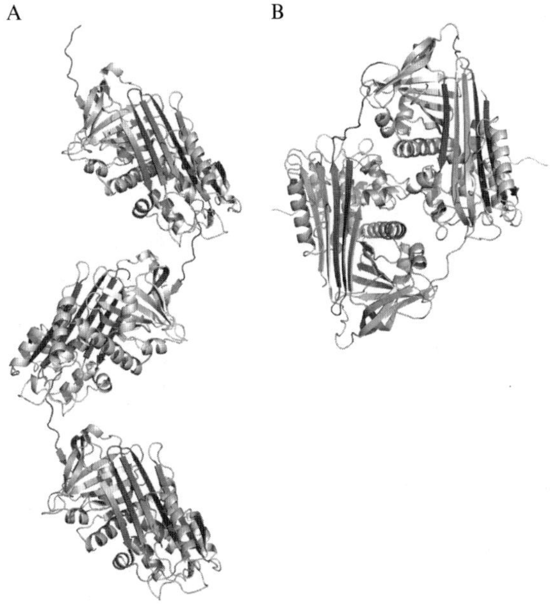

Figure 5.2 Polymerization of serpins. Colors as for Fig. 5.1. (A) Cleaved polymer: The structure of the cleaved antitrypsin polymer (PDB ID: 1D5S). (B) Domain-swapped dimer: The structure of antithrombin dimer showing swapping of both the RCL (red) and the fifth β-strand (blue) of one antithrombin molecule into the A-sheet of the neighboring antithrombin molecule (green) (PDB ID: 2ZNH). (See Color Insert.)

Recently, a new model for serpin polymerization has been proposed. The structure of a domain-swapped antithrombin dimer (Fig. 5.2B) revealed that both the fifth strand and the RCL can be incorporated into the A-sheet of another serpin molecule (Yamasaki *et al.*, 2008). This structure can be adapted to model long chain polymers. The new domain-swapped model has several properties consistent with polymers purified *in vivo*. In particular, the substantial interface buried in the domain swap is consistent with formation of a highly stable polymer, and the domain-swapping event itself is consistent with assembly from a folding intermediate.

10. Crystallization of Serpins and Serpin Complexes

Initial crystallization of serpins was reported over 100 years ago, including crystallization of ovalbumin from hen's eggs (Hopkins and Pinkus, 1898). Crystallization of serpins preceded technological advances such as X-ray crystallography, protein sequencing, gene sequencing, and recombinant DNA techniques. It was not until 1984 that the first serpin structure was determined (PDB ID: 7API; Loebermann *et al.*, 1984). This protein was purified from its native source, human plasma. Subsequent serpin structures in the 1980s and early 1990s also utilized material purified from their native source including plakalbumin from hen egg white (cleaved ovalbumin, no PDB entry (Wright *et al.*, 1990)), ovalbumin from hen egg whites (PDB ID: 1OVA (Stein *et al.*, 1990)), α1-antichymotrypsin from plasma (PDB ID: 2ACH (Baumann *et al.*, 1991)), and antithrombin III from plasma (PDB IDs: 1ATT (Mourey *et al.*, 1993), 1ANT (Carrell *et al.*, 1994), 1ATH (Schreuder *et al.*, 1994)). Interestingly, the first intracellular serpin to be structurally characterized was sourced from a native source as well: horse leukocytes were used to as a source for leukocyte elastase inhibitor (PDB ID: 1HLE (Baumann *et al.*, 1992)).

In 1991/1992, the structure of PAI-1 produced using recombinant DNA technology was presented (PDB ID: 1C5G (Mottonen *et al.*, 1992)). Since then, recombinant expression of protein has opened up the opportunities either to crystallize proteins that are difficult to purify from the native source or to create mutant serpins more amenable to crystallization (Sharp *et al.*, 1999). A range of different expression systems have subsequently been used including *Escherichia coli*, mammalian cell lines (BHK (PDB ID: 3KCG (Johnson *et al.*, 2010)) and CHO (PDB ID: 1A7C (Xue *et al.*, 1998))), baculovirus expression (PDB ID: 3H5C (Huang *et al.*, 2010)), *Drosophila melanogaster*, and yeast, *Pichia pastoris* (PDB ID: 2OAY (Beinrohr *et al.*, 2007)). In addition, the determination of mutant forms of serpins has also gone hand-in-glove with experiments to understand serpin conformational change (Yamasaki *et al.*, 2010).

11. Conclusions

To conclude, the structure of only a handful of human serpins remains to be determined—these include HSP47 as well as many members of the intracellular branch of the serpin superfamily. The recent structure of a domain-swapped serpin dimer (Yamasaki *et al.*, 2008, #129) provides insights into a possible mechanism for serpin polymerization. Many interesting structures remain to be determined, however, including the complex between serpins and receptor molecules as well as structures that will shed additional light on the specificity of serpin–protease interactions.

REFERENCES

Aertgeerts, K., De Bondt, H. L., *et al.* (1995). Mechanisms contributing to the conformational and functional flexibility of plasminogen activator inhibitor-1. *Nat. Struct. Biol.* **2**(10), 891–897.

Al-Ayyoubi, M., Gettins, P. G., *et al.* (2004). Crystal structure of human maspin, a serpin with antitumor properties: Reactive center loop of maspin is exposed but constrained. *J. Biol. Chem.* **279**(53), 55540–55544.

Baglin, T. P., Carrell, R. W., Church, F. C., Esmon, C. T., and Huntington, J. A. (2002). Crystal structures of native and thrombin-complexed heparin cofactor II reveal a multistep allosteric mechanism. *Proc. Natl. Acad. Sci. USA* **99,** 11079–11084.

Baumann, U., Huber, R., Bode, W., Grosse, D., Lesjak, M., and Laurell, C. B. (1991). Crystal structure of cleaved human alpha 1-antichymotrypsin at 2.7 A resolution and its comparison with other serpins. *J. Mol. Biol.* **218,** 595–606.

Baumann, U., Bode, W., Huber, R., Travis, J., and Potempa, J. (1992). Crystal structure of cleaved equine leucocyte elastase inhibitor determined at 1.95 A resolution. *J. Mol. Biol.* **226,** 1207–1218.

Beinrohr, L., Harmat, V., Dobo, J., Lorincz, Z., Gal, P., and Zavodszky, P. (2007). C1 inhibitor serpin domain structure reveals the likely mechanism of heparin potentiation and conformational disease. *J. Biol. Chem.* **282,** 21100–21109.

Briand, C., Kozlov, S. V., Sonderegger, P., and Grutter, M. G. (2001). Crystal structure of neuroserpin: a neuronal serpin involved in a conformational disease. *FEBS Lett.* **505,** 18–22.

Carrell, R. W., Stein, P. E., Fermi, G., and Wardell, M. R. (1994). Biological implications of a 3 A structure of dimeric antithrombin. *Structure* **2,** 257–270.

Dementiev, A., Simonovic, M., *et al.* (2003). Canonical inhibitor-like interactions explain reactivity of alpha1-proteinase inhibitor Pittsburgh and antithrombin with proteinases. *J. Biol. Chem.* **278**(39), 37881–37887.

Dementiev, A., Petitou, M., Herbert, J. M., and Gettins, P. G. (2004). The ternary complex of antithrombin-anhydrothrombin-heparin reveals the basis of inhibitor specificity. *Nat. Struct. Mol. Biol.* **11,** 863–867.

Dementiev, A., Dobo, J., and Gettins, P. G. (2006). Active site distortion is sufficient for proteinase inhibition by serpins: Structure of the covalent complex of alpha1-proteinase inhibitor with porcine pancreatic elastase. *J. Biol. Chem.* **281,** 3452–3457.

Dewilde, M., Strelkov, S. V., *et al.* (2009). High quality structure of cleaved PAI-1-stab. *J. Struct. Biol.* **165**(2), 126–132.

Di Giusto, D. A., Sutherland, A. P., *et al.* (2005). Plasminogen activator inhibitor-2 is highly tolerant to P8 residue substitution—Implications for serpin mechanistic model and prediction of nsSNP activities. *J. Mol. Biol.* **353**(5), 1069–1080.

Dunstone, M. A., Dai, W., Whisstock, J. C., Rossjohn, J., Pike, R. N., Feil, S. C., Le Bonniec, B. F., Parker, M. W., and Bottomley, S. P. (2000). Cleaved antitrypsin polymers at atomic resolution. *Protein Sci.* **9,** 417–420.

Elliott, P. R., Lomas, D. A., *et al.* (1996). Inhibitory conformation of the reactive loop of alpha 1-antitrypsin. *Nat. Struct. Biol.* **3**(8), 676–681.

Elliott, P. R., Abrahams, J. P., *et al.* (1998). Wild-type alpha 1-antitrypsin is in the canonical inhibitory conformation. *J. Mol. Biol.* **275**(3), 419–425.

Elliott, P. R., Pei, X. Y., *et al.* (2000). Topography of a 2.0 A structure of alpha1-antitrypsin reveals targets for rational drug design to prevent conformational disease. *Protein Sci.* **9**(7), 1274–1281.

Engh, R., Lobermann, H., *et al.* (1989). The S variant of human alpha 1-antitrypsin, structure and implications for function and metabolism. *Protein Eng.* **2**(6), 407–415.

Fulton, K. F., Buckle, A. M., *et al.* (2005). The high resolution crystal structure of a native thermostable serpin reveals the complex mechanism underpinning the stressed to relaxed transition. *J. Biol. Chem.* **280**(9), 8435–8442.

Gooptu, B., Hazes, B., Chang, W. S., Dafforn, T. R., Carrell, R. W., Read, R. J., and Lomas, D. A. (2000). Inactive conformation of the serpin alpha(1)-antichymotrypsin indicates two-stage insertion of the reactive loop: Implications for inhibitory function and conformational disease. *Proc. Natl. Acad. Sci. USA* **97,** 67–72.

Gooptu, B., Miranda, E., *et al.* (2009). Crystallographic and cellular characterisation of two mechanisms stabilising the native fold of alpha1-antitrypsin: Implications for disease and drug design. *J. Mol. Biol.* **387**(4), 857–868.

Harrop, S. J., Jankova, L., *et al.* (1999). The crystal structure of plasminogen activator inhibitor 2 at 2.0 A resolution: Implications for serpin function. *Structure* **7**(1), 43–54.

Hopkins, F. G., and Pinkus, S. N. (1898). Observations on the crystallization of animal proteids. *J. Physiol.* **23,** 130–136.

Horvath, A. J., Irving, J. A., *et al.* (2005). The murine orthologue of human antichymotrypsin: A structural paradigm for clade A3 serpins. *J. Biol. Chem.* **280**(52), 43168–43178.

Huang, X., Dementiev, A., Olson, S. T., and Gettins, P. G. (2010). Basis for the specificity and activation of the serpin protein Z-dependent proteinase inhibitor (ZPI) as an inhibitor of membrane-associated factor Xa. *J. Biol. Chem.* **285,** 20399–20409.

Huntington, J. A., Pannu, N. S., Hazes, B., Read, R. J., Lomas, D. A., and Carrell, R. W. (1999). A 2.6 A structure of a serpin polymer and implications for conformational disease. *J. Mol. Biol.* **293,** 449–455.

Huntington, J. A., Read, R. J., and Carrell, R. W. (2000a). Structure of a serpin-protease complex shows inhibition by deformation. *Nature* **407,** 923–926.

Huntington, J. A., McCoy, A., *et al.* (2000b). The conformational activation of antithrombin. A 2.85-A structure of a fluorescein derivative reveals an electrostatic link between the hinge and heparin binding regions. *J. Biol. Chem.* **275**(20), 15377–15383.

Huntington, J. A., Kjellberg, M., *et al.* (2003). Crystal structure of protein C inhibitor provides insights into hormone binding and heparin activation. *Structure* **11**(2), 205–215.

Im, H., Woo, M. S., *et al.* (2002). Interactions causing the kinetic trap in serpin protein folding. *J. Biol. Chem.* **277**(48), 46347–46354.

Irving, J. A., Pike, R. N., Lesk, A. M., and Whisstock, J. C. (2000). Phylogeny of the serpin superfamily: Implications of patterns of amino acid conservation for structure and function. *Genome Res.* **10,** 1845–1864.

Irving, J. A., Steenbakkers, P. J., Lesk, A. M., Op den Camp, H. J., Pike, R. N., and Whisstock, J. C. (2002). Serpins in prokaryotes. *Mol. Biol. Evol.* **19,** 1881–1890.

Irving, J. A., Cabrita, L. D., *et al.* (2003). The 1.5 A crystal structure of a prokaryote serpin: Controlling conformational change in a heated environment. *Structure* **11**(4), 387–397.

Jankova, L., Harrop, S. J., *et al.* (2001). Crystal structure of the complex of plasminogen activator inhibitor 2 with a peptide mimicking the reactive center loop. *J. Biol. Chem.* **276**(46), 43374–43382.

Jensen, J. K., and Gettins, P. G. (2008). High-resolution structure of the stable plasminogen activator inhibitor type-1 variant 14-1B in its proteinase-cleaved form: A new tool for detailed interaction studies and modeling. *Protein Sci.* **17**(10), 1844–1849.

Jin, L., Abrahams, J. P., Skinner, R., Petitou, M., Pike, R. N., and Carrell, R. W. (1997). The anticoagulant activation of antithrombin by heparin. *Proc. Natl. Acad. Sci. USA* **94,** 14683–14688.

Johnson, D. J., and Huntington, J. A. (2003). Crystal structure of antithrombin in a heparin-bound intermediate state. *Biochemistry* **42**(29), 8712–8719.

Johnson, D. J., and Huntington, J. A. (2004). The influence of hinge region residue Glu-381 on antithrombin allostery and metastability. *J. Biol. Chem.* **279**(6), 4913–4921.

Johnson, D. J., Li, W., Adams, T. E., and Huntington, J. A. (2006a). Antithrombin-S195A factor Xa-heparin structure reveals the allosteric mechanism of antithrombin activation. *EMBO J.* **25,** 2029–2037.

Johnson, D. J., Langdown, J., *et al.* (2006b). Crystal structure of monomeric native antithrombin reveals a novel reactive center loop conformation. *J. Biol. Chem.* **281**(46), 35478–35486.

Johnson, D. J., Langdown, J., and Huntington, J. A. (2010). Molecular basis of factor IXa recognition by heparin-activated antithrombin revealed by a 1.7-A structure of the ternary complex. *Proc. Natl. Acad. Sci. USA* **107,** 645–650.

Kaslik, G., Patthy, A., Balint, M., and Graf, L. (1995). Trypsin complexed with alpha 1-proteinase inhibitor has an increased structural flexibility. *FEBS Lett.* **370,** 179–183.

Kim, S., Woo, J., *et al.* (2001). A 2.1 A resolution structure of an uncleaved alpha(1)-antitrypsin shows variability of the reactive center and other loops. *J. Mol. Biol.* **306**(1), 109–119.

Klieber, M. A., Underhill, C., Hammond, G. L., and Muller, Y. A. (2007). Corticosteroid-binding globulin, a structural basis for steroid transport and proteinase-triggered release. *J. Biol. Chem.* **282,** 29594–29603.

Langdown, J., Belzar, K. J., *et al.* (2009). The critical role of hinge-region expulsion in the induced-fit heparin binding mechanism of antithrombin. *J. Mol. Biol.* **386**(5), 1278–1289.

Law, R. H., Irving, J. A., *et al.* (2005). The high resolution crystal structure of the human tumor suppressor maspin reveals a novel conformational switch in the G-helix. *J. Biol. Chem.* **280**(23), 22356–22364.

Law, R. H., Sofian, T., *et al.* (2008). X-ray crystal structure of the fibrinolysis inhibitor alpha2-antiplasmin. *Blood* **111**(4), 2049–2052.

Li, W., and Huntington, J. A. (2008). The heparin binding site of protein C inhibitor is protease-dependent. *J. Biol. Chem.* **283**(51), 36039–36045.

Li, J., Wang, Z., *et al.* (1999). The structure of active serpin 1K from Manduca sexta. *Structure* **7**(1), 103–109.

Li, W., Johnson, D. J., Esmon, C. T., and Huntington, J. A. (2004). Structure of the antithrombin-thrombin-heparin ternary complex reveals the antithrombotic mechanism of heparin. *Nat. Struct. Mol. Biol.* **11,** 857–862.

Li, W., Adams, T. E., *et al.* (2007). Structure of native protein C inhibitor provides insight into its multiple functions. *J. Biol. Chem.* **282**(18), 13759–13768.

Li, W., Adams, T. E., Nangalia, J., Esmon, C. T., and Huntington, J. A. (2008). Molecular basis of thrombin recognition by protein C inhibitor revealed by the 1.6-A structure of the heparin-bridged complex. *Proc. Natl. Acad. Sci. USA* **105,** 4661–4666.

Loebermann, H., Tokuoka, R., Deisenhofer, J., and Huber, R. (1984). Human alpha 1-proteinase inhibitor. Crystal structure analysis of two crystal modifications, molecular model and preliminary analysis of the implications for function. *J. Mol. Biol.* **177,** 531–557.

Lomas, D. A., Evans, D. L., Finch, J. T., and Carrell, R. W. (1992). The mechanism of Z alpha 1-antitrypsin accumulation in the liver. *Nature* **357,** 605–607.

Lukacs, C. M., Zhong, J. Q., *et al.* (1996). Arginine substitutions in the hinge region of antichymotrypsin affect serpin beta-sheet rearrangement. *Nat. Struct. Biol.* **3**(10), 888–893.

Lukacs, C. M., Rubin, H., *et al.* (1998). Engineering an anion-binding cavity in antichymotrypsin modulates the "spring-loaded" serpin-protease interaction. *Biochemistry* **37** (10), 3297–3304.

McCoy, A. J., Pei, X. Y., *et al.* (2003). Structure of beta-antithrombin and the effect of glycosylation on antithrombin's heparin affinity and activity. *J. Mol. Biol.* **326**(3), 823–833.

McGowan, S., Buckle, A. M., *et al.* (2006). X-ray crystal structure of MENT: Evidence for functional loop-sheet polymers in chromatin condensation. *EMBO J.* **25**(13), 3144–3155.

Mottonen, J., Strand, A., Symersky, J., Sweet, R. M., Danley, D. E., Geoghegan, K. F., Gerard, R. D., and Goldsmith, E. J. (1992). Structural basis of latency in plasminogen activator inhibitor-1. *Nature* **355,** 270–273.

Mourey, L., Samama, J. P., Delarue, M., Petitou, M., Choay, J., and Moras, D. (1993). Crystal structure of cleaved bovine antithrombin III at 3.2 A resolution. *J. Mol. Biol.* **232,** 223–241.

Nar, H., Bauer, M., *et al.* (2000). Plasminogen activator inhibitor 1. Structure of the native serpin, comparison to its other conformers and implications for serpin inactivation. *J. Mol. Biol.* **297**(3), 683–695.

Pearce, M. C., Morton, C. J., *et al.* (2008). Preventing serpin aggregation: The molecular mechanism of citrate action upon antitrypsin unfolding. *Protein Sci.* **17**(12), 2127–2133.

Pearce, M. C., Powers, G. A., *et al.* (2010). Identification and characterization of a misfolded monomeric serpin formed at physiological temperature. *J. Mol. Biol.* **403**(3), 459–467.

Renatus, M., Zhou, Q., *et al.* (2000). Crystal structure of the apoptotic suppressor CrmA in its cleaved form. *Structure* **8**(7), 789–797.

Ricagno, S., Caccia, S., *et al.* (2009). Human neuroserpin: Structure and time-dependent inhibition. *J. Mol. Biol.* **388**(1), 109–121.

Ryu, S. E., Choi, H. J., *et al.* (1996). The native strains in the hydrophobic core and flexible reactive loop of a serine protease inhibitor: Crystal structure of an uncleaved alpha1-antitrypsin at 2.7 A. *Structure* **4**(10), 1181–1192.

Schreuder, H. A., de Boer, B., Dijkema, R., Mulders, J., Theunissen, H. J., Grootenhuis, P. D., and Hol, W. G. (1994). The intact and cleaved human antithrombin III complex as a model for serpin-proteinase interactions. *Nat. Struct. Biol.* **1,** 48–54.

Sharp, A. M., Stein, P. E., Pannu, N. S., Carrell, R. W., Berkenpas, M. B., Ginsburg, D., Lawrence, D. A., and Read, R. J. (1999). The active conformation of plasminogen activator inhibitor 1, a target for drugs to control fibrinolysis and cell adhesion. *Structure* **7,** 111–118.

Simonovic, M., Gettins, P. G. W., *et al.* (2000). Crystal structure of viral serpin crmA provides insights into its mechanism of cysteine proteinase inhibition. *Protein Sci.* **9**(8), 1423–1427.

Simonovic, M., Gettins, P. G., *et al.* (2001). Crystal structure of human PEDF, a potent anti-angiogenic and neurite growth-promoting factor. *Proc. Natl. Acad. Sci. USA* **98**(20), 11131–11135.

Skinner, R., Abrahams, J. P., *et al.* (1997). The 2.6 A structure of antithrombin indicates a conformational change at the heparin binding site. *J. Mol. Biol.* **266**(3), 601–609.

Skinner, R., Chang, W. S., *et al.* (1998). Implications for function and therapy of a 2.9 A structure of binary-complexed antithrombin. *J. Mol. Biol.* **283**(1), 9–14.

Song, H. K., Lee, K. N., *et al.* (1995). Crystal structure of an uncleaved alpha 1-antitrypsin reveals the conformation of its inhibitory reactive loop. *FEBS Lett.* **377**(2), 150–154.

Stavridi, E. S., O'Malley, K., Lukacs, C. M., Moore, W. T., Lambris, J. D., Christianson, D. W., Rubin, H., and Cooperman, B. S. (1996). Structural change in alpha-chymotrypsin induced by complexation with alpha 1-antichymotrypsin as seen by enhanced sensitivity to proteolysis. *Biochemistry* **35,** 10608–10615.

Stein, P. E., Leslie, A. G., Finch, J. T., Turnell, W. G., McLaughlin, P. J., and Carrell, R. W. (1990). Crystal structure of ovalbumin as a model for the reactive centre of serpins. *Nature* **347,** 99–102.

Stein, P. E., Leslie, A. G., *et al.* (1991). Crystal structure of uncleaved ovalbumin at 1.95 A resolution. *J. Mol. Biol.* **221**(3), 941–959.

Stout, T. J., Graham, H., *et al.* (2000). Structures of active and latent PAI-1: A possible stabilizing role for chloride ions. *Biochemistry* **39**(29), 8460–8469.

Takehara, S., Onda, M., *et al.* (2009). The 2.1-A crystal structure of native neuroserpin reveals unique structural elements that contribute to conformational instability. *J. Mol. Biol.* **388**(1), 11–20.

Tucker, H. M., Mottonen, J., *et al.* (1995). Engineering of plasminogen activator inhibitor-1 to reduce the rate of latency transition. *Nat. Struct. Biol.* **2**(6), 442–445.

Wei, A., Rubin, H., *et al.* (1994). Crystal structure of an uncleaved serpin reveals the conformation of an inhibitory reactive loop. *Nat. Struct. Biol.* **1**(4), 251–258.

Wei, Z., Yan, Y., Carrell, R. W., and Zhou, A. (2009). Crystal structure of protein Z-dependent inhibitor complex shows how protein Z functions as a cofactor in the membrane inhibition of factor X. *Blood* **114,** 3662–3667.

Whisstock, J. C., and Bottomley, S. P. (2006). Molecular gymnastics: Serpin structure, folding and misfolding. *Curr. Opin. Struct. Biol.* **16,** 761–768.

Whisstock, J., Lesk, A. M., and Carrell, R. (1996). Modeling of serpin-protease complexes: Antithrombin-thrombin, alpha 1-antitrypsin (358Met– > Arg)-thrombin, alpha 1-antitrypsin (358Met– > Arg)-trypsin, and antitrypsin-elastase. *Proteins* **26,** 288–303.

Wright, H. T., and Scarsdale, J. N. (1995). Structural basis for serpin inhibitor activity. *Proteins* **22,** 210–225.

Wright, H. T., Qian, H. X., and Huber, R. (1990). Crystal structure of plakalbumin, a proteolytically nicked form of ovalbumin. Its relationship to the structure of cleaved alpha-1-proteinase inhibitor. *J. Mol. Biol.* **213,** 513–528.

Xue, Y., Bjorquist, P., Inghardt, T., Linschoten, M., Musil, D., Sjolin, L., and Deinum, J. (1998). Interfering with the inhibitory mechanism of serpins: Crystal structure of a complex formed between cleaved plasminogen activator inhibitor type 1 and a reactive-centre loop peptide. *Structure* **6,** 627–636.

Yamasaki, M., Arii, Y., *et al.* (2002). Loop-inserted and thermostabilized structure of P1-P1' cleaved ovalbumin mutant R339T. *J. Mol. Biol.* **315**(2), 113–120.

Yamasaki, M., Li, W., Johnson, D. J., and Huntington, J. A. (2008). Crystal structure of a stable dimer reveals the molecular basis of serpin polymerization. *Nature* **455,** 1255–1258.

Yamasaki, M., Sendall, T. J., Harris, L. E., Lewis, G. M., and Huntington, J. A. (2010). Loop-sheet mechanism of serpin polymerization tested by reactive center loop mutations. *J. Biol. Chem.* **285,** 30752–30758.

Ye, S., Cech, A. L., Belmares, R., Bergstrom, R. C., Tong, Y., Corey, D. R., Kanost, M. R., and Goldsmith, E. J. (2001). The structure of a Michaelis serpin-protease complex. *Nat. Struct. Biol.* **8,** 979–983.

Zhang, Q., Buckle, A. M., Law, R. H., Pearce, M. C., Cabrita, L. D., Lloyd, G. J., Irving, J. A., Smith, A. I., Ruzyla, K., Rossjohn, J., Bottomley, S. P., and Whisstock, J. C. (2007). The N terminus of the serpin, tengpin, functions to trap the metastable native state. *EMBO Rep.* **8,** 658–663.

Zhang, Q., Law, R. H., Bottomley, S. P., Whisstock, J. C., and Buckle, A. M. (2008). A structural basis for loop C-sheet polymerization in serpins. *J. Mol. Biol.* **376,** 1348–1359.

Zheng, B., Matoba, Y., *et al.* (2009). Crystal structure of SCCA1 and insight about the interaction with JNK1. *Biochem. Biophys. Res. Commun.* **380**(1), 143–147.

Zhou, A., Huntington, J. A., Pannu, N. S., Carrell, R. W., and Read, R. J. (2003a). How vitronectin binds PAI-1 to modulate fibrinolysis and cell migration. *Nat. Struct. Biol.* **10,** 541–544.

Zhou, A., Stein, P. E., *et al.* (2003b). Serpin polymerization is prevented by a hydrogen bond network that is centered on his-334 and stabilized by glycerol. *J. Biol. Chem.* **278**(17), 15116–15122.

Zhou, A., Stein, P. E., *et al.* (2004). How small peptides block and reverse serpin polymerisation. *J. Mol. Biol.* **342**(3), 931–941.

Zhou, A., Wei, Z., Read, R. J., and Carrell, R. W. (2006). Structural mechanism for the carriage and release of thyroxine in the blood. *Proc. Natl. Acad. Sci. USA* **103,** 13321–13326.

Zhou, A., Wei, Z., *et al.* (2008). The S-to-R transition of corticosteroid-binding globulin and the mechanism of hormone release. *J. Mol. Biol.* **380**(1), 244–251.

CHAPTER SIX

Serpins as Hormone Carriers: Modulation of Release

Robin Carrell, Xiaoqiang Qi,[1] *and* Aiwu Zhou

Contents

Abstract

The hormone-carrying serpins, thyroxine- and corticosteroid-binding globulins, TBG and CBG, provide a clear example of the way the serpin conformational mechanism can be adapted not only to give an irreversible switching-off of function but also more significantly to allow a constant dynamic modulation of activity. This is illustrated here with the demonstration that hormone release from both TBG and CBG is responsive to changes in ambient temperature and specifically to changes in body temperature. An exception to this adaptation of the serpin mechanism is seen with another family member, angiotensinogen, in which hormone release is modulated by a redox switch and is apparently independent of changes in the serpin framework.

1. Serpins and Allosteric Modulation

The recognition that all the members of the serpin family share the same conserved template structure (Huber and Carrell, 1989) brought with it two challenges. The first, which has been amply answered in other chapters in this volume, is how this single template has been adapted to provide such a diversity of functions. But the more puzzling question to those early in the field was why the serpins had been preferentially selected

Cambridge Institute for Medical Research, University of Cambridge, Cambridge, United Kingdom
[1] Current address: Department of Biochemistry, Nanjing Medical University, Nanjing, People's Republic of China

Methods in Enzymology, Volume 501
ISSN 0076-6879, DOI: 10.1016/B978-0-12-385950-1.00006-7

as controlling factors in the critical pathways of higher organisms, including coagulation, complement activation, and tissue development and growth? The answer to this question came with the understanding of the reason for the predominance of the serpins as protease inhibitors. This culminated with the structural demonstration of the profound conformational transition that results in the essentially irreversible inactivation by serpins of their target proteases (Huntington *et al.*, 2000). The ability of the serpins to undergo this remarkable change of fold had been indicated much earlier in the first of the serpin structures, that of reactive-loop cleaved α1-antitrypsin (Loebermann *et al.*, 1984) and then that of an intact serpin, ovalbumin (Stein *et al.*, 1990). The demonstration of the generality of this S-to-R conformational transition from a stressed to a relaxed form in the serpins led to the prediction of its likely adaptation to allow the modulation of serpin activity (Carrell and Owen, 1985; Carrell *et al.*, 1991; Stein *et al.*, 1990; Fig. 6.1).

Subsequent structures confirmed the predictions shown in Fig. 6.1C, with the presence of full loop insertion in latent PAI-1 (Mottonen *et al.*, 1992) and notably with the loop changes associated with the activation of antithrombin by the heparin pentasaccharide. Whereas the prior structures of antithrombin in the absence of heparin (Carrell *et al.*, 1994; Schreuder *et al.*, 1994) had shown the reactive loop to be partially inserted, a later structure showed that on complexing with heparin, an extension of the D-helix took place with an accompanying full expulsion and hence activation of the reactive-center loop (Jin *et al.*, 1997). It was at first assumed that this allosteric activation of antithrombin functioned as an on–off switch, but a subsequent structure of the heparin–antithrombin complex surprisingly showed, contrary to the previous structure of the same complex, a partial incorporation of the reactive loop (Johnson and Huntington, 2003). The demonstration of these two alternative conformations in the same complex provides direct structural evidence that the reactive loop can move dynamically into and out of the A-sheet, with the binding of heparin resulting in a shift of the equilibrated balance in favor of full exposure of the loop. In this way, antithrombin provides a paradigm for the allosteric regulation of serpin activity. It is this in-built potential of the serpins to allow a responsive modulation of activity that explains their selection and adaptation as the controlling factors in key metabolic and intracellular processes, as notably seen with the hormone-carrying globulins, thyroxine- and corticosteroid-binding globulins, TBG and CBG.

2. Hormone Carriage—TBG and CBG

Although the recognition of the serpins as a protein family was initiated with the homologous alignment of two protease inhibitors, α1-antitrypsin and antithrombin (Carrell *et al.*, 1979; Petersen *et al.*, 1979), there was much

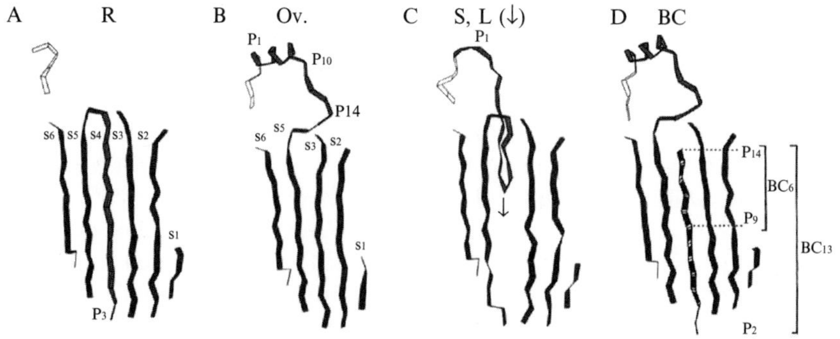

Figure 6.1 Schematic from 1991 showing predicted modulatory mechanisms. (A) Fully inserted reactive loop in cleaved antitrypsin, (B) fully exposed loop of ovalbumin, and (D) the six-stranded binary complex formed on incubation of antitrypsin with a synthetic 13-residue loop-peptide (Schulze *et al.*, 1990). (C) The proposed modulatory movement of the reactive loop into and out of the A-sheet specifically predicted to give the inactivation of latent PAI-1 and the heparin activation of antithrombin and now seen to modulate hormone release from TBG and CBG (reproduced with permission from Carrell *et al.*, 1991).

interest in the subsequent identification of noninhibitory serpins and most particularly of three hormone carriers—angiotensinogen (Doolittle, 1983), thyroxine-binding globulin (Flink *et al.*, 1986), and corticosteroid-binding globulin (Hammond *et al.*, 1987). The alignment of the sequences of the three hormone carriers with other serpins provided immediate clues as to their function. The presence of atypical reactive-center sequences made it unlikely that these newly identified serpins retained any inhibitory function and this was further excluded with angiotensinogen (Stein *et al.*, 1989) by a reactive-loop hinge-sequence P15–P12 that was incompatible with ready loop insertion into the A-sheet. TBG and CBG, however, although having lost any inhibitory activity, do retain the typical hinge sequence of an active serpin. Both were shown to undergo the typical S-to-R change (Pemberton *et al.*, 1988) as confirmed 20 years later by the structures of their cleaved reactive-loop conformers (Qi *et al.*, 2011; Zhou *et al.*, 2008).

The demonstration that proteolytic cleavage of the reactive loops of TBG and CBG resulted in hormone release (Hammond *et al.*, 1990; Jirasakuldech *et al.*, 2000; Pemberton *et al.*, 1988; Suda *et al.*, 2000) provided direct evidence of a conformational release mechanism compatible with hormone delivery to areas of proteolytic activity, as in sites of inflammation. The good sense that this mechanism makes was in a way misleading, as it led to the concept that the carriage of thyroxine and corticosteroids is essentially an on-and-off event, with the release being triggered by the S-to-R irreversible transition. The wider modulatory potential of the mechanism, however, has recently become evident with the solving of a series of structures of TBG and CBG (Klieber *et al.*, 2007; Qi *et al.*, 2011; Zhou *et al.*, 2006, 2008). The two binding globulins are closely homologous, with thyroxine and cortisol both binding to identical pockets on the serpin framework, between helices H and A and strands 3–5 of the B-sheet (Fig. 6.2). The hormones each form a series of hydrophobic and polar bonds with the residues lining the pockets, with specific stabilization by π-bonding with an arginine in TBG and with a homologous tryptophan in CBG. Although the first crystal structures of the TBG and CBG hormone complexes implied a tight specificity of binding, subsequent structures emphasize the plasticity of the pockets, with the interactions of the liganded hormones within the pocket being reflected in the affinity rather than precise specificity of binding. Thus both globulins can still bind their respective hormones even after cleavage and transition to the R-form, but

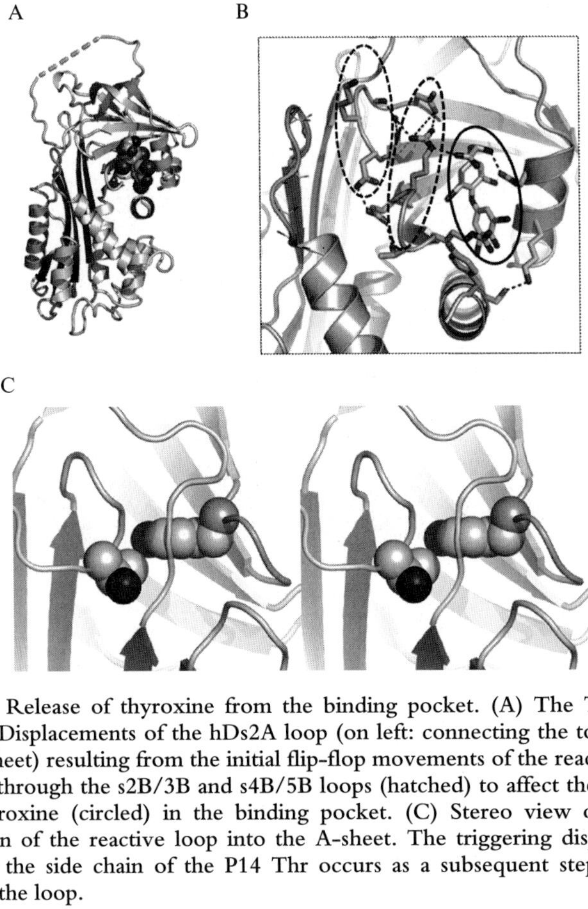

Figure 6.2 Release of thyroxine from the binding pocket. (A) The TBG binding pocket. (B) Displacements of the hDs2A loop (on left: connecting the top of helix D and the A-sheet) resulting from the initial flip-flop movements of the reactive loop are transmitted through the s2B/3B and s4B/5B loops (hatched) to affect the interactions holding thyroxine (circled) in the binding pocket. (C) Stereo view of the partial incorporation of the reactive loop into the A-sheet. The triggering displacement of Tyr 241 by the side chain of the P14 Thr occurs as a subsequent step, on further insertion of the loop.

with a substantially reduced affinity. Likewise, other structures show that the TBG pocket can readily accommodate and bind a range of thyroxine analogues and unrelated drugs and dyes, though again with a lesser affinity (Qi *et al.*, 2011).

Early evidence that movements of the reactive loop influenced the ligand affinity of the binding globulins came from the demonstration that a shortening of the reactive-center loop of TBG resulted in an increased affinity for thyroxine and conversely a lengthening of the loop gave a decreased affinity (Grasberger *et al.*, 2002). The explanation for this became apparent with the solving of the structure of the TBG–thyroxine complex (Zhou *et al.*, 2006). This, with ancillary evidence from natural variants of TBG (Bertenshaw *et al.*, 1991), clearly indicated the existence of a flip-flop modulatory mechanism, with the affinity of the binding of thyroxine being governed by a limited movement of the reactive loop into and out of the A-sheet, as demonstrably occurs with antithrombin. Similarly, TBG crystallized, as does active antithrombin, with its reactive loop partially incorporated in the A-sheet, to the level of a conserved threonine at P14. The complete insertion of the side chain of the P14 threonine is impeded by a conserved underlying tyrosine, Tyr 241 in the s2B/3B loop, which critically interacts with underlying peptide loops of the B-sheet that bind to and flank thyroxine in the binding pocket. In this way, the perturbation of Tyr241 on the complete insertion of the P14 threonine will initiate a chain of changes resulting in a decrease in binding affinity (Fig. 6.2C).

Although the displacements resulting from the complete insertion of the side chain of the P14 Thr 342 can trigger further hormone release, we now believe that the fine-tuning of the physiological balance between the bound and unbound hormones principally occurs with the initial entry of the loop into the A-sheet prior to this triggering event. The significance of this initial movement of the loop into and out of the sheet, to the level of P14, became apparent with the determination of the structure of CBG (Klieber *et al.*, 2007). This showed, as compared to the TBG structure, a fully exposed reactive loop and an accompanying unwinding of the D-helix, with the two structures, of TBG and CBG, providing a strikingly comparable identity with the antithrombin–heparin-bound and -unbound structures, respectively. Subsequent structural studies have confirmed the role of this limited flip-flop movement of the loop, in TBG as well as in CBG, in modifying the affinity of hormone binding. Entry of the reactive loop will result in an expansion of sheet A with a consequent reordering of the loop that connects strand 2 of the A-sheet with the top of the D-helix. This connecting loop is in close contact with loop s2/3B and it in turn is in contact with loop s4B/5B that flanks the binding pocket. As modeled in Fig. 6.2B, any slight changes in the packing of these loops will affect the plasticity of the pocket and perturb the polar and hydrophobic bonds that anchor the hormone in the binding site.

Thus contrary to earlier concepts of a simple on–off mechanism, it is now evident that the binding and release of hormones from the carrier globulins is a much more subtle process, primarily reflecting an allosteric and dynamic equilibrium in both CBG and TBG between high- and low-affinity conformers of the intact circulating proteins. Hence each of these carriers will have a characteristic dissociation curve (Schreiber, 2002), comparable to the similar allosteric dissociation curve of hemoglobin, as shown with CBG in Fig. 6.3. Inherent to this is that the uptake and release of the hormones will be dependent on tissue concentrations of the unbound hormones, with the carrier proteins acting as buffers to evenly maintain the optimal concentrations of the free hormones. For example, the tissue concentration of free thyroxine at 20 p*M*/L is just a fraction of the 100 n*M*/L of thyroxine bound to TBG. In this way, TBG functions as a circulating store of thyroxine that is amply able to buffer the tissue concentrations of free thyroxine. The binding capacity of TBG in the circulation is only 25% saturated, so small shifts in the K_d of TBG will

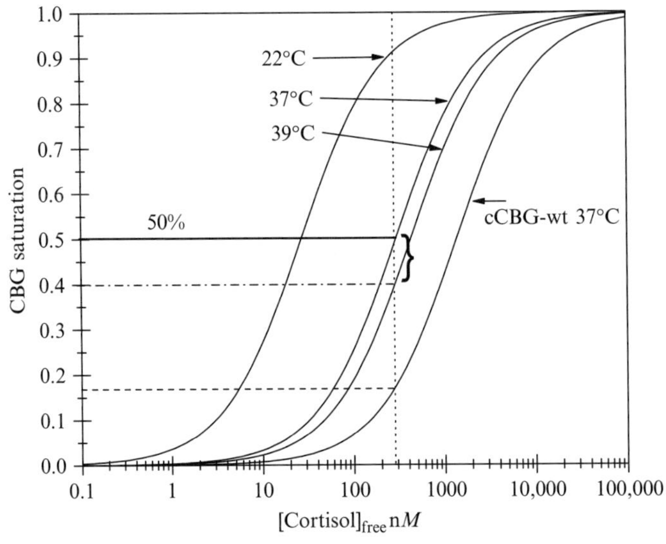

Figure 6.3 Temperature modulation of hormone release. The binding to recombinant (unglycosylated) CBG of cortisol. Curves plotted using the determined dissociation constant K_d at each temperature and the law of mass action (Schreiber, 2002). Saturations were plotted as follows: Saturation = [ligand]/([ligand] + K_d). The seemingly small shift that will occur in fevers (37–39 °C) results, at 50% saturation, in the bracketed release of a fifth of the carried corticosteroid. Reactive-loop cleavage (cCBG) gives a marked diminution but not complete loss of binding affinity (Qi *et al.*, 2011).

potentially result in major changes in the concentration of unbound thyroxine in the tissues. This, together with the flexibility provided by an allosteric release mechanism, led to our proposal that the hormone carriage by TBG and CBG would be responsive to changes in their environments or to ligand or receptor interactions (Zhou *et al.*, 2006, 2008).

3. Temperature Response: A Protein Thermocouple

A finding from antithrombin studies of special relevance to TBG and CBG is the temperature sensitivity of the flip-flop movements of the reactive loop, as evidenced by variant antithrombins that capture the over-insertion of the loop as the irreversible latent conformer (Beauchamp *et al.*, 1998). That this is also so with TBG is similarly indicated by the temperature-sensitive induction of the latent state in comparable loop-sheet variants (Bertenshaw *et al.*, 1991). With this mechanism in mind, we proposed that TBG and CBG could potentially function as protein thermocouples to allow a physiologically advantageous release of hormones in response to changes in body temperature (Zhou *et al.*, 2008). This has now been confirmed with CBG, with the demonstration of changes in K_d and the consequent release of cortisol in response to increases in the normal body temperature from 37 °C to that in fevers of 39 °C (Cameron *et al.*, 2010). Temperature-induced changes in the K_d of TBG are more difficult to assess because of its tight-binding affinity for T4 ($K_d \sim 0.1$ n*M*; Refetoff *et al.*, 1996), which makes measurement by direct fluorimetry impracticable. To overcome this, thyroxine fluorophores have recently been developed that allow the ready monitoring of changes in binding affinity not only in native TBG but also in its engineered mutants (Qi *et al.*, 2011).

The use of the thyroxine fluorophores has confirmed quantitatively what is also clearly observed structurally, that the complete S-to-R change in TBG results merely in a decreased thyroxine-binding affinity, from at 37 °C a K_d (with thyroxine) of 0.075 to that of 0.28 n*M* in the cleaved form. A practical implication is the fallibility of measurements made in the laboratory at room temperature of 22 °C with respect to physiological status at the body temperature of 37 °C. Of medical interest is the significant shift that takes place with the small rise in body temperature to 39 °C as characteristically occurs with infections, with a decrease in binding affinity that will predictably result in a 20% increase in the concentration of free plasma thyroxine. The consequent increased release in fevers of thyroxine from TBG and even more so of cortisol from CBG will in both cases be an advantageous response to infection and inflammation.

4. Angiotensinogen and Its Interaction with Renin

A serpin, angiotensinogen, is a starting point for the complex interactions of the renin–angiotensin system (RAS) that control blood pressure in mammals by influencing salt retention and vasoconstriction (Cleland and Reid, 1996; Dickson and Sigmund, 2006). The key mediator in the RAS system is the series of angiotensin peptides initiated by the highly specific release by the enzyme renin of the decapeptide angiotensin I from the amino-terminus of angiotensinogen. Although angiotensinogen is secreted from the liver and is present in the plasma in relatively high concentration (0.8 μ*M*), it is now believed that its primary action occurs at a tissue and cellular level, with the efficiency of cleavage being greatly increased by the interaction of renin with the membrane prorenin receptor (Nguyen *et al.*, 2002). The story of angiotensinogen illustrates the pitfalls that can result from the development of a field of studies in the absence of a starting structural basis. Despite genetic and other evidence (Kim *et al.*, 1995) indicating that angiotensinogen directly contributed to the regulation of blood pressure, the long-standing diagrammatic depiction of angiotensinogen as a blob with the decapeptide dangling from it reinforced the misapprehension that it was merely a passive substrate. The recent solving of the structure of angiotensinogen and of its complex with renin (Fig. 6.4) has now resolved this by showing that not only does angiotensinogen undergo

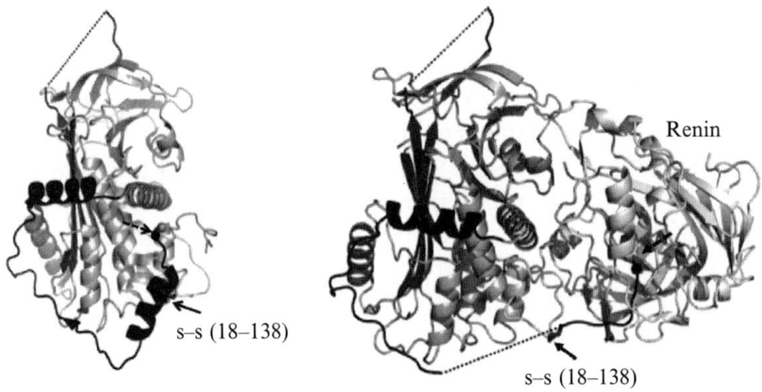

Figure 6.4 Angiotensinogen and its complex with renin: Showing (left) the amino-terminal superstructure anchored to the serpin framework of angiotensinogen by two helices (dark gray) with the hatched arrow indicating the buried renin-cleavage site. Interaction with renin, shown right, displaces a protruding CD peptide loop with the concerted movement of the angiotensin-containing terminal into the active-site cleft of renin.

an active conformational interaction with renin but that it also incorporates with this an elegant fine-tuning switch (Zhou *et al.*, 2010). The completion of a range of structures of angiotensinogen brought with it another surprise. As expected, the structures show that angiotensinogen retains the archetypal framework of a noninhibitory serpin with the A-sheet remaining tightly closed and the reactive loop fully exposed. But, the structures, including that of its complex with renin, showed no evidence of the confidently expected loop-sheet shifts and framework rotations that typify the serpin conformational change (Whisstock *et al.*, 2000). Thus, although there is a substantial modulatory control of the release of angiotensin, this is apparently independent of the framework structure of angiotensinogen, which remains essentially unchanged.

The solving of the angiotensinogen structures was the culmination of a 20-year effort and required the recombinant production and crystallization of rat and mouse angiotensinogens as well as that of the human (see Chapter 4). Diffraction datasets for each of these were obtained at up to 2.1 Å resolution at the Daresbury and Diamond synchrotron light sources. Solving the structures provided a challenging test of the advanced molecular replacement algorithms in *Phaser* (McCoy *et al.*, 2007) because there were no close homologues of known structure. A weak molecular replacement solution for human angiotensinogen, together with a low-resolution $GdCl_3$ derivative, gave electron density that was used to solve the C2 crystal form of rat angiotensinogen. After twofold averaging, electron density from the C2 crystal form of rat angiotensinogen then enabled the solving of the structure of the $P3_221$ crystal form of the same protein. An electron density map sufficient for model building was then obtained by fourfold multicrystal averaging across the two crystal forms. Finally, the mouse angiotensinogen structure was solved by molecular replacement, utilizing four copies of the model of the rat protein, and this in turn led to a high-resolution structure of human angiotensinogen. This shows (Fig. 6.4) that the 61-residue amino-terminal extension, containing the angiotensin sequence, forms an ordered superstructure bound to the body of angiotensinogen by two new helices.

The surprise was that the terminal angiotensin sequence was not freely exposed as expected but rather was firmly bound to the body of the molecule with the critical peptide cleavage site being inaccessibly buried. The question arising from this was how does renin gain access to this cleavage site? The answer to this became apparent with the crystallization and solving of the structure of the initiating complex formed by human angiotensinogen with recombinantly inactivated (Asp292Ala) human renin (Zhou *et al.*, 2010). Although only at 4.4 Å resolution, the complex clearly shows that the interaction with renin results in two major changes in angiotensinogen. The complementary binding of renin causes a 10 Å displacement of the protruding CD loop on angiotensinogen along with a concerted 10–20 Å extension and movement of the N-terminal substrate peptide of angiotensinogen into the active-site cleft of renin. These two activating shifts are seen to be critically linked by a conserved disulphide bond between Cys 138 in the CD loop and Cys 18 in the amino-terminal tail, close to the 10–11 Leu-Val renin-cleavage site.

Previous work (Gimenez-Roqueplo *et al.*, 1998; Streatfeild-James *et al.*, 1998) had focused attention on this conserved 18–138 disulphide bridge, which the structure now showed to be externally sited. Confirmation that this is a labile disulphide bond was demonstrated by the incubation of angiotensinogen over a range of physiological redox potentials. Unequivocal evidence of the presence in the circulation of the reduced unbridged form as well as the oxidized bridged form came from the examination of a series of normal plasma samples, showing the consistent presence of near 40% of angiotensinogen in the reduced form, apparently independent of age or gender. This finding of equilibrated reversibility, between the bridged and unbridged forms, immediately suggested the possibility of a control mechanism modulating the release of angiotensin. Confirmation of the physiological relevance of this bridge came from the redox titration of isolated angiotensinogen indicating a reduction potential for this dithiol/disulphide equilibrium of − 230 mV at 25 °C and pH 7. A crystal structure of reduced mouse angiotensinogen confirmed that the breaking of the 18–138 disulphide bridge does occur on reduction. Angiotensinogen will predictably be secreted from the endoplasmic reticulum in the oxidized form but will then be subject to reduction on interaction with tissue and plasma reducing systems. Although the ratio of the reduced to oxidized forms is remarkably consistent at near 40% in the plasma of healthy controls, the redox poise of the transition will readily allow the reversible switch from the oxidized to the reduced form in focal tissues or vascular beds.

To our initial puzzlement, the differences between the two forms, in terms of the efficiency of release by renin of angiotensin, were relatively small. When the assays of angiotensin release by renin were, however, repeated in the presence of the prorenin receptor, a clear fourfold increase in activity was evident in the oxidized bridged form, confirming that the 18–138 disulphide could indeed act as a redox switch capable of fine-tuning the release of angiotensin (Fig. 6.5). This corroboration of the significance of the receptor-bound activation of renin is in keeping with current understandings that the critical release of angiotensin takes place focally at a cellular level in renal and other tissues including the brain (Kobori *et al.*, 2007). In particular, our identification of an activating redox switch is in keeping with a body of observations by others indicating that oxidative stress can be a contributory cause to the onset of hypertension (Harrison and Gongora, 2009). The suspicion that increases in blood pressure could result from the oxidative conversion of angiotensinogen to its more active bridged form was difficult to confirm, as changes taking place at a focal tissue level are likely to have little effect on the pooled circulating angiotensinogen. An exception,

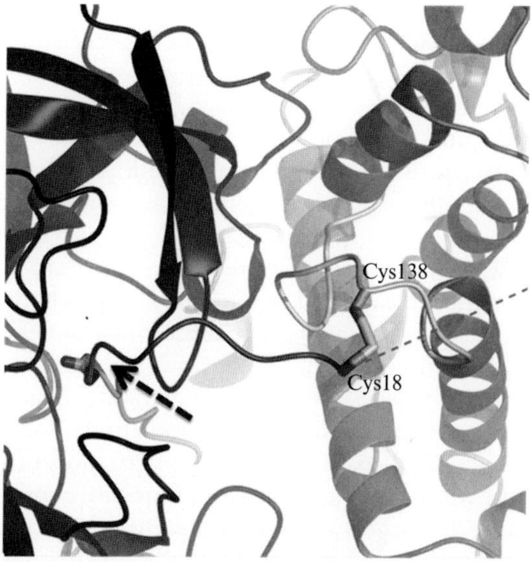

Figure 6.5 The redox switch: The complex with renin (left) highlighting the labile disulphide bond that bridges the two mobile elements: the displaced CD loop containing Cys 138 and the amino-terminus containing Cys 18 and the neighboring angiotensin cleavage site (hatched arrow).

however, is the more overt oxidative stress resulting from placental dysfunction (Burton and Jauniaux, 2004; Myatt, 2010). This is associated with the increase in blood pressure that in 2–7% of all pregnancies leads on to preeclampsia—the hypertensive crisis of pregnancy responsible globally for an estimated 50,000 maternal and 500,000 infant deaths each year.

To test at a pilot level whether conversion of angiotensinogen to its more active oxidized form occurred in preeclampsia, 24 plasma samples provided by Professor Broughton Pipkin in Nottingham were examined—12 from preeclamptics and 12 from matched normal pregnancies (Mistry *et al.*, 2008). Although mixed and with blinded identifiers, electrophoresis subsequent to SH-labeling showed clear differences between the two cohorts, with some of the preeclamptic samples being immediately recognizable by a marked decrease in reduced as compared to oxidized angiotensinogen. Evidence that the fourfold increase in activity that occurs on oxidation is a sufficient contributory cause of preeclampsia came from an earlier experiment of nature (Inoue *et al.*, 1995). Investigation of a 17-year-old female with preeclampsia revealed a mutation at the renin-cleavage site, Leu10Val with a consequent increase in the catalytic release of angiotensin from the mutant, comparable to that observed with the transition of angiotensinogen from the reduced to the oxidized form. Thus there is now a direct series of findings linking the oxidative changes in the placenta with the hypertension in pregnancy that leads on to the development of preeclampsia.

A further potential tissue interaction is likely to exacerbate this difference in activity between the reduced and oxidized forms of angiotensinogen. The reduction and subsequent modification of labile disulphides is now realized to be widely involved in the modulation of protein activity (Nagahara *et al.*, 2009). It is clear from the structure of angiotensinogen that any such modification of the reduced 18–138 sulphydryls would perturb its interaction with renin. Pertinently, it was shown that the reduced form of angiotensinogen can be blocked and hence stabilized by interaction with a known hypotensive agent, nitric oxide. As well as having a direct vasorelaxant effect, nitric oxide also has a potentially additional hypotensive action in being able to selectively *S*-nitrosylate-exposed thiols (Derakhshan *et al.*, 2007; Stamler *et al.*, 2001). Incubation of reduced angiotensinogen with the simple *S*-nitrosothiol donor SNAP (Zhou *et al.*, 2010) led to the stoichiometric loss of the 18 and 138 thiols and the formation of *S*-nitroso-angotensinogen, confirming the nitrosylation and hence blockage of both cysteines. Taken together with the structural evidence, these findings strongly indicate an as yet unrecognized modulatory mechanism at the commencement of the principal pathway controlling blood pressure.

Crystal structures more often than not provide unexpected surprises. We describe two such findings here: the conformational contribution of angiotensinogen to the interaction with renin and the presence of a redox-sensitive fine-tuning switch. But there was also an unexpected negative finding, with the absence of any apparent involvement of the serpin framework in the mechanism or modulation of angiotensin cleavage and release. As an archaic member of the serpin family, there were some differences from the typical serpin framework, with a shift in orientation of helix F and a nearly complete absence of helix H. Nevertheless, no apparent changes occurred in the serpin framework of angiotensinogen on formation of the complex. Is angiotensinogen an exception to the rule that the success of serpins as modulatory agents is dependent on adaptations of the characteristic serpin conformational transition? Does the serpin structure merely provide a frame for the added-on terminal tail? Before definitive answers can be given to these questions, further structures of angiotensinogen are required, notably of its complex at higher resolution, of its reactive-loop cleaved form, and of its post renin-cleavage form. Those close to the current structural studies of angiotensinogen suspect there may yet be more to this story.

ACKNOWLEDGMENTS

We gratefully acknowledge the support of the British Heart Foundation and the Isaac Newton Trust of Trinity College, University of Cambridge.

REFERENCES

Beauchamp, N. J., Pike, R. N., Daly, M., Butler, L., Makris, M., Dafforn, T. R., Zhou, A., Fitton, H. L., Preston, F. E., Peake, I. R., and Carrell, R. W. (1998). Antithrombins Wibble and Wobble (T85M/K): Archetypal conformational diseases with in vivo latent-transition, thrombosis, and heparin activation. *Blood* **92,** 2696–2706.

Bertenshaw, R., Takeda, K., and Refetoff, S. (1991). Sequencing of the variant thyroxine-binding globulin (TBG)-Quebec reveals two nucleotide substitutions. *Am. J. Hum. Genet.* **48,** 741–744.

Burton, G. J., and Jauniaux, E. (2004). Placental oxidative stress: From miscarriage to preeclampsia. *J. Soc. Gynecol. Invest.* **11,** 342–352.

Cameron, A., Henley, D., Carrell, R., Zhou, A., Clarke, A., and Lightman, S. (2010). Temperature-responsive release of cortisol from its binding globulin: A protein thermocouple. *J. Clin. Endocrinol. Metab.* **95,** 4689–4695.

Carrell, R. W., and Owen, M. C. (1985). Plakalbumin, alpha 1-antitrypsin, antithrombin and the mechanism of inflammatory thrombosis. *Nature* **317,** 730–732.

Carrell, R., Owen, M., Brennan, S., and Vaughan, L. (1979). Carboxy terminal fragment of human α-1-antitrypsin from hydroxylamine cleavage: Homology with antithrombin III. *Biochem. Biophys. Res. Commun.* **91,** 1032–1037.

Carrell, R. W., Evans, D. L., and Stein, P. E. (1991). Mobile reactive centre of serpins and the control of thrombosis. *Nature* **353,** 576–578.

Carrell, R. W., Stein, P. E., Fermi, G., and Wardell, M. R. (1994). Biological implications of a 3 Å structure of dimeric antithrombin. *Structure* **2,** 257–270.

Cleland, S. J., and Reid, J. L. (1996). The renin-angiotensin system and the heart: A historical review. *Heart* **76,** 7–12.

Derakhshan, B., Hao, G., and Gross, S. S. (2007). Balancing reactivity against selectivity: The evolution of protein S-nitrosylation as an effector of cell signaling by nitric oxide. *Cardiovasc. Res.* **75,** 210–219.

Dickson, M. E., and Sigmund, C. D. (2006). Genetic basis of hypertension: Revisiting angiotensinogen. *Hypertension* **48,** 14–20.

Doolittle, R. F. (1983). Angiotensinogen is related to the antitrypsin-antithrombin-ovalbumin family. *Science* **222,** 417–419.

Flink, I. L., Bailey, T. J., Gustafson, T. A., Markham, B. E., and Morkin, E. (1986). Complete amino acid sequence of human thyroxine-binding globulin deduced from cloned DNA: Close homology to the serine antiproteases. *Proc. Natl. Acad. Sci. USA* **83,** 7708–7712.

Gimenez-Roqueplo, A. P., Celerier, J., Schmid, G., Corvol, P., and Jeunemaitre, X. (1998). Role of cysteine residues in human angiotensinogen. Cys232 is required for angiotensinogen-pro major basic protein complex formation. *J. Biol. Chem.* **273,** 34480–34487.

Grasberger, H., Golcher, H. M., Fingerhut, A., and Janssen, O. E. (2002). Loop variants of the serpin thyroxine-binding globulin: Implications for hormone release upon limited proteolysis. *Biochem. J.* **365,** 311–316.

Hammond, G. L., Smith, C. L., Goping, I. S., Underhill, D. A., Harley, M. J., Reventos, J., Musto, N. A., Gunsalus, G. L., and Bardin, C. W. (1987). Primary structure of human corticosteroid binding globulin, deduced from hepatic and pulmonary cDNAs, exhibits homology with serine protease inhibitors. *Proc. Natl. Acad. Sci. USA* **84,** 5153–5157.

Hammond, G. L., Smith, C. L., Paterson, N. A., and Sibbald, W. J. (1990). A role for corticosteroid-binding globulin in delivery of cortisol to activated neutrophils. *J. Clin. Endocrinol. Metab.* **71,** 34–39.

Harrison, D. G., and Gongora, M. C. (2009). Oxidative stress and hypertension. *Med. Clin. North Am.* **93,** 621–635.

Huber, R., and Carrell, R. W. (1989). Implications of the three-dimensional structure of α_1-antitrypsin for structure and function of serpins. *Biochemistry* **28,** 8951–8966.

Huntington, J. A., Read, R. J., and Carrell, R. W. (2000). Structure of a serpin-protease complex shows inhibition by deformation. *Nature* **407,** 923–926.

Inoue, I., Rohrwasser, A., Helin, C., Jeunemaitre, X., Crain, P., Bohlender, J., Lifton, R. P., Corvol, P., Ward, K., and Lalouel, J. M. (1995). A mutation of angiotensinogen in a patient with preeclampsia leads to altered kinetics of the renin-angiotensin system. *J. Biol. Chem.* **270,** 11430–11436.

Jin, L., Abrahams, J. P., Skinner, R., Petitou, M., Pike, R. N., and Carrell, R. W. (1997). The anticoagulant activation of antithrombin by heparin. *Proc. Natl. Acad. Sci. USA* **94,** 14683–14688.

Jirasakuldech, B., Schussler, G. C., Yap, M. G., Drew, H., Josephson, A., and Michl, J. (2000). A characteristic serpin cleavage product of thyroxine-binding globulin appears in sepsis sera. *J. Clin. Endocrinol. Metab.* **85,** 3996–3999.

Johnson, D. J., and Huntington, J. A. (2003). Crystal structure of antithrombin in a heparin-bound intermediate state. *Biochemistry* **42,** 8712–8719.

Kim, H. S., Krege, J. H., Kluckman, K. D., Hagaman, J. R., Hodgin, J. B., Best, C. F., Jennette, J. C., Coffman, T. M., Maeda, N., and Smithies, O. (1995). Genetic control of blood pressure and the angiotensinogen locus. *Proc. Natl. Acad. Sci. USA* **92,** 2735–2739.

Klieber, M. A., Underhill, C., Hammond, G. L., and Muller, Y. A. (2007). Corticosteroid-binding globulin, a structural basis for steroid transport and proteinase-triggered release. *J. Biol. Chem.* **282,** 29594–29603.

Kobori, H., Nangaku, M., Navar, L. G., and Nishiyama, A. (2007). The intrarenal renin-angiotensin system: From physiology to the pathobiology of hypertension and kidney disease. *Pharmacol. Rev.* **59,** 251–287.

Loebermann, H., Tokuoka, R., Deisenhofer, J., and Huber, R. (1984). Human α_1-proteinase inhibitor. Crystal structure analysis of two crystal modifications, molecular model and preliminary analysis of the implications for function. *J. Mol. Biol.* **177,** 531–556.

McCoy, A. J., Grosse-Kunstleve, R. W., Adams, P. D., Winn, M. D., Storoni, L. C., and Read, R. J. (2007). Phaser crystallographic software. *J. Appl. Crystallogr.* **40,** 658–674.

Mistry, H. D., Wilson, V., Ramsay, M. M., Symonds, M. E., and Broughton Pipkin, F. (2008). Reduced selenium concentrations and glutathione peroxidase activity in pre-eclamptic pregnancies. *Hypertension* **52,** 881–888.

Mottonen, J., Strand, A., Symersky, J., Sweet, R. M., Danley, D. E., Geoghegan, K. F., Gerard, R. D., and Goldsmith, E. J. (1992). Structural basis of latency in plasminogen activator inhibitor-1. *Nature* **355,** 270–273.

Myatt, L. (2010). Review: Reactive oxygen and nitrogen species and functional adaptation of the placenta. *Placenta* **31**(Suppl), S66–S69.

Nagahara, N., Matsumura, T., Okamoto, R., and Kajihara, Y. (2009). Protein cysteine modifications: (1) Medical chemistry for proteomics. *Curr. Med. Chem.* **16,** 4419–4444.

Nguyen, G., Delarue, F., Burckle, C., Bouzhir, L., Giller, T., and Sraer, J. D. (2002). Pivotal role of the renin/prorenin receptor in angiotensin II production and cellular responses to renin. *J. Clin. Invest.* **109,** 1417–1427.

Pemberton, P. A., Stein, P. E., Pepys, M. B., Potter, J. M., and Carrell, R. W. (1988). Hormone binding globulins undergo serpin conformational change in inflammation. *Nature* **336,** 257–258.

Petersen, T. E., Dudek-Wojciechowska, G., Sottrup-Jensen, L., and Magnusson, S. (eds.), (1979). Primary Structure of Antithrombin III: Partial Homology with alpha1-Antitrypsin, Elsevier/North Holland, Amsterdam.

Qi, X., Loiseau, F., Chan, W. L., Yan, Y., Wei, Z., Milroy, L., Myers, R. M., Ley, S. V., Read, R. J., Carrell, R. W., and Zhou, A. (2011). Allosteric modulation of hormone release from thyroxine and corticosteroid binding-globulins. *J. Biol. Chem.* **286**(18), 16163–16173.

Refetoff, S., Murata, Y., Mori, Y., Janssen, O. E., Takeda, K., and Hayashi, Y. (1996). Thyroxine-binding globulin: Organization of the gene and variants. *Horm. Res.* **45,** 128–138.

Schreiber, G. (2002). The evolutionary and integrative roles of transthyretin in thyroid hormone homeostasis. *J. Endocrinol.* **175,** 61–73.

Schreuder, H. A., de Boer, B., Dijkema, R., Mulders, J., Theunissen, H. J., Grootenhuis, P. D., and Hol, W. G. (1994). The intact and cleaved human antithrombin III complex as a model for serpin-proteinase interactions. *Nat. Struct. Biol.* **1,** 48–54.

Schulze, A. J., Baumann, U., Knof, S., Jaeger, E., Huber, R., and Laurell, C. B. (1990). Structural transition of alpha 1-antitrypsin by a peptide sequentially similar to beta-strand s4A. *Eur. J. Biochem.* **194,** 51–56.

Stamler, J. S., Lamas, S., and Fang, F. C. (2001). Nitrosylation. The prototypic redox-based signaling mechanism. *Cell* **106,** 675–683.

Stein, P. E., Tewkesbury, D. A., and Carrell, R. W. (1989). Ovalbumin and angiotensinogen lack serpin S-R conformational change. *Biochem. J.* **262,** 103–107.

Stein, P. E., Leslie, A. G., Finch, J. T., Turnell, W. G., McLaughlin, P. J., and Carrell, R. W. (1990). Crystal structure of ovalbumin as a model for the reactive centre of serpins. *Nature* **347,** 99–102.

Streatfeild-James, R. M., Williamson, D., Pike, R. N., Tewksbury, D., Carrell, R. W., and Coughlin, P. B. (1998). Angiotensinogen cleavage by renin: Importance of a structurally constrained N-terminus. *FEBS Lett.* **436,** 267–270.

Suda, S. A., Gettins, P. G., and Patston, P. A. (2000). Linkage between the hormone binding site and the reactive center loop of thyroxine binding globulin. *Arch. Biochem. Biophys.* **384,** 31–36.

Whisstock, J. C., Skinner, R., Carrell, R. W., and Lesk, A. M. (2000). Conformational changes in serpins: I. The native and cleaved conformations of alpha(1)-antitrypsin. *J. Mol. Biol.* **296,** 685–699.

Zhou, A., Wei, Z., Read, R. J., and Carrell, R. W. (2006). Structural mechanism for the carriage and release of thyroxine in the blood. *Proc. Natl. Acad. Sci. USA* **103,** 13321–13326.

Zhou, A., Wei, Z., Stanley, P. L., Read, R. J., Stein, P. E., and Carrell, R. W. (2008). The S-to-R transition of corticosteroid-binding globulin and the mechanism of hormone release. *J. Mol. Biol.* **380,** 244–251.

Zhou, A., Carrell, R. W., Murphy, M. P., Wei, Z., Yan, Y., Stanley, P. L., Stein, P. E., Broughton Pipkin, F., and Read, R. J. (2010). A redox switch in angiotensinogen modulates angiotensin release. *Nature* **468,** 108–111.

CHAPTER SEVEN

Serpin–Glycosaminoglycan Interactions

Chantelle M. Rein,[*,†] Umesh R. Desai,[‡] *and* Frank C. Church[*,†,§,¶]

Contents

1. Quantitative Methods 106
 1.1. Fluorescence spectroscopy 106
 1.2. Kinetic studies using fluorescence spectroscopy 111
 1.3. Mass spectrometry 113
 1.4. Affinity co-electrophoresis 114
 1.5. Solid-phase binding 114
 1.6. Surface plasmon resonance 116
 1.7. Isothermal titration calorimetry 116
 1.8. Circular dichroism 117
 1.9. NMR 117
 1.10. X-ray crystallography 118
2. Qualitative Methods 121
 2.1. Affinity chromatography 121
 2.2. Electromobility shift assays 121
 2.3. Immunochemical methods 121
 2.4. *In silico* methods 122
3. Animal Models 127
 3.1. Antithrombin 127
 3.2. Heparin cofactor II 128
 3.3. Protein C inhibitor 130
Acknowledgments 131
References 131

* Department of Pathology and Laboratory Medicine, The University of North Carolina at Chapel Hill, Chapel Hill, North Carolina, USA
† Department of Medicine, The University of North Carolina at Chapel Hill, Chapel Hill, North Carolina, USA
‡ Department of Medicinal Chemistry and Institute for Structural Biology and Drug Discovery, Virginia Commonwealth University, Richmond, Virginia, USA
§ Department of Pharmacology, The University of North Carolina at Chapel Hill, Chapel Hill, North Carolina, USA
¶ UNC McAllister Heart Institute, The University of North Carolina at Chapel Hill, Chapel Hill, North Carolina, USA

Methods in Enzymology, Volume 501
ISSN 0076-6879, DOI: 10.1016/B978-0-12-385950-1.00007-9

Abstract

Serpins (serine protease inhibitors) have traditionally been grouped together based on structural homology. They share common structural features of primary sequence, but not all serpins require binding to cofactors in order to achieve maximal protease inhibition. In order to obtain physiologically relevant rates of inhibition of target proteases, some serpins utilize the unbranched sulfated polysaccharide chains known as glycosaminoglycans (GAGs) to enhance inhibition. These GAG-binding serpins include antithrombin (AT), heparin cofactor II (HCII), and protein C inhibitor (PCI). The GAGs heparin and heparan sulfate have been shown to bind AT, HCII, and PCI, while HCII is also able to utilize dermatan sulfate as a cofactor. Other serpins such as PAI-1, kallistatin, and α_1-antitrypsin also interact with GAGs with different endpoints, some accelerating protease inhibition while others inhibit it. There are many serpins that bind or carry ligands that are unrelated to GAGs, which are described elsewhere in this work. For most GAG-binding serpins, binding of the GAG occurs in a conserved region of the serpin near or involving helix D, with the exception of PCI, which utilizes helix H. The binding of GAG to serpin can lead to a conformational change within the serpin, which can lead to increased or tighter binding to the protease, and can accelerate the rates of inhibition up to 10,000-fold compared to the unbound native serpin. In this chapter, we will discuss three major GAG-binding serpins with known physiological roles in modulating coagulation: AT (*SERPINC1*), HCII (*SERPIND1*), and PCI (*SERPINA5*). We will review methodologies implemented to study the structure of these serpins and those used to study their interactions with GAG's. We discuss novel techniques to examine the serpin–GAG interaction and finally we review the biological roles of these serpins by describing the mouse models used to study them.

1. Quantitative Methods

1.1. Fluorescence spectroscopy

Fluorescence spectroscopy is an important technique in the study of serpin–GAG (glycosaminoglycan) interactions because of its high sensitivity, relative ease of applicability, and wide range of fluorescence labels and probes. Steady-state fluorescence spectroscopy, which relies on the change of quantum yield, extinction coefficient, emission wavelength, rotational mobility, or energy transfer, is the most exploited technique for monitoring ligand binding to proteins. The serpin–GAG system most commonly studied using fluorescence spectroscopy is antithrombin (AT)–heparin, for which the intrinsic serpin fluorescence increases $\sim$30–40% at saturation (Fig. 7.1). This fluorescence change forms the staple of most thermodynamic and kinetic studies involving AT. For affinity measurements, a

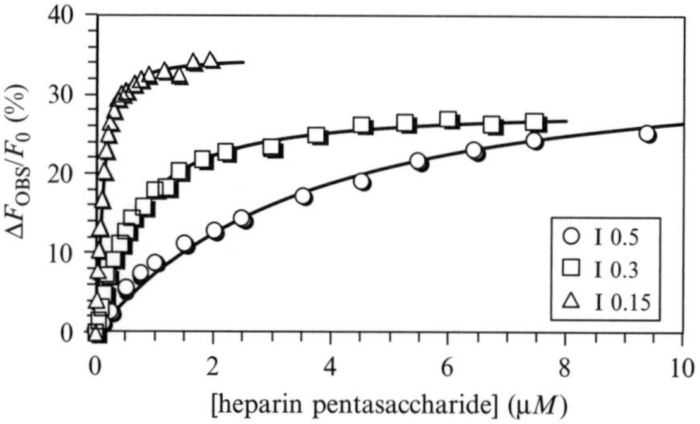

Figure 7.1 Fluorescence titration of heparin pentasaccharide binding to human AT in 20 m*M* sodium phosphate buffer, pH 7.4, containing 0.1 m*M* EDTA, 0.1% (w/v) PEG8000, and 100 m*M* NaCl (Δ), 250 m*M* NaCl (□), or 450 m*M* NaCl (○) at 25 °C. The antithrombin concentration in these titrations was 0.09 μ*M* (Δ), 0.30 μ*M* (□), or 0.92 μ*M* (○). Solid lines represent fits to the data using Eq. (7.1) assuming a stoichiometry of binding $n = 1$.

solution of AT (~10–500 n*M*) in an appropriate buffer is titrated with small aliquots of highly concentrated solution of the GAG, preferably in the same buffer, followed by measuring the intensity of light emitted at a specific wavelength between 330 and 350 nm ($\lambda_{EX} \sim 280$ nm). The change in serpin fluorescence (ΔF_{OBS}) at a fixed wavelength of emission reaches a plateau at high enough GAG concentration, which can be fitted by quadratic binding Eq. (7.1) to calculate the equilibrium dissociation constant K_D of serpin (P)–GAG (ligand L) complex, "n" the apparent stoichiometry of binding, and maximal change in fluorescence as a ratio of the initial fluorescence ($\Delta F_{MAX}/F_0$).

$$\frac{\Delta F_{OBS}}{F_0} = \frac{\Delta F_{MAX}}{F_0} \frac{n[P]_0 + [L]_0 + K_D - \sqrt{\left(n[P]_0 + [L]_0 + K_D\right)^2 - 4n[P]_0[L]_0}}{2[P]_0} \tag{7.1}$$

The K_D of interaction is most accurately measured in titrations where the serpin concentration is in the range of the affinity, while the stoichiometry of binding "n" is better determined when [P] is approximately 10- to 100-fold the K_D (Nordenman *et al.*, 1978). For AT–full-length heparin interaction under physiological conditions, these concentrations would be ~10 and ~500 n*M* for measurement of K_D and n, respectively. Care should be taken to ensure that the GAG titrant is transparent to the incident or emitted light, that serpin dilution following titration is a maximum of 10%, that serpin or GAG adsorption to cuvette surface is minimal, and that mixing does not induce serpin denaturation resulting in light scattering.

Fluorescence-based affinity measurement relies on environmental changes introduced on one or more tryptophans of the serpin by GAG binding. For AT, Trp^{49}, Trp^{189}, Trp^{225}, and Trp^{307} contribute 8%, 10%, 19%, and 63%, respectively, of the total fluorescence (Meagher *et al.*, 1998). Of these, Trp^{225} and Trp^{307} account for the majority of fluorescence enhancement induced by heparin binding most probably arising from the movement of helices D and H. Since the discovery of heparin-induced enhancement in AT's fluorescence in the late 1970s (Einarsson, 1978; Einarsson and Andersson, 1977; Nordenman *et al.*, 1978), the method has been extensively exploited for measuring the affinity of various forms of heparin under many conditions. In fact, it was the major tool in the identification of heparin residues critical for AT binding (Atha *et al.*, 1985, 1987; Desai *et al.*, 1998b). The affinities measured using this technique have ranged from ~10 n*M* to ~200 μ*M*, suggesting a dynamic range of ~4–5 log units.

Heparin oligosaccharides induce varying fluorescence enhancements in AT depending on the microscopic structure of the variants. Whereas pentasaccharide DEFGH (Fig. 7.2) induces 30% enhancement, a variant lacking the critical 3-*O*-sulfate group of residue F induced less than 5% ΔF_{MAX}. This raises the possibility of linear correlation between extent of intrinsic fluorescence enhancement and conformational activation of AT. For example, low-affinity heparin induces a ΔF_{MAX} of only 8% and activates the serpin only 60-fold (Streusand *et al.*, 1995). Likewise, under appropriate conditions 3-*O*-desulfated heparin pentasaccharide shows 25–30% fluorescence enhancement and almost unaffected conformational activation (Richard *et al.*, 2009). Yet, these correlations are likely to be approximate and not strictly linear.

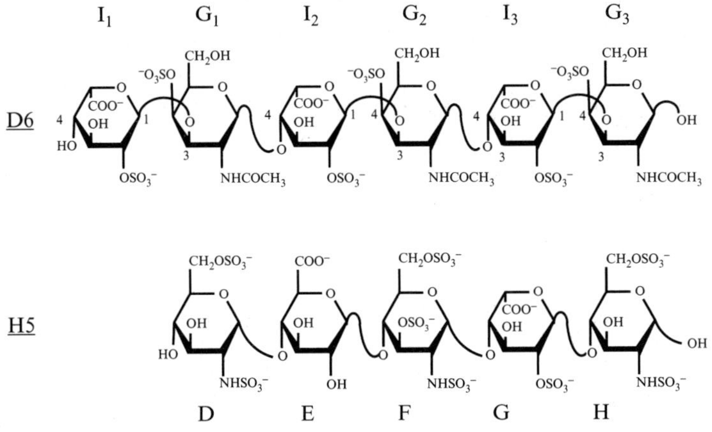

Figure 7.2 Structures of DS hexasaccharide D6, which binds to HCII, and heparin pentasaccharide H5, which binds to AT.

Other examples of intrinsic tryptophan fluorescence-based affinity measurement include low molecular weight heparins binding to plasma AT (Lin *et al.*, 2001; Liu *et al.*, 2007), full-length heparin and heparin pentasaccharide binding to AT mutants (Jairajpuri *et al.*, 2003; Monien *et al.*, 2005; Schedin-Weiss *et al.*, 2002a), AT glycoforms (Olson *et al.*, 1997), and covalent AT–heparin complex binding to free AT (Berry *et al.*, 1998). Each of these systems behaves in a manner similar to the plasma AT–heparin system. Some variances are possible as displayed by $Ser^{380} \rightarrow$ Trp AT, which exhibits a 17 nm red shifted λ_{EM} of 354 nm following heparin binding (Huntington *et al.*, 1996). It is also important to remember that a GAG may possess poor affinity for the target serpin, necessitating higher concentrations of the serpin in titrations for good signal-to-noise ratio. For example, titrations of low-affinity heparin with plasma AT were performed using 10 μ*M* inhibitor. In such circumstances, it may be necessary to use an excitation wavelength (e.g., 295 nm) that obviates inner filter corrections (Streusand *et al.*, 1995). Another alternative is to use a small path length on the excitation side or to introduce appropriate inner filter correction (Henry *et al.*, 2009). Finally, heparin binding to AT induces a blue shift of 15 nm in Trp^{49} fluorescence and a red shift of 5 nm in Trp^{225} fluorescence (Meagher *et al.*, 1998). Fluorescence lifetime changes for Trp^{225} and Trp^{307} are also observed upon heparin binding. However, these shifts in emission maxima or lifetime changes of individual tryptophans have not yet been exploited for AT thermodynamic or kinetic studies.

For situations where GAG binding does not affect the intrinsic fluorescence, extrinsic probes may be used. A particularly useful probe is TNS (2-(*p*-toluidinyl)-naphthalene-6-sulfonic acid). TNS is a hydrophobic, environment sensitive probe that binds weakly to AT ($K_D > 125$ μ*M*) to form a highly fluorescent AT–TNS complex with λ_{EM} of ~432. Heparin binding to AT results in 50–90% decrease in fluorescence at 432 nm and a blue shift of ~3 nm (Meagher *et al.*, 2000). A small amount of TNS (~5–40 μ*M*) is used to provide a good fluorescence signal at 432 nm and the GAG is titrated in a manner similar to the intrinsic fluorescence titration experiment. The resultant fluorescence change as a function of ligand concentration is analyzed using Eq. (7.1) to calculate the affinity of interaction. This protocol was adopted for heparin binding to AT mutants devoid of amino acids 134–137 (Meagher *et al.*, 2000). Likewise, the affinities of cleaved and latent AT for heparin and heparin pentasaccharide have been measured using TNS as extrinsic probe (Schedin-Weiss *et al.*, 2008).

Another extrinsic probe used for studying AT–GAG interaction is NBD (*N*,*N'*-dimethyl-*N*-acetyl-*N'*-(7-nitrobenz-2-oxa-1,3-diazol-4-yl) ethylenediamine). NBD is a hydrophobic, covalently bound probe for site-specific attachment to reactive cysteine residues on proteins. When introduced at the P1 position of $Arg^{393} \rightarrow$ Cys AT, the fluorescence of NBD at 542 nm shows dramatic increases of 32–105% upon interaction with dextran sulfate, low-affinity heparin and full-length heparin (Futamura *et al.*, 2001) corresponding to significant differences in the microenvironment of the exposed reactive center loop (RCL) following conformational change. Likewise, the fluorescence of the dansyl group (5-dimethylaminonapthalene-1-sulfonyl-) has also been used for AT–heparin affinity studies (Piepkorn, 1981; Piepkorn *et al.*, 1980). The interaction of dansylated heparin with AT results in an ~80% increase in fluorescence at ~510 nm, which serves as an excellent quantitative monitor of interaction.

The interaction of GAGs with other serpins has been less studied in comparison to that with AT, although GAG binding most probably induces some conformational changes in other serpins. For example, the intrinsic tryptophan fluorescence change induced by heparin binding to heparin cofactor II (HCII) is almost zero, while that induced by dermatan sulfate (DS) is −13% (Liaw *et al.*, 1999). This implies that differential changes are induced in HCII by DS and heparin. For a more accurate description of these differences, dansylated HCII (λ_{EM} = 520 nm; λ_{EX} = 335 nm) has been used (Liaw *et al.*, 1999). In an alternative protocol, the binding of heparin to HCII was monitored by TNS fluorescence (Schedin-Weiss *et al.*, 2008). The HCII–TNS system displayed some interesting characteristics. The λ_{EM} of HCII–TNS complex was found to be 448 nm, rather than 432 nm as found for AT (λ_{EX} = 330 nm). More interestingly, heparin chains between 8 and 12 units long displayed a 40–70% increase in TNS fluorescence and 1.5 nm blue shift, while chains 14 units and longer caused a 30–70% fluorescence quench and 7.5 nm red shift. Smaller oligosaccharide chains (<8-mer) did not affect HCII–TNS fluorescence. The chain length dependent change in TNS fluorescence for HCII appears to have no other parallels in GAG–serpin interaction. Other examples of the use of TNS include protein C inhibitor (PCI), which displayed a 39% quench at λ_{EM} of 448 nm with a 2–3 nm blue shift upon heparin binding (Li *et al.*, 2007). Cleaved PCI, on the other hand, showed only a 2% decrease in TNS fluorescence following heparin binding.

Although the above discussion may suggest that intrinsic fluorescence is not of much use for serpins other than AT, this is not necessarily the case. Squamous cell carcinoma antigens (SCCAs) interact with heparin (Higgins *et al.*, 2010). The binding of the two isoforms, SCCA-1 and SCCA-2, to heparin is associated with nearly 80–100% quench in the fluorescence of four tryptophans from which affinities of 4.2 and 2.0 μ*M*, respectively, were calculated. By contrast, the intrinsic fluorescence increases ~25% upon heparin binding to protease nexin-1 (PN1) (Arcone *et al.*, 2009). Yet it is difficult to accurately measure the K_D of interaction because the affinity is too high (~0.8 n*M*; Rovelli *et al.*, 1992) for fluorescence titrations. The sub-nanomolar affinity implies that concentrations used in experiments measurement would be in the stoichiometric range, rather than in the K_D range. This is an important limitation of fluorescence experimentation.

1.2. Kinetic studies using fluorescence spectroscopy

The enhancement in intrinsic fluorescence of AT following heparin binding has resulted in a major effort to understand the kinetics of GAG interaction with serpins at an atomic level. The 30–40% increase in fluorescence can be easily followed in a stopped-flow fluorometer to measure the rate of GAG binding to AT. In most stopped-flow instruments, solutions of AT and GAG are rapidly mixed in a microcell while continuously monitoring the light emitted using a filter of 300 nm (λ_{EX} = 280 nm). Assuming that the dead time of the instrument is small ($\leq$1 ms), reaction rates of the order of 500 s^{-1} can be measured with good precision. Typically, more than 10 reaction profiles are collected and averaged to improve the signal-to-noise ratio. Under pseudo-first-order conditions ([H] $\gg$ [AT]), the increase in intrinsic fluorescence follows a typical first-order profile (Fig. 7.3A), as observed for pentasaccharide DEFGH and full-length heparin, which can be computer fit by a single exponential equation to derive the observed rate constant of interaction (k_{OBS}) (Olson *et al.*, 1981). Some profiles may be more accurately fit using a double exponential equation, as

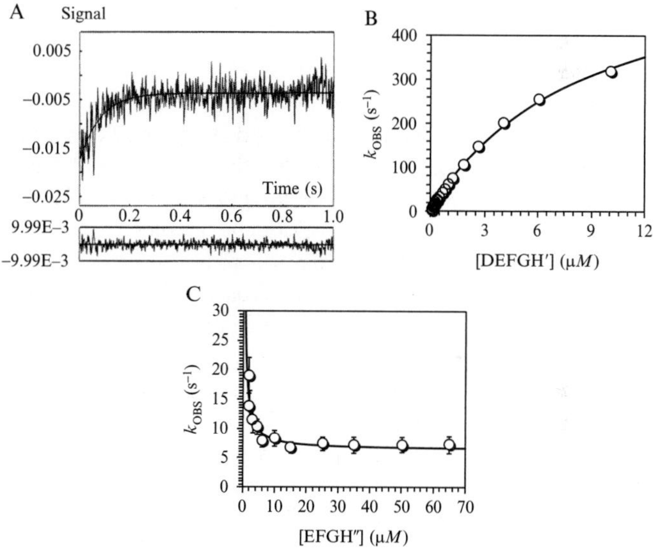

Figure 7.3 Stopped-flow fluorimetry based rapid kinetics of heparin oligosaccharides binding to AT. (A) A typical profile of increase in intrinsic fluorescence following rapid mixing of AT and oligosaccharide under pseudo-first-order conditions. Multiple profiles are averaged and fitted by a single (e.g., DEFGH′) or double (e.g., EFGH″) exponential equation to derive the observed rate constant (k_{OBS}) of interaction. (B, C) Oligosaccharide concentration dependence of k_{OBS} for the binding of pentasaccharide DEFGH′ (B) or tetrasaccharide EFGH″ (C) to AT at 25 °C. Solid lines are fits of the data by Eqs. (7.2) and (7.3), respectively.

observed with tetrasaccharide EFGH″ (Desai *et al.*, 1998b), which will give two observed rate constants.

For most heparin variants, the k_{OBS} of interaction with AT initially increases linearly with the concentration of ligand, followed by a gradual approach to a limiting rate (Fig. 7.3B). For others (e.g., tetrasaccharide EFGH″; Desai *et al.*, 1998b), the k_{OBS} decreases with increasing ligand concentration and reaches a constant rate (Fig. 7.3C), while for 3-*O*-desulfated pentasaccharide, the k_{OBS} first decreases and then increases with ligand concentration (Richard *et al.*, 2009). These profiles can be analyzed on the basis of a two-step, conformational activation process (Scheme 7.1). In the case of DEFGH, the recognition of AT in its native state results in a low-affinity AT–H complex that rapidly undergoes a major conformational change to give the activated AT⋆–H complex through the induced conformational activation pathway. Alternatively, a two-step, pre-equilibrium pathway may also occur with some ligands, for example, EFGH″ and 3-*O*-desulfated pentasaccharide (Richard *et al.*, 2009). In this pathway, the heparin variant binds specifically to the activated state of the serpin (AT⋆), rather than the native state, that exists in pre-equilibrium with the native state. This specific binding to AT⋆ drags the preexisting equilibrium toward the same AT⋆–H complex observed in the induced conformational activation pathway. The overall kinetics of heparin binding to AT may be described by either Eq. (7.2) or (7.3) or a combination of the two for simultaneous operation of both pathways (Richard *et al.*, 2009). In either pathway, it is assumed that the steps involved in binding are governed by rapid equilibrium.

$$k_{OBS} = \frac{k_2 \times [H]_0}{[H]_0 + K_1} + k_{-2} \tag{7.2}$$

$$k_{OBS} = \frac{k_{-3} \times K_4}{[H]_0 + K_4} + k_3 \tag{7.3}$$

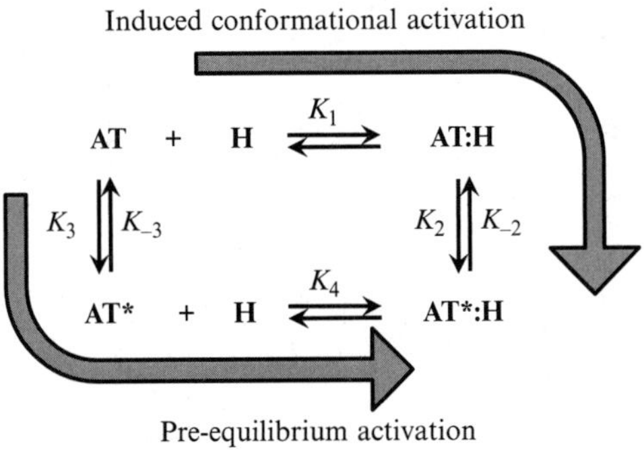

Scheme 7.1 Heparin-induced conformational activation process of AT.

Fluorescence-based stopped-flow fluorimetry has the capability to derive the microscopic parameters K_1, K_4, k_2, k_3, k_{-2}, and k_{-3} that fully describe the kinetics of GAG binding to serpin, although appropriate simplifications may have to be made to account for the interdependence of the parameters and the possibility of inadequate data for either of the two pathways. Using this technique, the role of pentasaccharide DEFGH in AT activation was deduced (Olson *et al.*, 1992). Likewise, the technique was also particularly useful in identifying the roles of individual residues of the DEFGH sequence (Desai *et al.*, 1998a,b,a). Further, site-specific AT mutants have been studied to elucidate the roles of several heparin-binding residues including Arg^{129} (Desai *et al.*, 2000), Lys^{125} (Schedin-Weiss *et al.*, 2002b), Lys^{114} (Arocas *et al.*, 2001), Lys^{11} (Schedin-Weiss *et al.*, 2004), Arg^{13} (Schedin-Weiss *et al.*, 2004), and Trp^{49} (Monien *et al.*, 2005).

The extensive use of stopped-flow fluorimetry to study AT–heparin interaction has led to its application to the corresponding HCII–heparin system. Yet, as described above, intrinsic HCII fluorescence does not provide a convenient signal for monitoring the interaction. Hence, the mechanism of heparin activation of HCII was elucidated using the change in fluorescence of TNS bound to the serpin (O'Keeffe *et al.*, 2004). The applicability of fluorescence-based stopped-flow fluorimetry to these two systems indicates that it is likely to be a powerful tool for investigation of the kinetics of GAG binding to other serpins.

1.3. Mass spectrometry

Both structural and dynamic properties of serpin–ligand interactions can be studied using mass spectrometry techniques. Hydrogen/deuterium exchange has been a useful technique for studying protein conformations in that amide hydrogens can be used to investigate solvent accessibility, protein lability, and secondary structure of the protein. In this technique, the protein backbone amide hydrogens are replaced with deuterium from a solvent containing D_2O. This exchange is dependent upon the accessibility of the hydrogen atoms dictated by the protein structure. Amide hydrogens that are surface exposed will exchange for deuterium more quickly that those buried within the protein or those that are hydrogen bonded (Busenlehner and Armstrong, 2005). Mass spectrometry is employed and the rate of exchange can be measured for the entire protein and can provide clues as to solvent accessibility, secondary structure, and protein stability. In a recent study, hydrogen/deuterium (H/D) exchange and hydroxyl radical-mediated footprinting were used to study the differences seen between X-ray crystal structures and dynamic protein conformations of serpins (Zheng *et al.*, 2008). This method has been used to study the conformational dynamics of α_1-antitrypsin (Tsutsui *et al.*, 2006) and has also been applied to studying the conformational dynamics of AT complexed with heparin (Boucher *et al.*, 2009).

Hydroxyl-mediated protein footprinting is a more recently developed technique to assess protein structure, assembly, and conformational change in solution. This technology is based on measurements of reactivity of amino acid side-chain groups (Takamoto and Chance, 2006). Hydroxyl radicals can be generated in a number of ways for these types of studies including Fenton reagent, electrical discharge or in the case of the above study, synchrotron footprinting. In this method, water will radiohydrolyze in response to X-rays, producing hydroxyl radicals, and these radicals then react with accessible side chains of proteins in solution to form stable oxidative modifications. Proteins are proteolytically degraded and subjected to quantitative liquid chromatography coupled MS and tandem MS. The results are especially helpful in examining conformational changes of proteins due to ligand binding. The combination of these two methods was used to compare static crystallographic data with dynamic data in α1-antitrypsin in solution and demonstrate novel structural features of this serpin, and will be useful in future studies on the structure of other ligand-bound serpins (Zheng *et al.*, 2008).

1.4. Affinity co-electrophoresis

Affinity co-electrophoresis is a technique developed for investigating the interaction between proteoglycans or GAGs and proteins and for determining the specificity of these reactions (Lee and Lander, 1991). This method involves the radiolabeling of a GAG of interest and subsequent electrophoresis in a nondenaturing agarose gel with or without differing concentrations of serpin. The electrophoretic pattern is visualized and allows for the determination of an apparent retardation coefficient. The dissociation constant (K_d) can be determined due to the dependence of the retardation coefficient on protein concentration. Affinity co-electrophoresis has many advantages over other methods looking at GAG binding, including the use of trace quantities of the interacting molecules, ability to study behaviors of native proteins and of radiolabeled GAGs that can be essentially unmodified, measurement of strengths of binding even for relatively weak interactions characteristic of GAG–serpin interactions, detection of protein binding heterogeneity in a GAG, and the ease of performing the experiment at relatively low costs (San Antonio and Lander, 2001). This method has been used to determine the K_d for heparin and DS binding to HCII, and for examining the binding of low- and high-affinity heparin molecules to AT (Lee and Lander, 1991; Shirk *et al.*, 2000).

1.5. Solid-phase binding

The interaction between serpins and their ligands has commonly been studied using solid-phase binding techniques. One binding partner is usually immobilized on a solid support such as beads, and the other binding partner is added in solution and equilibrium is established. Following separation of the solid phase from the solution phase, usually by centrifugation in the case of beads, the solid phase is washed to remove any unbound solution protein and the amount of solution protein remaining on the solid phase is measured. There are several techniques used to facilitate the quantitation of bound solution species. First, the solution species can be radiolabeled and the amount of radioactivity remaining in the solid phase can be measured. Alternatively, the solution species can also be fluorescently labeled using readily available labeling reagents and the amount of fluorescence in the solid fraction measured spectroscopically. Additionally, a fluorescently labeled or enzyme-conjugated antibody to the solution species may be used in the case that the solution species is a protein and immunoassays may be used for detection. The advantage of the final two techniques is the avoidance of the use of radioactive isotopes, and the use of solid-phase binding methods is popular do to the ease of which they are done. However, the requirement for immobilization of one of the molecules of interest can lead to problems if the exact orientation of the immobilized species is not known or is not uniform. The immobilization can affect the accessibility of the binding site for its binding partner or the overall conformation of the molecule. Although the first hurdle can be overcome by site-specific attachment of the species, it does not account for any conformational change induced by attachment. However, this method has remained popular and has been used recently to examine the binding of differentially sulfated heparin mimetic saccharides to AT and HCII (Kim and Kiick, 2007).

A variation of the solid-phase method that has been used extensively to study the affinity of serpin–ligand interactions employs a competitive binding of increasing concentrations of solution-phase species. Following binding of the solution-phase species to the solid-phase species, addition of increasing concentrations of the solid-phase species will dissociate the complex of immobilized-solution. This method allows for the quantitation of both specific and nonspecific binding interactions between binding partners. For example, AT bound to heparin immobilized on agarose was analyzed by displacing the AT bound to immobilized heparin with solution-phase heparin (Olson *et al.*, 1991). More recently, this technique was used to examine the binding of the heparin pentasaccharide, low-affinity heparin, and high-affinity heparin to latent and cleaved forms of AT. This study used heparin-Sepharose in order to bind the two forms of AT and monitored the displacement of AT from this matrix by heparin competitors, showing that the pentasaccharide and high-affinity heparin bound to both latent and cleaved AT with higher affinity than low-affinity heparin (Schedin-Weiss *et al.*, 2008).

1.6. Surface plasmon resonance

Surface plasmon resonance is a technique that has been used to study the specificity, rates, and affinities of intermolecular interactions. This technique also uses an immobilized species, but this species is immobilized onto a biosensor chip either directly or indirectly using antibodies or other coupling methods. A solution species is then flowed over the chip and changes in the refractive index at the surface of the chip, which is measured as a change in the intensity and angle of light reflected from the chip's surface due to the binding of the two species, is analyzed in real time. Competition assays with other possible binding partners or mutant binding partners can also be performed by flowing a competing solution species over the chip and looking for changes in refractive index, and dissociation of complex formation can be studied by flowing solution lacking interaction species over the chip. In this way, the kinetics, affinity, and thermodynamics of binding can be investigated in a single assay. However, using this technique does not solve the problem of immobilization of one of the species, although many new coupling chemistries have been developed that allow for immobilization in a site-specific manner with more ease than was once possible. Surface plasmon resonance has been used extensively to characterize the serpin–ligand interaction. The effects of calcium on the heparin/HCII interaction were studied using a heparin biochip (Zhang *et al.*, 2004). Additionally, the kinetics of thrombin binding to AT in the presence or absence of heparin were examined using biochips coated with antibodies against either thrombin or AT for immobilization (Elg and Deinum, 2002).

1.7. Isothermal titration calorimetry

Isothermal titration calorimetry is a thermodynamic method that measures changes in heat released or absorbed during a molecular binding event and is therefore a useful tool in measuring the interaction of serpins with GAGs. Measurement of the change in heat allows for the determination of binding parameters, including enthalpy (ΔH), entropy (ΔS), stoichiometry, Gibbs free energy changes (ΔG), and binding constants. Thermodynamic data provide insight into conformational changes, hydrophobic interactions, and charge–charge interactions, which are of particular interest in the binding of GAGs. The use of ITC allows for generation of true binding affinities that are often not able to be obtained using surface plasmon resonance or solid-phase binding methods. Additionally, all reactions occur in the solution phase, negating the need for immobilization and possible experimental consequences. This method has been used extensively in the past to study AT binding to high-affinity heparin species containing the heparin pentasaccharide (Toida *et al.*, 1996) and has more recently been used to study the effects of new heparin mimetics that were created to bind to AT (Kim and Kiick, 2007) and to determine the effect of extending the GAG chain of the heparin pentasaccharide on serpin binding (Guerrini *et al.*, 2008).

1.8. Circular dichroism

Circular dichroism (CD) spectroscopy measures differences in absorption of left-handed polarized light versus right-handed polarized light that occur due to conformational change of a macromolecule. CD is able to provide information about the bonds and structure of a protein and how these properties change once a ligand is bound by measuring the CD spectrum of a protein and determining how the spectra changes once a ligand is bound. The CD spectrum of this ligand–protein interaction depends on the CD spectrum of the ligand itself, and its intensity is determined by the strength of its binding to a protein. Thus, CD has been useful in studying how the interaction of serpins with GAGs can alter the structure of the serpin, particularly the binding of reactive loop peptides to sheet A of the serpin (Schulze *et al.*, 1992). While this technique has been used to study the interaction of many different polysaccharides with serpins, a problem arises in that this technique has failed to take into account the motion and spectroscopic properties of the GAG itself in the contribution to protein structural changes (Rudd *et al.*, 2008). This problem is particularly important to address when describing the interaction between GAGs such as heparan, HS, and DS and serpins. A recent study has shown that the CD spectra of heparin and heparin derivatives vary due to their carbohydrate pattern and conformation (Rudd *et al.*, 2007). Therefore, a more refined CD method known as vibration circular dichroism (VCD), which takes into account the individual spectra of the polysaccharides, has been used to investigate the interaction of heparin with several nonserpin proteins and can be used in future studies looking at GAG–serpin interactions (Rudd *et al.*, 2009).

1.9. NMR

NMR has been a very useful tool in the study of the binding of GAGs to serpins. Solution state NMR has become a widely accepted method for studying the structural changes of proteins after complexing with ligands in the solution phase and has yielded considerable amounts of data on the binding parameters of polysaccharides to proteins. By monitoring the changes in NMR properties such as chemical shifts, nuclear Overhauser effect (NOE), and transverse relaxation rate in the free- and ligand-bound state, conformational changes in the protein can be examined (Clarkson and Campbell, 2003). Two-dimensional NMR techniques can also be used to map the ligand binding site of a protein. While NMR experiments give a relatively detailed picture of the interaction between heparin and protein, several drawbacks exist. This technique requires milligram quantities of interacting species at high concentrations, raising solubility problems and making it difficult to accurately determine the association constant (K_a) (Capila and Linhardt, 2002). Augmenting the results obtained with NMR spectroscopy with other techniques such as CD will provide further details into the changes in both the serpin and the ligand upon formation of the serpin–ligand complex.

1.10. X-ray crystallography

A widely used method of examining ligand binding sites in a serpin and studying the conformational changes induced by ligand binding is comparison of the crystal structures of the unbound and ligand-bound forms of the serpin. In the case of AT, the binding of heparin is required for its biological function as an inhibitor of coagulation serine proteases. In the native state, the RCL of AT is partially inserted into β-sheet A, but upon heparin binding the loop is expelled. This expulsion coupled with the ensuing conformational change in the molecule increases the affinity of AT for its target proteases, such as factor Xa. When AT is crystallized in the presence of heparin, helix D becomes elongated and a new helix I is formed, causing conformational change throughout the molecule ultimately leading to the expulsion of the RCL (Pearce *et al.*, 2007). When the RCL is cleaved upon interaction with the target protease, it is incorporated into β sheet A, leading to global conformational changes. In the past decade, many new crystal structures have been solved for the GAG-binding serpins, which has allowed much more insight into the mechanisms governing serpin activity. Recent crystal structures of AT have revealed that the native form in the monomeric state has a different RCL conformation (Fig. 7.4A), including a novel salt bridge, that was not previously seen in the dimeric form (Carrell *et al.*, 1994). The RCL showed a 20° shift and made contact with the body of AT, including a salt bridge between Arg^{393} and Glu^{237} (Johnson *et al.*, 2006). This structure more closely represents the circulating form of AT and explains why native AT is a poor inhibitor of coagulation proteins. Moreover, the X-ray crystal structures of several binary and ternary complexes involving AT and heparin have recently been solved (Fig. 7.4A). Although the native conformation and the high-affinity heparin bound state of AT were characterized several years ago (Jin *et al.*, 1997), the mechanism of how the conformational change affects the affinity for heparin has just recently been elucidated. A crystal structure of a heparin pentasaccharide-AT intermediate has been characterized in which the complex has undergone all activating conformational changes except for expulsion of the RCL and elongation of helix D. This crystal structure has demonstrated that the

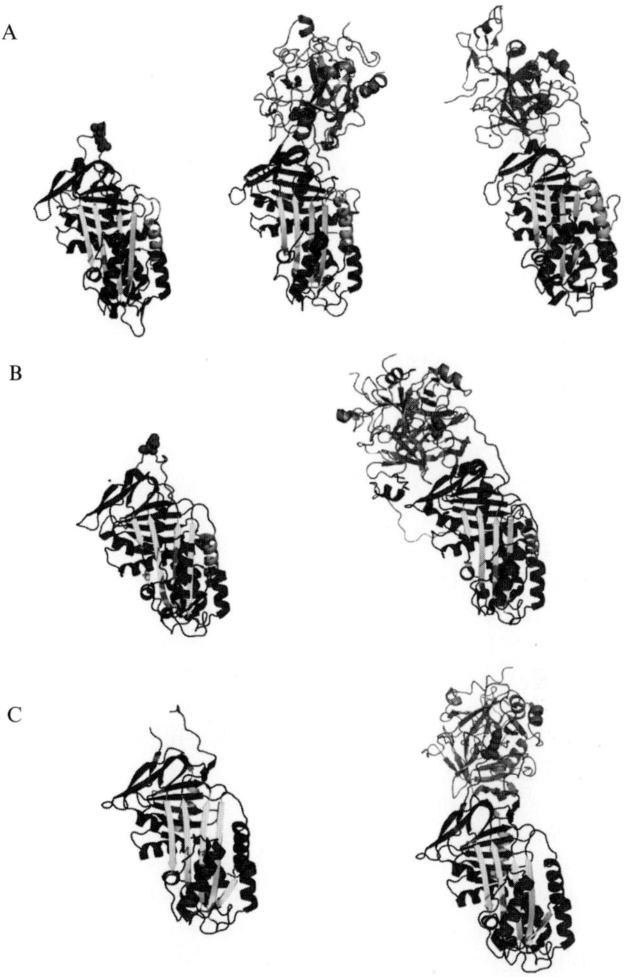

Figure 7.4 Native and serpin–GAG–protease complex structures. (A) Native AT (1E04), AT-heparin–thrombin (1TB6), and AT–heparin–factor Xa (2GD4) complexes shown as ribbon diagrams. (B) Native HCII (1JMJ) and HCII–heparin–thrombin (1JMO) complexes shown as ribbon diagrams. (C) Native PCI (2HI9) and PCI–heparin–thrombin (3B9F) complexes shown as ribbon diagrams. In all models, serpins are shown in blue with P1 residues as spheres, the RCL in magenta, regions involved in GAG binding in green, and β sheet A in yellow. Thrombin heavy and light chains are shown in orange and factor Xa is shown in red. (For interpretation of the references to color in this figure legend, the reader is referred to the Web version of this chapter.)

basis of the high-affinity heparin bound state of AT is not due to an enhanced interaction with the pentasaccharide, but instead a lower global free energy due to conformational changes elsewhere in the molecule

(Johnson and Huntington, 2003). These conformational changes upon pentasaccharide binding leading to a decrease in the global free energy were later determined to be caused by hinge region expulsion, the last step of the activation pathway of AT by heparin (Langdown *et al.*, 2009). A number of conformational intermediates involving heparin, AT, and thrombin as the target protease have been studied using X-ray crystallography. These intermediates provide evidence for a bridging mechanism where heparin is able to make contact with residues in both thrombin and AT, and highlight the importance of Arg and Lys residues in AT and the heparin-binding exosite of thrombin (Dementiev *et al.*, 2004; Li *et al.*, 2004). Additionally, a recent study has determined the crystal structure of the Michaelis complex of factor IXa with heparin-activated AT (Johnson *et al.*, 2010). Since AT inhibition of factor IXa is dependent upon exosite interactions (Langdown *et al.*, 2004), characterization of these interactions is important in understanding how AT recognizes factor IXa as a target protease. This study reveals why the heparin-induced conformational change in AT is required to allow active site and exosite interactions with factor IXa and the nature of the interactions. AT bound to the pentasaccharide in the canonical fashion leading to RCL expulsion and complementary exosites on AT and factor IXa stabilized the complex. These exosites explain the requirement for heparin-binding AT for in AT-mediated factor IXa inhibition. Surprisingly, a role for heparin as a bridge between AT and factor IXa was also suggested by these results.

X-ray crystallography studies have also shed light on the novel mechanism of HCII inhibition of thrombin. HCII is known to bind both heparin and DS, and this binding increases the affinity of HCII for its protease substrate, thrombin. The crystal structures of native and S195A thrombin complexed with HCII (Fig. 7.4B) have shown that in the absence of heparin or DS, the unique N-terminal acidic domain of native HCII is sequestered by interaction with the body of HCII (Baglin *et al.*, 2002). Upon GAG binding, conformational changes occur within HCII, including the expulsion of the RCL and release of the acidic domain. The free acidic domain is then able to bind to exosite 1 of thrombin, bringing the active site of thrombin in closer proximity to the RCL. Further, the GAG provides a bridging function in the interaction between thrombin and HCII, in a similar manner to that of heparin with thrombin and AT.

The crystal structures of PCI have provided much useful information on the unique structural elements found within this serpin. While all other members of the serpin family use basic residues within helix D for heparin binding, PCI is known to utilize an extended, nonlinear binding site with residues within helix H and a basic N-terminal extension (Huntington *et al.*, 2003; Phillips *et al.*, 1994), although recent evidence suggests that neither the N-terminus nor the residues adjacent to helix H play a role in heparin binding in PCI (Li *et al.*, 2007). Additionally, heparin acceleration of target protease inhibition by PCI, namely APC and thrombin, is thought to occur via a bridging mechanism (Li *et al.*, 2007, 2008), and not by an allosteric mechanism as with other serpins. Finally, these studies have shown a role for the elongated and highly flexible RCL of PCI (Fig. 7.4C) in that the ternary complex can only be formed by the alignment of the heparin-binding sites of both thrombin and PCI through exosite contacts and by exploiting the extra long RCL region (Li *et al.*, 2008).

2. Qualitative Methods

2.1. Affinity chromatography

Affinity chromatography is a useful tool for confirming whether or not a postulated binding partner actually interacts with a protein. This technique is performed by immobilizing the ligand, in this case a GAG, to a matrix of agarose or sepharose and packed into a column. A solution containing the protein of interest, here a serpin, is then applied to the column and if the protein interacts with the ligand, it will be retained in the column after elution with a physiological buffer. This technique has been widely used to show interactions between heparin and AT, HCII, and PCI (Gettins *et al.*, 1996). It has also been used more recently to identify and characterize new serpin family members (Majdoub *et al.*, 2009; Mansour *et al.*, 2010).

2.2. Electromobility shift assays

Binding of a ligand to a protein can easily be visualized by electrophoresis of a mixture of the two binding partners on a nondenaturing gel. Changes in the mobility of either the protein or the serpin to a higher molecular weight complex are indicative of molecular interaction. This technique has been used to show binding of reactive loop peptides to β sheet A of serpins (Schulze *et al.*, 1992). Additionally, EMSA has been used to identify regulatory RNA aptamers that bind to particular serpins (Madsen *et al.*, 2010), as well as to verify interactions of putative drug targets with serpins such as AT and HCII (De Fatima *et al.*, 2009).

2.3. Immunochemical methods

Immunochemical methods are longstanding techniques used to identify and characterize serpin–ligand interactions. The utility of these methods depends heavily on the specificity of antibodies against the particular serpin of interest. Recent descriptions of newly generated antibodies against specific conformations of serpins will provide a valuable tool in analyzing the conformational changes in serpins that occur upon binding to their GAG ligands (Kjellberg *et al.*, 2006; Strandberg *et al.*, 2001). The most quantitative of the immunochemical methods, ELISA, involves immobilizing a serpin–ligand and detecting binding of the serpin with a specific antibody. Binding of a ligand to a serpin can also be deduced using co-immunoprecipitation studies where the serpin–ligand mixture is pulled down by a specific serpin antibody conjugated to a bead, the beads washed, and the bound ligands eluted and identified. Crossed immunoelectrophoresis in which the serpin is electrophoresed in one dimension in the presence of ligand and then electrophoresed in a second dimension in the presence of an antibody incorporated into the gel provides qualitative information regarding serpin–ligand interactions. This technique has been used to examine the binding of heparin to AT in plasma (Lane *et al.*, 1993), to determine the conformational state of AT (Corral *et al.*, 2003), to examine the binding of heparin to naturally occurring AT mutants (Martinez-Martinez *et al.*, 2010), and to determine the effects of competing substances on AT inhibitory function (Martinez-Martinez *et al.*, 2009). Additionally, HCII deficiencies have been identified using this method (Kanagawa *et al.*, 2001).

2.4. *In silico* methods

Nearly all serpins of the coagulation and fibrinolytic systems including AT, HCII, plasminogen activator inhibitor–1, PCI, and PN1 interact with GAGs (Gettins, 2002). It is commonly assumed that these interactions arise from an optimal combination of structural features on both GAGs and serpins that engineer affinity. Whether these interactions are specific, or more accurately selective, remains an open question, except for the AT–heparin interaction, which is known to be highly specific. A major reason for the poor description of these interactions at an atomic level is the structural diversity of GAGs. Challenges also exist on the serpin side, since most GAG-binding sites on proteins contain multiple arginine and lysine residues that are typically surface exposed.

To resolve the limited structural knowledge on GAG–protein interactions and to utilize the power of nature's massive GAG sequence library, computational approaches have been developed. With respect to serpins in particular, molecular dynamics and docking approaches were developed to derive the heparin pentasaccharide binding site on AT (Bitomsky and Wade, 1999; Grootenhuis *et al.*, 1991), but the modeled geometries were significantly different from the cocrystal structure. This is not surprising because modeling GAGs is challenging due to their high negative charge density, which induces recognition of practically any collection of positively charged residues, especially on shallow, surface exposed sites of proteins (Carter *et al.*, 2005; Faham *et al.*, 1996; Pellegrini *et al.*, 2000; Schlessinger *et al.*, 2000). A molecular dynamics study of AT–heparin pentasaccharide interaction has correctly predicted the induced-fit mechanism as well as the preferred conformations of iduronic acid residues in the complex (Verli and Guimaraes, 2005), yet computational modeling typically suffers from the lack of understanding about the specificity of serpin–GAG interactions.

To resolve this problem, a novel approach of predicting high-specificity GAG sequences has been developed using a genetic algorithm-based docking and scoring technique (Raghuraman *et al.*, 2006). The approach relies on the application of a dual-filter strategy, with "simulated affinity" being the first step and "consistency of binding geometry" being the second step, to sort a combinatorially generated library of "specific" and "nonspecific" GAG sequences (Fig. 7.5). The phenomenal potential of this approach originates from its power to mimic the solution-phase interaction of GAG sequences with a target binding site on a protein using genetic algorithm-based computation, to utilize a large number of GAG sequences, natural as well as unnatural, to comprehensively screen the structural landscape, and to identify few distinct sequences that are predicted to be highly specific. Thus, the approach attempts to identify needle(s) in the haystack of GAG sequences that bind to the serpin with high affinity.

The genetic algorithm-based combinatorial virtual library screening (CVLS) approach involves the construction of GAG virtual library using programming scripts used in standard molecular modeling software such as SYBYL®. Typically disaccharide sequences, for example, →4)-GlcN*p*2S6S-(1 → 4) IdoA*p*2S-(1→ and its desired structural variants, are

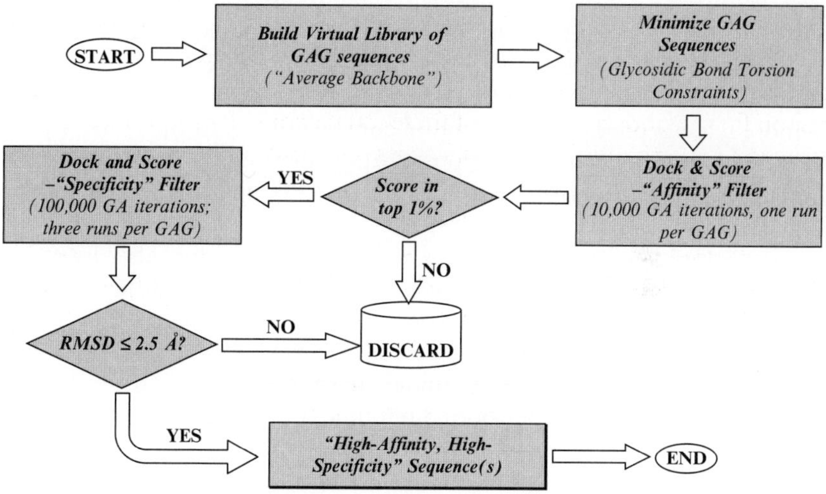

Figure 7.5 Genetic algorithm-based CVLS algorithm used to study the interaction of GAG sequences with AT and HCII. A GAG sequence library is constructed using the average backbone hypothesis, minimized using standard molecular modeling routines and each library member is docked onto the target binding site on a serpin. Analysis using the interaction score and RMSD yields sequences that bind the serpin in the target binding site with "high affinity and high specificity." See text for details.

built using the disaccharide geometry reported in the crystal structures (Jin *et al.*, 1997; Johnson and Huntington, 2003; Li *et al.*, 2004; Dementiev *et al.*, 2004). These sequences are then combinatorially conjoined to produce an oligosaccharide library. In the first development of this approach, 6859 heparin hexasaccharide sequences were generated from 19 different disaccharides ($19 \times 19 \times 19 = 6859$). A key factor to consider in the construction of combinatorial oligosaccharide library is the ϕ_H and ψ_H interglycosidic bond angles (Raghuraman *et al.*, 2006). Analysis of multiple oligosaccharide structures reported in the literature indicated that ϕ_H/ψ_H vary within $\pm 30°$, a relatively small range considering the large number of possibilities for a structurally diverse molecule (Bitomsky and Wade, 1999; Dementiev *et al.*, 2004; Grootenhuis *et al.*, 1991; Jin *et al.*, 1997; Johnson and Huntington, 2003; Li *et al.*, 2004). Thus, an "average backbone" structure, wherein the ϕ_H/ψ_H interglycosidic bond angles are held constant at the mean of known solution values, is typically used to reliably simulate physiological GAG–serpin interaction.

Following the construction of the combinatorial library, genetic algorithm-based docking of each library member onto the putative binding site on the serpin is initiated. GOLD is one of the genetic algorithm-based search engines that have the capability of scoring each GAG–serpin solution, although such a search protocol can be designed in-house also. In this approach, a population of arbitrarily docked ligand orientations is evaluated using a scoring function, following which an iterative optimization procedure attempts to improve the score through modification in binding geometry. This process mimics the solution binding phenomenon, wherein molecules find the best fit through numerous collisions. As the initial population is selected at random, several genetic algorithm runs are required to more reliably predict correct bound conformations. At the conclusion of the first phase of search, each serpin–GAG structure is scored and the best sequences, as defined by a scoring filter, for example, the top 1% of all sequences, are selected.

The "affinity"-filtered sequences are then screened for self-consistency of docking through a more rigorous docking and scoring study. In these experiments, the genetic algorithm is run for a much longer time period allowing the geometric search to take place over greater conformational space and enabling a more determined selection of the best geometries. Docking experiments are performed multiple times to ensure reproducibility and reduce false positives. The solutions are evaluated for consistent binding geometry, for example, RMSD of less than 2.5 Å. Experiments with the AT–heparin system has shown that this exhaustive process identifies sequences with "high affinity and high specificity" (Fig. 7.6; Raghuraman *et al.*, 2006).

The CVLS approach described above facilitates the extraction of a "pharmacophore," key interactions that drive binding specificity. For example, the sequences map developed from library screening identifies groups that display minimal variation, for example, $3S_F$, $2S_F$, and $6S_D$ in DEF (Fig. 7.6), which defines the pharmacophore. In contrast, domains with higher RMSD, and therefore possessing significant movement, define locations in which structural modification can be introduced to design new sequences or ligands.

Another application of CVLS approach reported in the literature is the elucidation of the putative binding geometry of a high-affinity DS

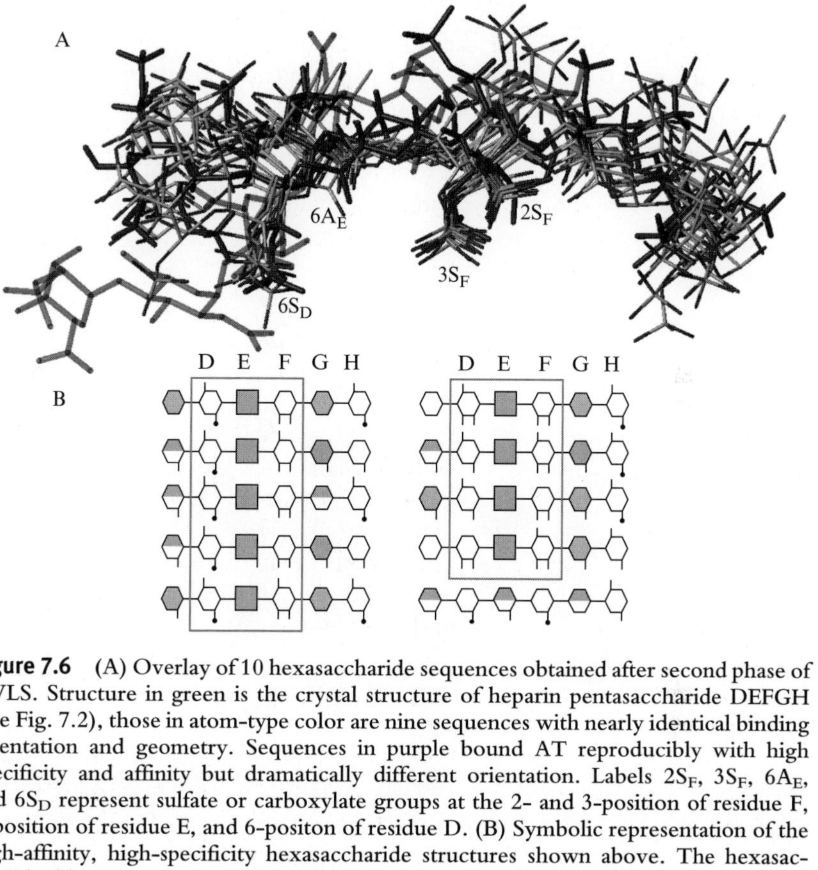

Figure 7.6 (A) Overlay of 10 hexasaccharide sequences obtained after second phase of CVLS. Structure in green is the crystal structure of heparin pentasaccharide DEFGH (see Fig. 7.2), those in atom-type color are nine sequences with nearly identical binding orientation and geometry. Sequences in purple bound AT reproducibly with high specificity and affinity but dramatically different orientation. Labels $2S_F$, $3S_F$, $6A_E$, and $6S_D$ represent sulfate or carboxylate groups at the 2- and 3-position of residue F, 6-position of residue E, and 6-positon of residue D. (B) Symbolic representation of the high-affinity, high-specificity hexasaccharide structures shown above. The hexasaccharide library sequence runs {UAp(1 → 4)GlcNp(1 → 4)}$_3$, where UA is either IdoAp (shaded hexagon) or GlcAp (shaded square). Sulfated substitution at 2-, 3-, or 6-positions of either UAp or GlcNp is indicated with a line (—), while acetate substitution at the 2-position of GlcNp is indicated with a line-dot (—•). Iduronic acid residues in $^2S_0$ conformation are shown as fully shaded hexagons, while those in $^1C_4$ conformation are shown as half-filled hexagons. Figure taken from Raghuraman *et al.* (2006), and reproduced with permission. (For interpretation of the references to color in this figure legend, the reader is referred to the Web version of this chapter.)

hexasaccharide onto HCII (Raghuraman *et al.*, 2010). It has been known that a rare DS hexasaccharide D6 (Fig. 7.2) binds HCII with high affinity, while sequences with higher levels of sulfation bind poorly, suggesting significant specificity of interaction (Maimone and Tollefsen, 1990; Pavão *et al.*, 1995). Taking cue from the CVLS results with the AT–H system, all possible topologies of D6 were constructed. This included three possible helical folds (2_1-, 3_2-, or 8_3-helices), four possible major conformers ($^1C_4$, $^4C_1$, $^2S_0$, and $^0S_2$) for IdoA*p* and the most favored conformer for GalN*p* ($^4C_1$). The total of 192 topologies were screened using the CVLS approach and the activated form of HCII was extracted from the S195A thrombin–HCII Michaelis complex (PDB entry 1JMO) (Baglin *et al.*, 2002). Analysis of the best topologies led to the identification of only one topology that satisfied all known biochemical data from mutagenesis studies (Fig. 7.7). The interesting aspect of this binding geometry is that it exhibits an ~60° angle with helix D, which is radically different from that of pentasaccharide H5 binding to AT, despite a strong degree of structural and sequence similarity between the two serpins. It remains to be seen whether this

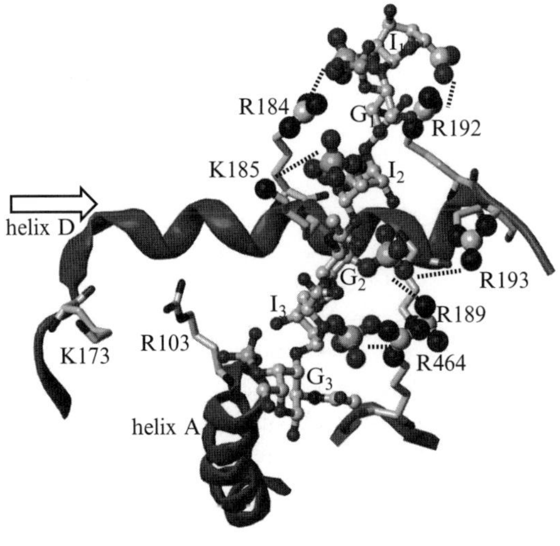

Figure 7.7 Predicted binding geometry of D6 onto HCII. Helices D and A are shown in magenta. Basic residues of HCII are shown as sticks, and the D6 sequence is rendered as ball-and-stick. D6 "hit" topology is 2_1-$^0S_2$·$^1C_4$·$^1C_4$, which binds at ~60° to helix D axis. Amino acid, sulfate, and carboxylate atoms involved in putative D6–HCII interactions are shown using an increased van der Waals radius. Interactions are indicated using dotted lines. Labels I_1 through I_3 and G_1 through G_3 are saccharide residue labels (see Fig. 7.2). The direction of the helix D axis is shown by an arrow. Figure taken from Raghuraman *et al.* (2010) and reprinted with permission. (For interpretation of the references to color in this figure legend, the reader is referred to the Web version of this chapter.)

computationally predicted binding geometry and topology is observed in solution experiments, for example, by NMR or crystallography.

Overall, the combinatorial virtual screening procedure is a powerful approach that identifies GAG sequences with high specificity for interacting with serpins. It affords the elucidation of the "pharmacophore" for possible drug discovery applications and is capable of identifying GAG binding geometry onto serpins. The novel approach is expected to be generally useful for many GAG–serpin interactions.

3. Animal Models

Established mouse models in which expression of AT, HCII, and PCI have been either abolished or altered have provided important information regarding the physiological roles of GAG-binding serpins. Additionally, several studies have highlighted the importance of heparin, HS, and DS, in regulating serpin function by altering or ablating the enzymes responsible for GAG chain synthesis. This section will highlight these mouse model methods with respect to their phenotype and how they have aided in the understanding of serpin–GAG interaction and function.

3.1. Antithrombin

AT deficient mice were generated by replacing exon 2 of the AT gene with a neomycin resistance gene (Ishiguro *et al.*, 2000). Breeding between heterozygote $AT^{+/-}$ animals yielded no live homozygous null $AT^{-/-}$ mice, suggesting that loss of AT is incompatible with prenatal life. Analysis of $AT^{-/-}$ embryos showed that these mice died in late gestation (embryonic days 15.5–16.5). There was prevalent fibrinogen deposition in the myocardium and liver of the null embryos, with subsequent tissue degradation in these areas. These embryos also showed signs of subcutaneous and intracranial hemorrhage, although fibrin deposition was absent in these tissues, suggesting either a consumption of fibrinogen or hepatic insufficiency in the $AT^{-/-}$ embryos. Heterozygous $AT^{+/-}$ offspring were observed at the expected Mendelian frequency, suggesting that reduced AT levels are compatible with normal prenatal development. $AT^{+/-}$ mice were shown to have approximately 50% of the normal plasma AT antigen and activity levels and were morphologically similar to their wild-type littermates. $AT^{+/-}$ mice showed no evidence of spontaneous thrombosis up to 14 months of age, but lipopolysaccharide challenge led to increased fibrin deposition in the kidney, liver, and small myocardial blood vessels (Yanada *et al.*, 2002). Administration of human AT prior to LPS challenge partially rescued the glomerular fibrin deposition phenotype in the $AT^{+/-}$ animals suggesting that AT plays a key role in prevention of renal thrombosis.

HS chains containing 3-*O*-sulfated glucosamine residues, such as those found on the cell surface of endothelial cells and in the subendothelial basement membrane, are known to activate AT. The rate-limiting step for biosynthesis of the 3-*O*-sulfated glucosamine residues is glucosaminyl-3-*O*-sulfotransferase-1 (3-OST-1), which is encoded by the *Hs3st1* gene. Replacement of exon 8 of the *Hs3st1* gene with a neomycin resistance gene yielded mice deficient in 3-OST-1 (Hajmohammadi *et al.*, 2003). 3-OST-1 homozygous null mice displayed no 3-OST-1 activity in plasma or tissues compared to wild-type animals. Tissues from these mice also showed a decreased ability to accelerate inhibition of factor Xa by AT. Surprisingly, these mice did not have increased tissue fibrin deposition under normal or hypoxic conditions and showed no change in the rate of thrombus formation in a ferric chloride carotid artery injury model. Interestingly, the 3-OST-1 null mice showed intrauterine growth retardation and postnatal death, which the heterozygotes did not manifest. Investigation into a possible gross coagulopathic or placental thrombotic phenotype showed no evidence of thrombosis, hemorrhage, fetal anatomic abnormalities or placental vessel defects. The lack of a prothrombotic phenotype in 3-OST-1 knockout mice questions the role of AT interaction with 3-*O*-sulfated GAGs in maintaining hemostasis in mice.

The binding of GAGs to the heparin-binding site of AT was examined by creating mice harboring an R48C mutation. This corresponds to residue R47 in human AT, which is important in the AT/heparin interaction and is frequently mutated in AT deficiency (Lane *et al.*, 1996). The R48C mutation eliminated the enhancing effect of heparin like molecules on the inhibition of coagulation both *in vivo* and *in vitro* (Dewerchin *et al.*, 2003). Mice homozygous for the R48C mutation developed spontaneous life threatening thrombosis in early postnatal life, including thromboses in the heart, liver, and ocular vessels. Neonates that did not survive until birth suffered from severe thrombosis in the heart. These animals also exhibited low heparin cofactor activity, and AT-mediated inhibition of thrombin by heparin, heparin pentasaccharide, or a heparin mimetic was completely abolished. These results suggest a crucial role for heparin binding to AT in regulating blood coagulation.

3.2. Heparin cofactor II

The initial HCII knockout mouse model was generated by inserting a PGK-neomycin resistance gene in place of exon 1 of the *SERPIND1* gene (He *et al.*, 2002). HCII $^{-/-}$ mice were born at expected Mendelian frequency and had normal growth and survival when assessed at 12 months of age. No spontaneous thrombosis was noted in the $HCII^{-/-}$ mice;

however, these mice did have a faster time to occlusion in a carotid artery photochemical injury model compared to wild-type littermates. The shortened occlusion time of the HCII$^{-/-}$ mice in this model was corrected by administration of purified HCII prior to injury. More recently, HCII null mice have been generated using the same gene targeting strategy (Aihara *et al.*, 2007), but in contrast to the results of He *et al.*, these HCII$^{-/-}$ mice were embryonic lethal. Analysis of embryos showed death occurred between embryonic days 6.5–8.5, although further analysis to reveal cause of lethality was not undertaken. It was hypothesized that strain differences between the mice generated by He *et al.*, and the mice generated by Aihara *et al.* led to this discrepancy in results. Due to the embryonic lethality of the HCII$^{-/-}$ in this study, HCII$^{+/-}$ mice that produced 50% of the normal HCII levels were used for analysis of thrombotic tendency. HCII$^{+/-}$ mice did not show spontaneous thrombosis, although there was evidence of enhanced platelet aggregation in response to ADP, suggesting that HCII has a suppressive role in platelet function. Additionally, in a wire insertion injury model, femoral arteries of HCII$^{+/-}$ mice had a higher incidence of occlusion due to thrombosis compared to wild-type mice. Prior to injury, administration of purified human HCII to the HCII$^{+/-}$ mice decreased the incidence of occlusion to wild-type levels. Moreover, a role for HCII in thrombin dependent vascular remodeling was established, with HC$^{+/-}$ mice exhibiting prominent hyperplasia and increased PAR1 expression in the vasculature. The increased expression of PAR1 in HCII$^{+/-}$ mice, along with enhanced platelet aggregation suggests a role for HCII in platelet function.

DS and heparin are both known to bind and to accelerate the antithrombotic activities of HCII. In HCII$^{-/-}$ mice generated by He *et al.*, DS inhibits thrombosis in carotid artery injury models in wild-type mice but has no effect in mice lacking HCII (Vicente *et al.*, 2004). These results suggest that HCII is required for the antithrombotic effects of DS. Additionally, HCII$^{-/-}$ mice injected with a mutant HCII with decreased affinity for DS (R189H) did not restore thrombotic occlusion whereas either wild-type HCII or HCII deficient in heparin binding were able to restore occlusion time to normal (He *et al.*, 2008). These results highlight the importance of DS interaction with HCII function to regulate thrombus formation. Moreover, the antithrombotic effects of DS are correlated with *N*-acetylgalactosamine-4-*O*-sulfate residues of DS. Future studies in mice deficient in D4ST-1, the sulfotransferase responsible for 4-*O*-sulfation of GalNAc, may shed light on this dependency. Finally, studies on DS epimerase, the enzyme responsible for glucuronic acid to iduronic acid changes in the synthesis of chondroitin and DS, have hinted at the importance of DS sulfation patterns in maintaining hemostasis. A DS epimerase knockout mouse was generated through replacement of exon 2 of the DS epimerase-1 gene with a neomycin resistance gene (Maccarana *et al.*, 2009). While the hemostatic parameters of these mice were not measured, it was noted that DSE$^{-/-}$ mice produced a smaller litter size compared to wild-type animals. This was hypothesized to be due to the maternal/fetal placenta harboring an HCII mediated thrombotic imbalance, but further studies will be required to determine the molecular mechanism behind the small litter size.

3.3. Protein C inhibitor

PCI studies in mice have long been complicated by the fact that mice and humans have differential organ expression profiles of PCI. In humans, PCI is expressed in many organs notably in the liver, kidney, and reproductive organs. In contrast, mouse PCI is only detected in reproductive organs and neither is it expressed in the mouse liver nor in the mouse plasma by both antigen and activity assay. Therefore, several strategies have been employed to study PCI in a more human-specific organ environment.

Disruption of the mouse PCI gene was accomplished by replacing exons 2–5 of mouse PCI with a neomycin resistance gene (Uhrin *et al.*, 2000). Homozygous null PCI$^{-/-}$ mice were born at the expected Mendelian frequency, suggesting that PCI deficiency is compatible with embryonic life. PCI$^{-/-}$ mice showed no evidence of altered hemostasis, however PCI$^{-/-}$ male mice were infertile. Sperm from PCI$^{-/-}$ mice were morphologically abnormal and were unable to fertilize wild-type oocytes *in vitro*. This defect was shown to be due to the requirement of PCI in the movement of spermatogenic cells in the blood–testis barrier and maturation of spermatids in the seminiferous tubules (Uhrin *et al.*, 2007). The action of urokinase plasminogen activator (uPA) and tissue plasminogen activator (tPA), major serine proteases in the semen of rodents, in PCI$^{-/-}$ mice was hypothesized to play a role in excessive proteolysis and contribute to the infertility phenotype. However, this hypothesis was ruled out when double knockout PCI$^{-/-}$/uPA$^{-/-}$ or PCI$^{-/-}$/tPA$^{-/-}$ mice failed to rescue the infertility of PCI$^{-/-}$ mice (Uhrin *et al.*, 2007).

In order to study the actions of PCI in the plasma where it is not normally found in mice, two strategies have been implemented. First, the human PCI cDNA was placed under the control of the mouse albumin enhancer/promoter region in order to obtain liver specific expression of human PCI (Wagenaar *et al.*, 2000). Transgenic homozygous mice were born and developed normally with no hemostatic abnormalities. Plasma levels of human PCI in mice were approximately twofold higher than in human plasma in the homozygous transgenics. Following endotoxin challenge, there were no differences in cytokine production or mortality between transgenic and wild-type mice. A second study generated mice expressing human PCI by inserting the entire human PCI genomic sequence into fertilized mouse oocytes (Hayashi *et al.*, 2004). Transgenic mice expressed human PCI in a variety of tissues including liver, kidney, brain, lung and reproductive organs. Plasma PCI levels were shown to be approximately four times higher than human plasma PCI levels. Human PCI expressed in transgenic mice was able to inhibit both exogenous human APC and endogenous mouse APC. In response to endotoxin, transgenic mice expressing human PCI showed a prolonged aPTT, and decreased fibrinogen and AT levels, suggestive of a consumption of coagulation factors typically found in disseminated intravascular coagulopathy.

ACKNOWLEDGMENTS

The authors would like to thank Dr. Dougald Monroe for providing critical feedback in molecular modeling tutoring. Stipend support for CMR is in part through NIH grant T32 HL007149-34. This work was supported in part by the National Institutes of Health (National Institute of Aging and Heart, Lung and Blood) R21AG031068 to FCC and HL090586 and HL099420 to URD, and by an Established Investigator Award 0640053N from the American Heart Association National Center to URD.

REFERENCES

Aihara, K., Azuma, H., Akaike, M., Ikeda, Y., Sata, M., Takamori, N., Yagi, S., Iwase, T., Sumitomo, Y., Kawano, H., Yamada, T., Fukuda, T., *et al.* (2007). Strain-dependent embryonic lethality and exaggerated vascular remodeling in heparin cofactor II deficient mice. *J. Clin. Invest.* **117,** 1514–1526.

Arcone, R., Chinali, A., Pozzi, N., Parafati, M., Maset, F., Pietropaolo, C., and Filippis, V. D. (2009). Conformational and biochemical characterization of a biologically active rat recombinant Protease Nexin-1 expressed in E. coli. *Biochim. Biophys. Acta—Proteins & Proteomics* **1794,** 602–614.

Arocas, V., Bock, S. C., Raja, S., Olson, S. T., and Björk, I. (2001). Lysine 114 of antithrombin is of crucial importance for the affinity and kinetics of heparin pentasaccharide binding. *J. Biol. Chem.* **276,** 43809–43817.

Atha, D. H., Lormeau, J. C., Petitou, M., Rosenberg, R. D., and Choay, J. (1987). Contribution of 3-O- and 6-O-sulfated glucosamine residues in the heparin-induced conformational change in antithrombin III. *Biochemistry (NY)* **26,** 6454–6461.

Atha, D. H., Lormeau, J. C., Petitou, M., Rosenberg, R. D., and Choay, J. (1985). Contribution of monosaccharide residues in heparin binding to antithrombin III. *Biochemistry (NY)* **24,** 6723–6729.

Baglin, T. P., Carrell, R. W., Church, F. C., Esmon, C. T., and Huntington, J. A. (2002). Crystal structures of native and thrombin-complexed heparin cofactor II reveal a multistep allosteric mechanism. *Proc. Natl. Acad. Sci. USA* **99,** 11079–11084.

Berry, L., Stafford, A., Fredenburgh, J., O'Brodovich, H., Mitchell, L., Weitz, J., Andrew, M., and Chan, A. K. C. (1998). Investigation of the anticoagulant mechanisms of a covalent antithrombin-heparin complex. *J. Biol. Chem.* **273,** 34730–34736.

Bitomsky, W., and Wade, R. C. (1999). Docking of glycosaminoglycans to heparin-binding proteins: Validation for aFGF, bFGF, and antithrombin and application to IL-8. *J. Am. Chem. Soc.* **121,** 3004–3013.

Boucher, L., de, l.C., Bock, S., and Wintrode, P. L. (2009). Conformational dynamics of antithrombin III with its allosteric activator heparin. *Biophys. J.* **96,** 69a.

Busenlehner, L. S., and Armstrong, R. N. (2005). Insights into enzyme structure and dynamics elucidated by amide H/D exchange mass spectrometry. *Arch. Biochem. Biophys.* **433,** 34.

Capila, I., and Linhardt, R. J. (2002). Heparin/protein interactions. *Angew. Chem. Int. Ed.* **41,** 390–412.

Carrell, R. W., Stein, P. E., Fermi, G., and Wardell, M. R. (1994). Biological implications of a 3 Å structure of dimeric antithrombin. *Structure* **2,** 257–270.

Carter, W. J., Cama, E., and Huntington, J. A. (2005). Crystal structure of thrombin bound to heparin. *J. Biol. Chem.* **280,** 2745–2749.

Clarkson, J., and Campbell, I. D. (2003). Studies of protein-ligand interactions by NMR. *Biochem. Soc. Trans.* **31,** 1006–1009.

Corral, J., Rivera, J., Martinez, C., Gonzalez-Conejero, R., Minano, A., and Vicente, V. (2003). Detection of conformational transformation of antithrombin in blood with crossed immunoelectrophoresis: New application for a classical method. *J. Lab. Clin. Med.* **142,** 298–305.

De Fatima, A., Fernandes, S. A., and Sabino, A. A. (2009). Calixarenes as new platforms for drug design. *Curr. Drug Discov. Technol.* **6,** 151–170.

Dementiev, A., Petitou, M., Herbert, J., and Gettins, P. G. W. (2004). The ternary complex of antithrombin-anhydrothrombin-heparin reveals the basis of inhibitor specificity. *Nat. Struct. Mol. Biol.* **11,** 863.

Desai, U. R., Petitou, M., Bjork, I., and Olson, S. T. (1998a). Mechanism of heparin activation of antithrombin: Evidence for an induced-fit model of allosteric activation involving two interaction subsites. *Biochemistry (NY)* **37,** 13033–13041.

Desai, U. R., Petitou, M., Björk, I., and Olson, S. T. (1998b). Mechanism of heparin activation of antithrombin. *J. Biol. Chem.* **273,** 7478–7487.

Desai, U. R., Swanson, S. C., Bock, I. Björk, and Olson, S. T. (2000). Role of arginine 129 in heparin binding and activation of antithrombin. *J. Biol. Chem.* **275,** 18976–18984.

Dewerchin, M., Herault, J., Wallays, G., Petitou, M., Schaeffer, P., Millet, L., Weitz, J. I., Moons, L., Collen, D., Carmeliet, P., and Herbert, J. (2003). Life-threatening thrombosis in mice with targeted Arg48-to-Cys mutation of the heparin-binding domain of antithrombin. *Circ. Res.* **93,** 1120–1126.

Einarsson, R. (1978). The increase in human antithrombin III tryptophan fluorescence produced by heparin. *Biochim. Biophys. Acta—Protein, Structure* **534,** 165.

Einarsson, R., and Andersson, L. (1977). Binding of heparin to human antithrombin III as studied by measurements of tryptophan fluorescence. *Biochim. Biophys. Acta—Protein, Structure* **490,** 104–111.

Elg, S., and Deinum, J. (2002). The interaction between captured human thrombin and antithrombin studied by surface plasmon resonance, and the effect of melagatran. *Spectroscopy* **16,** 257–270.

Faham, S., Hileman, R. E., Fromm, J. R., Linhardt, R. J., and Rees, D. C. (1996). Heparin structure and interactions with basic fibroblast growth factor. *Science* **271,** 1116–1120.

Futamura, A., Beechem, J. M., and Gettins, P. G. W. (2001). Conformational equilibrium of the reactive center loop of antithrombin examined by steady state and time-resolved fluorescence measurements: Consequences for the mechanism of factor Xa inhibition by antithrombin-heparin complexes. *Biochemistry (NY)* **40,** 6680–6687.

Gettins, P. G. W., Patston, P. A., and Olson, S. T. (1996). Serpins: Structure, Function and Biology. R. G. Landes Co., Austin.

Gettins, P. G. W. (2002). Serpin structure, mechanism, and function. *Chem. Rev.* **102,** 4751–4804.

Grootenhuis, P. D. J., Van Boeckel, and Constant, A. A. (1991). Constructing a molecular model of the interaction between antithrombin III and a potent heparin analog. *J. Am. Chem. Soc.* **113,** 2743–2747.

Guerrini, M., Guglieri, S., Casu, B., Torri, G., Mourier, P., Boudier, C., and Viskov, C. (2008). Antithrombin-binding octasaccharides and role of extensions of the active pentasaccharide sequence in the specificity and strength of interaction. *J. Biol. Chem.* **283,** 26662–26675.

Hajmohammadi, S., Enjyoji, K., Princivalle, M., Christi, P., Lech, M., Beeler, D., Rayburn, H., Schwartz, J. J., Barzegar, S., de Agostini, A. I., Post, M. J., Rosenberg, R. D., *et al.* (2003). Normal levels of anticoagulant heparan sulfate are not essential for normal hemostasis. *J. Clin. Invest.* **111,** 989.

Hayashi, T., Nishioka, J., Kamada, H., Asanuma, K., Kondo, H., Gabazza, E. C., Ido, M., and Suzuki, K. (2004). Characterization of a novel human protein C inhibitor (PCI) gene transgenic mouse useful for studying the role of PCI in physiological and pathological conditions. *J. Thromb. Haemost.* **2,** 949.

He, L., Giri, T. K., Vicente, C. P., and Tollefsen, D. M. (2008). Vascular dermatan sulfate regulates the antithrombotic activity of heparin cofactor II. *Blood* **111,** 4118–4125.

He, L., Vicente, C. P., Westrick, R. J., Eitzman, D. T., and Tollefsen, D. M. (2002). Heparin cofactor II inhibits arterial thrombosis after endothelial injury. *J. Clin. Invest.* **109,** 213.

Henry, B. L., Connell, J., Liang, A., Krishnasamy, C., and Desai, U. R. (2009). Interaction of antithrombin with sulfated, low molecular weight lignins. *J. Biol. Chem.* **284,** 20897–20908.

Higgins, W. J., Fox, D. M., Kowalski, P. S., Nielsen, J. E., and Worrall, D. M. (2010). Heparin enhances serpin inhibition of the cysteine protease cathepsin L. *J. Biol. Chem.* **285,** 3722–3729.

Huntington, J. A., Kjellberg, M., and Stenflo, J. (2003). Crystal structure of protein c inhibitor provides insights into hormone binding and heparin activation. *Structure* **11,** 205.

Huntington, J. A., Olson, S. T., Fan, B., and Gettins, P. G. W. (1996). Mechanism of heparin activation of antithrombin. Evidence for reactive center loop preinsertion with expulsion upon heparin binding. *Biochemistry (NY)* **35,** 8495–8503.

Ishiguro, K., Kojima, T., Kadomatsu, K., Nakayama, Y., Takagi, A., Suziki, M., Takeda, N., Ito, M., Yamamoto, K., Matsushita, T., Kusugami, K., Muramatsu, T., *et al.* (2000). Complete antithrombin deficiency in mice results in embryonic lethality. *J. Clin. Invest.* **106,** 873.

Jairajpuri, M. A., Lu, A., Desai, U., Olson, S. T., Bjork, I., and Bock, S. C. (2003). Antithrombin III phenylalanines 122 and 121 contribute to its high affinity for heparin and its conformational activation. *J. Biol. Chem.* **278,** 15941–15950.

Jin, L., Abrahams, J. P., Skinner, R., Petitou, M., Pike, R. N., and Carrell, R. W. (1997). The anticoagulant activation of antithrombin by heparin. *Proc. Natl. Acad. Sci. USA* **94,** 14683–14688.

Johnson, D. J. D., and Huntington, J. A. (2003). Crystal structure of antithrombin in a heparin-bound intermediate state. *Biochemistry (NY)* **42,** 8712–8719.

Johnson, D. J. D., Langdown, J., and Huntington, J. A. (2010). Molecular basis of factor IXa recognition by heparin-activated antithrombin revealed by a 1.7-Å structure of the ternary complex. *Proc. Natl. Acad. Sci. USA* **107,** 645–650.

Johnson, D. J. D., Langdown, J., Li, W., Luis, S. A., Baglin, T. P., and Huntington, J. A. (2006). Crystal structure of monomeric native antithrombin reveals a novel reactive center loop conformation. *J. Biol. Chem.* **281,** 35478–35486.

Kanagawa, Y., Shigekiyo, T., Aihara, K., Akaike, M., Azuma, H., and Matsumoto, T. (2001). Molecular mechanism of Type I congenital heparin cofactor (HC) II deficiency caused by a missense mutation at reactive P2 site: HC II Tokushima. *Thromb. Haemost.* **85,** 101.

Kim, S. H., and Kiick, K. L. (2007). Heparin-mimetic sulfated peptides with modulated affinities for heparin-binding peptides and growth factors. *Peptides* **28,** 2125–2136.

Kjellberg, M., Ikonomou, T., and Stenflo, J. (2006). The cleaved and latent forms of antithrombin are normal constituents of blood plasma: A quantitative method to measure cleaved antithrombin. *J. Thromb. Haemost.* **4,** 168–176.

Lane, D. A., Olds, R. J., Boisclair, M., Chowdhury, V., Thein, S. L., Cooper, D. N., Blajchman, M., Perry, D., Emmerich, J., and Aiach, M. (1993). Antithrombin III mutation database: First update. For the Thrombin and its Inhibitors Subcommittee of the Scientific and Standardization Committee of the International Society on Thrombosis and Haemostasis. *Thromb. Haemost.* **70,** 361.

Lane, D. A., Kunz, G., Olds, R. J., and Thein, S. L. (1996). Molecular genetics of antithrombin deficiency. *Blood Rev.* **10,** 59.

Langdown, J., Belzar, K. J., Savory, W. J., Baglin, T. P., and Huntington, J. A. (2009). The critical role of hinge-region expulsion in the induced-fit heparin binding mechanism of antithrombin. *J. Mol. Biol.* **386,** 1278.

Langdown, J., Johnson, D. J. D., Baglin, T. P., and Huntington, J. A. (2004). Allosteric activation of antithrombin critically depends upon hinge region extension. *J. Biol. Chem.* **279,** 47288–47297.

Lee, M. K., and Lander, A. D. (1991). Analysis of affinity and structural selectivity in the binding of proteins to glycosaminoglycans: Development of a sensitive electrophoretic approach. *Proc. Natl. Acad. Sci. USA* **88,** 2768–2772.

Li, W., Adams, T. E., Kjellberg, M., Stenflo, J., and Huntington, J. A. (2007). Structure of native protein c inhibitor provides insight into its multiple functions. *J. Biol. Chem.* **282,** 13759–13768.

Li, W., Adams, T. E., Nangalia, J., Esmon, C. T., and Huntington, J. A. (2008). Molecular basis of thrombin recognition by protein C inhibitor revealed by the 1.6-Å structure of the heparin-bridged complex. *Proc. Natl. Acad. Sci. USA* **105,** 4661–4666.

Li, W., Johnson, D. J. D., Esmon, C. T., and Huntington, J. A. (2004). Structure of the antithrombin-thrombin-heparin ternary complex reveals the antithrombotic mechanism of heparin. *Nat. Struct. Mol. Biol.* **11,** 857.

Liaw, P. C. Y., Austin, R. C., Fredenburgh, J. C., Stafford, A. R., and Weitz, J. I. (1999). Comparison of heparin- and dermatan sulfate-mediated catalysis of thrombin inactivation by heparin cofactor II. *J. Biol. Chem.* **274,** 27597–27604.

Lin, P., Sinha, U., and Betz, A. (2001). Antithrombin binding of low molecular weight heparins and inhibition of factor Xa. *Biochim. Biophys. Acta—General Subjects* **1526,** 105.

Liu, L., Mushero, N., Hedstrom, L., and Gershenson, A. (2007). Short-lived protease-serpin complexes: Partial disruption of the rat trypsin active site. *Protein Sci.* **16,** 2403–2411.

Maccarana, M., Kalamajski, S., Kongsgaard, M., Magnusson, S. P., Oldberg, A., and Malmstrom, A. (2009). Dermatan sulfate epimerase 1-deficient mice have reduced content and changed distribution of iduronic acids in dermatan sulfate and an altered collagen structure in skin. *Mol. Cell. Biol.* **29,** 5517–5528.

Madsen, J. B., Dupont, D. M., Andersen, T. B., Nielsen, A. F., Sang, L., Brix, D. M., Jensen, J. K., Broos, T., Hendrickx, M. L. V., Christensen, A., Kjems, J., and Andreasen, P. A. (2010). RNA aptamers as conformational probes and regulatory agents for plasminogen activator inhibitor-1. *Biochemistry* **49,** 4103–4115.

Maimone, M. M., and Tollefsen, D. M. (1990). Structure of a dermatan sulfate hexasaccharide that binds to heparin cofactor II with high affinity. *J. Biol. Chem.* **265,** 18263–18271.

Majdoub, H., Mansour, M. B., Chaubet, F., Roudesli, M. S., and Maaroufi, R. M. (2009). Anticoagulant activity of a sulfated polysaccharide from the green alga Arthrospira platensis. *Biochim. Biophys. Acta—General Subjects* **1790,** 1377.

Mansour, M. B., Dhahri, M., Hassine, M., Ajzenberg, N., Venisse, L., Ollivier, V., Chaubet, F., Jandrot-Perrus, M., and Maaroufi, R. M. (2010). Highly sulfated dermatan sulfate from the skin of the ray Raja montagui: Anticoagulant activity and mechanism of action. *Comp. Biochem. Physiol. B Biochem. Mol. Biol.* **156,** 206.

Martinez-Martinez, I., Ordonez, A., Guerrero, J. A., Pedersen, S., Minano, A., Teruel, R., Velazquez, L., Kristensen, S. R., Vicente, V., and Corral, J. (2009). Effects of acrolein, a natural occurring aldehyde, on the anticoagulant serpin antithrombin. *FEBS Lett.* **583,** 3165.

Martinez-Martinez, I., Ordonez, A., Navarro-Fernandez, J., Perez-Lara, A., Gutierrez-Gallego, R., Giraldo, R., Martinez, C., Llop, E., Vicente, V., and Corral, J. (2010). Antithrombin Murcia (K241E) causing antithrombin deficiency: A possible role for altered glycosylation. *Haematologica* **95,** 1358–1365.

Meagher, J. L., Beechem, J. M., Olson, S. T., and Gettins, P. G. W. (1998). Deconvolution of the fluorescence emission spectrum of human antithrombin and identification of the tryptophan residues that are responsive to heparin binding. *J. Biol. Chem.* **273,** 23283–23289.

Meagher, J. L., Olson, S. T., and Gettins, P. G. W. (2000). Critical role of the linker region between helix D and strand 2A in heparin activation of antithrombin. *J. Biol. Chem.* **275,** 2698–2704.

Monien, B. H., Krishnasamy, C., Olson, S. T., and Desai, U. R. (2005). Importance of tryptophan 49 of antithrombin in heparin binding and conformational activation. *Biochemistry (NY)* **44,** 11660–11668.

Nordenman, B., Danielsson, Å., and Björk, I. (1978). The binding of low-affinity and high-affinity heparin to antithrombin. *Eur. J. Biochem.* **90,** 1–6.

O'Keeffe, D., Olson, S. T., Gasiunas, N., Gallagher, J., Baglin, T. P., and Huntington, J. A. (2004). The heparin binding properties of heparin cofactor II suggest an antithrombin-like activation mechanism. *J. Biol. Chem.* **279,** 50267–50273.

Olson, S. T., Björk, I., Sheffer, R., Craig, P. A., Shore, J. D., and Choay, J. (1992). Role of the antithrombin-binding pentasaccharide in heparin acceleration of antithrombin-proteinase reactions. Resolution of the antithrombin conformational change contribution to heparin rate enhancement. *J. Biol. Chem.* **267,** 12528–12538.

Olson, S. T., Srinivasan, K. R., Björk, I., and Shore, J. D. (1981). Binding of high affinity heparin to antithrombin III. Stopped flow kinetic studies of the binding interaction. *J. Biol. Chem.* **256,** 11073–11079.

Olson, S. T., Bock, P. E., and Sheffer, R. (1991). Quantitative evaluation of solution equilibrium binding interactions by affinity partitioning: Application to specific and nonspecific protein-heparin interactions. *Arch. Biochem. Biophys.* **286,** 533.

Olson, S. T., Frances-Chmura, A. M., Swanson, R., Björk, I., and Zettlmeissl, G. (1997). Effect of individual carbohydrate chains of recombinant antithrombin on heparin affinity and on the generation of glycoforms differing in heparin affinity. *Arch. Biochem. Biophys.* **341,** 212.

Pavão, M. S. G., Mourão, P. A. S., Mulloy, B., and Tollefsen, D. M. (1995). A unique dermatan sulfate-like glycosaminoglycan from ascidian. *J. Biol. Chem.* **270,** 31027–31036.

Pearce, M., Bottomley, S., Pike, R., and Lesk, A. (2006). Serpin conformations. *In* "Molecular and cellular aspects of the serpinopathies and disorders in serpin activity," (D. Lomas and G. Silverman, eds.). World Scientific, Singapore.

Pellegrini, L., Burke, D. F., von Delft, F., Mulloy, B., and Blundell, T. L. (2000). Crystal structure of fibroblast growth factor receptor ectodomain bound to ligand and heparin. *Nature* **407,** 1029–1034.

Phillips, J. E., Cooper, S. T., Potter, E. E., and Church, F. C. (1994). Mutagenesis of recombinant protein C inhibitor reactive site residues alters target proteinase specificity. *J. Biol. Chem.* **269,** 16696–16700.

Piepkorn, M. W. (1981). Dansyl (5-dimethylaminonaphthalene-1-sulphonyl)-heparin binds antithrombin III and platelet factor 4 at separate sites. *Biochem. J.* **196,** 649–651.
Piepkorn, M. W., Lagunoff, D., and Schmer, G. (1980). Binding of heparin to antithrombin III: The use of dansyl and rhodamine labels. *Arch. Biochem. Biophys.* **205,** 315.
Raghuraman, A., Mosier, P. D., and Desai, U. R. (2010). Understanding dermatan sulfate heparin cofactor II interaction through virtual library screening. *ACS Med. Chem Lett.* **1,** 281–285.
Raghuraman, A., Mosier, P. D., and Desai, U. R. (2006). Finding a needle in a haystack: Development of a combinatorial virtual screening approach for identifying high specificity heparin/heparan sulfate sequence(s). *J. Med. Chem.* **49,** 3553–3562.
Richard, B., Swanson, R., and Olson, S. T. (2009). The signature 3-O-sulfo group of the anticoagulant heparin sequence is critical for heparin binding to antithrombin but is not required for allosteric activation. *J. Biol. Chem.* **284,** 27054–27064.
Rovelli, G., Stone, S. R., Guidolin, A., Sommer, J., and Monard, D. (1992). Characterization of the heparin-binding site of glia-derived nexin/protease nexin-1. *Biochemistry (NY)* **31,** 3542–3549.
Rudd, T. R., Yates, E. A., and Hricovini, M. (2009). Spectroscopic and theoretical approaches for the determination of heparin saccharide structure and the study of protein-glycosaminoglycan complexes in solution. *Curr. Med. Chem.* **16,** 4750–4766.
Rudd, T. R., Guimond, S. E., Skidmore, M. A., Duchesne, L., Guerrini, M., Torri, G., Cosentino, C., Brown, A., Clarke, D. T., Turnbull, J. E., Fernig, D. G., and Yates, E. A. (2007). Influence of substitution pattern and cation binding on conformation and activity in heparin derivatives. *Glycobiology* **17,** 983–993.
Rudd, T. R., Nichols, R. J., and Yates, E. A. (2008). Selective detection of protein secondary structural changes in solution protein polysaccharide complexes using vibrational circular dichroism (VCD). *J. Am. Chem. Soc.* **130,** 2138–2139.
San Antonio, J. D., and Lander, A. D. (2001). Affinity coelectrophoresis of proteoglycan-protein complexes. *Methods Mol. Biol.* **171,** 401–414.
Schedin-Weiss, S., Arocas, V., Bock, S. C., Olson, S. T., and Björk, I. (2002a). Specificity of the basic side chains of Lys114, Lys125, and Arg129 of antithrombin in heparin binding. *Biochemistry (NY)* **41,** 12369–12376.
Schedin-Weiss, S., Desai, U. R., Bock, S. C., Gettins, P. G. W., Olson, S. T., and Björk, I. (2002b). Importance of lysine 125 for heparin binding and activation of antithrombin. *Biochemistry (NY)* **41,** 4779–4788.
Schedin-Weiss, S., Desai, U. R., Bock, S. C., Olson, S. T., and Björk, I. (2004). Roles of N-terminal region residues Lys11, Arg13, and Arg24 of antithrombin in heparin recognition and in promotion and stabilization of the heparin-induced conformational change. *Biochemistry* **43,** 675–683.
Schedin-Weiss, S., Richard, B., Hjelm, R., and Olson, S. T. (2008). Antiangiogenic forms of antithrombin specifically bind to the anticoagulant heparin sequence. *Biochemistry (NY)* **47,** 13610–13619.
Schlessinger, J., Plotnikov, A. N., Ibrahimi, O. A., Eliseenkova, A. V., Yeh, B. K., Yayon, A., Linhardt, R. J., and Mohammadi, M. (2000). Crystal structure of a ternary FGF-FGFR-heparin complex reveals a dual role for heparin in FGFR binding and dimerization. *Mol. Cell* **6,** 743–750.
Schulze, A. J., Frohnert, P. W., Engh, R. A., and Huber, R. (1992). Evidence for the extent of insertion of the active site loop of intact.alpha.1 proteinase inhibitor in.beta.-sheet A. *Biochemistry (NY)* **31,** 7560–7565.
Shirk, R. A., Parthasarathy, N., San Antonio, J. D., Church, F. C., and Wagner, W. D. (2000). Altered dermatan sulfate structure and reduced heparin cofactor II-stimulating activity of biglycan and decorin from human atherosclerotic plaque. *J. Biol. Chem.* **275,** 18085–18092.

Strandberg, K., Bhiladvala, P., Holm, J., and Stenflo, J. (2001). A new method to measure plasma levels of activated protein C in complex with protein C inhibitor in patients with acute coronary syndromes. *Blood Coag. Fibrinol.* **12,** 503–510.

Streusand, V. J., Björk, I., Gettins, P. G. W., Petitou, M., and Olson, S. T. (1995). Mechanism of acceleration of antithrombin-proteinase reactions by low affinity heparin. *J. Biol. Chem.* **270,** 9043–9051.

Takamoto, K., and Chance, M. R. (2006). Radiolytic protein footprinting with mass spectrometry to probe the structure of macromolecular complexes. *Annu. Rev. Biophys. Biomol. Struct.* **35,** 251–276.

Toida, T., Hileman, R. E., Smith, A. E., Vlahova, P. I., and Linhardt, R. J. (1996). Enzymatic preparation of heparin oligosaccharides containing antithrombin III binding sites. *J. Biol. Chem.* **271,** 32040–32047.

Tsutsui, Y., Liu, L., Gershenson, A., and Wintrode, P. L. (2006). The conformational dynamics of a metastable serpin studied by hydrogen exchange and mass spectrometry. *Biochemistry (NY)* **45,** 6561–6569.

Uhrin, P., Dewerchin, M., Hilpert, M., Chrenek, P., Schöfer, C., Zechmeister-Machhart, M., Krönke, G., Vales, A., Carmeliet, P., Binder, B. R., and Geiger, M. (2000). Disruption of the protein C inhibitor gene results in impaired spermatogenesis and male infertility. *J. Clin. Invest.* **106,** 1531.

Uhrin, P., Schöfer, C., Zaujec, J., Lubos, R., Hilpert, M., Weipoltshammer, K., Jerabeck, I., Pertzkall, I., Furtmüller, M., Dewerchin, M., Binder, B. R., and Geiger, M. (2007). Male fertility and protein C inhibitor/plasminogen activator inhibitor-3 (PCI): Localization of PCI in mouse testis and failure of single plasminogen activator knockout to restore spermatogenesis in PCI-deficient mice. *Fertil. Steril.* **88,** 1049.

Verli, H., and Guimaraes, J. A. (2005). Insights into the induced-fit mechanism in antithrombin—Heparin interaction using molecular dynamics simulations. *J. Mol. Graph. Modell.* **24,** 203.

Vicente, C. P., He, L., Pavao, M. S. G., and Tollefsen, D. M. (2004). Antithrombotic activity of dermatan sulfate in heparin cofactor II-deficient mice. *Blood* **104,** 3965–3970.

Wagenaar, G. T. M., van Vuuren, A. J. H., Girma, M., Tiekstra, M. J., Kwast, L., Koster, J. G., Rijneveld, A. W., Elisen, M. G. L. M., van der Poll, T., and Meijers, J. C. M. (2000). Characterization of transgenic mice that secrete functional human protein C inhibitor into the circulation. *Thromb. Haemost.* **83,** 93.

Yanada, M., Kojima, T., Ishiguro, K., Nakayama, Y., Yamamoto, K., Matsushita, T., Kadomatsu, K., Nishimura, M., Muramatsu, T., and Saito, H. (2002). Impact of antithrombin deficiency in thrombogenesis: Lipopolysaccharide and stress-induced thrombus formation in heterozygous antithrombin-deficient mice. *Blood* **99,** 2455–2458.

Zhang, F., Wu, Y., Ma, Q., Hoppensteadt, D., Fareed, J., and Linhardt, R. J. (2004). Studies on the effect of calcium in interactions between heparin and heparin cofactor II using surface plasmon resonance. *Clin. Appl. Thromb. Hemost.* **10,** 249–257.

Zheng, X., Wintrode, P. L., and Chance, M. R. (2008). Complementary structural mass spectrometry techniques reveal local dynamics in functionally important regions of a metastable serpin. *Structure* **16,** 38–51.

CHAPTER EIGHT

Targeting Serpins in High-Throughput and Structure-Based Drug Design

Yi-Pin Chang,* Ravi Mahadeva,† Anathe O. M. Patschull,‡ Irene Nobeli,‡ Ugo I. Ekeowa,§ Adam R. McKay,¶ Konstantinos Thalassinos,¶ James A. Irving,§ Imran Haq,§ Mun Peak Nyon,‡ John Christodoulou,¶ Adriana Ordóñez,§ Elena Miranda,|| *and* Bibek Gooptu‡

Contents

1. Introduction 140
2. Targeting the s4A Site with Peptides in a Pathogenic Variant of α_1-antitrypsin 142
   2.1. Interrogation of peptide sequences—Initial library synthesis 144
   2.2. PAGE-based readouts: Nondenaturing and urea-native PAGE 144
   2.3. Combinatorial approach 145
   2.4. Analysis of biomolecular interactions using surface plasmon resonance 147
3. Computational Approaches 148
   3.1. Choosing a starting model 148
   3.2. Site mapping 149
   3.3. Docking 151
   3.4. Evaluation of fit 153
   3.5. *In silico* approaches targeting the hydrophobic pocket flanking β-sheet A in α_1-antitrypsin 154
4. *In vitro* Screening of Small Molecules 155
   4.1. Fluorescence-based thermal shift assays 156
   4.2. Use of mass spectrometry for drug design in serpins 156

* Chemistry Research Laboratory, Department of Chemistry, University of Oxford, Oxford, United Kingdom
† Respiratory Medicine Division, Department of Medicine, University of Cambridge School of Clinical Medicine, Addenbrooke's and Papworth Hospitals, Cambridge, United Kingdom
‡ ISMB/Birkbeck, Crystallography, Department of Biological Sciences, Birkbeck College, London, United Kingdom
§ Department of Medicine, Cambridge Institute for Medical Research, University of Cambridge, Cambridge, United Kingdom
¶ ISMB/UCL, Research Department of Structural & Molecular Biology, University College London, London, United Kingdom
|| Dipartimento di Biologia e Biotecnologie 'Charles Darwin', Università di Roma La Sapienza, Piazzale Aldo Moro 5, Roma, Italy

Methods in Enzymology, Volume 501
ISSN 0076-6879, DOI: 10.1016/B978-0-12-385950-1.00008-0

Abstract

Native, metastable serpins inherently tend to undergo stabilizing conformational transitions in mechanisms of health (e.g., enzyme inhibition) and disease (serpinopathies). This intrinsic tendency is modifiable by ligand binding, thus structure-based drug design is an attractive strategy in the serpinopathies. This can be viewed as a labor-intensive approach, and historically, its intellectual attractiveness has been tempered by relatively limited success in development of drugs reaching clinical practice. However, the increasing availability of a range of powerful experimental systems and higher-throughput techniques is causing academic and early-stage industrial pharmaceutical approaches to converge. In this review, we outline the different systems and techniques that are bridging the gap between what have traditionally been considered distinct disciplines. The individual methods are not serpin-specific. Indeed, many have only recently been applied to serpins, and thus investigators in other fields may have greater experience of their use to date. However, by presenting examples from our work and that of other investigators in the serpin field, we highlight how techniques with potential for automation and scaling can be combined to address a range of context-specific challenges in targeting the serpinopathies.

1. Introduction

The metastable structural scaffold to which members of the serpin (*ser*ine *p*rotease *in*hibitor) superfamily of proteins fold allows them to function as adjustable rheostats regulating critical physiological processes. The dynamic serpin functional mechanism utilizes stabilizing conformational transitions, but it is intrinsically vulnerable to subversion by point mutations that facilitate the formation of aberrant conformers. As structural insights into these related processes have developed, an obvious implication has been the potential to design and/or screen ligands to modulate physiological or block pathological mechanisms.

There is a creative tension between two approaches to drug design: high-throughput and structure-based. High-throughput approaches have required significant investment to screen very large libraries of compounds against targets in automated procedures. Ligand selection and optimization are based upon readout criteria rather than the degree to which they can be demonstrated to act in accordance with a guiding mechanistic hypothesis. Structure-based approaches tailor ligands to bind structurally characterized target sites. Pharmaceutical companies have tended to favor high-throughput approaches coupled with readouts that model human physiology. In contrast, academic centers with expertise in mechanistic characterization have tended to focus upon structure-based strategies.

Put crudely, rational design is driven by highly detailed inputs, while high-throughput screening relies upon functional outputs that ideally are as biologically relevant as possible. Both approaches have inherent strengths and weaknesses but they are not necessarily mutually exclusive. As structural methodologies become increasingly robust, as processing power reduces in cost, and as automation technologies diffuse to the wider research market, it is increasingly feasible for any group to operate hybrid strategies. This chapter therefore outlines a range of techniques that can provide mechanistic insights but can also be scaled and, in many cases, automated. The aim is to review both well-established and novel approaches that we have found useful in studying serpin–ligand interactions. Our experience predominantly relates to development of novel therapeutics for the prototypic serpinopathy, α_1-antitrypsin deficiency. However, key questions are likely to be generally relevant across the serpin field. For example, how do we define targets—what sites in what conformational states? Or, how do we tell if the drugs might work—what readouts do we use at what stage? The optimal answers to such questions will vary according to context; thus in our experience, flexibility and the ability to approach questions using multiple techniques (in-house or through collaborations) are invaluable. This review is not intended to be exhaustive, but to describe the principles and potential of different methods, with illustrative "case studies" where possible of how these have been applied and modified.

To date, the most successful drug development in the serpin field has been in developing synthetic heparin-like oligosaccharides for therapeutic activation of the anticoagulant serpin antithrombin. This built upon insights from crystallographic and biochemical data into the mechanism of allosteric activation by a pentasaccharide sequence within the physiological ligand heparan sulfate proteoglycan (Jin *et al.*, 1997). Other serpins that play roles in regulating coagulation and fibrinolysis are also regulated by allosteric glycosaminoglycan (GAG) interactions (Huntington, 2003). The fibrinolytic serpin PAI-1 is also an important modulator of cellular pathways and may play roles in the biology of a number of diseases (Gramling and Church, 2010); its activated state is maintained by binding the GAG vitronectin. The development of vitronectin-mimetic compounds is therefore of similar pharmacological interest (Zhou *et al.*, 2003).

For α_1-antitrypsin, the physiological "ligand" used as a starting point for drug development is the reactive loop itself. After describing the rationale and techniques developed for this work, we go on to detail methods that may be used in small molecule screening programs *in silico*, *in vitro*, and in more complex physiological systems.

2. Targeting the s4A Site with Peptides in a Pathogenic Variant of α_1-Antitrypsin

Formation of pathogenic serpin polymers, like formation of the latent and cleaved conformers, involves the insertion of the residues of the serpin reactive site loop into β-sheet A to form a central, 4th β-strand (s4A; Dunstone *et al.*, 2000; Gooptu *et al.*, 2000; Huntington *et al.*, 1999; Lomas *et al.*, 1992; Sivasothy *et al.*, 2000; Yamasaki *et al.*, 2008). The use of peptide analogues of the reactive site loop to block polymerization by direct competition for binding in the s4A position is therefore logical. While this is effective *in vitro* (Lomas *et al.*, 1992), the use of 11–13-mer peptides to fill the entire length of the s4A site is not attractive therapeutically as they are difficult to deliver as drugs and are not entirely serpin-specific (Björk *et al.*, 1992; Lomas *et al.*, 1992; Schulze *et al.*, 1990; Skinner *et al.*, 1998). Moreover, in the case of the most clinically significant serpinopathy α_1-antitrypsin deficiency, those tested bind and inactivate wildtype (M) even more readily than the pathological mutant (Z) species (Chang *et al.*, 1996; Skinner *et al.*, 1998).

These and other data (Fitton *et al.*, 1997; Gooptu *et al.*, 2000) led to the hypothesis that the s4A site in the polymerogenic intermediate state populated by Z α_1-antitrypsin was partially occupied by insertion of proximal hinge residues, facilitating opening of the lower s4A site. This predicts that the upper s4A site in the Z variant would be more resistant to, and the lower s4A site more accepting of, peptide insertion compared to M α_1-antitrypsin. The finding that a 12-mer peptide annealed to Z α_1-antitrypsin at a slower rate compared with M α_1-antitrypsin strongly supported this hypothesis (Mahadeva *et al.*, 2002). To test it further, a 6-mer (FLEAIG) peptide mimicking the P7–P2 reactive loop residues that anneal to the lower s4A site in the latent and cleaved conformers was incubated with M and with Z α_1-antitrypsin. This region is nonconserved (Gettins, 2002) and likely plays a pivotal role in both binding affinity and specificity. As predicted, the FLEAIG peptide bound far more readily to Z than to M α_1-antitrypsin (Mahadeva *et al.*, 2002). These findings indicated that specificity could be optimized and so a program of further studies has built upon this result to define the optimal, minimal active sequence, with regard to binding affinity, specificity, and kinetics. A number of the techniques that were applied may be more generally applicable for drug development against serpin targets using both peptide and small molecule-led approaches. Figure 8.1A shows an overview of this process.

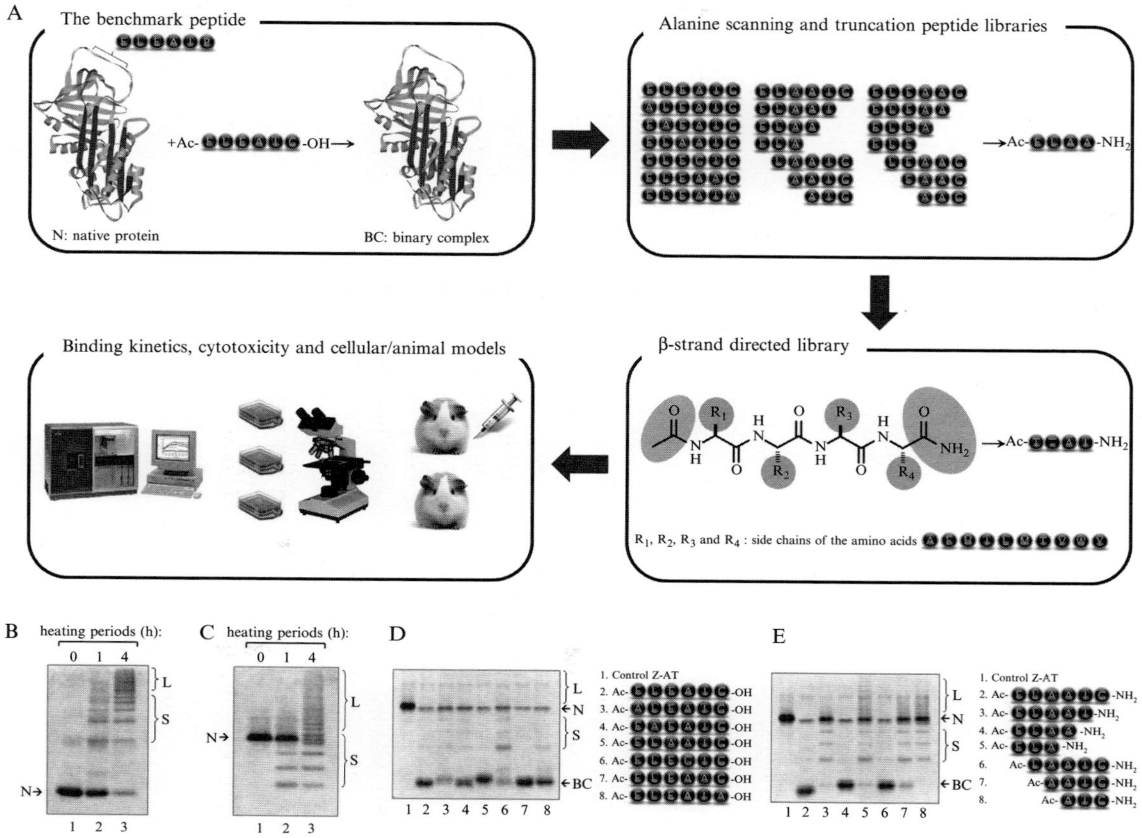
A
The benchmark peptide
+Ac- -OH
N: native protein
BC: binary complex
Alanine scanning and truncation peptide libraries
Ac- -NH2
β-strand directed library
R1, R2, R3 and R4 : side chains of the amino acids
Binding kinetics, cytotoxicity and cellular/animal models
B
heating periods (h):
0 1 4
L
S
N
1 2 3
C
heating periods (h):
0 1 4
L
S
N
1 2 3
D
L
N
S
BC
1 2 3 4 5 6 7 8
1. Control Z-AT
2. Ac- -OH
3. Ac- -OH
4. Ac- -OH
5. Ac- -OH
6. Ac- -OH
7. Ac- -OH
8. Ac- -OH
E
L
N
S
BC
1 2 3 4 5 6 7 8
1. Control Z-AT
2. Ac- -NH2
3. Ac- -NH2
4. Ac- -NH2
5. Ac- -NH2
6. Ac- -NH2
7. Ac- -NH2
8. Ac- -NH2

2.1. Interrogation of peptide sequences—Initial library synthesis

The FLEAIG peptide was initially probed by sequential alanine scanning and truncation to identify key residues and the minimum length for preferential binding to Z α_1-antitrypsin (Chang *et al.*, 2006) (Fig. 8.1B). The chemical synthesis of the requisite peptide library was carried out on PAL (Advanced ChemTech, Inc., Louisville, KY, USA) or Wang resin (AnaSpec, Inc., San Jose, CA, USA) using standard *N*-(9-fluorenyl)methoxycarbonyl (Fmoc) peptide synthesis techniques.

2.2. PAGE-based readouts: Nondenaturing and urea-native PAGE

Nondenaturing PAGE is an excellent technique for demonstrating polymerization of α_1-antitrypsin, as formation of dimer species from native protein is reported by a significant and discrete cathodal bandshift (Lomas *et al.*, 1993) (Fig. 8.1B). Recent data indicate that formation of the polymerogenic intermediate species is also reported by smaller cathodal bandshift from the native protein (Ekeowa *et al.*, 2010). It is therefore a natural choice of readout for polymerization blockade *in vitro* (Chang *et al.*, 1996; Fitton *et al.*, 1997; Lomas *et al.*, 1993). However, in the case of assessing binding of peptides to the s4A site in α_1-antitrypsin in a medium-to-high-throughput manner, it has limitations. Binding of small peptides to α_1-antitrypsin tends to result in subtle bandshifts (and may cause no bandshift if the peptide carries no net charge) relative to the native conformer. It can therefore be hard to quantify the occurrence of the binding event (i.e., the change from monomer to protein:peptide complex) directly.

Figure 8.1 (A) Overview of combinatorial approach. The benchmark peptide was derived from the reactive loop of native α_1-antitrypsin. The N-terminal proline was replaced by glycine due to synthetic considerations. The structure of binary complex (BC) is illustrated with α_1-antitrypsin and an imaginary peptide. The screening of alanine scanning and truncation peptide libraries identified the 4-mer peptide Ac-FLAA-NH_2 that binds to Z α_1-antitrypsin. Binding kinetics, cytotoxicity, and cellular/animal models were also studied. (B and C) 8% (w/v) nondenaturing PAGE (B) and (C) 8% (w/v) 8 *M* urea-native PAGE demonstrating the patterns seen with heat-induced polymerization of M α_1-antitrypsin (4 μg, heated at 58 °C) at 0 (lane 1), 1 (lane 2), and 4 (lane 3) h. The native conformer (N), short-chain (S), and long chain (L) polymers are indicated. (D and E) 8% (w/v) nondenaturing PAGE containing 8 *M* urea demonstrating the effect of alanine scanning of Ac-FLEAIG-OH and the derived truncation peptides on its binding to Z α_1-antitrypsin. Z α_1-antitrypsin was incubated with a 100-fold molar excess of the peptides as shown in the figure at 37 °C for 3 d. All lanes contain 2.5 μg of AT. (For color version of this figure, the reader is referred to the Web version of this chapter.)

To address this limitation, nondenaturing gels containing 8 *M* urea have been used to good effect (urea-native PAGE; Mahadeva *et al.*, 2002). Migration of native and polymeric samples on 8 *M* urea-native PAGE is shown in Fig. 8.1C. While native α_1-antitrypsin unfolds fully in 8 *M* urea conditions, peptide binding at the s4A site is associated with hyperstabilization and so, like heat-induced polymers, the fold structure maintains its integrity. In 8 *M* urea-native PAGE (Fig. 8.1D and E), binary-complexed α_1-antitrypsin can be readily distinguished as it clearly migrates more anodally than either polymers or residual native monomer (same mass but unfolded). Its formation can therefore be quantified by comparison with appropriate control samples (starting material and material incubated in the absence of peptide) followed by scanning digitization and use of densitometric software (e.g., Gel-Pro, Media Cybernetic, MD, USA). Normalization of the summed values for gel bands from each experiment allows quantitative comparison of the degree of binding seen with different peptides under identical conditions. Figure 8.1D and E demonstrates initial peptide screening steps. Native (N), short polymeric (S, migrating anodal to denatured native), and long polymeric (L, defined by migration slower than denatured native) as well as peptide-complexed (binary complexed, BC) conformers are indicated.

The use of this conformation-sensitive readout assay to assess peptide binding allowed confident identification of two additional 6-mer peptides (Ac-FLAAIG-OH and Ac-FLEAAG-OH) that showed similar binding to Z α_1-antitrypsin compared to Ac-FLEAIG-OH (Chang *et al.*, 2006). Further, truncation variants were identified (Ac-FLEAA-NH_2 and Ac-FLAA-NH_2) that also demonstrated similar binding.

2.3. Combinatorial approach

The binding avidity and effectiveness of the FLAA peptide in blocking polymerization encouraged a systematic attempt to expand the molecular diversity of further potential 4-mer peptide ligands for testing against α_1-antitrypsin. A peptide library scaled up by three orders of magnitude was generated by a combinatorial approach. To assess the best strategy for optimization, the reactive loops of inhibitory serpins were analyzed in terms of the Chou–Fasman secondary structure propensities of constituent amino acids (Chou and Fasman, 1978). 60.1% (69.0% if restricted to P12–P3′) of these residues are comprised by members of the top 10 β-sheet forming amino acids. For comparison, only 38.3% loop residues in noninhibitory serpins were from this set. Therefore, the top 10 β-sheet prone amino acids (A, F, H, I, L, M, T, V, W, and Y) were selected as the building blocks of the β-strand directed library.

The split-and-mix method was employed to construct a series of β-strand-directed libraries (Furka *et al.*, 1991). A simple example of a 3 × 3 × 3 library giving all 27 possible combinations of tripeptides is illustrated in Fig. 8.2A. To screen all possible combinations of the 10 candidate

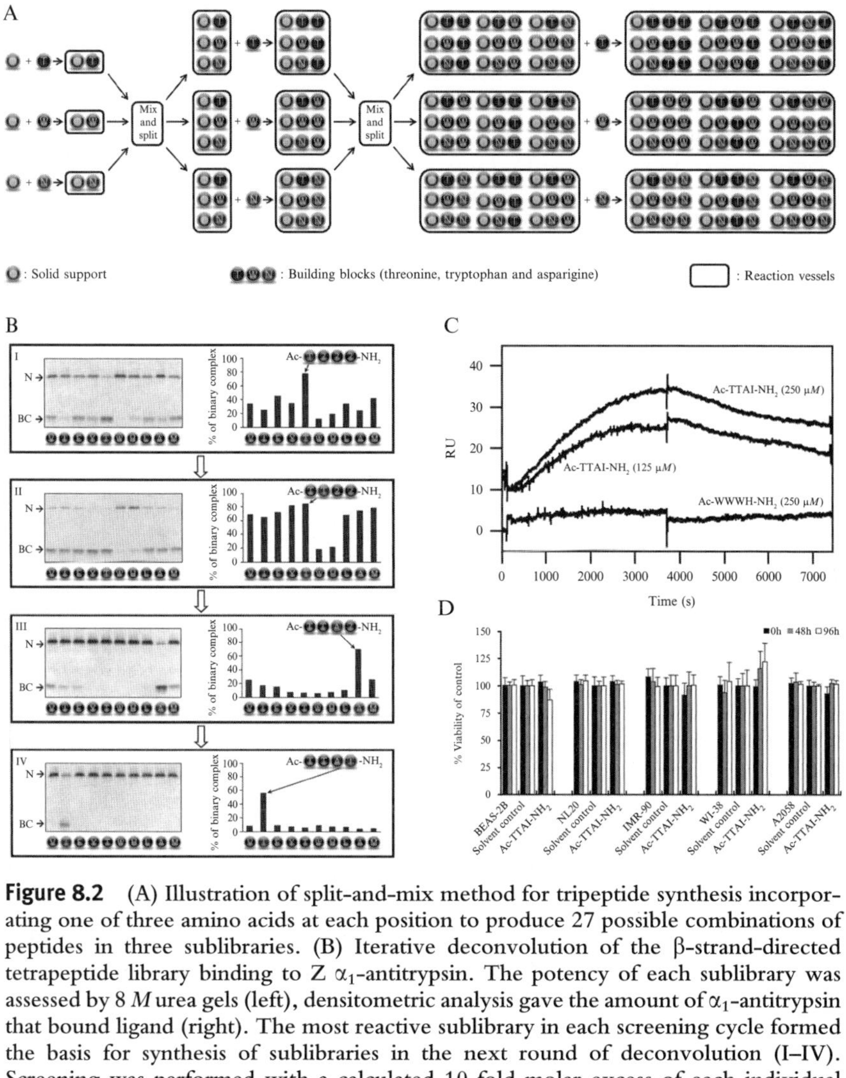

Figure 8.2 (A) Illustration of split-and-mix method for tripeptide synthesis incorporating one of three amino acids at each position to produce 27 possible combinations of peptides in three sublibraries. (B) Iterative deconvolution of the β-strand-directed tetrapeptide library binding to Z α_1-antitrypsin. The potency of each sublibrary was assessed by 8 *M* urea gels (left), densitometric analysis gave the amount of α_1-antitrypsin that bound ligand (right). The most reactive sublibrary in each screening cycle formed the basis for synthesis of sublibraries in the next round of deconvolution (I–IV). Screening was performed with a calculated 10-fold molar excess of each individual peptide in each sublibrary compared with Z α_1-antitrypsin. Incubations were performed at 37 °C for 2 h, or 1 h for the last round of screening. (C) Surface plasmon resonance demonstrating binding of the TTAI peptide to immobilized Z α_1-antitrypsin. (D) Lack of cytotoxicity of TTAI peptide against a panel of different cell types, assessed by formazan assay and colorimetric analysis. (For color version of this figure, the reader is referred to the Web version of this chapter.)

amino acids in a 4-mer peptide ($Ac\text{-}X_1X_2X_3X_4\text{-}NH_2$) required a 4 × 10 matrix, resulting in synthesis of 1 × 10^4 peptides using the same solid-phase method outlined above. These peptides were assayed as 10 sublibrary batteries, based upon their N-terminal (X_1) amino-acids—the only ones that are known by this approach. The most potent sublibrary was determined by the percentage of α_1-antitrypsin–peptide complex formation as assessed by 8 *M* urea-native PAGE (Fig. 8.2B, I). This set, containing Ac-$TX_2X_3X_4$ peptides, was then further analyzed by deconvolution (Fig. 8.2B, II-IV). Ten new sublibraries were synthesized by the split-and-mix method up to the X_2 residue with T as the N-terminal (X1) residue. Each therefore contained a mixture of 100 peptide combinations Ac-$TX_2X_3X_4$ in which X_2 was known. After retesting each sublibrary the process was repeated until an optimal tetramer of known sequence was identified (iterative deconvolution). The optimal residues (X_1 = T, X_2 = T, X_3 = A and X_4 = I) for α1-antitrypsin binding were therefore determined from four generations of libraries (Ac-$X_1X_2X_3X_4$-NH_2, Ac-$TX_2X_3X_4$-NH_2, Ac-TTX_3X_4-NH_2, Ac-TTAX_4-NH_2), respectively. The size of the sublibraries reduced by a factor of 10 with each iteration, i.e. 10^4, 10^3, 10^2, and ultimately 10 individual peptides with known sequences were assessed. The optimal X_4 site was determined to be phenylalanine for binding to M α_1-antitrypsin or isoleucine for binding to Z α_1-antitrypsin. The deconvoluted sequence was therefore Ac-TTAF/I-NH_2 (Chang *et al.*, 2009). The final peptides were screened by far more stringent-binding criteria than those used in previous peptide-binding studies (10-fold molar excess over α_1-antitrypsin and 1 h, compared with 50 or 100-fold molar excess over 24–48 h).

2.4. Analysis of biomolecular interactions using surface plasmon resonance

The result of library screening was further validated by surface plasmon resonance (SPR; Chang *et al.*, 2009). SPR-based ligand detection relies upon changes in surface electromagnetic waves at an interface between metal and an external medium upon binding of a new material to that interface. Z α_1-antitrypsin (50 μg/mL, 10 m*M* acetate, pH 4.5) was immobilized onto the SPR sensorchip (CM5 sensor chip linked to BIAcore 3000 optical biosensor) by a standard amine coupling procedure according to the manufacturer's recommendations. Ac-TTAI-NH_2 (125 and 250 μ*M*) and a negative control peptide Ac-WWWH-NH_2 (250 μ*M*) were injected (2 μl/min, 37 °C 20 m*M* Na_2HPO_4, 150 m*M* NaCl, pH 7.4) over the immobilized Z α_1-antitrypsin to give the binding sensorgrams. Multiple washes with 10 m*M* glycine buffer (pH 3.0) were performed between the injection cycles. Sensorgrams were recorded as a plot of binding response (resonance unit, RU) versus time and confirmed the specificity and tight-binding of Ac-TTAI-NH_2 to Z α_1-antitrypsin (Fig. 8.2C).

Chip-based SPR is a powerful technology for biomolecular interaction analyses in real time without the need of tags or labels (Rich and Myszka, 2006). It is highly scalable as a screening tool for drug discovery purposes as appropriate washing of the protein chip allows its reuse to assay multiple ligands in an automated screen (Neumann *et al.*, 2007). Since the interaction had already been well characterized *in vitro*, in this case, it was used as a binding validation assay. Nevertheless, its successful use here demonstrates its potential in screening TTAI-mimetic and other small-molecule drugs. Figure 8.2D reports a complementary cell survival assay demonstrating the lack of cytotoxicity of the TTAI peptide against five cell lines (see Section 5).

3. Computational Approaches

Computational (*in silico*) approaches to drug design combine high throughput and rational, structure-based elements. Where active molecules are known, a purely ligand-based approach is possible, with properties of the known ligands guiding design of new inhibitors. Where ligand data are sparse, a structure-based approach relying on at least one structural model of the target is required. Most commonly, this method involves identification of potential drug binding sites on the molecule (An *et al.*, 2005; Nayal and Honig, 2006; Perot *et al.*, 2010) and docking of small molecules to these sites (Brooijmans and Kuntz, 2003a; Zhou *et al.*, 2007). Alternatively, *de novo* design of inhibitory molecules to complement the characteristics of defined binding sites could be considered truly rational drug design. However, *in silico* modeling of the interaction of any particular compound with a target site remains unlikely to translate to an equivalent interaction *in vitro*. It is therefore advisable to test a large range of such hits *in vitro*. Synthesizing novel entities is expensive and technically nontrivial, so computational docking of existing libraries of small molecule structures is currently favored. These structures explore large ranges of chemical space and the compounds are typically purchasable at low cost, allowing selection of multiple potential "hits" for *in vitro* testing.

3.1. Choosing a starting model

There are a wealth of X-ray crystallographic data on serpins representing various clades (Silverman *et al.*, 2001) of the superfamily. For α_1-antitrypsin, 10 crystal structures of the native conformer (wildtype and mutant) have been deposited with the RCSB PDB. Homology modeling could, theoretically, provide initial structural models for serpins from underrepresented clades. However, as binding sites present on one serpin are not necessarily found on others (Elliott *et al.*, 2000), it is currently not clear whether such models would be sufficiently reliable for confident definition of potential-binding sites.

Use of the highest resolution structure available as a starting model provides the most reliable information on relative atomic positions within the structure.

However, lower resolution structures may also indicate valuable information on protein dynamics relating to binding sites in solution (Furnham *et al.*, 2006). Further, crystallographic starting models may bias toward "snapshot" structural features that are favored under crystallization conditions (e.g., lattice contacts) but not in solution physiologically. This could be addressed by using solution NMR, however, the relatively large size of serpins presents technical challenges (Section 4.4). Alternatively, an ensemble of native-like states can be generated *in silico* from an initial crystal structure. Molecular dynamics approaches explore pathways of conformational change *in silico* based upon energetic favourability, but are computationally very intensive. Simpler alternatives include Normal Mode Analysis (Brooks and Karplus, 1985; Ma, 2005) and distance constraints-based approaches adopted by the programs CONCOORD (de Groot *et al.*, 1997) and tCONCOORD (Seeliger *et al.*, 2007). CONCOORD randomly perturbs the coordinates of the protein atoms, thus stochastically producing a very large number of hypothetical "structures." Unrealistic structures are filtered out according to a list of distance constraints derived from the initial input structure. Figure 8.3A illustrates the use of CONCOORD in α_1-antitrypsin. Site-mapping using the conformers generated by this program may provide an indication of the persistence of sites identified in the starting crystal structure.

For serpins, the inherent potential for significant conformational change adds complexity to the choice of starting structural model. Cleaved or latent conformers may be used to model the polymeric protomer as they are biochemically analogous in their hyperstabilized (relaxed) characteristics. The hydrophobic pocket flanking β-sheet A in α_1-antitrypsin was identified by comparison of crystal structures of the native and cleaved conformers (Elliott *et al.*, 2000). Further, the concept of a branched pathway from native to polymeric or latent states has led to the development of a structural model of a common intermediate or M* species (Gooptu *et al.*, 2009).

3.2. Site mapping

Site mapping can be performed with or without prior assumptions specific to the system being studied. The s4A and β-sheet A flanking hydrophobic pocket targeting approaches (Fig. 8.3B) are case studies of predefined target sites based upon crystallographic and biochemical data. Global site mapping can be performed according to geometric or energy-based criteria. Geometric approaches define cavities as possible binding sites based upon volume and shape. They tend to be less computationally intensive and were used in early site mapping programs such as SURFNET (Laskowski, 1995), POCKET (Levitt and Banaszak, 1992), and LIGSITE (Hendlich *et al.*, 1997). SURFNET and POCKET identify sites by fitting virtual spheres into the solvent-accessible space between protein atoms. LIGSITE scans the protein surface from cubic grid points to define pockets.

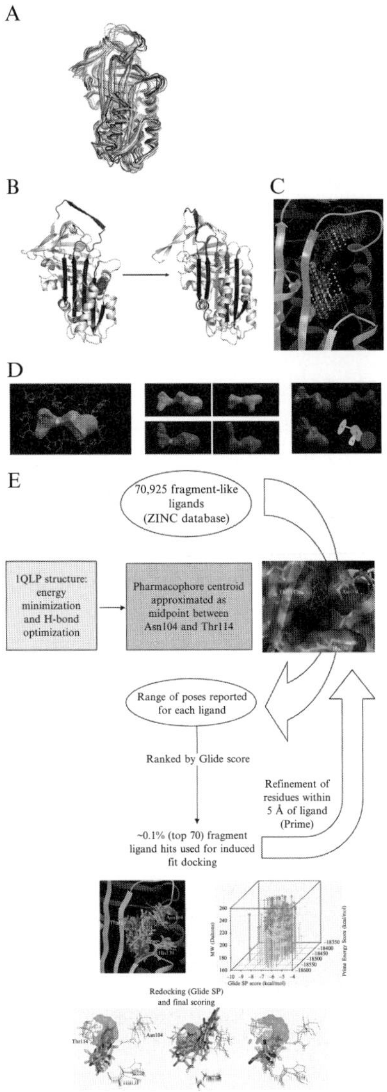

Figure 8.3 (A) Computationally generated, native-like conformers of α_1-antitrypsin. A hundred conformers were generated using the program CONCOORD. A representative sample of 8 (from 100) conformers is shown. All 100 structures were screened for persistence of binding sites observed in crystallographic structures. (B) Screening targets identified through interpretation of prior crystallographic and biochemical data. The s4A site (see Section 2) believed to be patent in the polymerogenic intermediate of α_1-antitrypsin (right) is highlighted in orange. The hydrophobic pocket flanking β-sheet A is shown on the native conformer (1QLP) as identified by SURFNET (purple mesh). As this cavity is abolished by expansion of β-sheet A, compounds

Site volume is generally the main criterion for binding site prediction, as most experimentally identified protein-drug complexes are observed in large sites (Nayal and Honig, 2006; Perot *et al.*, 2010). However, ligand-binding capacity is not accurately predicted by geometry alone. Geometry-based methods may add simple terms for atom, amino acid, and molecular-binding site properties, for example, proportions of hydrogen donor or acceptor groups (Perot *et al.*, 2010). More recent site mapping programs such as PocketFinder (An *et al.*, 2005) and SiteMap (Halgren, 2007) use probe and energy-based methods to estimate the interaction energy between the probe and the protein. Both of these programs are able to define potentially druggable sites in α_1-antitrypsin although PocketFinder requires manipulation of two residues to define the cavity flanking β-sheet A (Mallya *et al.*, 2007). Energies are calculated along a three-dimensional grid sampling the space around the structural model. These values are combined with geometry-based scoring to give a druggability scoring function. For example, SiteMap scores sites and their druggability using a function of volume, site enclosure (solvent exclusion) with a penalty factor for hydrophilicity (Halgren, 2009). Other calculated parameters giving further information about the site are tabulated in the SiteMap output (Fig. 8.3C).

3.3. Docking

Once potential-binding sites have been defined, they can be used as targets for virtual docking of compound structure databases (libraries). Pharmaceutical companies typically compile such libraries in-house. However, compound

binding here should prevent polymerization. (C) Hydrophobic pocket flanking β-sheet A as identified using SiteMap. Each site is highlighted using white spheres (i.e., site points), around which hydrophobic (yellow grid), hydrogen acceptor (red grid), and hydrogen donor (blue grid) maps are highlighted. (D) The same site shown in (B) and (C), as identified by PocketFinder (following relaxation of Asn104). Changes in pocket dimensions are demonstrated by comparison of volumes pre- (blue, PocketFinder only) and post- (green) ligand-induced changes modeled using the program SCARE. Division of this cavity into subvolumes defined chemically and computationally is shown (right). (E) Upper panel; overview of process in which a cavity subvolume (shown as blue mesh within the pocket flanking β-sheet A) was identified by comparison of the high-resolution crystal structures of Thr114Phe (PDB code: 3DRM) and wildtype (1QLP) α_1-antitrypsin. This was then used as the target for a computational screen of >70,000 fragment-like compounds. Asn104, Thr114, and His139 were used as coordinating residues. Lower panel; The best hits (ensemble top left) were classified (top right) by two docking scores, molecular weight and proximity to target centroid (heatmap coloring $\geq$5.0 Å, yellow; 2.5–5.0 Å, orange; $\leq$2.5 Å, red). Three ensembles of five fragments that scored highly by different criteria are shown (lower row, target in cyan, coordinating α_1-antitrypsin residues shown as per induced-fit, color coded with appropriate ligand). (See Color Insert.)

libraries may be available commercially (e.g., Available Chemicals Directory and ChemNavigator Database) or through public access initiatives such as ZINC (Irwin and Shoichet, 2005), DrugBank (Wishart *et al.*, 2006), ChEMBL (Warr and Overington, 2009), and PubChem (Wang *et al.*, 2009). Compounds can be selected on the basis of various characteristics. Popular criteria include compliance with the Lipinski "Rule of 5." This observes that success in drug development is highly correlated with: $M_r < 500$ Da, lipophilic index (log P) < 5, ≤ 5 hydrogen bond donors, ≤ 10 hydrogen bond acceptors. A modified version of this, the "Rule of 3," has been proposed to guide drug design via exploration of a greater diversity of chemical space using still smaller (<300 Da) molecular fragments to identify potential hits.

In silico docking models the optimal binding of a ligand to a pocket. The simplest docking algorithms dock ligands by rigid body fitting (Brooijmans and Kuntz, 2003a). However, most modern docking programs incorporate ligand flexibility in the calculations, allowing sampling of many different ligand conformations during docking (Totrov and Abagyan, 2008). Docking programs such as GOLD (Verdonk *et al.*, 2003), Autodock (Park *et al.*, 2006), DOCK (Lang *et al.*, 2009), and Glide (Friesner *et al.*, 2004) scan the potential energy landscape to find the global energy minimum for the protein:ligand complex. They then rank potential ligand poses and score those selected according to predefined criteria using two different functions. To account for potential flexibility of the binding pocket in solution or local conformational changes induced by proximity of ligand and pocket groups, a range of ligand protonation states and ligand or receptor conformers may be trialed (Halgren, 2009; Irwin and Schoichet, 2005).

Alternatively, protein side-chain flexibility in solution and ligand-induced changes can be modeled by newer docking algorithms (Totrov and Abagyan, 2008), though this requires significantly greater computational resources. In this approach, single structural starting models of ligand and protein are used and the program identifies the conformations giving the most favorable interaction in an "induced-fit" manner. An example of such an approach is the Induced Fit Protocol, a method that combines protein energy minimization using the program Prime with ligand docking using the program Glide within the Schrödinger Suite (Schrödinger LLP, NY, USA).

The internal coordinate mechanics (ICM) algorithm also allows flexible fitting of the ligand into the receptor site. It adjusts torsion and phase angles as variables according to the Monte Carlo method (Metropolis and Ulam, 1949). The receptor is represented by precalculated grid potentials incorporating the shape and specificity of multiple energy terms (e.g., electrostatics, directional hydrogen bonding, hydrophobic interactions). Truncated van der Waals repulsion terms are used to limit adverse effects of minor steric clashes. These make minimization more efficient (soft docking) while acknowledging some receptor flexibility. Docking to this weighted grid model is fast and avoids evaluation of each atomic interaction between the ligand and the protein (Carlson and McCammon, 2000).

3.4. Evaluation of fit

Docking programs rely on scoring functions to evaluate the fit between protein and ligand. Usually simple molecular mechanics functions are used to select reasonable poses for a single protein–ligand complex during the docking run (Halperin *et al.*, 2002) while more sophisticated functions are required to rank final poses of different ligands. Scoring functions may be categorized as physics-based, empirical, or knowledge-based (Bohm, 2003). Physics (force field)-based functions (Bohm, 2003; Brooijmans and Kuntz, 2003b) involve nonbonded terms borrowed from popular intermolecular force fields, such as electrostatic and van der Waals terms. These effectively ignore entropy and solvation and attempt to calculate the gas-phase enthalpy of binding. Empirical scoring functions (Eldridge *et al.*, 1997; Friesner *et al.*, 2004; Rarey *et al.*, 1996) decompose the free energy of binding into a sum of weighted contributions (such as hydrogen bonds, hydrophobic interactions, ionic interactions, and entropic terms), dependant on the protein and ligand coordinates. Weighting terms are decided by fitting to a large dataset of known experimental binding free energies. Empirical scoring functions are very popular in docking but their success is often a result of cancelation of errors, thus making it very hard to improve them by rational addition or improvement of terms. Knowledge-based methods (Gohlke *et al.*, 2000; Mitchell *et al.*, 1999; Muegge and Martin, 1999) utilize Boltzmann's law to translate the probability of observing a pair-wise interaction (based upon statistical analysis of available protein–ligand complex structures) to an energy estimate for that interaction. Once derived, these potentials are very fast to evaluate and can account for interactions that may be neglected or misrepresented in common force fields. However, the theoretical basis of these potentials has been heavily criticized, (Ben-Naim, 1997) and practical issues such as the low numbers of observations for certain atom–atom pairs make their reliability questionable.

Efforts to improve scoring functions concentrate on better solvation models (Mysinger and Shoichet, 2010), rescoring of docked poses using more sophisticated methods (Graves *et al.*, 2008; Kalyanaraman *et al.*, 2005), and consensus scoring (Bissantz *et al.*, 2000; Wang and Wang, 2001). Ultimately, however, there is a limit to the accuracy of methods that ignore or oversimplify the contribution of entropy to binding, that ignore the unbound state for the protein and ligand, and that break down free energy into a sum of additive terms, thus ignoring cooperative effects. Taken together with the narrow window of affinity that these scoring functions have to predict (approximately 4.5 kcal/mol Leach *et al.*, 2006), these factors make accurate affinity prediction a tall order for any scoring function.

3.5. *In silico* approaches targeting the hydrophobic pocket flanking β-sheet A in α_1-antitrypsin

The hydrophobic pocket flanking β-sheet A in α_1-antitrypsin was first identified and characterized (Elliott *et al.*, 2000) using the program SURFNET (Laskowski, 1995) (Fig. 8.3B; grid separations of 0.8 Å, probe sphere radii 2.0–4.0 Å). Its potential as an allosteric target for polymerization blockade was based upon comparison of native and cleaved structures. This analysis implied that filling of this surface accessible cavity would prevent the expansion of β-sheet A required for polymerization. Experimental data supported this hypothesis (Parfrey *et al.*, 2003), and so *in silico* docking was performed using the ICM (Totrov and Abagyan, 1997) algorithm. The original crystallographic coordinates of the lateral hydrophobic pocket of the crystal structure of native wildtype α_1-antitrypsin were screened against ~1.2 million commercially available drug-like compounds. Next the screen was repeated following an assessment of local side-chain flexibility and modeling of the effects of ligand-induced changes by a biased probability Monte Carlo simulation of the 17 side chains surrounding the pocket. Lowest energy representatives were selected for conformations where side-chain torsion RMSD was $<15°$. Optimal docking interactions (poses) were ranked by the ICM scoring function. Compounds ranked in the top 1% were selected for biological testing if the ligand was clearly located in the pocket and made at least one hydrogen bond with the protein.

This approach can be iterated. SCARE (Bottegoni *et al.*, 2008) is an ICM-based approach that models induced fit effects involving backbone as well as side-chain rearrangements to improve pocket representations during docking. Figure 8.3D shows changes in the hydrophobic pocket flanking β-sheet A between the volume defined by PocketFinder using the 1QLP crystal structure of α_1-antitrypsin as a starting point and that defined using SCARE during ligand docking. This pocket can then be defined in terms of linked subvolumes allowing a rational approach to optimization of different parts of the ligand moiety (Fig. 8.3D, right).

Experimental data have also been used to generate a smaller target volume within the cavity where pharmacophore binding may confer resistance to polymerization without abolishing inhibitory activity (Parfrey *et al.*, 2003). This volume (Fig. 8.3E) was therefore attractive for targeting of fragment compounds, but since it was not tightly enclosed, it was unclear whether it could be defined sufficiently for effective *in silico* docking. A proof-of-principle *in silico* screen was performed.

The fragment-like subset labeled "usual" of the ZINC database (Irwin and Schoichet, 2005) was selected for initial screening. This ligand was prepared using programs from the Schrödinger Suite 2007 (Schrödinger LLP, New York, NY, USA) and the protein was prepared using Maestro version 8.0 (Protein Preparation Wizard). The potential energy grid was precalculated using a box centered on the centroid of residues Asn104 and Thr114. The box was 39 Å in each dimension, but the ligand center was constrained to lie within 14 Å of the box center. No additional constraints were used.

Initial screening using Glide version 4.5 generated a range of poses. The best pose for each ligand calculated using Emodel was ranked using the Glide SP score. The top 70 (0.1%) ligands resulting from this Glide SP run were then submitted to the Induced Fit Docking protocol. Preliminary docking by Glide SP used a scaling factor for van der Waals radii of 0.5 for both the ligand and the receptor. For each ligand, the best 20 poses were retained. Prime version 1.6 was used to optimize receptor side chains within 5.0 Å of each posed ligand. Finally, ligands were redocked into the optimized structures.

The resulting poses were ranked using a composite of the Glide SP (binding-site energy) and the Prime (molecular mechanics + solvation energy function for the protein–ligand complex) scores as previously described (Sherman *et al.*, 2006). The process, summarized in Fig. 8.3E, successfully generated numerous hits with docking scores theoretically equivalent to K_d values in the submicromolar range.

4. *In vitro* Screening of Small Molecules

As discussed in Section 2.2, nondenaturing and urea-native PAGE can usefully report polymerization blockade or increased protein stability where this is a direct consequence of ligand binding. As well as reporting peptide binding, nondenaturing PAGE has been used to report binding of small molecules targeted against the cavity flanking β-sheet A (Mallya *et al.*, 2007). However, such techniques may not report weaker binding of small molecules that nevertheless might be useful as lead compounds and they are not highly automatable. A range of biophysical and structural techniques is therefore likely to be useful in future higher throughput screening programs to identify small molecule binding to serpins.

In this review, scalability refers to a technique's potential for screening many compounds using the same experimental setup without a disproportionate increase (and ideally a proportional reduction) in technical requirements. A powerful aid to this is to assay compounds within subsets or "cocktails" rather than individually. The identity of hit ingredients responsible for positive results can then be dissected out. In effect, this is similar to the peptide deconvolution process described in Section 2, but in small molecule cocktails, identification of hit constituents does not generally require further rounds of synthesis.

4.1. Fluorescence-based thermal shift assays

Thermal shift assays are based upon the propensity of some dyes, structurally related to 8-anilino-1-naphthalene sulfonate (ANS), to bind to partially denatured, though not folded or fully denatured material and fluoresce upon binding. Fluorescence therefore peaks when there is maximal population of a partially denatured state. For a two state unfolding transition, this will approximate the T_m of this transition. Binding of a ligand that stabilizes the complex against denaturation will therefore be reported by an increase in the temperature at which peak fluorescence is observed (Cimmperman *et al.*, 2008). The significance of binding that results in an apparent drop in T_m is harder to interpret.

A popular method utilizing this approach is the ThermoFluor assay. Protein unfolding is monitored by fluorescence of the solvatochromic fluorescent dye SYPRO Orange using a Real Time PCR machine. It is robust, sensitive, highly scalable, and automatable though hits require validation by other techniques. It is therefore used in high-throughput screening for drug development (Lavinder *et al.*, 2009). A simple protocol is as follows:

1. Dissolve compounds in 100% Dimethyl sulfoxide (DMSO) to give a stock solution of 20 m*M*.
2. Assays are performed in 25 μl reaction volume in individual wells within a 96 well plate. This includes 1 μl SYPRO Orange (1:200 dilution). The protein concentration is 1 mg/ml, compound concentration is 1 m*M* (i.e., final DMSO 5%).

Duplicate each assay in six wells, including controls.

Increase temperature in 0.5 °C steps from 10 to 95 °C and measure fluorescence at each step (excitation wavelength ~492 nm and emission wavelength ~580 nm).

Outputs can be plotted as change in fluorescence intensity (arbitrary fluorescence units, RFU), or as the rate of change in fluorescence respect to temperature (dRFU/d*t*). The apparent T_m is the temperature at which peak fluorescence, in the former representation, or the corresponding trough dRFU/d*t* value, occur.

4.2. Use of mass spectrometry for drug design in serpins

Mass spectrometry (MS) measures the mass-to-charge (m/z) ratio of ionized molecules in the gas phase. In the past decade, it has emerged as a powerful complementary technique for characterization of protein complexes and protein–ligand interactions in structural biology (Heck, 2008; McCammon and Robinson, 2004).

The "soft" electrospray ionization (ESI) technique has proven to be the ionization method of choice for the analysis of large proteins. It produces gaseous ions from sample solutions flowing (1–20 μl/min) through a metal capillary maintained at high voltage (2–5 kV) and atmospheric pressure. Ions are driven by the electric field causing positive charge to build at the capillary tip (Taylor, 1964). At high field strength, a fine spray of highly charged droplets emerges. These conditions are not optimized to preserve noncovalent assemblies. Nevertheless, ESI has been used in industrial, high-throughput programs (often preceded by a chromatography step to clean the sample and then used in a tandem MS experiment). Such screens can process 10^5–10^6 compounds per week (Allen Annis *et al.*, 2007; Muckenshnabel *et al.*, 2004). The limitations inherent in conventional ESI can be significantly overcome by miniaturization (nanoflow (or nano-)ESI) (Wilm and Mann, 1994). Borosilicate or quartz capillaries (tip diameters 1–4 μm), made conductive by coating with gold or inserting a thin metal wire down the length, allow the use of far lower flow rates (~20 nl/min) and lower voltage (~1.5 kV). Longer analysis times from small samples (~30 min from 1–5 μl) become possible. Further, the smaller droplets produced (<200 nm in diameter) easily evaporate (Wilm and Mann, 1996) negating the use of heating for aqueous solutions. Such conditions allow proteins to remain in "native-like" conformation and preserve non-covalent intramolecular and ligand interactions.(Juraschek *et al.*, 1999; Karas *et al.*, 2000).

4.2.1. Nano-ESI of noncovalent assemblies: Sample and mass spectrometer considerations

Solution conditions must maintain fragile noncovalent interactions yet be suitable for MS. Commonly, 20–70 μl samples are buffer exchanged from storage conditions into aqueous solutions of ammonium acetate or ammonium bicarbonate pH 7.0 using microcentrifugal membrane concentrators or gel filtration columns (e.g., MicroBiospin, Bio-Rad, Hercules, CA).

The most commonly used mass spectrometer for studies of noncovalent assemblies is the quadrupole–time of flight (Q–ToF). In addition to conventional MS analyzers, this hybrid instrument contains both a quadrupole and a time-of-flight analyzer separated by a collision cell. Ions of interest are selected by m/z characteristics in the first (quadrupole) analyzer and subjected to collisions in the gas-filled collision cell, resulting in their dissociation. The products are then analyzed with the second (ToF) analyzer. ToF analyzers are simple in their design and have a theoretically unlimited mass range. Q–ToF designs which utilize a heated transfer capillary should be avoided, as in our experience this induces some dissociation of complexes.

Large ions acquire a large amount of kinetic energy upon transfer into the vacuum of the mass spectrometer. This leads to ion beam broadening and transmission losses as ions traverse the instrument. The use of higher pressures in the initial stages of the mass spectrometer can compensate for this through "collisional cooling"—the damping of energy as a result of collisions with residual gas molecules (Chernushevich and Thomson, 2004;

Krutchinsky *et al.*, 1998). Adduction of residual solvent and buffer molecules can also cause peak broadening. Optimizing desolvation may improve this, however, there is often a trade-off since increasing desolvation may disrupt noncovalent interactions in the gas phase.

4.2.2. Characterization of ligand binding

The use of MS to study protein–ligand interactions has been reported extensively in the literature (Daniel *et al.*, 2002). MS techniques are sensitive and rapid, provide stoichiometry data, and can assay a wide range of masses (Fig. 8.4). Gas-phase complexes can retain the structural features seen in solution, even after relatively long periods of time (15–30 ms; Ruotolo *et al.*, 2005). However, noncovalent interactions may vary between solution and gas phase. Thus in a vacuum, two point charges separated by 5 Å will demonstrate a bond strength that is 3.5 kJ/mol higher than in water at 25 °C. For hydrogen bonds, such effects can result in bond strengths of 17 kJ/mol for peptides in the gas phase corresponding to bond strengths of 2–6 kJ/mol in water (Sheu *et al.*, 2003). Conversely, the strength of hydrophobic effects in aqueous environments is driven by entropically unfavorable effects of solvent exposure of hydrophobic moieties. In the gas phase, minimal solvation will therefore weaken hydrophobic interactions. Fortunately, this consideration appears to be less critical for small ligands binding to proteins of similar size to serpins (McCammon *et al.*, 2002; Rostom *et al.*, 2000). This may be explained in part by the fact that large proteins often contain binding pockets and channels, which act to bury the ligand and prevent the depletion of solvent.

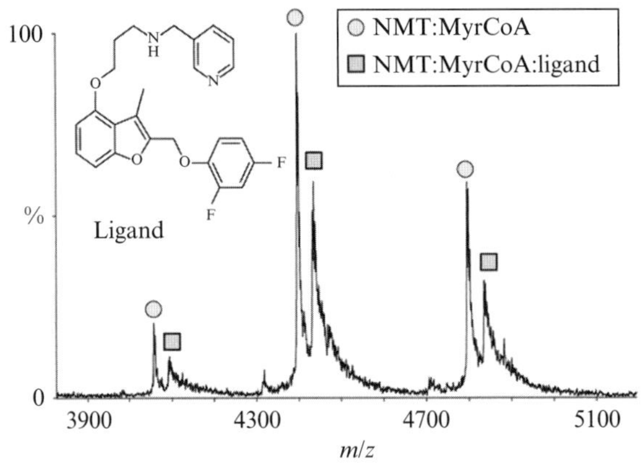

Figure 8.4 Nano-ESI mass spectra of the products of incubating the complex between *N*-myristoyl transferase and its cofactor myristoyl coenzyme A with excess of an antifungal ligand. The ligand binds to the enzyme complex to form a (1:)1:1 ternary complex. (For color version of this figure, the reader is referred to the Web version of this chapter.)

4.2.3. Solution based titration protocol

The simplest use of nano-ESI–MS in drug development is to monitor a solution-phase titration for the determination of the equilibrium of a protein ligand pair (Lim *et al.*, 1995). This assumes that the ionization process does not affect the equilibrium. Generally, the host concentration is kept constant whereas the guest concentration is varied over a range of about two orders of magnitude. The intensity of the complex is then compared to the intensity of the free protein for every ligand concentration.

Sample preparation

1. Prepare 20–70 μl of 5–20 μ*M* protein solution buffer exchanged into aqueous ammonium acetate using either microcentrifugal gel filtration columns or membrane concentrators.
2. Prepare ligand at a concentration suitable for titration experiments so that minimal dilution is required. Organic solvents such as DMSO, often used to solubilize small molecules, may cause structural perturbations of the protein. We have found that a limiting DMSO concentration to $\leq$10% (v/v) in the reaction mixture is usually sufficient to prevent any changes in structure as evident from the charge state distribution.

Titration: Infusion and Ionization

3. Carefully, trim ends of glass capillaries open under a microscope (some commercial needle suppliers provide ionization needles that are already open).
4. Mount capillary in source and load with 1–4 μl of sample using gel loader pipette tips.

Optimization of instrumental parameters

5. Calibrate instrument using an appropriate standard; we favor 33 mg/ml cesium iodide in a 50:50 solution of MeCN and water.
6. Determine minimal capillary voltage and backing gas pressure for ionization and stable spray.
7. Optimize pressure in the initial stages of the vacuum chamber to maximize the transmission of the ions of interest. This is instrument specific and should be optimized for each experiment.
8. Screen carefully for minimal effective desolvation conditions within the instrument before recording *m/z* data.

Data analysis

9. Data recorded from MS titration experiments allow the determination of the relative concentrations of different states to be plotted as a function of

total ligand concentration. Assuming equal ionization efficiencies for the apo- and ligand-bound forms of the protein, the fractions of each corresponding species in solution can be quantified by comparison of peak heights or areas. Where peak widths differ across charge state series, peak areas may be more accurate. The number of binding sites can be deduced from the stoichiometry at saturation. Taken together, these data allow K_d values to be calculated using appropriate theoretical models. MS methods of K_d estimation were recently shown to compare favorably with SPR and circular dichroism spectroscopy for the calmodulin–melittin system (Mathur *et al.*, 2007). However, at present, we advise that, although K_d values derived from MS data are internally consistent, they should not be directly compared with K_ds for other interactions derived from more robustly validated methods.

Alternatively, "gas-phase methods" involve dissociation of the complex. Such methods can produce a rapid and straightforward evaluation of gas-phase stabilities. Typically, the energy at which 50% of the complex dissociates is taken as a measure of gas-phase stability.

4.2.4. Automation

Nano-ESI MS can now be automated and will likely allow academic centers to use it for medium- to high-throughput (10^4–10^5 compounds) screening of small molecules against target proteins and characterization of interactions. Robotic ionization devices such as the TriVersa NanoMate (Advion Biosciences, Ithaca, NY) enable rapid, inexpensive screening and K_d estimation for protein–ligand complexes with speeds up to 50 times faster and 100 times less expensive than NMR. The devices contain a microfluidics chip comprising an array of nanoelectrospray nozzles etched in a silicon wafer. Although the same ionization processes are involved, the fabrication method used is highly reproducible and the field strength created by the nanoelectrospray nozzles allows for a more efficient and stable spray.

4.2.5. Mass spectrometry of peptide fragments

In addition to the techniques described for studying protein–ligand interactions directly using intact protein above, many MS approaches provide information by studying cleaved peptide fragments; such studies may localize binding sites. For drug development, this can involve compounds designed to form covalent cross-links with the protein target upon binding (Erlanson and Hansen, 2004). The observed fragment sizes allow identification of sequence and the general binding site of covalently bound compound. The binding of the CG molecule to plasma derived Z α_1-antitrypsin was investigated by this approach in case it involved covalent linkage, with and without PNGase deglycanation (Mallya *et al.*, 2007). No such linkage was observed in the 70% of primary sequence that was detected. However, this highlights a limitation of fragment-based approaches: the challenge of interpreting negative results in the context of subtotal sequence coverage. Nevertheless, MS fragment analysis of hydrogen–deuterium exchange and hydroxyl free radical footprinting profiles have been characterized for various α_1-antitrypsin conformers (Sengupta *et al.*, 2009; Tsutsui *et al.*, 2006; Zheng *et al.*, 2008). Changes in such profiles could therefore give a readout of serpin-compound interactions and residues affected by binding in screening programs.

4.3. Crystallographic approaches

The crystallographic structure of a protein in complex with a small molecule ligand can provide high-resolution insights into binding mode, stoichiometry, mechanism of action, and conformational effects. These data inform subsequent modeling and chemical synthesis steps. Additionally, crystallographic methods provide an excellent means of detecting and localizing the binding of fragment-like (<250 Da) compounds. These are believed to be invaluable as a first stage of drug development, as lead compound optimization tends to modifications that increase molecular size (Carr *et al.*, 2005). The probability of a drug reaching market correlates inversely with molecular weight—relatively few drug molecules are >500 Da (Lipinski *et al.*, 1997)—so it is attractive to start the process with initial leads that are as small as possible. However, smaller size correlates with lower binding affinity. Consequently, relatively few *in vitro*, high-throughput approaches are sufficiently sensitive to detect fragment screen hits. Crystallographic (and NMR spectroscopic) methods can provide the required sensitivity.

Advances in high-throughput nanocrystalization of macromolecules reduce the amount of material required for crystallization experiments, permitting a wider range of buffer conditions to be evaluated. Advances in data processing and structure solution software facilitate the progression from data to crystallographic model. In this section, reference will be made to liquid-handling equipment capable of dispensing 100–250 nl drops in a 96-well format crystallization plate, with a reservoir volume of 50–150 μl. A format for mounting trays by hand is a 24-well plate, with drops usually 0.5–5 μl in volume and a reservoir volume of 250–1000 μl.

Pivotal to the crystallographic characterization of a ligand is the ability to obtain diffracting crystals. For the greatest chance of success, the protein preparation should have the following features:

- *Homogeneity*: single species on SDS-PAGE and/or denaturing MS, for serpins conformational homogeneity state should be assessed by native-PAGE or size exclusion chromatography.
- *A low buffer and salt concentration*: ensures pH and salt composition of the protein:buffer drop is dominated by the buffer components.
- *Minimal organic solvents/detergents*

- *High protein concentration*: values of 5–20 mg/ml are typical and generally have to be determined empirically by conducting several crystallization experiments.

The small molecule used in the crystallization experiment should be:

- *Homogenous*
- *High concentration, in a buffered, protein-compatible solvent*: The contribution of solvents other than water to the final buffer composition should be minimal.

The three typical scenarios for undertaking crystallization experiments depend on the level of *a priori* knowledge about which buffer conditions favor crystal growth, the characteristics of the crystals themselves and the effect of the ligand on the protein configuration.

4.3.1. The target protein has not yet been crystallized

This necessitates the broadest initial strategy to be taken and favors the use of a high-throughput approach to test as many conditions as possible. If sufficient quantities of the ligand are available, the target protein is mixed with a 2–10-fold molar concentration of ligand (or at a concentration 5–10-times the K_d value for the interaction, whichever is greater). If not, attempts can first be made to crystallize the protein in the absence of ligand.

Many commercial "sparse" screens are available (so-called because they canvas a wide range of buffer conditions), that can provide useful information for subsequent fine-tuning steps. A typical vapor–diffusion experiment proceeds as follows: for each well on the crystallization plate containing a specified volume of crystallization buffer "mother liquor," equal volumes of protein solution and crystallization buffer are mixed in a droplet. The well is then sealed with the droplet sitting or suspended above the buffer. The plate is stored at constant temperature, usually at 20–22 or 4 °C. As the protein drop has an initial 50% concentration of buffer components with respect to the reservoir, vapor diffusion causes the drop to slowly shrink over time, increasing both the protein and buffer component concentrations. This equilibration process occurs more rapidly for small drops than larger ones.

With luck, a condition is found in which well-defined crystals form. If not, the behavior of the protein in the drops, such as the formation of needles or microcrystalline precipitate, can guide subsequent screening steps. A "fine-screen" approach is then used, in which buffer constituents are varied by small increments and decrements around the lead condition, until a condition yielding crystals is found.

4.3.2. The target protein crystallizes in known conditions—soaking

Even if a condition favoring crystal formation is known, it is a good idea to perform a "fine-screen" of buffer components around this condition, to allow for experimental variability such as differences in reagents, protein solution, humidity, and temperature. Care should be taken to replicate the concentration of constituents in the protein solution, the protein:buffer drop volume, reservoir volume, and temperature. A lead condition identified in a 200 nl drop does not necessarily directly scale up to a 1 μl volume.

Protein crystals contain pockets and channels of disordered solvent that run throughout the crystal. Consequently, it is often possible for small molecules to diffuse into the lattice of preformed crystals. Once a condition is found that produces well-diffracting crystals (a resolution less than 2.2 Å is ideal), a series of soaking experiments can be undertaken. Usually, this involves soaking crystals for 12–24 h in drops with a range of small molecule concentrations, collecting diffraction data from each crystal, and processing the data for the presence of density corresponding with the ligand.

4.3.3. The target protein has one or more known crystallization conditions, but soaking has failed

Three scenarios can confound a soaking experiment: inaccessibility of the ligand-binding pocket due to interactions between protein molecules in the crystal lattice, the occurrence of a ligand-binding site at a protein–protein interface, or a change in the protein upon ligand binding that is not compatible with the packing of protein molecules within the lattice. In the first instance, it is worth attempting a cocrystallography experiment where the ligand is introduced at different concentrations prior to the crystallization experiment. In the latter two scenarios, a survey of conditions that yield different crystal forms is warranted. Failing this, a *de novo* cocrystallization experiment using sparse screens, detailed above, should be used to produce novel crystallization leads.

We have tried the latter two approaches in an attempt to crystallize CG and derivative compounds in complex with α_1-antitrypsin. Thus far, soaking experiments have indicated that the preformed α_1-antitrypsin crystals used show no adverse effects by light microscopy on exposure to 5% DMSO or ethanol. However, beyond 6 h of soaking, we have observed complete loss of diffraction data in such conditions indicating that the crystal lattice is adversely affected by these conditions. Cocrystallization attempts have so far resulted only in crystals of the apo-, native conformer.

4.4. Using NMR spectroscopy for drug discovery in the serpinopathies

Nuclear magnetic resonance (NMR) spectroscopy is a powerful method for determination of protein structure and dynamics (Lindorff-Larsen *et al.*, 2005). Techniques have progressed such that it is now feasible, though still challenging, to apply it to proteins as large as serpins (~45 kDa when nonglycosylated). It is increasingly used as an aid to drug discovery.

Here, we introduce strategies and challenges relevant to this application in serpins based upon our experience to date with α_1-antitrypsin.

To use NMR spectroscopy to provide detailed information of the interactions of a serpin with ligands, a characterization of NMR spectra of the isolated protein is desirable. For larger biomolecules (>25 kDa), the residue-specific cross-peaks that define 2D NMR spectra broaden due to slower molecular tumbling in solution. Further, the identification of individual cross-peaks is hampered by the increase in overlapping that occurs due to population of a relatively small spectral region with a greater number of cross-peaks. Nevertheless, successful assignment of cross-peaks to specific residues in the protein sequence, when combined with complementary high-resolution crystallographic data, provides a powerful tool in detecting and localizing ligand interactions. For 2D NMR spectroscopy, isotopic labeling with ^{15}N is usual. When coupled with site-directed mutagenesis, it may possible to assign some residues individually within a ^{1}H–^{15}N heteronuclear single quantum coherence (HSQC) spectrum for a uniformly ^{15}N-labeled protein. A similar approach has assigned a few alanine residues in α_1-antitrypsin using protein produced such that only alanine residues were isotopically labeled, resulting in a greatly simplified 2D spectrum (Gettins *et al.*, 2004). For maximal sequence assignment, uniformly triple-labeled ($^{2}H/^{13}C/^{15}N$) sample may be required.

4.4.1. Preparation of isotopically labeled α_1-antitrypsin for NMR studies

The strategies for recombinant production of many serpins are well established. We have adapted an existing protocol (Parfrey *et al.*, 2003) to produce uniformly-labeled recombinant wildtype α_1-antitrypsin. We culture BL21 (DE3) *E. coli* in M9 minimal media in which the only nitrogen source is ^{15}N-ammonium chloride. If ^{13}C-labeling is required, then ^{13}C-glucose is used as the sole carbon source. This results in reasonably good expression yields of 3–10 mg/l. Of importance, deuteration/perdeuteration (e.g., uniform triple-labeling with $^{2}H/^{15}N/^{13}C$) may be used to reduce spectral complexity and improve relaxation properties. However, the necessary use of deuterated water (D_2O) for culture media slows cell growth, tends to reduce yields, and greatly increases costs.

4.4.2. NMR spectra of serpins

The early application of heteronuclear NMR methods to serpins showed the significant promise of this method in their study (Peterson and Gettins, 2001). Improvement in NMR technology and methodology has increased the sensitivity and resolution of the spectrum. The use of a cryoprobe enables three to fourfold enhancement of the detection sensitivity in the high-resolution NMR compared to conventional probes (Kovacs *et al.*, 2005). The observation of nearly all amino acids in fingerprint ^{1}H–^{15}N

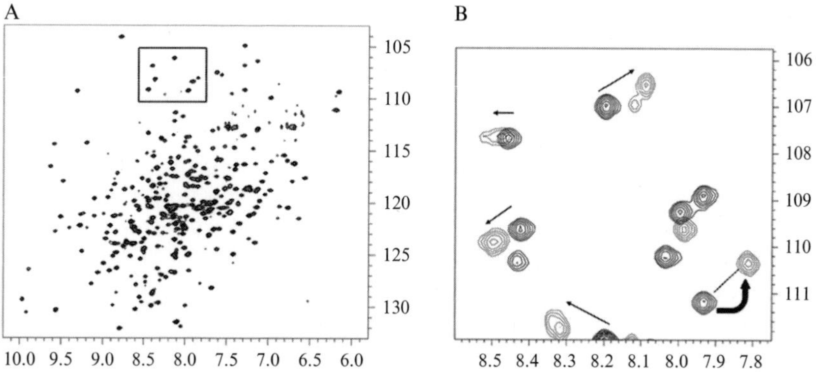

Figure 8.5 (A) [^{1}H–^{15}N]-TROSY-HSQC spectrum of native α_1-antitrypsin at 25 °C (Nyon *et al.*, manuscript submitted). The ^{1}H chemical shift (ppm) data are plotted on the *x*-axis, ^{15}N chemical shift data (ppm) on the *y*-axis. (B) Changes seen in one part of the spectrum (boxed in A) upon ligand binding. Comparison of [^{1}H–^{15}N]-TROSY-HSQC spectra (at 37 °C) of native α_1-antitrypsin (black) and α_1-antitrypsin when saturated with ligand (red) shows changes in chemical shift (straight arrows). Two possible patterns of chemical shift change are indicated in the lower right corner. Gradual migration of cross-peaks from the apo- to ligand-saturated states (e.g., along the path of the dashed line) is the characteristic pattern of chemical shift change for weak binding with rapid conformational exchange. Discrete jumps in cross-peaks (indicated by curved arrow) indicate tight binding with very slow rates of conformational exchange between the bound and apo-state. (See Color Insert.)

TROSY-HSQC spectrum has recently been achieved (Fig. 8.5A; Nyon *et al.*, submitted for publication) using modern triple-resonance strategies (Kay, 2005). This has resulted in a near complete backbone resonance assignment. The assignment opens the way for many studies of the solution behavior of α_1-antitrypsin including identification and characterization of potential inhibitors of polymerization.

4.4.3. NMR strategies for studying ligand binding to the serpins

NMR spectroscopy is exquisitely sensitive and quantitative in assessing changes in solution behavior (Heller and Kessler, 2001; Pellecchia *et al.*, 2008). Several methods for investigating such interactions have been proposed. Compound screening can be performed based on the unique structural information on the binding site and analyzing the chemical shift perturbation. The process may be simplified by use of selective side-chain labeling. However, the resonance assignment of fingerprint spectra of α_1-antitrypsin provides ca. 300 potential probes of changes reporting ligand binding even without selective labeling. Figure 8.5B shows how residue-specific changes can be detected in 2D NMR spectra on ligand binding to α_1-antitrypsin.

NMR signals can be detected and measured across a wide concentration range (10^{-2}–10^{-6} *M*) and can be used to detect weak ligand–protein interactions. However, such interactions are difficult to characterize because of the short lifetime of the complex. Saturation transfer difference (STD), a double resonance nuclear Overhauser effect (NOE) method, can overcome this problem. Here, a saturated proton signal rapidly spreads through spin diffusion to all protons associated with the protein, including those of bound ligands (Mayer and Meyer, 1999). Changes in ligand signal from the dissociated free state identify ligand binding. STD has been widely used to probe low affinity interactions between small compounds and has been used for lead generation in drug discovery (Mayer *et al.*, 2006). The resonances of small molecules are typically sharp and this allows simultaneous screening of multiple molecules.

WaterLOGSY experiments exploit the different pathways of magnetization transfer to bulk water to identify protein–ligand interactions (Dalvit *et al.*, 2000). Moreover, paramagnetic relaxation enhancement (PRE) experiments can complement NOE experiments, providing information across long ranges (15–24 Å). In screens, the incorporation of a spin label (e.g., nitroxide radical) into a ligand, or alternatively the target binding site of a protein, can identify ligand binding (Jahnke *et al.*, 2001) through sensitive T_2 relaxation experiments.

Optimization of lead compounds to generate high-affinity binding compound or analogues may be performed by a "SAR (structure–activity relationships) by NMR" strategy. This is a linked-fragment approach where the high-affinity ligands are constructed through stepwise optimization of ligands for individual protein subsites (Oltersdorf *et al.*, 2005).

5. Mammalian Cell Models and Beyond

Mammalian cell models of α_1-antitrypsin and neuroserpin polymerization are evaluated in chapter 18 of this journal issue. In addition to the recent development of a further stably transfected inducible model (also described), the following cell models of α_1 antitrypsin deficiency have been reported:

(i) human skin fibroblast cell lines (CJZ12B) from PIZZ individuals engineered for stable expression of Z α_1-antitrypsin (Burrows *et al.*, 2000);
(ii) CHO-K1 cell lines expressing M, Z, Saar, and Saar/Z variants of α_1-antitrypsin (Lin *et al.*, 2001);
(iii) human epidermal cell line HeLa expressing the same α_1-antitrypsin variants as in (ii), in a Tet-Off inducible system (Hidvegi *et al.*, 2005);

(iv) COS-7 cell transient transfection with expression of M, Z, Thr114Phe/Z, and Gly117Phe/Z α_1-antitrypsin (Gooptu *et al.*, 2009; Kroeger *et al.*, 2009);
(v) murine hepatoma Hepa1-6 cell line with transient expression of Z α_1-antitrypsin (Hidvegi *et al.*, 2005);
(vi) human hepatoma cell transfected for transient expression of human α_1-antitrypsin $\text{Null}_{\text{Hong Kong}}$ (Cameron *et al.*, 2009);
(vii) human inducible pluripotent stem cell (iPS) lines generated from Z α_1-antitrypsin homozygote human dermal fibroblasts, and hepatocyte lines derived from these (Rashid *et al.*, 2010).

The use of mammalian cell models, like many of the techniques described in this chapter, can be used in focused or high-throughput ways. A focused approach uses the cell model to "bridge" between *in vitro* and *in vivo* testing. Compounds identified from readouts in other systems (urea-native gels, MS, NMR, etc.) can be tested for cytotoxicity and/or efficacy prior to assessment in animal models. The TTAI peptide (described in Section 2) was assessed for cytotoxicity in assays using two normal lung epithelial cell lines (BEAS-2B and NL20), two normal lung fibroblast cell lines (WI-38 and IMR-90), and one cancer cell line (A2058) (Chang *et al.*, 2009). These cell lines were not specifically models of α_1-antitrypsin deficiency. However, the COS-7 transient transfection system has been used to evaluate the capacity of stabilizing mutations to rescue polymerization phenotype on a Z background (Gooptu *et al.*, 2009). This system could be used similarly for compound testing. The readouts were Western blot and ELISA-based analyses of intracellular and culture medium supernatant material using both polymer specific (2C1 mAb) and conformer nonspecific primary antibodies. Restoration of functional activity was demonstrated by a colorimetric sandwich technique using 96-well plates coated with target protease (bovine α-chymotrypsin). Incubation of these wells with media supernatants captured active material, and the proportion of chymotrypsin inhibition was assessed by ELISA (Gooptu *et al.*, 2009).

The Hepa1 cell line has been used to test efficacy of the CG compound targeted at the hydrophobic pocket flanking β-sheet A (Mallya *et al.*, 2007). This has the advantage of being liver derived, and this may be relevant as polymer handling may differ between cell types. Hundred micromolar of the CG compound was incubated with cells expressing Z α_1-antitrypsin and effects monitored by pulse-chase experiments following radiolabeled methionine incorporated into the protein. This demonstrated that the CG compound reduced the intracellular retention half-time of Z α_1-antitrypsin by 70%. The strength of cell models in providing information of physiological interest was shown by the absence of a concomitant increase in α_1-antitrypsin secreted into the culture medium and the lack of glycan maturation in intracellular material. These observations indicated that CG interactions were causing folding intermediate species to be targeted for degradation rather than performing a chaperoning function.

The use of mammalian cell models of serpinopathies as direct targets of high-throughput screens has not yet been described in the literature. However, such an approach will be attractive to many drug developers as the lack of readouts informing upon structural mechanisms is offset by readouts that may be more closely related to effects in animal models or humans.

Recently, the use of mouse models to evaluate the effects of drugs that can alter the disposal of the retained mutant protein, such as rapamycin (Kaushal *et al.*, 2010) or carbamazepine (Hidvegi *et al.*, 2010), has been reported. Currently, animal models are likely to be too expensive to scale up for high-throughput screening. However, the tissue-based readouts used may well be valuable in informing future drug-design strategies in the serpinopathies to take account of increased complexity at the level of organs. However, animal models do not necessarily recapitulate human disease optimally. Therefore, the use of recently pioneered iPS approaches (Rashid *et al.*, 2010) are likely to be an invaluable complement to this. In addition, this work raises the exciting prospect of therapeutic use of patient-specific iPS cell lines in which the genetic defect has been corrected.

6. Conclusion

The development of synthetic heparinoids was an excellent example of how academic and pharmaceutical industry approaches could combine to produce a class of drugs in clinical use within a relatively short (10–15 year) timescale (Koopman and Buller, 2003). Two general factors aided this process: it involved refining and modifying a physiological ligand to produce compounds that were inherently well tolerated systemically, and it was aimed at a subset of a large clinical market. Rarer conditions such as the serpinopathies have traditionally been less attractive targets for the pharmaceutical industry. Moreover, strategies that develop novel ligands are likely to face greater challenges than modification of a physiological ligand. These factors have contributed to low levels of translation of structural insights from academia into drugs developed industrially.

The methods described in this chapter may redress this balance. Academic research groups can deploy techniques that provide high-resolution structural data on serpins, yet are sufficiently sensitive, scalable, and automatable to allow higher throughput. *In vitro* and computational approaches can be used iteratively to enrich screening of small molecules and their derivatives. Studies in mammalian cell models may indicate safety and effectiveness *in vivo*. Taken together with orphan disease legislation, it is entirely feasible that such programs can optimize lead compounds making them sufficiently attractive for further development by the pharmaceutical industry.

REFERENCES

Allen Annis, D., Nickbarg, E., Yang, X., Ziebell, M. R., and Whitehurst, C. E. (2007). Affinity selection-mass spectrometry screening techniques for small molecule drug discovery. *Curr. Opin. Chem. Biol.* **11,** 518–526.

An, J., Totrov, M., and Abagyan, R. (2005). Pocketome via comprehensive identification and classification of ligand binding envelopes. *Mol. Cell. Proteomics* **4,** 752–761.

Ben-Naim, A. (1997). Statistical potentials extracted from protein structures: Are these meaningful potentials? *J. Chem. Phys.* **107,** 3698–3706.

Bissantz, C., Folkers, G., and Rognan, D. (2000). Protein-based virtual screening of chemical databases. 1. Evaluation of different docking/scoring combinations. *J. Med. Chem.* **43,** 4759–4767.

Björk, I., Ylinenjärvi, K., Olson, S. T., and Bock, P. E. (1992). Conversion of antithrombin from an inhibitor of thrombin to a substrate with reduced heparin affinity and enhanced conformational stability by binding of a tetradecapeptide corresponding to the P1-P14 region of the putative reactive bond loop of the inhibitor. *J. Biol. Chem.* **267,** 1976–1982.

Bohm, H.-J. (2003). Prediction of non-bonded interactions in drug design. *In* "Protein-Ligand Interactions," (H.-J. Bohm and G. Schneider, eds.), pp. 3–20. Wiley-VCH Verlag, Weinheim.

Bottegoni, G., Kufareva, I., Totrov, M., and Abagyan, R. (2008). A new method for ligand docking to flexible receptors by dual alanine scanning and refinement (SCARE). *J. Comput. Aided Mol. Des.* **22,** 311–325.

Brooijmans, N., and Kuntz, I. D. (2003a). Molecular recognition and docking algorithms. *Annu. Rev. Biophys. Biomol. Struct.* **32,** 35–373.

Brooijmans, N., and Kuntz, I. D. (2003b). Molecular recognition and docking algorithms. *Annu. Rev. Biophys. Biomol. Struct.* **32,** 335–373.

Brooks, B., and Karplus, M. (1985). Normal modes for specific motions of macromolecules: application to the hinge-bending mode of lysozyme. *Proc. Natl. Acad. Sci. USA* **82,** 4995–4999.

Burrows, J. A. J., Willis, L. K., and Perlmutter, D. H. (2000). Chemical chaperones mediate increased secretion of mutant α_1-antitrypsin (α_1-AT) Z: A potential pharmacologcial strategy for prevention of liver injury and emphysema. *Proc. Natl. Acad. Sci. USA* **97,** 1796–1801.

Cameron, P. H., Chevet, E., Pluquet, O., Thomas, D. Y., and Bergeron, J. J. (2009). Calnexin phosphorylation attenuates the release of partially misfolded alpha1-antitrypsin to the secretory pathway. *J. Biol. Chem.* **284,** 34570–34579.

Carlson, H. A., and McCammon, J. A. (2000). Accommodating protein flexibility in computational drug design. *Mol. Pharmacol.* **57,** 213–218.

Carr, R. A. E., Congreve, M., Murray, C. W., and Rees, D. C. (2005). Fragment-based lead discovery: Leads by design. *Drug Discov. Today* **10,** 987–992.

Chang, W.-S. W., Wardell, M. R., Lomas, D. A., and Carrell, R. W. (1996). Probing serpin reactive loop conformations by proteolytic cleavage. *Biochem. J.* **314,** 647–653.

Chang, Y. P., Mahadeva, R., Chang, W. S. W., Shukla, A., Dafforn, T. R., and Chu, Y. H. (2006). Identification of a 4-mer peptide inhibitor that effectively blocks the polymerization of pathogenic Z α_1-antitrypsin. *Am. J. Respir. Cell Mol. Biol.* **35,** 540–548.

Chang, Y. P., Mahadeva, R., Chang, W. S., Lin, S. C., and Chu, Y. H. (2009). Small-molecule peptides inhibit Z alpha1-antitrypsin polymerization. *J. Cell. Mol. Med.* **13,** 2304–2316.

Chernushevich, I. V., and Thomson, B. A. (2004). Collisional cooling of large ions in electrospray mass spectrometry. *Anal. Chem.* **76,** 1754–1760.

Chou, P. Y., and Fasman, G. D. (1978). Empirical predictions of protein conformation. *Annu. Rev. Biochem.* **47,** 251–276.

Cimmperman, P., Baranauskiene, L., Jachimoviciute, S., Jachno, J., Torresan, J., Michailoviene, V., Matuliene, J., Serekaite, J., Burnelis, V., and Matulis, D. (2008). A quantitative model of thermal stabilization and destabilization of proteins by ligands. *Biophys. J.* **95,** 3222–3231.

Dalvit, C., Pevarello, P., Tato, M., Veronesi, M., Vulpetti, A., and Sundstrom, M. (2000). Identification of compounds with binding affinity to proteins via magnetization transfer from bulk water. *J. Biomol. NMR* **18,** 65–68.

Daniel, J. M., Friess, S. D., Rajagopalan, S., Wendt, S., and Zenobi, R. (2002). Quantitative determination of noncovalent binding interactions using soft ionization mass spectrometry. *Int. J. Mass Spectrom.* **216,** 1–27.

de Groot, B. L., van Aalten, D. M. F., Scheek, R. M., Amadei, A., Vriend, G., and Berendsen, H. J. C. (1997). Prediction of protein conformational freedom from distance constraints. *Proteins Struct. Funct. Genet.* **29,** 240–251.

Dunstone, M. A., Dai, W., Whisstock, J. C., Rossjohn, J., Pike, R. N., Feil, S. C., Le Bonniec, B. F., Parker, M. W., and Bottomley, S. P. (2000). Cleaved antitrypsin polymers at atomic resolution. *Protein Sci.* **9,** 417–420.

Ekeowa, U. I., Freeke, J., Miranda, E., Gooptu, B., Bush, M. F., Perez, J., Teckman, J. H., Robinson, C. V., and Lomas, D. A. (2010). Defining the mechanism of polymerization in the serpinopathies. *Proc. Natl. Acad. Sci. USA* **107,** 17146–17151.

Eldridge, M. D., Murray, C. W., Auton, T. R., Paolini, G. V., and Mee, R. P. (1997). Empirical scoring functions: I. The development of a fast empirical scoring function to estimate the binding affinity of ligands in receptor complexes. *J. Comput. Aided Mol. Des.* **11,** 425–445.

Elliott, P. R., Pei, X. Y., Dafforn, T. R., and Lomas, D. A. (2000). Topography of a 2.0Å structure of α1-antitrypsin reveals targets for rational drug design to prevent conformational disease. *Protein Sci.* **9,** 1274–1281.

Erlanson, D. A., and Hansen, S. K. (2004). Making drugs on proteins: Site-directed ligand discovery for fragment-based lead assembly. *Curr. Opin. Chem. Biol.* **8,** 399–406.

Fitton, H. L., Pike, R. N., Carrell, R. W., and Chang, W.-S. W. (1997). Mechanisms of antithrombin polymerisation and heparin activation probed by insertion of synthetic reactive loop peptides. *Biol. Chem.* **378,** 1059–1063.

Friesner, R. A., Banks, J. L., Murphy, R. B., Halgren, T. A., Klicic, J. J., Mainz, D. T., Repasky, M. P., Knoll, E. H., Shelley, M., Perry, J. K., *et al.* (2004). Glide: A new approach for rapid, accurate docking and scoring. 1. Method and assessment of docking accuracy. *J. Med. Chem.* **47,** 1739–1749.

Furka, A., Sebesteyen, M., Asgedom, M., and Dibo, G. (1991). General method for rapid synthesis of multicomponent peptide mixtures. *Int. J. Pept. Protein Res.* **37,** 487–493.

Furnham, N., Blundell, T. L., DePristo, M. A., and Terwilliger, T. C. (2006). Is one solution good enough? *Nat. Struct. Mol. Biol.* **13,** 184–185.

Gettins, P. G. (2002). The F-helix of serpins plays an essential, active role in the proteinase inhibition mechanism. *FEBS Lett.* **523,** 2–6.

Gettins, P. G., Backovic, M., and Peterson, F. C. (2004). Use of NMR to study serpin function. *Methods* **32,** 120–129.

Gohlke, H., Hendlich, M., and Klebe, G. (2000). Knowledge-based scoring function to predict protein-ligand interactions. *J. Mol. Biol.* **295,** 337–356.

Gooptu, B., Hazes, B., Chang, W.-S. W., Dafforn, T. R., Carrell, R. W., Read, R., and Lomas, D. A. (2000). Inactive conformation of the serpin α_1-antichymotrypsin indicates two stage insertion of the reactive loop; implications for inhibitory function and conformational disease. *Proc. Natl. Acad. Sci. USA* **97,** 67–72.

Gooptu, B., Miranda, E., Nobeli, I., Mallya, M., Purkiss, A., Leigh Brown, S. C., Summers, C., Phillips, R. L., Lomas, D. A., and Barrett, T. E. (2009). Crystallographic and cellular characterisation of two mechanisms stablising the native fold of α_1-antitrypsin: Implications for disease and drug design. *J. Mol. Biol.* **387,** 857–868.

Gramling, M. W., and Church, F. C. (2010). Plasminogen activator inhibitor-1 is an aggregate response factor with pleiotropic effects on cell signaling in vascular disease and the tumor microenvironment. *Thromb. Res.* **125,** 377–381.

Graves, A. P., Shivakumar, D. M., Boyce, S. E., Jacobson, M. P., Case, D. A., and Shoichet, B. K. (2008). Rescoring docking hit lists for model cavity sites: Predictions and experimental testing. *J. Mol. Biol.* **377,** 914–934.

Halgren, T. A. (2007). New method for fast and accurate binding-site identification and analysis. *Chem. Biol. Drug Des.* **69,** 146–148.

Halgren, T. A. (2009). Identifying and characterizing binding sites and assessing druggability. *J. Chem. Inf. Model.* **49,** 377–389.

Halperin, I., Ma, B., Wolfson, H., and Nussinov, R. (2002). Principles of docking: An overview of search algorithms and a guide to scoring functions. *Proteins* **47,** 409–443.

Heck, A. J. (2008). Native mass spectrometry: A bridge between interactomics and structural biology. *Nat. Methods* **5,** 927–933.

Heller, M., and Kessler, H. (2001). NMR spectroscopy in drug design. *Pure Appl. Chem.* **73,** 1429–1436.

Hendlich, M., Rippmann, F., and Barnickel, G. (1997). LIGSITE: Automatic and efficient detection of potential small molecule-binding sites in proteins. *J. Mol. Graph. Model.* **15,** 359–363.

Hidvegi, T., Schimdt, B. Z., Hale, P., and Perlmutter, D. H. (2005). Accumulation of mutant alpha1-antitrypsin Z in the endoplasmic reticulum activates caspases-4 and -12, NFkappaB, and BAP31 but not the unfolded protein response. *J. Biol. Chem.* **280,** 39002–39015.

Hidvegi, T., Ewing, M., Hale, P., Dippold, C., Beckett, C., Kemp, C., Maurice, N., Mukherjee, A., Goldbach, C., Watkins, S., *et al.* (2010). An autophagy-enhancing drug promotes degradation of mutant alpha1-antitrypsin Z and reduces hepatic fibrosis. *Science* **329,** 229–232.

Huntington, J. A. (2003). Mechanisms of glycosaminoglycan activation of the serpins in hemostasis. *J. Thromb. Haemost.* **1,** 1535–1549.

Huntington, J. A., Pannu, N. S., Hazes, B., Read, R., Lomas, D. A., and Carrell, R. W. (1999). A 2.6Å structure of a serpin polymer and implications for conformational disease. *J. Mol. Biol.* **293,** 449–455.

Irwin, J. J., and Shoichet, B. K. (2005). ZINC—A free database of commercially available compounds for virtual screening. *J. Chem. Inf. Model.* **45,** 177–182.

Jahnke, W., Rudisser, S., and Zurini, M. (2001). Spin label enhanced NMR screening. *J. Am. Chem. Soc.* **123,** 3149–3150.

Jin, L., Abrahams, J.-P., Skinner, R., Petitou, M., Pike, R., and Carrell, R. W. (1997). The anticoagulant activation of antithrombin by heparin. *Proc. Natl. Acad. Sci. USA* **94,** 14683–14688.

Juraschek, R., Dülcks, T., and Karas, M. (1999). Nanoelectrospray—More than just a minimized electrospray ionization source. *J. Am. Soc. Mass. Spectrom.* **10,** 300–308.

Kalyanaraman, C., Bernacki, K., and Jacobson, M. P. (2005). Virtual screening against highly charged active sites: Identifying substrates of alpha-beta barrel enzymes. *Biochemistry* **44,** 2059–2071.

Karas, M., Bahr, U., and Dulcks, T. (2000). Nano-electrospray ionization mass spectrometry: Addressing analytical problems beyond routine. *Fresenius J. Anal. Chem.* **366,** 669–676.

Kaushal, S., Annamali, M., Blomenkamp, K., Rudnick, D., Halloran, D., Brunt, E. M., and Teckman, J. H. (2010). Rapamycin reduces intrahepatic alpha-1-antitrypsin mutant Z protein polymers and liver injury in a mouse model. *Exp. Biol. Med. (Maywood)* **235,** 700–709.

Kay, L. E. (2005). NMR studies of protein structure and dynamics. *J. Magn. Reson.* **173,** 193–207.

Koopman, M. M. W., and Buller, H. R. (2003). Short- and long-acting synthetic pentasaccharides. *J. Intern. Med.* **254,** 335–342.

Kovacs, H., Moskau, D., and Spraul, M. (2005). Cryogenically cooled probes—A leap in NMR technology. *Prog. Nucl. Magn. Reson. Spectrosc.* **46,** 131–155.

Kroeger, H., Miranda, E., MacLeod, I., Perez, J., Crowther, D. C., Marciniak, S. J., and Lomas, D. A. (2009). Endoplasmic reticulum-associated degradation (ERAD) and autophagy cooperate to degrade polymerogenic mutant serpins. *J. Biol. Chem.* **284,** 22793–22802.

Krutchinsky, I. V., Spicer, V. L., Ens, W., and Standing, K. G. (1998). Collisional damping interface for an electrospray ionization time-of-flight mass spectrometer. *J. Am. Soc. Mass Spectrom.* **9,** 569–579.

Lang, P. T., Brozell, S. R., Mukherjee, S., Pettersen, E. F., Meng, E. C., Thomas, V., Rizzo, R. C., Case, D. A., James, T. L., and Kuntz, I. D. (2009). DOCK 6: Combining techniques to model RNA-small molecule complexes. *RNA* **15,** 1219–1230.

Laskowski, R. A. (1995). SURFNET: A program for visualizing molecular surfaces, cavities and intermolecular interactions. *J. Mol. Graph.* **13,** 323–330.

Lavinder, J. J., Hari, S. B., Sullivan, B. J., and Magliery, T. J. (2009). High-throughput thermal scanning: A general, rapid dye-binding thermal shift screen for protein engineering. *J. Am. Chem. Soc.* **131,** 3794–3795.

Leach, A. R., Shoichet, B. K., and Peishoff, C. E. (2006). Prediction of protein-ligand interactions. Docking and scoring: Successes and gaps. *J. Med. Chem.* **49,** 5851–5855.

Levitt, D. G., and Banaszak, L. J. (1992). POCKET: A computer graphics method for identifying and displaying protein cavities and their surrounding amino acids. *J. Mol. Graph. Model.* **10,** 229–234.

Lim, H. K., Hsieh, Y. L., Ganem, B., and Henion, J. (1995). Recognition of cell-wall peptide ligands by vancomycin group antibiotics—Studies using ion-spray mass-spectrometry. *J. Mass Spectrom.* **30,** 708–714.

Lin, L., Schmidt, B., Teckman, J., and Perlmutter, D. H. (2001). A naturally occurring nonpolymerogenic mutant of α1-antitypsin charcaterized by prolonged retention in the endoplasmic reticulum. *J. Biol. Chem.* **276,** 33893–33898.

Lindorff-Larsen, K., Best, R. B., DePristo, M. A., Dobson, C. M., and Vendruscolo, M. (2005). Simultaneous determination of protein structure and dynamics. *Nature* **433,** 128–132.

Lipinski, C. A., Lombardo, F., Dominy, B. W., and Feeney, P. J. (1997). Experimental and computational approaches to estimate solubility and permeability in drug discovery and development settings. *Adv. Drug Deliv. Rev.* **23,** 3–25.

Lomas, D. A., Evans, D. L., Finch, J. T., and Carrell, R. W. (1992). The mechanism of Z α_1-antitrypsin accumulation in the liver. *Nature* **357,** 605–607.

Lomas, D. A., Evans, D. L., Stone, S. R., Chang, W.-S. W., and Carrell, R. W. (1993). Effect of the Z mutation on the physical and inhibitory properties of α_1-antitrypsin. *Biochemistry* **32,** 500–508.

Ma, J. (2005). Usefulness and limitations of normal mode analysis in modeling dynamics of biomolecular complexes. *Structure* **13,** 373–380.

Mahadeva, R., Dafforn, T. R., Carrell, R. W., and Lomas, D. A. (2002). Six-mer peptide selectively anneals to a pathogenic serpin conformation and blocks polymerisation: Implications for the prevention of Z α_1-antitrypsin related cirrhosis. *J. Biol. Chem.* **277,** 6771–6774.

Mallya, M., Phillips, R. L., Saldanha, S. A., Gooptu, B., Leigh Brown, S. C., Termine, D. J., Shirvani, A. M., Wu, Y., Sifers, R. N., Abagyan, R., and Lomas, D. A. (2007). Small molecules block the polymerization of Z alpha1-antitrypsin and increase the clearance of intracellular aggregates. *J. Med. Chem.* **50,** 5357–5363.

Mathur, S., Badertscher, M., Scott, M., and Zenobi, R. (2007). Critical evaluation of mass spectrometric measurement of dissociation constants: Accuracy and cross-validation against surface plasmon resonance and circular dichroism for the calmodulin-melittin system. *Phys. Chem. Chem. Phys.* **9,** 6187–6198.

Mayer, M., and Meyer, B. (1999). Characterization of ligand binding by saturation transfer difference NMR spectroscopy. *Angew. Chem. Int. Ed.* **38,** 1784–1788.

Mayer, M., Lang, P. T., Gerber, S., Madrid, P. B., Pinto, I. G., Guy, R. K., and James, T. L. (2006). Synthesis and testing of a focused phenothiazine library for binding to HIV-1 TAR RNA. *Chem. Biol.* **13,** 993–1000.

McCammon, M. G., and Robinson, C. V. (2004). Structural change in response to ligand binding. *Curr. Opin. Chem. Biol.* **8,** 60–65.

McCammon, M. G., Scott, D. J., Keetch, C. A., Greene, L. H., Purkey, H. E., Petrassi, H. M., Kelly, J. W., and Robinson, C. V. (2002). Screening transthyretin amyloid fibril inhibitors: Characterization of novel multiprotein, multiligand complexes by mass spectrometry. *Structure* **10,** 851–863.

Metropolis, N., and Ulam, S. (1949). The Monte Carlo method. *J. Am. Stat. Assoc.* **44,** 335–341.

Mitchell, J. B. O., Laskowski, R. A., Alex, A., and Thornton, J. M. (1999). BLEEP-potential of mean force describing protein-ligand interactions. I. Generating potential. *J. Comput. Chem.* **20,** 1165–1176.

Muckenshnabel, I., Falchetto, R., Mayr, L. M., and Filipuzzi, I. (2004). SpeedScreen: Label-free liquid chromatography-mass spectrometry-based high-throughput screening for the discovery of orphan protein ligands. *Anal. Biochem.* **324,** 241–249.

Muegge, I., and Martin, Y. C. (1999). A general and fast scoring function for protein-ligand interactions: A simplified potential approach. *J. Med. Chem.* **42,** 791–804.

Mysinger, M. M., and Shoichet, B. K. (2010). Rapid context-dependent ligand desolvation in molecular docking. *J. Chem. Inf. Model.* **50,** 1561–1573.

Nayal, M., and Honig, B. (2006). On the nature of cavities on protein surfaces: Application to the identification of drug-binding sites. *Proteins Struct. Funct. Bioinform.* **63,** 892–906.

Neumann, T., Junker, H. D., Schmidt, K., and Sekul, R. (2007). SPR-based fragment screening: Advantages and applications. *Curr. Top. Med. Chem.* **7,** 1630–1642.

Nyon, M. P., Kirkpatrick, J., Cabrita, L. D., Christodoulou, J., and Gooptu, B. 1H, 15N and 13C backbone resonance assignments of the archetypal serpin α1-antitrypsin. Manuscript submitted, 2011.

Oltersdorf, T., Elmore, S. W., Shoemaker, A. R., Armstrong, R. C., Augeri, D. J., Belli, B. A., Bruncko, M., Deckwerth, T. L., Dinges, J., Hajduk, P. J., *et al.* (2005). An inhibitor of Bcl-2 family proteins induces regression of solid tumours. *Nature* **435,** 677–681.

Parfrey, H., Mahadeva, R., Ravenhill, N., Zhou, A., Dafforn, T. R., Foreman, R. C., and Lomas, D. A. (2003). Targeting a surface cavity of α_1-antitrypsin to prevent conformational disease. *J. Biol. Chem.* **278,** 33060–33066.

Park, H., Lee, J., and Lee, S. (2006). Critical assessment of the automated AutoDock as a new docking tool for virtual screening. *Proteins* **65,** 549–554.

Pellecchia, M., Bertini, I., Cowburn, D., Dalvit, C., Giralt, E., Jahnke, W., James, T. L., Homans, S. W., Kessler, H., Luchinat, C., *et al.* (2008). Perspectives on NMR in drug discovery: A technique comes of age. *Nat. Rev. Drug Discov.* **7,** 738–745.

Perot, S., Sperandio, O., Miteval, M. A., Camproux, A.-C., and Villoutreix, B. O. (2010). Druggable pockets and binding site centric chemical space: A paradigm shift in drug discovery. *Drug Discov. Today* **15,** 656–667.

Peterson, F. C., and Gettins, P. G. (2001). Insight into the mechanism of serpin-proteinase inhibition from 2D [1H-15N] NMR studies of the 69kDa alpha 1-proteinase inhibitor Pittsburght-trypsin covalent complex. *Biochemistry* **29,** 6284–6292.

Rarey, M., Kramer, B., Lengauer, T., and Klebe, G. (1996). A fast flexible docking method using an incremental construction algorithm. *J. Mol. Biol.* **261,** 470–489.

Rashid, S. T., Corbineau, S., Hannan, N., Marciniak, S. J., Miranda, E., Alexander, G. J., Huang-Doran, I., Griffin, J., Ahrlund-Richter, L., Skepper, J., *et al.* (2010). Modeling inherited metabolic disorders of the liver using human induced pluripotent stem cells. *J. Clin. Invest.* **120,** 3127–3136.

Rich, R. L., and Myszka, D. G. (2006). Commercial optical biosensor literature. *J. Mol. Recognit.* **19,** 478–534.

Rostom, A. A., Tame, J. R. H., Ladbury, J. E., and Robinson, C. V. (2000). Specificity and interactions of the protein OppA: Partitioning solvent binding effects using mass spectrometry. *J. Mol. Biol.* **296,** 269–279.

Ruotolo, B. T., Giles, K., Campuzano, I., Sandercock, A. M., Bateman, R. H., and Robinson, C. V. (2005). Evidence for macromolecular protein rings in the absence of bulk water. *Science* **310,** 1658–1661.

Schulze, A. J., Baumann, U., Knof, S., Jaeger, E., Huber, R., and Laurell, C.-B. (1990). Structural transition of α_1-antitrypsin by a peptide sequentially similar to β-strand s4A. *Eur. J. Biochem.* **194,** 51–56.

Seeliger, D., Haas, J., and de Groot, B. L. (2007). Geometry-based sampling of conformational transitions in proteins. *Structure* **15,** 1482–1492.

Sengupta, T., Tsutsui, Y., and Wintrode, P. L. (2009). Local and global effects of a cavity filling mutation in a metastable serpin. *Biochemistry* **48,** 8233–8240.

Sherman, W., Day, T., Jacobson, M. P., Freisner, R. A., and Farid, R. (2006). Novel procedure for modeling ligand/receptor induced fit effects. *J. Med. Chem.* **49,** 534–553.

Sheu, S. Y., Yang, D. Y., Selzle, H. L., and Schlag, E. W. (2003). Energetics of hydrogen bonds in peptides. *Proc. Natl. Acad. Sci. USA* **100,** 12683–12687.

Silverman, G. A., Bird, P. I., Carrell, R. W., Church, F. C., Coughlin, P. B., Gettins, P., Irving, J., Lomas, D. A., Moyer, R. W., Pemberton, P., *et al.* (2001). The serpins are an expanding superfamily of structurally similar but functionally diverse proteins. Evolution, novel functions, mechanism of inhibition and a revised nomenclature. *J. Biol. Chem.* **276,** 33293–33296.

Sivasothy, P., Dafforn, T. R., Gettins, P. G. W., and Lomas, D. A. (2000). Pathogenic α_1-antitrypsin polymers are formed by reactive loop-β-sheet A linkage. *J. Biol. Chem.* **275,** 33663–33668.

Skinner, R., Chang, W.-S. W., Jin, L., Pei, X., Huntington, J. A., Abrahams, J.-P., Carrell, R. W., and Lomas, D. A. (1998). Implications for function and therapy of a 2.9Å structure of binary-complexed antithrombin. *J. Mol. Biol.* **283,** 9–14.

Taylor, G. (1964). Disintegration of water drops in an electric field. *Proc. R. Soc. Lond. A* **280,** 383–385.

Totrov, M., and Abagyan, R. (1997). Flexible protein-ligand docking by global energy optimization in internal coordinates. *Proteins* (Suppl. 1), 215–220.

Totrov, M., and Abagyan, R. (2008). Flexible ligand docking to multiple receptor conformations: A practical alternative. *Curr. Opin. Struct. Biol.* **18,** 178–184.

Tsutsui, Y., Lu, L., Gershenson, A., and Wintrode, P. L. (2006). The conformational dynamics of a metastable serpin studied by hydrogen exchange and mass spectrometry. *Biochemistry* **45,** 6561–6569.
Verdonk, M. L., Cole, J. C., Hartshorn, M. J., Murray, C. W., and Taylor, R. D. (2003). Improved protein-ligand docking using GOLD. *Proteins* **52,** 609–623.
Wang, R., and Wang, S. (2001). How does consensus scoring work for virtual library screening? An idealized computer experiment. *J. Chem. Inf. Comput. Sci.* **41,** 1422–1426.
Wang, Y., Xiao, J., Suzek, T. O., Zhang, J., Wang, J., and Bryant, S. H. (2009). PubChem: A public information system for analyzing bioactivities of small molecules. *Nucleic Acids Res.* **37,** 623–633.
Warr, W. A., and Overington, J. (2009). ChEMBL. An interview with John Overington, team leader, chemogenomics at the European Bioinformatics Institute Outstation of the European Molecular Biology Laboratory (EMBL-EBI). *J. Comput. Aided Mol. Des.* **23,** 195–198.
Wilm, M., and Mann, M. (1994). Electrospray and Taylor—Cone theory. Dole's beam of macromolecules at last? *Int. J. Mass Spectrom. Ion Proc.* **136,** 167–180.
Wilm, M., and Mann, M. (1996). Analytical properties of the nanoelectrospray ion source. *Anal. Chem.* **68,** 1–8.
Wishart, D. S., Knox, C., Chi Guo, A., Shrivastava, S., Hassanali, M., Stothard, P., Chang, Z., and Woolsey, J. (2006). DrugBank: A comprehensive resource for in silico drug discovery and exploration. *Nucleic Acids Res.* **34,** 668–672.
Yamasaki, M., Li, W., Johnson, D. J., and Huntington, J. A. (2008). Crystal structure of a stable dimer reveals the molecular basis of serpin polymerization. *Nature* **455,** 1255–1258.
Zheng, X., Wintrode, P. L., and Chance, M. R. (2008). Complementary structural mass spectrometry techniques reveal local dynamics in functionally important regions of a metastable serpin. *Structure* **16,** 38–51.
Zhou, A., Huntington, J. A., Pannu, N. S., Carrell, R. W., and Read, R. J. (2003). How vitronectin binds PAI-1 to modulate fibrinolysis and cell migration. *Nat. Struct. Biol.* **10,** 541–544.
Zhou, Z., Felts, A. K., Friesner, R. A., and Levy, R. M. (2007). Comparative performance of several flexible docking programs and scoring functions: Enrichment studies for a diverse set of pharmaceutically relevant targets. *J. Chem. Inf. Model.* **47,** 1599–1608.

CHAPTER NINE

Development of Inhibitors of Plasminogen Activator Inhibitor-1

Shih-Hon Li* *and* Daniel A. Lawrence†

Contents

1. Introduction 178
2. Serpins as Drug Targets 179
   2.1. Biochemistry of canonical inhibitors 179
   2.2. Uniqueness of the serpin fold 179
   2.3. Biologic functions of the serpin mechanism beyond protease inhibition 180
   2.4. Physiologic ligand-binding can affect the antiproteolytic function of serpins 182
3. Development of PAI-1 Inhibitors 184
   3.1. Efforts before screens 184
   3.2. High-throughput screening (HTS) and the screen reaction 184
   3.3. Validation 197
   3.4. Work-up of mechanisms of action 198
4. Concluding Remarks 200
References 200

Abstract

Plasminogen activator inhibitor-1 (PAI-1) belongs to the *ser*ine *p*rotease *in*hibitor super family (*serpin*) and is the primary inhibitor of both the tissue-type (tPA) and urokinase-type (uPA) plasminogen activators. PAI-1 has been implicated in a wide range of pathological processes where it may play a direct role in a variety of diseases. These observations have made PAI-1 an attractive target for small molecule drug development. However, PAI-1's structural plasticity and its capacity to interact with multiple ligands have made the identification and development of such small molecule PAI-1 inactivating agents challenging. In the following pages, we discuss the difficulties associated with screening for small molecule inactivators of PAI-1, in particular, and of serpins, in general. We discuss strategies for high-throughput screening (HTS)

* Department of Pathology, University of Michigan Medical School, Ann Arbor, Michigan, USA
† Division of Cardiovascular Medicine, Department of Internal Medicine, University of Michigan Medical School, Ann Arbor, Michigan, USA

Methods in Enzymology, Volume 501
ISSN 0076-6879, DOI: 10.1016/B978-0-12-385950-1.00009-2

of chemical and natural product libraries, and validation steps necessary to confirm identified hits. Finally, we describe steps essential to confirm specificity of active compounds, and strategies to examine potential mechanisms of compound action.

1. Introduction

Plasminogen activator inhibitor-1 (PAI-1) is the physiological inhibitor of tissue and urokinase plasminogen activators (tPA and uPA, respectively). Numerous studies over the past two decades have demonstrated PAI-1's role in normal physiologic processes, including thrombolysis and wound healing (Yepes *et al.*, 2006). PAI-1 has also been associated with many disease processes, including both acute diseases such as sepsis and myocardial infarction (Colucci *et al.*, 1985; Hamsten *et al.*, 1985), and chronic disorders such as cancer, atherosclerosis, and type 2 diabetes (De Taeye *et al.*, 2005; Durand *et al.*, 2004). The association of PAI-1 with these diseases has led to the suggestion that PAI-1 may contribute to their pathologies. However, the mechanistic role that PAI-1 plays in development of these diseases is not clear and is likely to be complex since PAI-1 can act through multiple pathways, such as modulating fibrinolysis through the regulation of plasminogen activators or by influencing tissue remodeling through the direct regulation of cell migration (Cale and Lawrence, 2007; Dupont *et al.*, 2009). In cardiovascular disease, for example, PAI-1 expression is significantly increased in severely atherosclerotic vessels (Schneiderman *et al.*, 1992), and PAI-1 protein levels rise consistently during disease progression from normal vessels to fatty streaks to atherosclerotic plaques (Robbie *et al.*, 1996). Studies in animal models have shown that blocking PAI-1 with antibodies (Berry *et al.*, 1998; Biemond *et al.*, 1995; Rupin *et al.*, 2001; van Giezen *et al.*, 1997) or dominant-negative PAI-1 mutants (Huang *et al.*, 2008, 2003; McMahon *et al.*, 2001; Wu *et al.*, 2009), or with small molecule inactivators of PAI-1 (Abderrahmani *et al.*, 2009; Baxi *et al.*, 2008; Crandall *et al.*, 2006; Hennan *et al.*, 2005, 2008; Leik *et al.*, 2006; Lijnen *et al.*, 2006; Smith *et al.*, 2006; Weisberg *et al.*, 2005) can ameliorate the severity of various disease models. Because these studies have suggested that the inactivation of PAI-1 may be of therapeutic benefit, PAI-1 has been a molecule of significant clinical interest as a drug target. However, after decades of research, only one compound, PAZ-417 (Jacobsen *et al.*, 2008), has been evaluated in clinical trials (ClinicalTrials.gov Identifier: NCT00739037), underscoring the difficulty that PAI-1 presents as a therapeutic target in drug discovery and especially for high-throughput screening (HTS) protocols. Several expert reviews are available about the design and implementation of HTS, and therefore, this review focuses on the specific challenges of identifying anti-PAI-1 agents.

2. Serpins as Drug Targets

2.1. Biochemistry of canonical inhibitors

Protease inhibition is not a complicated problem to solve in biochemical terms. Since the serine protease mechanism can be simplified into a two-step process, a putative inhibitor merely has to disrupt either the binding of a substrate molecule to the protease or a protease's catalytic activity. Canonical inhibitors accomplish the first method very efficiently. They are small proteins with inhibitor domains of ~3–21 kDa with structures that can vary significantly (Krowarsch *et al.*, 2003). However, they all have in common a solvent-exposed loop whose amino acid composition complements the active site of their target protease. During typical cleavage of a substrate, the C-terminal or P′ side of a substrate leaves the active site of the acyl-enzyme intermediate, making hydrolysis of the P1–P1′ bond ostensibly irreversible. However, with canonical serine protease inhibitors, the entire solvent-exposed loop is held firmly in place and cleavage results in very little conformational change in the loop (Betzel *et al.*, 1993; Shaw *et al.*, 1995). Consequently, the P′ side does not dissociate, allowing a protease to complex tightly with either an intact or cleaved canonical inhibitor. Thus, canonical inhibitors in complex with their target proteases occupy the active site and prevent binding and cleavage of true substrates.

2.2. Uniqueness of the serpin fold

Canonical inhibitors have no moving parts, and if one were to design a protease inhibitor *de novo*, a large complicated molecule with moving parts would most likely not be the first choice. Large molecules require more resources to synthesize, and moving parts need to work precisely. Serpins are such molecules, with their conserved core structure generally containing about 350 residues. Serpins utilize a multistep mechanism that is inherently a race (Lawrence *et al.*, 2000). Like the canonical inhibitors, serpins have an exposed loop, referred to as the reactive center loop (RCL; Fig. 9.1), and they rely on target proteases to initiate hydrolysis of the RCL and the formation of an acyl-enzyme complex via the first catalytic cycle of the serine protease mechanism (Lawrence *et al.*, 1995). The serpin machinery then needs to intercede to prevent deacylation, the second catalytic cycle, from taking place, or else the cleaved serpin is irreversibly inactivated. This is accomplished by a dramatic conformational change in the serpin where the RCL inserts into the central β-sheet of the serpin (Fig. 9.1) and in the process tightly docks the protease to the surface of the serpin and distorts the protease active site (Dementiev *et al.*, 2006; Huntington *et al.*, 2000; Lawrence *et al.*, 1995, 1990). This mechanism requires a protein scaffold

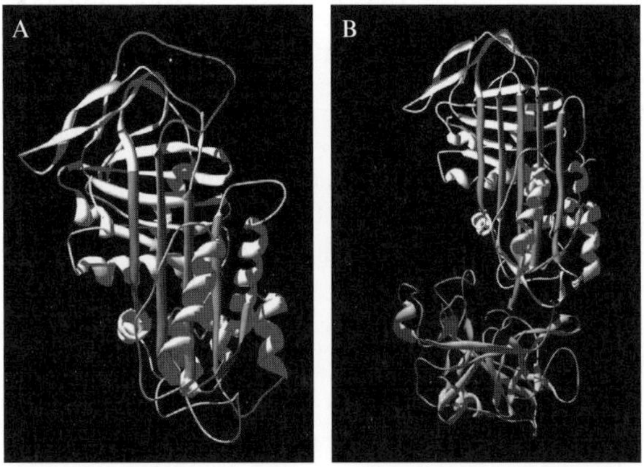

Figure 9.1 Native and protease-bound serpin structures. An archetypal inhibitory serpin shown in its (A) active conformation (PDB accession 1QLP) (Elliott *et al.*, 2000) and (B) in complex with a cognate protease (PDB accession 2D26) (Dementiev *et al.*, 2006). The RCL (red) serves as a substrate loop and inserts into the central β-sheet (green) upon acylation by a protease (light blue), stabilizing the covalent protease–serpin complex. (See Color Insert.)

that is particularly flexible during the initial interaction with a protease, coupled with an endpoint that is thermodynamically very stable so as to prevent the reverse reaction, deacylation. These aspects of the mechanism make serpins prone to off-pathway conformational states, including a latent conformation where the RCL is inserted into the central β-sheet without cleavage (Mottonen *et al.*, 1992), and to serpin polymerization (Belorgey *et al.*, 2007). This latter off-pathway conformation can have devastating pathobiological consequences (Belorgey *et al.*, 2007). Nonetheless, serpins account for approximately 10% of all proteins in blood, suggesting the existence of an evolutionary advantage for such a complex answer to the seemingly simple problem of protease inhibition.

2.3. Biologic functions of the serpin mechanism beyond protease inhibition

The structural lability of serpins is what sets them apart from the canonical serine protease inhibitors and is what gives serpins the potential to exert remarkable regulation over biological processes. Serpins have the ability to act as sensors for proteases of the extracellular environment and to report their presence via the conformational change inherent in the serpin-inhibitory mechanism. Two good examples of this conformational sensor and switch are found in PAI-1 and the hormone-binding serpins, corticosteroid-binding globulin (CBG) and thyroxine-binding globulin (TBG).

PAI-1 interacts with several nonprotease ligands, and these associations play an essential role in PAI-1 function (Dupont *et al.*, 2009; Yepes *et al.*, 2006). PAI-1 binds with high affinity to the cell adhesion protein vitronectin and to members of the endocytic low-density lipoprotein receptor (LDL-r) family, such as the lipoprotein receptor-related protein (LRP), and the very low-density lipoprotein receptor (VLDL-r). These nonprotease interactions are important for both PAI-1 localization and function, and they are largely conformationally controlled through structural changes associated with RCL insertion (Lawrence *et al.*, 1997; Stefansson *et al.*, 1998). When a protease covalently complexes with PAI-1, the subsequent conformational change in PAI-1 results in a greater than 100-fold decrease in affinity for vitronectin (Lawrence *et al.*, 1997). Conversely, native inhibitory PAI-1 binds weakly to the endocytic receptor family members, whereas the PAI-1-protease complex is recognized with single nanomolar K_D, leading to its internalization and degradation (Stefansson *et al.*, 1998, 2004). This switch in affinity from vitronectin to the endocytic receptor family members is mediated by RCL insertion, and the concomitant structural changes in the serpin structure as β-strands 3A and 5A slide apart to accommodate the intercalating RCL. These changes involve movement of two tertiary portions of PAI-1 relative to each other, known as the small serpin fragment (β-strands 1A–3A and α-helix F) and the large serpin fragment (the rest of the serpin; Stein and Chothia, 1991; Whisstock *et al.*, 2000). As the two fragments move apart, they remain connected via the flexible joint region, consisting of α-helices D and E. The surface of the vitronectin-binding site, consisting of residues located in β-strand 1A, the β-strand 2A/α-helix E loop, and the N-terminal ends of α-helices E and F, changes upon RCL insertion, leading to the disruption of hydrophobic and ionic interactions between PAI-1 and vitronectin, such as the critical interaction between PAI-1 residue, Arg^{101}, and Asp^{22} in vitronectin (Zhou *et al.*, 2003). The endocytic receptor family member-binding site is less well defined, but several studies suggest that it partially overlaps with the heparin-binding site in PAI-1 along α-helices D and E (Horn *et al.*, 1998; Rodenburg *et al.*, 1998; Skeldal *et al.*, 2006; Stefansson *et al.*, 1998). Since no X-ray crystal structure is available of PAI-1 in covalent complex with a protease, it is not clear how this site changes upon loop insertion. However, biochemical data suggest that a cryptic LRP-binding site is exposed upon acyl-enzyme complex formation with any protease and that the affinities of these complexes for endocytic receptor family members are similar or identical, consistent with the binding arising from a common protease-bound loop-inserted PAI-1 conformer (Stefansson *et al.*, 2004). Complexation with proteases, even ones that do not have intrinsic affinity toward LRP, causes a greater than 100-fold enhancement of binding between PAI-1 and receptor as compared to native PAI-1 alone. Thus, when PAI-1 that is localized to vitronectin in the extracellular matrix engages a target protease, the ensuing conformational change that accompanies protease inhibition induces a shift in the relative affinity from vitronectin to the endocytic receptor family members of 10,000- to 100,000-fold, repartitioning PAI-1 from the matrix to the cell surface (Yepes *et al.*, 2006). This effectively makes PAI-1 a conformationally driven biological sensor for the presence of specific enzymes (as defined by its target specificity), regulating cellular adhesion, migration, and proliferation.

This conformational switch is utilized in other serpins as well. The hormone-carriage serpins, TBG and CBG, rely on differing degrees of loop insertion to modulate the affinity for their respective hormones. Unlike other serpins, the top of β-sheet A in TBG appears to be obligately open, allowing the RCL to possibly toggle between a fully expelled state and one in which it is partially inserted up to the P12 residue (Grasberger *et al.*, 2002; Zhou *et al.*, 2006). Entry of the RCL up to residue P12 into the β-strand 3A–5A space is hypothesized to force complete insertion of the conserved P14 threonyl group, displacing a tyrosyl group into the closely packed hydrophobic core, possibly closing the thyroxine-binding site by 4–5 Å and decreasing the affinity for the hormone ligand (Zhou *et al.*, 2006). A prolyl side chain at the P8 position, a unique feature of TBG, prevents full insertion of the RCL and potentially allows for the reversible release of thyroxine from TBG in a manner that is regulated by the conformation of the RCL. CBG is thought to function in an analogous manner. As with TBG, RCL expulsion in CBG corresponds to a high-affinity state for cortisol, while partial-loop insertion into the central β-sheet corresponds to a low-affinity state (Klieber *et al.*, 2007; Zhou *et al.*, 2008). Unlike TBG, the RCL of CBG completely inserts into its central sheet upon loop cleavage by proteases, causing irreversible loss of cortisol-binding. The CBG RCL is recognized by elastases, and it has been proposed that cleavage of CBG by neutrophil elastase at sites of inflammation may trigger the irreversible release of its bound hormone (Hammond *et al.*, 1990). Recently, reversible hormone binding in CBG was reported to be affected by temperature, with the affinity for cortisol and progesterone being reduced by 16-fold when reaction temperatures were raised from 35 to 42 °C (Braun *et al.*, 2010), well within the physiological range for fever. Thus, at least two aspects of inflammation may regulate the local release of cortisol from CBG, while a physiologic trigger has yet to be found for the conformational regulation of thyroxin carriage by TBG.

2.4. Physiologic ligand-binding can affect the antiproteolytic function of serpins

Binding of physiologic ligands to serpins can also affect their RCL conformations. Two examples of this are the activation of antithrombin by heparin and the modulation of PAI-1 by vitronectin. Antithrombin is activated toward thrombin, and factors Xa and XIa by heparin and heparan sulfate glycosaminoglycans, by both a template mechanism in which the serpin and its cognate protease are brought in proximity and an allosteric change induced by a specific pentasaccharide (Bjork and Olson, 1997). Native antithrombin, in the absence of cofactor, demonstrates a reactive loop that is partially inserted into the top of β-sheet A (Skinner *et al.*, 1997). Binding of the pentasaccharide to a site involving the N-terminus, α-helices A, and D causes extension of helix D, closure of β-sheet A, and concomitant expulsion of the RCL, as revealed by X-ray crystallography (Jin *et al.*, 1997; Johnson *et al.*, 2006), relieving a repulsive interaction between the serpin and cognate proteases as well as making the RCL more accessible to active sites (Gettins and Olson, 2009). PAI-1 similarly exhibits cofactor-dependent modulation of its RCL conformation. While no X-ray crystal structures of wild-type PAI-1 in its native conformation are available, biochemical data suggest that the RCL of PAI-1 is also partially inserted into β-sheet A, as evidenced by cross-linking studies of the RCL with β-strands 3A and 5A (Hagglof *et al.*, 2004), and by the ability of native PAI-1 to bind a monoclonal antibody, MA-33B8, that specifically recognizes the top of β-sheet A and helix D in a loop-inserted conformation (Gorlatova *et al.*, 2003). Upon the binding of vitronectin to PAI-1, the RCL becomes expelled from the top of β-sheet A (Fa *et al.*, 1995; Hagglof *et al.*, 2004; Li *et al.*, 2008). However, in contrast to the heparin-induced closure of the antithrombin β-sheet A, vitronectin makes the β-strand 3A–5A cleft in PAI-1 more solvent accessible (Li *et al.*, 2008), either by expulsion of the RCL from the central sheet or by direct stabilizing effects on the partially open sheet.

These examples illustrate the advantage of the conformational flexibility of the serpin fold over the rigidity of canonical serine protease inhibitors. Although serpins are large, dynamic machines with increased susceptibility for dysfunction, their unique structural flexibility also allows for the modulation of binding sites. Serpins can act as biological sensors for the presence of specific proteolytic activities, utilizing the conformational changes that occur during protease inhibition to transduce that information via interactions with receptors in a protease-dependent context. Likewise, changes in the RCL conformation can alter binding sites for macromolecules as well as small compounds such as alteration of the common organochemical compound-binding pocket bordered by β-strands 2A, helices D, and E (Bjorquist *et al.*, 1998; Egelund *et al.*, 2001). Pertinent to the development of small inactivating compounds against serpins, ligands can also affect the conformation of the RCL and β-sheet A, two structural features critical for serpin protease inhibition. In this light, serpins are prime targets for the allosteric modulation of their inhibitory as well as ligand-binding properties.

3. Development of PAI-1 Inhibitors

3.1. Efforts before screens

The initial identification of agents as PAI-1 inhibitors revolved around incidental findings on the effects of reaction conditions and buffer additives on PAI-1 activity. Amphipathic compounds, such as sodium dodecyl sulfate, Triton X-100, Tween-80, and Nonidet P40, have been reported to inactivate PAI-1 by inducing substrate behavior or nonreactivity with plasminogen activators (Andreasen *et al.*, 1999; Gils and Declerck, 1998). Studies of the effects of these compounds on the binding of conformationally sensitive monoclonal antibodies to PAI-1 suggest that their mechanisms of action are to induce structural changes in the serpin (Kjoller *et al.*, 1996), even converting PAI-1 to the latent form under certain conditions. While neither binding sites nor stoichiometries have been identified for these amphipathic agents, in regard to future anti-PAI-1 drug development, the value of these studies lies in highlighting the labile nature of PAI-1 and in foreshadowing the importance of a thorough biochemical characterization of the potential mechanisms of action of any identified PAI-1 inactivating compounds.

3.2. High-throughput screening (HTS) and the screen reaction

Directed, purposeful development of anti-PAI-1 agents has for the most part required HTS of candidate molecules (Bryans *et al.*, 1996; Cale *et al.*, 2010; Elokdah *et al.*, 2004; Neve *et al.*, 1999). This is certainly the case for small compounds, but also held true for the PAI-1-inactivating monoclonal antibodies, though these were initially created as tools to study PAI-1 biochemistry. HTS can be a high-yield approach. However, careful consideration must enter into every step of the process, taking into account the selection of the drug target, the biochemical behavior of the candidate compounds, and how to verify the positive hits. This is only the beginning though, since positive hits often become lead structures for second- and third-generation synthetic compounds, while much work remains in defining mechanisms of action.

Screening for inactivators of PAI-1 is usually accomplished indirectly. A common setup compares the residual enzymatic activity of a protease after incubation with a serpin in the presence of a candidate-inactivating agent versus in the absence of such agents. Enzymatic activity is typically measured using small chromogenic or fluorogenic substrates. Incubation of an enzyme with PAI-1 followed by such a substrate decreases the amount of cleavage product as compared to the enzyme alone. In HTS, the inactivation of PAI-1 by a compound prevents the inhibition of the enzyme by PAI-1.

This results in an increase in the amount of substrate cleavage detected compared to the enzyme with PAI-1 but without compound (100% PAI-1 activity), and little or no change in amount of cleaved substrate detected compared to the enzyme alone (0% PAI-1 activity).

Reaction conditions deserve careful consideration, as they will affect the stability and behavior of all of the reaction components. The ionic strength and pH should be kept within physiological ranges. Small amounts of organic solvent might be used to aid in the solubility of compounds. Detergents can be added as well in order to prevent loss of protein to reaction vessels. However, any reagents added for the stability of compounds or of proteins should be checked for their effect on the activities of PAI-1 and the enzyme selected for the HTS. For example, dimethyl sulfoxide, or DMSO, is commonly used as a short- to medium-term storage solvent for concentrated small molecule stocks and is often included in HTS reactions. However, concentrations of DMSO greater than 10% can have adverse effects on PAI-1 activity, increasing substrate behavior, and also reduce the cleavage rate of small peptidyl substrates by tPA and uPA, depending on the protease and substrate pair. Other common organic solvents should likewise be tested. The effect of detergents on PAI-1 activity is well known. Nonionic detergents can induce substrate behavior and accelerate the latency transition in PAI-1, with Triton X-100, Nonidet P40, and Tween-80 being among such offenders (Andreasen *et al.*, 1999).

The reaction temperature will likely either be at 25 or 37 °C. Most published HTS for PAI-1 inactivators appears to have been carried out at room temperature (Bryans *et al.*, 1996; Cale *et al.*, 2010; Elokdah *et al.*, 2004; Neve *et al.*, 1999). This should be no surprise since PAI-1 transition to the latent conformation is much slower at 25 °C than at 37 °C, and given the size of current compound libraries and therefore the time needs to carry out HTS, the transition to latent PAI-1 becomes a real concern. The trend to miniaturize HTS has driven the shrinking of reaction volumes in order to expand the number of reactions run per microplate and raises the challenge of preventing volume loss due to evaporation. While automated HTS instruments have incorporated environmental controls, performing screens at room temperature can significantly reduce evaporation.

3.2.1. Selection of target PAI-1

As both subtle and large-scale conformational changes in PAI-1 can alter their affinities to many binding partners, it stands to reason that the type of PAI-1 to be used as a target should be carefully weighed. PAI-1 has been extensively studied and many variants and conformers of the inhibitor have been made available, including native, latent, cleaved, protease-complexed, as well as numerous mutations thereof. While this presents a wide array of tools for PAI-1 research, it also creates a challenge.

If the ultimate goal is to find a potent inactivator of PAI-1, certainly the appropriate choice for a target molecule is a PAI-1 variant that is inhibitory. There are, however, some things to consider. The structure of active PAI-1 is inherently unstable, slowly transitioning to the latent form. At room temperature and physiologic pH and ionic strength, PAI-1 auto-inactivates with a half-life of about 22 h (Blouse *et al.*, 2003), making active wild-type PAI-1 stable enough to use in large-scale or prolonged screening assays. It should be kept in mind, though, that the different rates of PAI-1 auto-inactivation, and inactivation by RCL-mimicking peptides or inhibitory monoclonal antibodies at 25 versus 37 °C, suggest that conformations of native PAI-1 at room temperature and physiologic temperature are very likely not identical. Rather, native PAI-1 may exist in equilibrium among multiple active conformations (Li *et al.*, 2008), with changes in environmental conditions causing that equilibrium to shift among states of differing stability.

There are a few conformationally stabilized PAI-1 mutants available that might be considered as alternative targets. PAI-$1_{14\text{-}1B}$ and PAI-1_{W175F} are two inhibitory mutants with prolonged half-lives as compared to wild-type PAI-1 (Berkenpas *et al.*, 1995; Blouse *et al.*, 2003). While they may be more resistant to auto-inactivation under physiological conditions, these variants have slight mechanistic differences from the wild-type molecule. Specifically, the mobility of the RCL in these stabilized variants may be limited compared to wild-type PAI-1. In the case of PAI-$1_{14\text{-}1B}$ specifically, the mobility of the central β-sheet is also limited in the active conformation (Sharp *et al.*, 1999). Consequently, PAI-$1_{14\text{-}1B}$ exhibits resistance to some amphipathic and organochemical inactivators of wild-type PAI-1 (Pedersen *et al.*, 2003). Mutants stabilized at the proximal hinge of the RCL, such as PAI-1_R (Lawrence *et al.*, 1994), may be useful for finding therapeutic agents that block PAI-1 from forming physiologic complexes, such as noncovalent Michaelis complexes with cognate proteases or complexes with vitronectin. A caveat that exists with the use of any mutant is that candidate-inactivating agents that are effective against the variant may be differentially potent against wild-type PAI-1, especially if the binding sites of the agents involved the altered residues.

One of the most effective and mechanistically intriguing PAI-1 inactivators arose from a screening assay that used PAI-1 in an inactive conformation. The monoclonal antibody MA-33B8 was raised against PAI-1 covalently complexed to tPA (Debrock and Declerck, 1997), and its mechanism of action involves binding to and stabilizing a conformational intermediate between active PAI-1 and latent PAI-1, making what is normally a reversible step in the transition to latency into an irreversible one, thus accelerating auto-inactivation (Gorlatova *et al.*, 2003; Verhamme *et al.*, 1999). It is very difficult to gauge how fruitful using a nonnative form of PAI-1 as a target protein could be; however, this approach would appear to limit potential candidate agents to mechanisms of action similar to that of

MA-33B8, capitalizing on the presence of unique conformational states that are in reversible equilibrium with the active form.

The manner in which PAI-1 is produced may also affect the results of a screen. Bacterial expression systems are capable of yields on the order of 100s of milligrams, while mammalian cells are generally capable of yields in micrograms. Recently, insect cells have been used to generate quantities of expressed proteins nearing those of bacterial systems, but with post-translational modifications. O-linked glycosylation is similar but nonidentical between insect and mammalian cells. N-linked glycosylation is generally of the high-mannose type in insects, while in mammals, complex N-linked glycosylation with mannose, galactose, *N*-acetylglucosamine, and neuramic acid with terminal sialic acids predominate (Brooks, 2004). Insect cells transfected with mammalian glycosyltransferases do allow for terminal sialization, though the moieties are still nonidentical and post-translational efficiencies could still be improved (Shi and Jarvis, 2007). PAI-1 is N-glycosylated at two of three potential sites, and while variants expressed in bacteria, insect cells, or mammalian cultures have all been found to be functional, even minor differences in post-translational modification may alter the binding of potential inactivators.

3.2.2. Choice of protease

In vitro, PAI-1 inhibits several trypsin-like serine proteases. Known second-order rate constants are shown in Table 9.1. PAI-1 inhibits tPA, uPA, and trypsin most rapidly (Blouse *et al.*, 2003; Olson *et al.*, 2001). While thrombin and activated protein C (APC) can be inactivated by PAI-1 on the order of $10^5\ M^{-1}\ s^{-1}$, this rate of inhibition requires the presence of the cofactor, vitronectin (Rezaie, 2001; Van Meijer *et al.*, 1997). PAI-1 is also more often a substrate for thrombin than an inhibitor, as shown by the high stoichiometries of inhibition. Plasmin can be targeted by PAI-1 as well (Hekman and Loskutoff, 1988). Theoretically, it should not matter which protease is used in the screen because presumably PAI-1 inactivates each of these proteases via the same serpin mechanism. A compound that disrupts the serpin mechanism should be revealed regardless of the protease used. In fact, important differences do exist in how PAI-1 interacts with these proteases, and some of these differences can directly affect the outcome of HTS. Practically, the choice of protease will also be based on availability of an amount of enzyme appropriate to screen the libraries of interest, and the robustness and availability of small protease-specific fluorogenic or chromogenic substrates.

TPA has been more often used in anti-PAI-1 screens than uPA, and this may have been related to the availability of highly purified tPA secondary to its use as a therapeutic agent. There are some practical obstacles to utilizing tPA, however. Only three proteins are known to be physiologically recognized by the tPA active site: plasminogen, PAI-1 and platelet-derived

Table 9.1 Inhibitory rate constants

Protease	k_{inh} (M^{-1} s^{-1})	Stoichiometry	Reference
tPA	2.3×10^7	1	Olson *et al.* (2001)
uPA	5.0×10^6	1	Blouse *et al.* (2003)
Trypsin	2.1×10^6	1	Olson *et al.* (2001)
Plasmin	2.7×10^4	1	Keijer *et al.* (1991)
Thrombin	9.8×10^2	3	Van Meijer *et al.* (1997)
+Vitronectin	2.1×10^5	5	
+Heparin	6.4×10^4	7	
APC	5.7×10^2	1	Rezaie (2001)
+Vitronectin	1.8×10^5	1	

growth factor C (PDGF-C) (Fredriksson *et al.*, 2004; Madison *et al.*, 1995). Occupancy requirements of the tPA subsite pockets S4 to S2′ severely narrow the specificity of the enzyme (Coombs *et al.*, 1996). Additionally, much of its catalytic efficiency comes from contributions of exosite interactions with potential substrates. Fibrin allosterically activates tPA and also bridges tPA to plasminogen (Chmielewska *et al.*, 1988; Hoylaerts *et al.*, 1982). TPA has direct exosite contacts with plasminogen and PAI-1 as well (Ke *et al.*, 1997). Thus small peptide substrates, which lack the capacity to engage in exosite interactions, are not expected to be robustly cleaved by tPA. HTS utilizing tPA often uses the enzyme at supraphysiologic concentrations for this reason, trying to increase the signal associated with substrate cleavage.

UPA also has limited active-site specificity, including plasminogen, PAI-1, and its own cell-surface receptor. Unlike tPA, uPA does not appear to be allosterically activated and may yield steadier baseline activity when exposed to substrate in the presence of a wide array of compounds. While uPA seems to have an exosite interaction with PAI-1, it does not appear to be inhibited by PAI-1 in exactly the manner as tPA. The 37-loop of the protease domains of the plasminogen activators contain basic residues that enhance their second-order rate constants of inhibition by PAI-1, presumably by providing an exosite to facilitate noncovalent Michaelis complex formation (Adams *et al.*, 1991; Madison *et al.*, 1989, 1990). The exosite interaction was additionally thought to decrease the acylation rate constant between tPA and PAI-1, slowing the conversion of the Michaelis complex to the acyl-enzyme complex (Olson *et al.*, 2001). This is not the case, however, for uPA or trypsin. More relevant to the design of the HTS reaction is the fact that this exosite on tPA may give tPA the strongest affinity for PAI-1 among these proteases, suggesting that lower concentrations of these reactants could be used. However, for reasons noted previously, the tPA concentration might in fact be inflated due to the unavailability of highly sensitive tPA-specific reporter substrates.

Not only must PAI-1 rapidly inhibit the selected protease, the resultant covalent complex must be stable on the timescale of the HTS. While complexes of PAI-1 with either tPA or uPA are long-lived, serpin complexes with either trypsin or thrombin are less so. Complex dissociation arises from several nonexclusive mechanisms. First, incomplete disruption of the protease active site may allow enough enzymatic activity to catalytically deacylate the serpin (Plotnick *et al.*, 2002). Second, nucleophilic agents or free base from the solvent can directly attack the acyl-ester bond between the protease and the serpin (Calugaru *et al.*, 2001). And third, the serpin–protease complex is a good substrate for free uninhibited protease molecules. This third mechanism is particularly true for highly processive proteases like trypsin and thrombin, as evidenced by the elimination of complex dissociation when reactions of serpins and these enzymes are mixed with chemical inhibitors such as phenylmethylsulfonyl fluoride, diisopropyl fluorophosphate, or chloromethyl ketones (Rezaie, 1998). Complex dissociation can lead to false-negative findings in the HTS setup described under Section 3.2 and can be reduced by using near physiologic reaction conditions and by keeping the PAI-1 concentration in excess of the protease concentration to promote engagement and inhibition of all protease molecules.

3.2.3. Compounds and compound libraries

Traditionally compound libraries take thousands of man-hours of work to assemble, including collection, purification, identification, and preservation (Mayr and Bojanic, 2009). As would be expected, the early development of compound libraries occurred within pharmaceutical companies, with HTS commonly being run on libraries several tens of thousands to a few hundred thousand compounds in size. These libraries quickly reached the million-compound mark, via continued synthesis and discovery, or through acquisition of smaller libraries. Soon though, the logistical obstacles of running HTS on millions of compounds prompted the development of intelligent workflow systems. Thus, focused libraries, internal hit validation, and artifact-reducing counter screens are common steps in producing quality hits that can serve as lead compounds.

Library screens for candidate PAI-1-inactivating agents have been ongoing for the past two decades. These screens included a wide array of molecules, including small chemicals, peptides, and nucleic acids. The small molecules encompass previously undescribed natural compounds and well-characterized compounds with existing therapeutic indications and therefore known safety profiles and pharmacokinetics (Cale *et al.*, 2010; Elokdah *et al.*, 2004; Neve *et al.*, 1999). Peptide and nucleic acid libraries are created much more rapidly than those of small compounds, utilizing phage-display and molecular biology techniques (Madsen *et al.*, 2010; Mathiasen *et al.*, 2008). As a historical note, some of earliest screens for PAI-1-binding molecules were for monoclonal antibodies (Debrock and Declerck, 1997). These antibody libraries often were made of a few hundred candidates, as this was a typical yield for studies involving animal immunization and splenocyte immortalization. However, the anti-PAI-1 monoclonal antibodies created over a decade ago still often serve as first-line agents to neutralize PAI-1 in cell culture.

The manner in which compounds are introduced merits some discussion. This is perhaps the first decision that will affect whether a compound will be deemed a positive hit. Initial screens are usually qualitative rather than quantitative, and so a single final concentration is often used for all compounds in a particular screen or from a particular library. Practically, this is the case because many libraries are maintained such that the stock compound solutions are of the same concentration. When a dilutional scheme for the library compounds is determined, a decision is implied that any potential effect of a compound on the target protein at compound concentrations above the final reaction concentration is disregarded. For example, if a compound happens to inactivate a target protein with an IC_{50} 100 μ*M*, while the final compound concentration in the screen is only 10 μ*M*, this compound would legitimately not be considered a hit within the parameters of the screen. The dilutional scheme and the final compound concentration may not be a decision that is solely based on forethought of the compound-target interaction. In fact, often the final concentration of compounds will be determined by the assay system, especially in automated platforms. Commonly, the compound concentration will be in vast excess over that of the target protein.

3.2.4. The significance of compound concentration

The concentration of compounds is typically in such excess over the target protein that binding remains pseudo-first order and is thus independent of the target protein concentration. In the most basic model, the formation of complexes between candidate compounds and PAI-1 can be described by the standard receptor–ligand-binding equation,

$$A = (A_{\max} \times I)/(K_D + I), \tag{9.1}$$

where A is the amount of complex formed between PAI-1 and a compound, A_{max} is the theoretical maximal possible amount of compound:PAI-1 complex, K_D is the affinity between a compound and PAI-1, and I is the compound concentration in the HTS reaction. In addition to the requirement of keeping the compounds in excess over PAI-1, this model also assumes that the binding of a compound and PAI-1 is stoichiometric, with a one-to-one interaction and the binding is reversible. If the compound concentration, I, in all of the screening reactions is fixed at 10 μM, and A_{max} is simply defined by the total PAI-1 concentration, then the dependence of compound:PAI-1 complex formation on the affinity of the interaction under the conditions described is shown in Fig. 9.2. As the affinity is decreased, fewer complexes are formed. What is less intuitive is that the fraction of PAI-1 that is bound by compound is relatively insensitive to the total PAI-1 concentration so long as the concentrations of the reactants are such that pseudo-first order conditions are preserved. Thus, the sensitivity of the screen, which we will take as the ability to identify compounds that inactivate a certain proportion of PAI-1, is independent of the total PAI-1

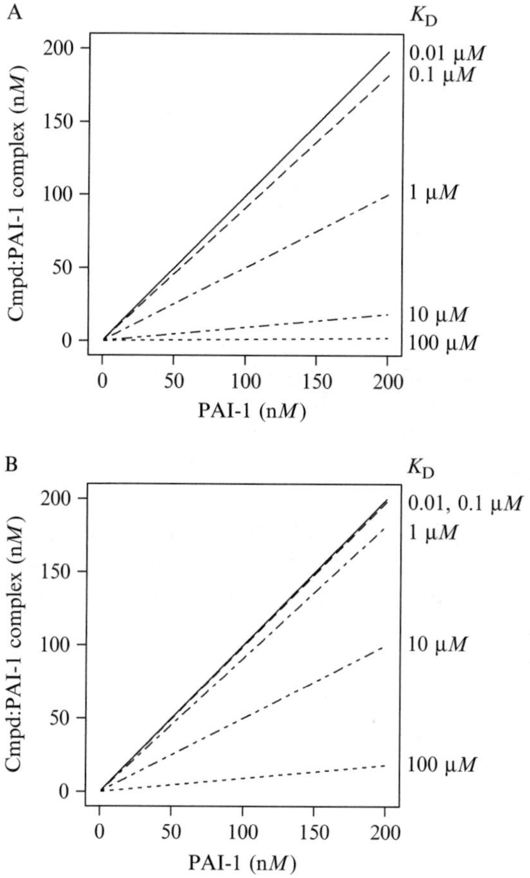

Figure 9.2 Dependence of compound-binding on compound concentration. The predicted concentration of compound:PAI-1 complex formation was calculated from the standard receptor–ligand-binding equation (Eq. (9.1)) as discussed in the text for theoretical compounds having a range of K_D as shown, in the presence of (A) 1 μM or (B) 10 μM compound.

concentration, be it 1, or 10, or 100 nM, so long as the compound concentration remains in vast excess. The linearity of each plot indicates that the proportion of PAI-1 that is inactivated is constant through the PAI-1 concentration range shown. Reduction of the compound concentration diminishes the ability of the screen to detect mid- and low-affinity compounds. Figure 9.2A shows the dependence of the formation of compound: PAI-1 complexes on the concentration of PAI-1 with different dissociation constants in the presence of 1 μM compound. Compared to the same reactions in the presence of 10 μM compound (Fig. 9.2B), there is less thermodynamic drive for the compounds to bind to PAI-1. Thus given any arbitrary threshold of what can be deemed a hit, the higher the compound concentration, the more likely a compound will be a hit.

3.2.5. The molar ratio of PAI-1 to protease

More important than the concentration of PAI-1 is the molar ratio of PAI-1 to protease. Simply stated, if PAI-1 is in excess over the protease, more PAI-1 will have to be inactivated by a compound before any significant relief of inhibition of the subsequent protease is to be detected. If the molar ratio of PAI-1 to protease is 2:1, and the stoichiometry of inhibition for the protease-PAI-1 interaction is 1, then 50% of the PAI-1 must first be inactivated by a candidate compound before PAI-1-mediated protease inhibition is affected. This increases the stringency of the HTS, eliminating some candidates from ever even being picked up as weak hits. Figure 9.3 shows the predicted amount of inhibition of 10 nM PAI-1 in the presence of 10 μM as calculated from the standard receptor–ligand-binding equation (Eq. (9.1)), given the affinities between a compound and PAI-1 as shown. The horizontal black lines represent the expected proportion of PAI-1 that is required to be inactivated by compounds just for such inactivation to be detected under the molar ratios of PAI-1 to protease as shown on the right axis. Under a twofold excess of PAI-1 over protease, compounds that failed to bind and inhibit PAI-1 with a K_D of greater than 10 μM would not be predicted to be hits. Further, if the molar ratio of PAI-1 to protease were to be increased to 10:1, then Fig. 9.3 shows that compounds that bind PAI-1 with a K_D of greater than 1 μM would not be expected to be hits.

The above model is constructed based on simple assumptions, such as the reversible binding between one compound molecule and one target protein. While these assumptions may not hold true for all compounds, the model is supported by the results of a 2009 HTS of the MicroSource SPECTRUM Collection (Cale *et al.*, 2010), wherein PAI-1 was kept at twofold molar excess over uPA. As the IC_{50} values were close to the K_D values for several of the HTS hits and synthetic derivatives thereof, the expected amount of PAI-1 inactivation was calculated in the presence of 10 μM of each hit as shown in Fig. 9.4 using the IC_{50} *in lieu* of the true K_D according to Eq. (9.1). The model would predict that all hits should

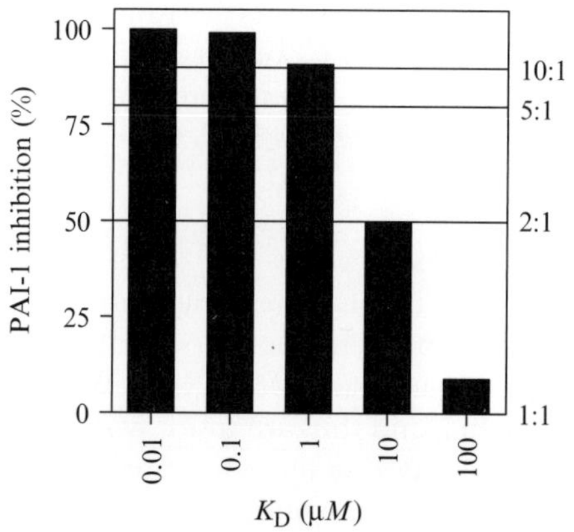

Figure 9.3 Effect of the molar ratio of PAI-1 to protease on HTS stringency. The amount of PAI-1 inhibition was predicted from the standard receptor–ligand-binding equation (Eq. (9.1)) for theoretical compounds with the affinities shown, and using 10 n*M* PAI-1 and 10 μ*M* compounds. If the amount of protease is varied according to the molar ratios of PAI-1 to protease as shown on the right, the amount of PAI-1 inhibition must exceed a certain threshold for each molar ratio as depicted by corresponding cutoff lines in order to be detectable on a screen. Otherwise, enough PAI-1 will remain active and completely inhibit the protease.

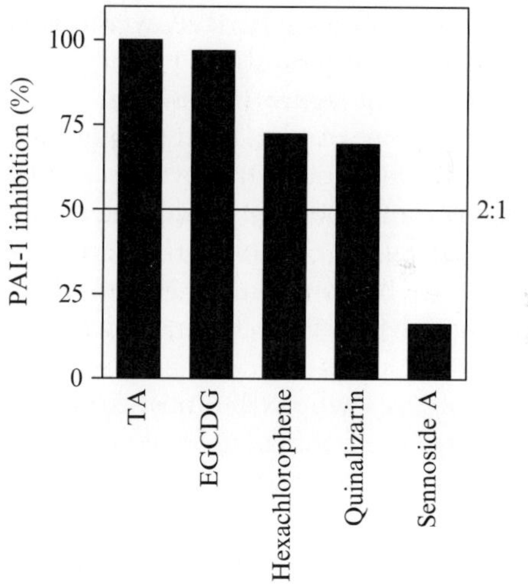

Figure 9.4 Hit compounds from a small-scale screen. A screen of the MicroSource SPECTRUM collection using 10 n*M* PAI-1 and 5 n*M* uPA yielded the hits as shown. The expected amount of PAI-1 inhibition under the HTS conditions was predicted from the empirical IC_{50}s of these compounds and Eq. (9.1). All but sennoside A were predicted to inactivate more than 50% of PAI-1 under the HTS conditions.

inactivate more than 50% of PAI-1 in order to be detected in the presence of twofold excess PAI-1 over uPA, and for four of five hits, this is exactly the case. The outlier, sennoside A, was found to not be stably soluble in DMSO, and this may have inflated its IC_{50} value.

3.2.6. Reporter substrates

Particularly useful for HTS involving proteases are small peptidyl substrates that are conjugated to reporter molecules. The most common substrate-bound chromophore is *p*-nitroanalide (*p*NA). While chromogenic substrates were developed first and thus are more widely available and perhaps less expensive than the fluorogenic substrates, they have nevertheless become less utilized in HTS. Chromophores have smaller signal-to-noise ratios than fluorophores. Many compounds absorb at the same wavelength as *p*NA. Additionally, compounds that precipitate will scatter light, which may be detected as false-positive absorbance.

Fluorophores have benefits and drawbacks as well. HTS of natural compound libraries have been attempted using substrates conjugated to 7-amido-4-methylcoumarin (AMC). While robust AMC-coupled substrates are commercially available for uPA and tPA, which can also be used for trypsin, interference is not uncommon between natural compounds and AMC. Many natural compounds absorb and fluoresce at wavelengths overlapping with one of the most commonly used fluorescent probes, fluorescein (Fowler *et al.*, 2002). Unfortunately, the spectrometric profile of fluorescein overlaps with that of AMC (Eldaw and Khalfan, 1988; Khalfan *et al.*, 1986), and so natural compounds have the tendency to quench AMC as well as generate fluorescent signals themselves.

An alternative fluorophore that has been conjugated to peptidyl protease substrates is rhodamine110 (Rhod110) (Leytus *et al.*, 1983a,b). The excitation and emission spectra of Rhod110 are identical to those of fluorescein, making the fluorescence signal of Rhod110 theoretically as susceptible as fluorescein to interference by compounds. However, the quantum yield of Rhod110 is greater than that of fluorescein, which overcomes the intrinsic fluorescence of many compounds (Chen *et al.*, 2001). While Rhod110-conjugated substrates are as a whole the most expensive of the fluorescent reporters discussed here, their remarkable sensitivity requires several fold lower concentrations of substrate that those linked to either *p*NA or AMC.

It must be understood that no one type of substrate will completely eliminate false positives, given the size of current compound libraries. One strategy that can be adaptable to an anti-PAI-1 screen is to use two differently conjugated substrates simultaneously in the same reaction, such as one peptidyl substrate linked to AMC and another linked to Rhod110 (Grant *et al.*, 2002). To be considered a true hit then, a compound must cause an appropriate recovery of fluorescent signal of both reporters, relying on the lower probability that one compound might interfere with the spectroscopic properties of two very different fluorophores.

3.2.7. Two sample HTS

Recently, we compared two screens for PAI-1-inactivating agents in the same compound libraries, varying reaction conditions such as PAI-1 concentration, enzyme concentration, enzyme identity, and substrate. These screens were run on a combination of the MicroSource SPECTRUM Collection (Kocisko *et al.*, 2003) and the BioFocus NIH Clinical Collection (NCC). The SPECTRUM library consists of 2000 compounds selected by medicinal chemists and biologists, 50% of which are off-patent drugs whose pharmacological and toxicological profiles are well known, 30% of which are natural products of unknown biological activities but which represent an array of structural diversity, and the remaining 20% of which are compounds that have known biological activities but were not further developed as drugs for a variety of reasons including toxicity. The NCC comprises 446 small molecules that have been used in clinical trials and were developed as drugs for an array of therapeutic indications.

PAI-1 at 10 n*M* was incubated with the small molecules before mixing with 5 n*M* uPA, followed by 50 μ*M* zGly-Gly-Arg-AMC (Fig. 9.5A). Alternatively, 10 n*M* PAI-1 was incubated with the compounds before the addition of 1 n*M* trypsin, followed by 2.5 μ*M* (CBZ-Ala-Arg)$_2$-Rhod110 (Fig. 9.5B). The signal of the uninhibited enzyme alone is shown in red, while samples containing PAI-1 alone without enzyme are shown in blue. Potential hit compounds prevent PAI-1 from completely inhibiting the proteases, and thus samples containing compound, PAI-1, and protease together (green) should never yield a signal greater than uninhibited protease alone (red) or less than nonprotease samples (blue). Nevertheless, four compounds from the screen utilizing an AMC-conjugated substrate exhibited fluorescence signal greater than that of uninhibited protease, and at least six compounds yielded signals lower than the nonprotease controls, consistent with the interference of these compounds with the AMC signal. In contrast, a screen using a Rhod110-conjugated substrate yielded only one compound that exhibited a signal greater than the theoretical maximum, with no compounds associated with a signal below the theoretical minimum.

The subtle differences between the two screens led to the different number of hits picked up under each set of conditions. Because the screen utilizing trypsin and a Rhod110 substrate generated a signal over four times greater than that by the screen utilizing uPA and an AMC-linked substrate, the Rhod110-based screen might be expected to have reported a greater number of hits simply based on the sensitivity of the HTS reaction scheme as defined by ratio of cleaved substrate signal to background signal. If all else were comparable, this very well might be the case, as many of the

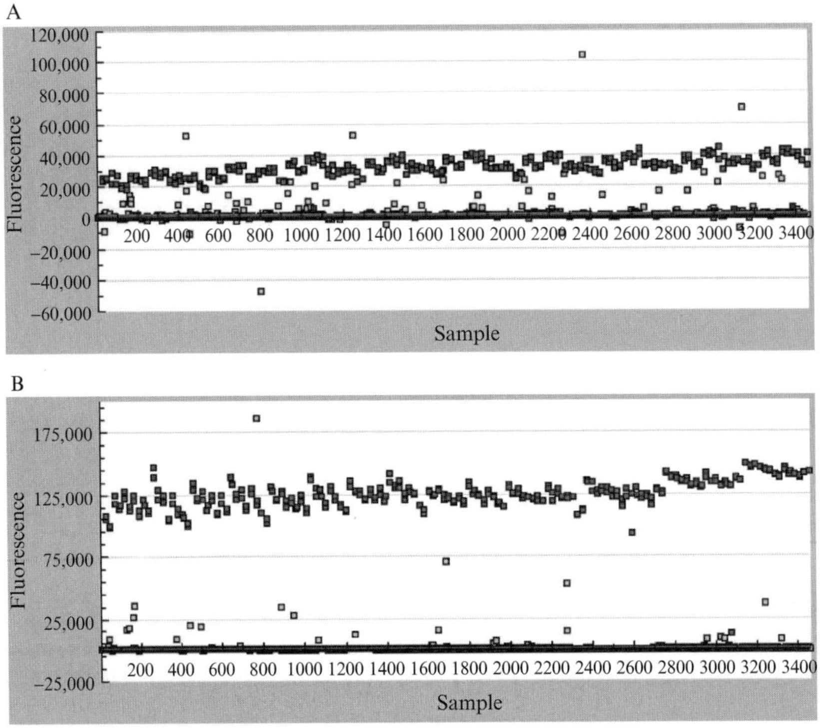

Figure 9.5 Two small-scale screens. The MicroSource SPECTRUM and the BioFocus NIH Clinical collections were screened under different conditions. (A) PAI-1 at 10 n*M* was mixed with 10 μ*M* of each compound, before the addition of 5 n*M* uPA followed by 50 μ*M* zGly-Gly-Arg-AMC. (B) Alternatively, 10 n*M* PAI-1 was mixed with 10 μ*M* of each compound before the addition of 1 n*M* trypsin followed by 2.5 μ*M* $(CBZ\text{-}Ala\text{-}Arg)_2$-Rhod110. In each case, HTS was performed in 40 m*M* HEPES, pH 7.8, 100 m*M* NaCl, and 0.005% Tween-20 at room temperature. Reactions containing PAI-1 alone are in *blue* while those containing uPA and substrate are in *red*. Mixtures of PAI-1 with compound and uPA are shown in *green*. (See Color Insert.)

compounds that yielded signals outside of the expected fluorescence range of AMC might fall within the wider expected range for Rhod110. However, as there is a 10-fold molar excess of PAI-1 over trypsin in the Rhod110-based HTS and only twofold excess of PAI-1 over uPA in the AMC-based screen, the screen utilizing Rhod110 is more stringent as discussed previously. Comparison of these two screens reiterates that the change in substrates predominantly affects the signal interference by small molecules while changing the PAI-1 to protease ratio moderately affects the stringency of the assay. While more sensitive fluorophores should overcome the intrinsic fluorescence of some small molecules and thus facilitate their interpretation as HTS hits, it is also possible to identify false positives during subsequent validation.

3.3. Validation

The validation process can vary but generally involves sorting through the hits to eliminate false positives. Commonly, a rescreen of the hits will be performed, either by using multiple replicates of the original HTS reaction conditions and/or by employing a limited dose-response wherein the compound concentration is varied. In the dose-response, a change in assay signal that is appropriate in relation to the change in small molecule concentration is reassuring. Then, HTS hits can be filtered for desirable or against undesirable chemical, structural, or safety properties. Exclusion criteria can include known toxicities or adverse affects, lack of specificity, difficulties with synthesis or production, and insolubility or other biochemical problems. Also, compound solutions can be inspected for intrinsic colorimetric or fluorometric properties that might interfere with assay interpretation. In screens that utilize a colorimetric readout, the precipitation of reactants and resultant increase in spectroscopic absorption can be mistaken for the proteolytic liberation of chromophores. If the number of hits is too large to inspect manually, plate assays can be helpful. For example, reading colorimetric reactions in a plate reader at wavelengths outside of the absorption spectrum of the chromophore may help to rule out the presence of precipitants.

A validation step that is more labor intensive but very informative is the analysis of reactions of hit compound, PAI-1, and a protease by SDS-PAGE. First, all true positive compounds must be able to prevent PAI-1 from forming covalent complexes with a protease. The presence of a PAI-1-protease band suggests that the signal in the HTS reaction did not correlate to protease activity. After the absence of a covalent complex has been verified, clues to the actual mechanism of the compound might be drawn from the fate of PAI-1. PAI-1 present in a reaction with compound and protease might electrophoretically migrate as a single band at a slightly faster rate than PAI-1 alone, suggesting the presence of RCL-cleaved serpin and that the compound caused PAI-1 to behave as a substrate (Knudsen and Nachman, 1988). If PAI-1 migrated in smaller molecular weight fragments, then the compound might have destabilized the structure of the serpin, allowing it to be partially digested by the protease. PAI-1 from a reaction mixture migrating alongside PAI-1 alone suggests that the compound has made PAI-1 proteolytically inert to the protease, either by preventing PAI-1 and the protease from interacting altogether or by preventing the protease from exerting catalytic activity on PAI-1 while the two are noncovalently bound in the Michaelis complex. Clues from this simple validation experiment can guide work-up of potential mechanisms of action of true positive HTS hits.

3.4. Work-up of mechanisms of action

Once compound hits have been validated, specificity is one of the three parameters that should be explored with all potential drugs. Because of their commercial availability, the effects of compounds on several well-described enzyme–serpin pairs can be assayed, including trypsin or elastase with α_1-proteinase inhibitor, chymotrypsin with α_1-antichymotrypsin, plasmin with α_2-antiplasmin, and thrombin with antithrombin. Different active PAI-1 variants can also be assayed. The effects of glycosylation on compound-susceptibility can be tested on eukaryotically expressed PAI-1. PAI-1 from different species can be compared. This information can be important if animal model experiments are planned. Also, should differences in compound-susceptibility be detected, comparison of primary amino acid sequences may yield information on the potential bind sites. The direct effect of compounds on the plasminogen activators can also be assessed. Some compounds can potentially also interact with the small peptidyl substrates or their reporter moieties, and this can be tested by comparing compound activity toward PAI-1 in assays without peptidyl substrates but analyzing actual cleavage and activation of plasminogen.

The second obligatory question is whether the inactivation of PAI-1 is reversible, or if PAI-1 activity is restored upon dissociation of the compound from PAI-1. Small and large molecules are routinely separated by dialysis or gel filtration. The reversibility of PAI-1 inactivation can be monitored by a simple dilutional experiment as well, wherein the proportion of antiproteolytic activity of PAI-1 in the presence of compound is compared to that in the absence of compound as the two samples are serially diluted in parallel. The ability to recover PAI-1 activity by any of these methods suggests that a compound is not chemically or covalently modifying PAI-1. It also suggests that the compound is not inducing latency. However, if PAI-1 activity cannot be recovered, compound-induced PAI-1 precipitation should be ruled out by centrifugation and pellet analysis on SDS-PAGE. Acceleration of the latency transition can also be considered and can be assessed via chemical or thermal denaturation studies (Lawrence *et al.*, 1994; Wang *et al.*, 1996) or with latency-sensitive fluorescently labeled PAI-1 variants (Shore *et al.*, 1994). If chemical or covalent modification of PAI-1 is suspected, compound-binding sites or sites of modification can be potentially identified via partial proteolysis and mass spectrometry.

The third question involves the stoichiometry of the compound:PAI-1 interaction. This is a very difficult question to answer and may involve highly sensitive biophysical techniques, such as surface plasmon resonance. Small molecules, especially natural ones, may bind promiscuously. A stoichiometry greater than 1 begs the question of whether there is more than one functional binding site. Multiple small molecules may bind to a protein, even while engagement of only one binding site leads to inactivation.

Concerns such as these may not be resolved until putative binding sites are defined by X-ray crystallography and mutagenesis.

Further work-up of the mechanisms of compound hits will likely be quite variable. A few considerations should be kept in mind though, given the biological and biochemical properties of PAI-1. For example, though most HTS might be carried out at room temperature, drugs will ultimately have to function at physiologic temperature. Also, as PAI-1 circulates bound to vitronectin, efficacy against the vitronectin:PAI-1 complex should be demonstrated. The framework of the serpin mechanism is also a good guide to direct dissection of compound mechanisms of action.

If it was not carried out during the validation phase, an SDS-PAGE analysis should be performed. As discussed above, this very basic experiment is potentially high yield. Several additional avenues of investigation could arise from the SDS-PAGE results. Should a hit compound cause accumulation of cleaved PAI-1, it could be parsed out by mass spectrometry or amino-terminal sequencing whether the cleavage site is still at the Met^{345}-Arg^{346} scissile bond. Did cleaved PAI-1 arise from substrate behavior or from accelerated dissociation of the protease-PAI-1 acyl-ester complex? Did the compound merely change the stoichiometry of inhibition between PAI-1 and a protease? If so, the effect of compounds on stoichiometries of inhibition could vary depending on the identity of the protease or presence of a cofactor. However, if a compound leads to the accumulation of unreacted PAI-1 in the presence of a cognate protease, one could ask whether PAI-1 is prevented from interacting at all with the protease or whether the compound has blocked PAI-1 acylation, thereby freezing the serpin mechanism at the noncovalent Michaelis complex. Standard methods of analyzing protein–protein interactions can distinguish among these possibilities, including size-exclusion chromatography by gel filtration or nondenaturing gel electrophoresis, as well as surface plasmon resonance and fluorescence polarization. If a compound is found to block noncovalent complex formation between PAI-1 and a protease, compound-induced serpin polymerization could be a potential drug mechanism. PAI-1 polymers are usually analyzed by nondenaturing gel electrophoresis, though any size-discriminating method can be employed.

A much sought-after piece of data is the drug-binding site, and X-ray crystallography is the standard method for defining such sites. Obtaining PAI-1 crystal structures is not trivial, given its structural lability. The only available X-ray crystal structures of PAI-1 in an active conformation were of the stabilized variant, PAI-$1_{14\text{-}1B}$, which carries four point mutations that may or may not affect compound-binding (Sharp *et al.*, 1999; Stout *et al.*, 2000). Potential binding sites should be confirmed via mutagenesis of native wild-type PAI-1. This controls for artifacts of X-ray crystallography, which by necessity uses very high protein and compound concentrations under nonphysiologic buffer conditions. This also verifies binding-site residues on wild-type PAI-1 as opposed to PAI-$1_{14\text{-}1B}$.

4. Concluding Remarks

PAI-1 is likely to remain of medical interest due to the continued accumulation of evidence for its association with human diseases. As its effects on pathological processes further unfold, the drive to develop pharmacologic interventions against PAI-1 will only continue to grow. HTS for anti-PAI-1 agents have just recently begun to be fruitful, with several new classes of potential lead compounds having been identified in just the past 5 years. Development of small molecule PAI-1 therapeutics will continue to be challenging on several fronts. Its conformational lability and environmental sensitivity may prove to be a double-edged sword, potentially leading to novel mechanisms of action while the window of drug efficacy might be narrowed depending on where PAI-1 is located physiologically. However, a concrete understanding of the serpin mechanism coupled with intimate knowledge of PAI-1 biology and biochemistry can guide the design, implementation, and interpretation of anti-PAI-1 HTS, the subsequent verification of hit compounds, as well as the identification of mechanisms of action of potential PAI-1 inhibitors.

REFERENCES

Abderrahmani, R., Francois, A., Buard, V., Benderitter, M., Sabourin, J. C., Crandall, D. L., and Milliat, F. (2009). Effects of pharmacological inhibition and genetic deficiency of plasminogen activator inhibitor-1 in radiation-induced intestinal injury. *Int. J. Radiat. Oncol. Biol. Phys.* **74,** 942–948.

Adams, D. S., Griffin, L. A., Nachajko, W. R., Reddy, V. B., and Wei, C. M. (1991). A synthetic DNA encoding a modified human urokinase resistant to inhibition by serum plasminogen activator inhibitor. *J. Biol. Chem.* **266,** 8476–8482.

Andreasen, P. A., Egelund, R., Jensen, S., and Rodenburg, K. W. (1999). Solvent effects on activity and conformation of plasminogen activator inhibitor-1. *Thromb. Haemost.* **81,** 407–414.

Baxi, S., Crandall, D. L., Meier, T. R., Wrobleski, S., Hawley, A., Farris, D., Elokdah, H., Sigler, R., Schaub, R. G., Wakefield, T., and Myers, D. (2008). Dose-dependent thrombus resolution due to oral plaminogen activator inhibitor (PAI)-1 inhibition with tiplaxtinin in a rat stenosis model of venous thrombosis. *Thromb. Haemost.* **99,** 749–758.

Belorgey, D., Hagglof, P., Karlsson-Li, S., and Lomas, D. A. (2007). Protein misfolding and the serpinopathies. *Prion* **1,** 15–20.

Berkenpas, M. B., Lawrence, D. A., and Ginsburg, D. (1995). Molecular evolution of plasminogen activator inhibitor-1: Functional stability. *EMBO J.* **14,** 2969–2977.

Berry, C. N., Lunven, C., Lechaire, I., Girardot, C., and O'Connor, S. E. (1998). Antithrombotic activity of a monoclonal antibody inducing the substrate form of plasminogen activator inhibitor type 1 in rat models of venous and arterial thrombosis. *Br. J. Pharmacol.* **125,** 29–34.

Betzel, C., Dauter, Z., Genov, N., Lamzin, V., Navaza, J., Schnebli, H. P., Visanji, M., and Wilson, K. S. (1993). Structure of the proteinase inhibitor eglin c with hydrolysed reactive centre at 2.0 A resolution. *FEBS Lett.* **317,** 185–188.

Biemond, B. J., Levi, M., Coronel, R., Janse, M. J., ten Cate, J. W., and Pannekoek, H. (1995). Thrombolysis and reocclusion in experimental jugular vein and coronary artery thrombosis. Effects of a plasminogen activator inhibitor type 1-neutralizing monoclonal antibody. *Circulation* **91,** 1175–1181.

Bjork, I., and Olson, S. T. (1997). Antithrombin. A bloody important serpin. *Adv. Exp. Med. Biol.* **425,** 17–33.

Bjorquist, P., Ehnebom, J., Inghardt, T., Hansson, L., Lindberg, M., Linschoten, M., Stromqvist, M., and Deinum, J. (1998). Identification of the binding site for a low-molecular-weight inhibitor of plasminogen activator inhibitor type 1 by site-directed mutagenesis. *Biochemistry* **37,** 1227–1234.

Blouse, G. E., Perron, M. J., Kvassman, J. O., Yunus, S., Thompson, J. H., Betts, R. L., Lutter, L. C., and Shore, J. D. (2003). Mutation of the highly conserved tryptophan in the serpin breach region alters the inhibitory mechanism of plasminogen activator inhibitor-1. *Biochemistry* **42,** 12260–12272.

Braun, B. C., Meyer, H. A., Reetz, A., Fuhrmann, U., and Kohrle, J. (2010). Effect of mutations of the human serpin protein corticosteroid-binding globulin on cortisol-binding, thermal and protease sensitivity. *J. Steroid Biochem. Mol. Biol.* **120,** 30–37.

Brooks, S. A. (2004). Appropriate glycosylation of recombinant proteins for human use: Implications of choice of expression system. *Mol. Biotechnol.* **28,** 241–255.

Bryans, J., Charlton, P., Chicarelli-Robinson, I., Collins, M., Faint, R., Latham, C., Shaw, I., and Trew, S. (1996). Inhibition of plasminogen activator inhibitor-1 activity by two diketopiperazines, XR330 and XR334 produced by Streptomyces sp. *J. Antibiot. (Tokyo)* **49,** 1014–1021.

Cale, J. M., and Lawrence, D. A. (2007). Structure-function relationships of plasminogen activator inhibitor-1 and its potential as a therapeutic agent. *Curr. Drug Targets* **8,** 971–981.

Cale, J. M., Li, S. H., Warnock, M., Su, E. J., North, P. R., Sanders, K. L., Puscau, M. M., Emal, C. D., and Lawrence, D. A. (2010). Characterization of a novel class of polyphenolic inhibitors of plasminogen activator inhibitor-1. *J. Biol. Chem.* **285,** 7892–7902.

Calugaru, S. V., Swanson, R., and Olson, S. T. (2001). The pH dependence of serpin-proteinase complex dissociation reveals a mechanism of complex stabilization involving inactive and active conformational states of the proteinase which are perturbable by calcium. *J. Biol. Chem.* **276,** 32446–32455.

Chen, Y., Muller, J. D., Eid, J. S., and Gratton, E. (2001). Two-photon fluorescence fluctuation microscopy. *In* "New Trends in Fluorescence Spectroscopy: Application to Chemical and Life Sciences," (B. Valeur and J. C. Brochon, eds.), pp. 277–296. Springer, New York.

Chmielewska, J., Rånby, M., and Wiman, B. (1988). Kinetics of the inhibition of plasminogen activators by the plasminogen-activator inhibitor. *Biochem. J.* **251,** 327–332.

Colucci, M., Paramo, J. A., and Collen, D. (1985). Generation in plasma of a fast-acting inhibitor of plasminogen activator in response to endotoxin stimulation. *J. Clin. Invest.* **75,** 818–824.

Coombs, G. S., Dang, A. T., Madison, E. L., and Corey, D. R. (1996). Distinct mechanisms contribute to stringent substrate specificity of tissue-type plasminogen activator. *J. Biol. Chem.* **271,** 4461–4467.

Crandall, D. L., Quinet, E. M., El, A. S., Hreha, A. L., Leik, C. E., Savio, D. A., Juhan-Vague, I., and Alessi, M. C. (2006). Modulation of adipose tissue development by pharmacological inhibition of PAI-1. *Arterioscler. Thromb. Vasc. Biol.* **26,** 2209–2215.

De Taeye, B., Smith, L. H., and Vaughan, D. E. (2005). Plasminogen activator inhibitor-1: A common denominator in obesity, diabetes and cardiovascular disease. *Curr. Opin. Pharmacol.* **5,** 149–154.

Debrock, S., and Declerck, P. J. (1997). Neutralization of plasminogen activator inhibitor-1 inhibitory properties: Identification of two different mechanisms. *Biochim. Biophys. Acta* **1337,** 257–266.

Dementiev, A., Dobo, J., and Gettins, P. G. (2006). Active site distortion is sufficient for proteinase inhibition by serpins: Structure of the covalent complex of alpha1-proteinase inhibitor with porcine pancreatic elastase. *J. Biol. Chem.* **281,** 3452–3457.

Dupont, D. M., Madsen, J. B., Kristensen, T., Bodker, J. S., Blouse, G. E., Wind, T., and Andreasen, P. A. (2009). Biochemical properties of plasminogen activator inhibitor-1. *Front. Biosci.* **14,** 1337–1361.

Durand, M. K., Bodker, J. S., Christensen, A., Dupont, D. M., Hansen, M., Jensen, J. K., Kjelgaard, S., Mathiasen, L., Pedersen, K. E., Skeldal, S., Wind, T., and Andreasen, P. A. (2004). Plasminogen activator inhibitor-I and tumour growth, invasion, and metastasis. *Thromb. Haemost.* **91,** 438–449.

Egelund, R., Einholm, A. P., Pedersen, K. E., Nielsen, R. W., Christensen, A., Deinum, J., and Andreasen, P. A. (2001). A regulatory hydrophobic area in the flexible joint region of plasminogen activator inhibitor-1, defined with fluorescent activity-neutralizing ligands. Ligand-induced serpin polymerization. *J. Biol. Chem.* **276,** 13077–13086.

Eldaw, A., and Khalfan, H. A. (1988). Aminomethyl coumarin acetic acid and fluorescein isothiocyanate in detection of leishmanial antibodies: A comparative study. *Trans. R. Soc. Trop. Med. Hyg.* **82,** 561–562.

Elliott, P. R., Pei, X. Y., Dafforn, T. R., and Lomas, D. A. (2000). Topography of a 2.0 A structure of alpha1-antitrypsin reveals targets for rational drug design to prevent conformational disease. *Protein Sci.* **9,** 1274–1281.

Elokdah, H., Abou-Gharbia, M., Hennan, J. K., McFarlane, G., Mugford, C. P., Krishnamurthy, G., and Crandall, D. L. (2004). Tiplaxtinin, a novel, orally efficacious inhibitor of plasminogen activator inhibitor-1: Design, synthesis, and preclinical characterization. *J. Med. Chem.* **47,** 3491–3494.

Fa, M., Karolin, J., Aleshkov, S., Strandberg, L., Johansson, L. B.Å., and Ny, T. (1995). Time-resolved polarized fluorescence spectroscapy studies of plasminogen activator inhibitor type 1: Conformational changes of the reactive center upon interations with target proteases, vitronectin and heparin. *Biochemistry* **34,** 13833–13840.

Fowler, A., Swift, D., Longman, E., Acornley, A., Hemsley, P., Murray, D., Unitt, J., Dale, I., Sullivan, E., and Coldwell, M. (2002). An evaluation of fluorescence polarization and lifetime discriminated polarization for high throughput screening of serine/threonine kinases. *Anal. Biochem.* **308,** 223–231.

Fredriksson, L., Li, H., Fieber, C., Li, X., and Eriksson, U. (2004). Tissue plasminogen activator is a potent activator of PDGF-CC. *EMBO J.* **23,** 3793–3802.

Gettins, P. G., and Olson, S. T. (2009). Activation of antithrombin as a factor IXa and Xa inhibitor involves mitigation of repression rather than positive enhancement. *FEBS Lett.* **583,** 3397–3400.

Gils, A., and Declerck, P. J. (1998). Modulation of plasminogen activator inhibitor 1 by Triton X-100—Identification of two consecutive conformational transitions. *Thromb. Haemost.* **80,** 286–291.

Gorlatova, N. V., Elokdah, H., Fan, K., Crandall, D. L., and Lawrence, D. A. (2003). Mapping of a conformational epitope on plasminogen activator inhibitor-1 by random mutagenesis. Implications for serpin function. *J. Biol. Chem.* **278,** 16329–16335.

Grant, S. K., Sklar, J. G., and Cummings, R. T. (2002). Development of novel assays for proteolytic enzymes using rhodamine-based fluorogenic substrates. *J. Biomol. Screen.* **7,** 531–540.

Grasberger, H., Golcher, H. M., Fingerhut, A., and Janssen, O. E. (2002). Loop variants of the serpin thyroxine-binding globulin: Implications for hormone release upon limited proteolysis. *Biochem. J.* **365,** 311–316.

Hagglof, P., Bergstrom, F., Wilczynska, M., Johansson, L. B., and Ny, T. (2004). The reactive-center loop of active PAI-1 is folded close to the protein core and can be partially inserted. *J. Mol. Biol.* **335,** 823–832.

Hammond, G. L., Smith, C. L., Paterson, N. A., and Sibbald, W. J. (1990). A role for corticosteroid-binding globulin in delivery of cortisol to activated neutrophils. *J. Clin. Endocrinol. Metab.* **71,** 34–39.

Hamsten, A., Wiman, B., de Faire, U., and Blombäck, M. (1985). Increased plasma levels of a rapid inhibitor of tissue plasminogen activator in young survivors of myocardial infarction. *N. Engl. J. Med.* **313,** 1557–1563.

Hekman, C. M., and Loskutoff, D. J. (1988). Bovine plasminogen activator inhibitor 1: Specificity determinations and comparison of the active, latent, and guanidine-activated forms. *Biochemistry* **27,** 2911–2918.

Hennan, J. K., Elokdah, H., Leal, M., Ji, A., Friedrichs, G. S., Morgan, G. A., Swillo, R. E., Antrilli, T. M., Hreha, A., and Crandall, D. L. (2005). Evaluation of PAI-039 [{1-benzyl-5-[4-(trifluoromethoxy)phenyl]-1H-indol-3-yl}(oxo)acetic acid], a novel plasminogen activator inhibitor-1 inhibitor, in a canine model of coronary artery thrombosis. *J. Pharmacol. Exp. Ther.* **314,** 710–716.

Hennan, J. K., Morgan, G. A., Swillo, R. E., Antrilli, T. M., Mugford, C., Vlasuk, G. P., Gardell, S. J., and Crandall, D. L. (2008). Effect of tiplaxtinin (PAI-039), an orally bioavailable PAI-1 antagonist, in a rat model of thrombosis. *J. Thromb. Haemost.* **6,** 1558–1564.

Horn, I. R., van den Berg, B. M., Moestrup, S. K., Pannekoek, H., and van Zonneveld, A. J. (1998). Plasminogen activator inhibitor 1 contains a cryptic high affinity receptor binding site that is exposed upon complex formation with tissue-type plasminogen activator. *Thromb. Haemost.* **80,** 822–828.

Hoylaerts, M., Rijken, D. C., Lijnen, H. R., and Collen, D. (1982). Kinetics of the activation of plasminogen by human tissue plasminogen activator. *J. Biol. Chem.* **257,** 2912–2919.

Huang, Y., Haraguchi, M., Lawrence, D. A., Border, W. A., Yu, L., and Noble, N. A. (2003). A mutant, noninhibitory plasminogen activator inhibitor type 1 decreases matrix accumulation in experimental glomerulonephritis. *J. Clin. Invest.* **112,** 379–388.

Huang, Y., Border, W. A., Yu, L., Zhang, J., Lawrence, D. A., and Noble, N. A. (2008). A PAI-1 mutant, PAI-1R, slows progression of diabetic nephropathy. *J. Am. Soc. Nephrol.* **19,** 329–338.

Huntington, J. A., Read, R. J., and Carrell, R. W. (2000). Structure of a serpin-protease complex shows inhibition by deformation. *Nature* **407,** 923–926.

Jacobsen, J. S., Comery, T. A., Martone, R. L., Elokdah, H., Crandall, D. L., Oganesian, A., Aschmies, S., Kirksey, Y., Gonzales, C., Xu, J., Zhou, H., Atchison, K., *et al.* (2008). Enhanced clearance of Abeta in brain by sustaining the plasmin proteolysis cascade. *Proc. Natl. Acad. Sci. USA* **105,** 8754–8759.

Jin, L., Abrahams, J. P., Skinner, R., Petitou, M., Pike, R. N., and Carrell, R. W. (1997). The anticoagulant activation of antithrombin by heparin. *Proc. Natl. Acad. Sci. USA* **94,** 14683–14688.

Johnson, D. J., Li, W., Adams, T. E., and Huntington, J. A. (2006). Antithrombin-S195A factor Xa-heparin structure reveals the allosteric mechanism of antithrombin activation. *EMBO J.* **25,** 2029–2037.

Ke, S. H., Tachias, K., Lamba, D., Bode, W., and Madison, E. L. (1997). Identification of a hydrophobic exosite on tissue type plasminogen activator that modulates specificity for plasminogen. *J. Biol. Chem.* **272,** 1811–1816.

Keijer, J., Linders, M., Wegman, J. J., Ehrlich, H. J., Mertens, K., and Pannekoek, H. (1991). On the target specificity of plasminogen activator inhibitor 1: the role of heparin, vitronectin, and the reactive site. *Blood.* **78,** 1254–1261.

Khalfan, H., Abuknesha, R., Rand-Weaver, M., Price, R. G., and Robinson, D. (1986). Aminomethyl coumarin acetic acid: A new fluorescent labelling agent for proteins. *Histochem. J.* **18,** 497–499.

Kjoller, L., Martensen, P. M., Sottrup-Jensen, L., Justesen, J., Rodenburg, K. W., and Andreasen, P. A. (1996). Conformational changes of the reactive-centre loop and beta-strand 5A accompany temperature-dependent inhibitor-substrate transition of plasminogen-activator inhibitor 1. *Eur. J. Biochem.* **241,** 38–46.

Klieber, M. A., Underhill, C., Hammond, G. L., and Muller, Y. A. (2007). Corticosteroid-binding globulin, a structural basis for steroid transport and proteinase-triggered release. *J. Biol. Chem.* **282,** 29594–29603.

Knudsen, B. S., and Nachman, R. L. (1988). Matrix plasminogen activator inhibitor: Modulation of the extracellular proteolytic environment. *J. Biol. Chem.* **263,** 9476–9481.

Kocisko, D. A., Baron, G. S., Rubenstein, R., Chen, J., Kuizon, S., and Caughey, B. (2003). New inhibitors of scrapie-associated prion protein formation in a library of 2000 drugs and natural products. *J. Virol.* **77,** 10288–10294.

Krowarsch, D., Cierpicki, T., Jelen, F., and Otlewski, J. (2003). Canonical protein inhibitors of serine proteases. *Cell. Mol. Life Sci.* **60,** 2427–2444.

Lawrence, D. A., Strandberg, L., Ericson, J., and Ny, T. (1990). Structure-function studies of the SERPIN plasminogen activator inhibitor type 1: Analysis of chimeric strained loop mutants. *J. Biol. Chem.* **265,** 20293–20301.

Lawrence, D. A., Olson, S. T., Palaniappan, S., and Ginsburg, D. (1994). Serpin reactive-center loop mobility is required for inhibitor function but not for enzyme recognition. *J. Biol. Chem.* **269,** 27657–27662.

Lawrence, D. A., Ginsburg, D., Day, D. E., Berkenpas, M. B., Verhamme, I. M., Kvassman, J.-O., and Shore, J. D. (1995). Serpin-protease complexes are trapped as stable acyl-enzyme intermediates. *J. Biol. Chem.* **270,** 25309–25312.

Lawrence, D. A., Palaniappan, S., Stefansson, S., Olson, S. T., Francis-Chmura, A. M., Shore, J. D., and Ginsburg, D. (1997). Characterization of the binding of different conformational forms of plasminogen activator inhibitor-1 to vitronectin: Implications for the regulation of pericellular proteolysis. *J. Biol. Chem.* **272,** 7676–7680.

Lawrence, D. A., Olson, S. T., Muhammad, S., Day, D. E., Kvassman, J. O., Ginsburg, D., and Shore, J. D. (2000). Partitioning of serpin-proteinase reactions between stable inhibition and substrate cleavage is regulated by the rate of serpin reactive center loop insertion into beta-sheet A. *J. Biol. Chem.* **275,** 5839–5844.

Leik, C. E., Su, E. J., Nambi, P., Crandall, D. L., and Lawrence, D. A. (2006). Effect of pharmacologic PAI-1 inhibition on cell motility and tumor angiogenesis. *J. Thromb. Haemost.* **4,** 2710–2715.

Leytus, S. P., Melhado, L. L., and Mangel, W. F. (1983a). Rhodamine-based compounds as fluorogenic substrates for serine proteinases. *Biochem. J.* **209,** 299–307.

Leytus, S. P., Patterson, W. L., and Mangel, W. F. (1983b). New class of sensitive and selective fluorogenic substrates for serine proteinases. Amino acid and dipeptide derivatives of rhodamine. *Biochem. J.* **215,** 253–260.

Li, S. H., Gorlatova, N. V., Lawrence, D. A., and Schwartz, B. S. (2008). Structural differences between active forms of plasminogen activator inhibitor type 1 revealed by conformationally sensitive ligands. *J. Biol. Chem.* **283,** 18147–18157.

Lijnen, H. R., Alessi, M. C., Frederix, L., Collen, D., and Juhan-Vague, I. (2006). Tiplaxtinin impairs nutritionally induced obesity in mice. *Thromb. Haemost.* **96,** 731–737.

Madison, E. L., Goldsmith, E. J., Gerard, R. D., Gething, M. H., and Sambrook, J. F. (1989). Serpin-resistant mutants of human tissue-type plasminogen activator. *Nature* **339,** 721–724.

Madison, E. L., Goldsmith, E. J., Gerard, R. D., Gething, M. H., Sambrook, J. F., and Bassel-Duby, R. S. (1990). Amino acid residues that affect interaction of tissue-type plasminogen activator with plasminogen activator inhibitor 1. *Proc. Natl. Acad. Sci. USA* **87,** 3530–3533.

Madison, E. L., Coombs, G. S., and Corey, D. R. (1995). Substrate specificity of tissue type plasminogen activator. Characterization of the fibrin independent specificity of t-PA for plasminogen. *J. Biol. Chem.* **270,** 7558–7562.

Madsen, J. B., Dupont, D. M., Andersen, T. B., Nielsen, A. F., Sang, L., Brix, D. M., Jensen, J. K., Broos, T., Hendrickx, M. L., Christensen, A., Kjems, J., and Andreasen, P. A. (2010). RNA aptamers as conformational probes and regulatory agents for plasminogen activator inhibitor-1. *Biochemistry* **49,** 4103–4115.

Mathiasen, L., Dupont, D. M., Christensen, A., Blouse, G. E., Jensen, J. K., Gils, A., Declerck, P. J., Wind, T., and Andreasen, P. A. (2008). A peptide accelerating the conversion of plasminogen activator inhibitor-1 to an inactive latent state. *Mol. Pharmacol.* **74,** 641–653.

Mayr, L. M., and Bojanic, D. (2009). Novel trends in high-throughput screening. *Curr. Opin. Pharmacol.* **9,** 580–588.

McMahon, G. A., Petitclerc, E., Stefansson, S., Smith, E., Wong, M. K., Westrick, R. J., Ginsburg, D., Brooks, P. C., and Lawrence, D. A. (2001). Plasminogen activator inhibitor-1 regulates tumor growth and angiogenesis. *J. Biol. Chem.* **276,** 33964–33968.

Mottonen, J., Strand, A., Symersky, J., Sweet, R. M., Danley, D. E., Geoghegan, K. F., Gerard, R. D., and Goldsmith, E. J. (1992). Structural basis of latency in plasminogen activator inhibitor-1. *Nature* **355,** 270–273.

Neve, J., Leone, P. A., Carroll, A. R., Moni, R. W., Paczkowski, N. J., Pierens, G., Bjorquist, P., Deinum, J., Ehnebom, J., Inghardt, T., Guymer, G., Grimshaw, P., *et al.* (1999). Sideroxylonal C, a new inhibitor of human plasminogen activator inhibitor type-1, from the flowers of Eucalyptus albens. *J. Nat. Prod.* **62,** 324–326.

Olson, S. T., Swanson, R., Day, D., Verhamme, I., Kvassman, J., and Shore, J. D. (2001). Resolution of Michaelis complex, acylation, and conformational change steps in the reactions of the serpin, plasminogen activator inhibitor-1, with tissue plasminogen activator and trypsin. *Biochemistry* **40,** 11742–11756.

Pedersen, K. E., Einholm, A. P., Christensen, A., Schack, L., Wind, T., Kenney, J. M., and Andreasen, P. A. (2003). Plasminogen activator inhibitor-1 polymers, induced by inactivating amphipathic organochemical ligands. *Biochem. J.* **372,** 747–755.

Plotnick, M. I., Samakur, M., Wang, Z. M., Liu, X., Rubin, H., Schechter, N. M., and Selwood, T. (2002). Heterogeneity in serpin-protease complexes as demonstrated by differences in the mechanism of complex breakdown. *Biochemistry* **41,** 334–342.

Rezaie, A. R. (1998). Reactivities of the S2 and S3 subsite residues of thrombin with the native and heparin-induced conformers of antithrombin. *Protein Sci.* **7,** 349–357.

Rezaie, A. R. (2001). Vitronectin functions as a cofactor for rapid inhibition of activated protein C by plasminogen activator inhibitor-1. Implications for the mechanism of profibrinolytic action of activated protein C. *J. Biol. Chem.* **276,** 15567–15570.

Robbie, L. A., Booth, N. A., Brown, A. J., and Bennett, B. (1996). Inhibitors of fibrinolysis are elevated in atherosclerotic plaque. *Arterioscler. Thromb. Vasc. Biol.* **16,** 539–545.

Rodenburg, K. W., Kjoller, L., Petersen, H. H., and Andreasen, P. A. (1998). Binding of urokinase-type plasminogen activator-plasminogen activator inhibitor-1 complex to the endocytosis receptors alpha2-macroglobulin receptor/low-density lipoprotein receptor-related protein and very-low- density lipoprotein receptor involves basic residues in the inhibitor. *Biochem. J.* **329**(Pt 1), 55–63.

Rupin, A., Martin, F., Vallez, M. O., Bonhomme, E., and Verbeuren, T. J. (2001). Inactivation of plasminogen activator inhibitor-1 accelerates thrombolysis of a platelet-rich thrombus in rat mesenteric arterioles. *Thromb. Haemost.* **86,** 1528–1531.

Schneiderman, J., Sawdey, M. S., Keeton, M. R., Bordin, G. M., Bernstein, E. F., Dilley, R. B., and Loskutoff, D. J. (1992). Increased type 1 plasminogen activator inhibitor gene expression in atherosclerotic human arteries. *Proc. Natl. Acad. Sci. USA* **89,** 6998–7002.

Sharp, A. M., Stein, P. E., Pannu, N. S., Carrell, R. W., Berkenpas, M. B., Ginsburg, D., Lawrence, D. A., and Read, R. J. (1999). The active conformation of plasminogen activator inhibitor 1, a target for drugs to control fibrinolysis and cell adhesion. *Structure* **7,** 111–118.

Shaw, G. L., Davis, B., Keeler, J., and Fersht, A. R. (1995). Backbone dynamics of chymotrypsin inhibitor 2: Effect of breaking the active site bond and its implications for the mechanism of inhibition of serine proteases. *Biochemistry* **34,** 2225–2233.

Shi, X., and Jarvis, D. L. (2007). Protein N-glycosylation in the baculovirus-insect cell system. *Curr. Drug Targets* **8,** 1116–1125.

Shore, J. D., Day, D. E., Francis-Chmura, A. M., Verhamme, I., Kvassman, J., Lawrence, D. A., and Ginsburg, D. (1994). A fluorescent probe study of plasminogen activator inhibitor-1: Evidence for reactive center loop insertion and its role in the inhibitory mechanism. *J. Biol. Chem.* **270,** 5395–5398.

Skeldal, S., Larsen, J. V., Pedersen, K. E., Petersen, H. H., Egelund, R., Christensen, A., Jensen, J. K., Gliemann, J., and Andreasen, P. A. (2006). Binding areas of urokinase-type plasminogen activator-plasminogen activator inhibitor-1 complex for endocytosis receptors of the low-density lipoprotein receptor family, determined by site-directed mutagenesis. *FEBS J.* **273,** 5143–5159.

Skinner, R., Abrahams, J. P., Whisstock, J. C., Lesk, A. M., Carrell, R. W., and Wardell, M. R. (1997). The 2.6 A structure of antithrombin indicates a conformational change at the heparin binding site. *J. Mol. Biol.* **266,** 601–609.

Smith, L. H., Dixon, J. D., Stringham, J. R., Eren, M., Elokdah, H., Crandall, D. L., Washington, K., and Vaughan, D. E. (2006). Pivotal role of PAI-1 in a murine model of hepatic vein thrombosis. *Blood* **107,** 132–134.

Stefansson, S., Muhammad, S., Cheng, X. F., Battey, F. D., Strickland, D. K., and Lawrence, D. A. (1998). Plasminogen activator inhibitor-1 contains a cryptic high affinity binding site for the low density lipoprotein receptor-related protein. *J. Biol. Chem.* **273,** 6358–6366.

Stefansson, S., Yepes, M., Gorlatova, N., Day, D. E., Moore, E. G., Zabaleta, A., McMahon, G. A., and Lawrence, D. A. (2004). Mutants of plasminogen activator inhibitor-1 designed to inhibit neutrophil elastase and cathepsin G are more effective in vivo than their endogenous inhibitors. *J. Biol. Chem.* **279,** 29981–29987.

Stein, P., and Chothia, C. (1991). Serpin tertiary structure transformation. *Mol. Biol.* **221,** 615–621.

Stout, T. J., Graham, H., Buckley, D. I., and Matthews, D. J. (2000). Structures of active and latent PAI-1: A possible stabilizing role for chloride ions. *Biochemistry* **39,** 8460–8469.

van Giezen, J. J., Wahlund, G., Nerme, G., and Abrahamsson, T. (1997). The Fab-fragment of a PAI-1 inhibiting antibody reduces thrombus size and restores blood flow in a rat model of arterial thrombosis. *Thromb. Haemost.* **77,** 964–969.

Van Meijer, M., Smilde, A., Tans, G., Nesheim, M. E., Pannekoek, H., and Horrevoets, A. J. (1997). The suicide substrate reaction between plasminogen activator inhibitor 1 and thrombin is regulated by the cofactors vitronectin and heparin. *Blood* **90,** 1874–1882.

Verhamme, I., Kvassman, J. O., Day, D., Debrock, S., Vleugels, N., Declerck, P. J., and Shore, J. D. (1999). Accelerated conversion of human plasminogen activator inhibitor-1 to its latent form by antibody binding. *J. Biol. Chem.* **274,** 17511–17517.

Wang, Z., Mottonen, J., and Goldsmith, E. J. (1996). Kinetically controlled folding of the serpin plasminogen activator inhibitor 1. *Biochemistry* **35,** 16443–16448.

Weisberg, A. D., Albornoz, F., Griffin, J. P., Crandall, D. L., Elokdah, H., Fogo, A. B., Vaughan, D. E., and Brown, N. J. (2005). Pharmacological inhibition and genetic deficiency of plasminogen activator inhibitor-1 attenuates angiotensin II/salt-induced aortic remodeling. *Arterioscler. Thromb. Vasc. Biol.* **25,** 365–371.

Whisstock, J. C., Skinner, R., Carrell, R. W., and Lesk, A. M. (2000). Conformational changes in serpins: I. The native and cleaved conformations of alpha(1)-antitrypsin. *J. Mol. Biol.* **296,** 685–699.

Wu, J., Peng, L., McMahon, G. A., Lawrence, D. A., and Fay, W. P. (2009). Recombinant plasminogen activator inhibitor-1 inhibits intimal hyperplasia. *Arterioscler. Thromb. Vasc. Biol.* **29,** 1565–1570.

Yepes, M., Loskutoff, D. J., and Lawrence, D. A. (2006). Plasminogen activator inhibitor-1. *In* "Hemostasis and Thrombosis: Basic Principles and Clinical Practice," (R. W. Colman, V. J. Marder, A. W. Clowes, J. N. George, and S. Z. Goldhaber, eds.), pp. 365–380. Lippincott Williams & Wilkins, Philadelphia, PA.

Zhou, A., Huntington, J. A., Pannu, N. S., Carrell, R. W., and Read, R. J. (2003). How vitronectin binds PAI-1 to modulate fibrinolysis and cell migration. *Nat. Struct. Biol.* **10,** 541–544.

Zhou, A., Wei, Z., Read, R. J., and Carrell, R. W. (2006). Structural mechanism for the carriage and release of thyroxine in the blood. *Proc. Natl. Acad. Sci. USA* **103,** 13321–13326.

Zhou, A., Wei, Z., Stanley, P. L., Read, R. J., Stein, P. E., and Carrell, R. W. (2008). The S-to-R transition of corticosteroid-binding globulin and the mechanism of hormone release. *J. Mol. Biol.* **380,** 244–251.

CHAPTER TEN

Bioinformatic Approaches for the Identification of Serpin Genes with Multiple Reactive Site Loop Coding Exons

Stefan Börner *and* Hermann Ragg[1]

Contents

1. Introduction 210
2. Procedure for Identification of Serpin Genes with mRSL Cassette Exons 211
   2.1. Construction of HMMs 213
   2.2. Processing of genomic sequences 213
   2.3. Identification of serpin genes with mRSL cassette exons 213
   2.4. Exon–intron structures of serpin genes 215
   2.5. Searching for mRSL serpin genes in the genome of a model organism 215
   2.6. Step-by-step protocol 217
   2.7. Limitations 218
3. Conclusion 219
Acknowledgment 220
References 220

Abstract

In several branches of the tree of life, alternative splicing of a single primary transcript may give rise to multiple serpin isoforms exhibiting different target enzyme specificities. Though the continuously increasing number of genome sequencing projects has been paralleled by a rapidly rising number of serpin genes, the full spectrum of isoforms that some of these genes can encode has often not been recognized in routine database searches. In this chapter, we introduce procedures that enable the systematic extraction of multi-isoform generating serpin genes from genomic sequences. Spot checking of a model organism demonstrates that the phyletic distribution of such genes appears

Department of Biotechnology, Faculty of Technology, Bielefeld University, Bielefeld, Germany
[1] Corresponding author.

Methods in Enzymology, Volume 501
ISSN 0076-6879, DOI: 10.1016/B978-0-12-385950-1.00010-9

to be largely underestimated. The bioinformatic approach presented here may help to dissect the complete antiproteolytic spectrum of a genome's serpin complement and to register the occurrence of multitasking serpin genes in eukaryotes for functional and evolutionary studies.

1. Introduction

Inhibitory serpins act as a counterbalance to several different clans of serine and cysteine peptidases through a unique reaction mechanism that involves large-scale rearrangements of the inhibitor molecule (Gettins 2002; Silverman *et al.*, 2001; Whisstock *et al.*, 2010). One of the factors crucially contributing to target choice selectivity and inhibitory efficiency is the reactive site loop (RSL), a flexible sequence of about 20 amino acids close to the C-terminal end of serpins presented as a bait to proteolytic enzymes. The RSL residues, notably between positions P6 to P2′ (nomenclature according to Schechter and Berger, 1967), are crucially important for a selective and efficient interaction with proteases (Dufour *et al.*, 2001), though many serpins also use auxiliary sequences, known as exosites, in order to do their job properly. Exosites may be located at various locations on the surface of the serpin scaffold or in the exterior parts of the RSL. The dramatic effects, which such additional contact sites may exercise on target specificity and on the kinetics of inhibition, clearly reveal that exosites represent a successful means of the serpins to antagonize proteases (Ragg *et al.*, 1990; Whisstock *et al.*, 2010).

From the viewpoint of genetics, two ways of enzyme control by serpins have evolved. A well-known mode of adapting the antiproteolytic repertoire is gene duplication followed by subfunctionalization or neofunctionalization of the progeny. In rodents, for instance, expansion of entire α_1-*antitrypsin*- and α_1-*antichymotrypsin*-like genes has led to the emergence of multiple paralogs sharing similar molecule scaffolds, but exhibiting strongly differing RSL sequences (Forsyth *et al.*, 2003; Hill and Hastie, 1987). A different strategy to increase inhibitor diversity rests upon the use of multiple dissimilar RSL cassettes that are individually hooked to one and the same serpin scaffold. These RSL cassettes are encoded in individual exons lined up like beads on a string in the genome. Through alternative splicing, one specimen of the exon set is chosen and connected to the invariable part of the transcript followed by translation into a variant displaying an individual target enzyme spectrum.

Multiple serpin isoforms, generated through alternative splicing of RSL cassette exons, were first described in the tobacco hornworm, *Manduca sexta* (Jiang *et al.*, 1994), and subsequently in various metazoans, mostly insects (Börner and Ragg, 2008; Brandt *et al.*, 2004; Danielli *et al.*, 2003; Hegedus *et al.*, 2008; Jiang *et al.*, 1996; Krüger *et al.*, 2002; Zou *et al.*, 2009), but also in nematodes (Krüger *et al.*, 2002; Luke *et al.*, 2006) and in the urochordate *Ciona intestinalis* (Ragg, 2007). As yet, no examples with plants are known. Serpin genes expressing more than 10 RSL variants are known to exist, and experimental validation has shown that the combined isoforms can inhibit a wide spectrum of proteolytic enzymes (Brüning *et al.*, 2007; Danielli *et al.*, 2003), underlining the evolutionary success of the strategy. The physiological interaction partners for a few of these isoforms have been identified, revealing that some of the multitasking serpin genes play a role in innate immunity (Ragan *et al.*, 2010; Silverman *et al.*, 2010).

Occasionally, multifunctional serpin genes were discerned following detection of mRNAs having identical sequences over most of their lengths, but differing in their 3′-parts that code for the RSL sequences (Sasaki, 1991). Subsequent analysis of the appropriate genomic sequences revealed the origin of these mRNAs from a common primary transcript. Database searches using signature motifs located close to the C-terminal end have also contributed to the identification of some of these serpin genes.

Standard gene identification programs may allow detection of mRSL encoding serpin genes, when appropriate search motifs are used for screening. However, in our experience, often some or even the majority of RSL cassette encoding exons escape detection. A major reason for this failure is the lack of considering additional downstream exons once a serpin sequence that includes an RSL at the 3′-end and that is properly terminated by a stop codon has been found. Therefore, we sought to develop a generally applicable methodology appropriate for unbiased, automated identification of such genes. The procedures given below may help to resolve this deficit.

2. Procedure for Identification of Serpin Genes with mRSL Cassette Exons

Our concept of identifying serpin genes with their complete set of RSL encoding cassette exons rests upon the following approach:

1. Construct hidden Markov models (HMMs) for peptide segments along the entire serpin sequence, including the RSL region, based on a large set of verified serpins covering a phylogenetically diverse set of eukaryotes.
2. Search adequately prepared genomic sequences for HMM matches.
3. The sections below outline the procedure, including its application to a model genome. At the end, a step-by-step user instruction is provided. Table 10.1 lists all databases and software programs used in this study.

Table 10.1 List of software programs and databases used

Software	Application	Web site
AUGUSTUS	Gene prediction program	http://augustus.gobics.de/
BLAST+	Sequence similarity search	ftp.ncbi.nih.gov/blast/executables/LATEST
CLUSTALW2	Multiple sequence alignments	http://www.ebi.ac.uk/clustalw/
EMBOSS	Software package for data evaluation and editing	http://emboss.sourceforge.net/
GENEDOC	Alignment viewer and editor	http://www.nrbsc.org/gfx/genedoc/
GenBank	Extraction of ESTs	http://www.ncbi.nlm.nih.gov/genbank/
HMMER2	Creation and application of HMMs	http://hmmer.janelia.org/
LOGOMAT-M	Creation of profile HMM logos	http://www.sanger.ac.uk/resources/software/logomat-m/
MAFFT	Multiple sequence alignments	http://mafft.cbrc.jp/alignment/software/
Python 2.7	Scripting	http://python.org
RefSeq	Extraction of verified serpins	http://www.ncbi.nlm.nih.gov/RefSeq/

2.1. Construction of HMMs

Serpin sequences were loaded from the NCBI reference sequence (RefSeq) database. The extracted sequences were aligned with five prototype serpins, including human α_1-antitrypsin (Hsap-A1AT, Accession Number: NP_001121177.1), human SERPIN B5 (Accession Number: NP_002630.2), srp-3 from *Caenorhabditis elegans* (Accession Number: NP_503528.2), Spn4 (isoform F) from *Drosophila melanogaster* (Accession Number: NP_995759.1), and serpin AT2G35580.1p from *Arabidopsis thaliana* (Accession Number: NP_181101.1), using NEEDLE (Needleman–Wunsch algorithm implemented in the EMBOSS package (Rice *et al.*, 2000)). The settings used were: amino acid substitution matrix: EBLOSUM80; gap open penalty: 8; gap extend penalty: 2. Only sequences having $\geq$40% similarity to at least one of the reference serpins were included. The remaining sequences were aligned with MAFFT using the G-INS-i strategy as described (Katoh and Toh, 2008), and all segments $\geq$10 residues sharing $\geq$60% similarity in at least six residues were identified. The selected segments were realigned with CLUSTALW2 (Larkin *et al.*, 2007), using the GONNET250 substitution matrix (gap open penalty: 25; gap extend penalty: 5). The resulting alignments were manually edited in GENEDOC (Nicholas *et al.*, 1997), discarding all columns having gaps in more than 80% of the positions. Using HMMBUILD from the HMMER2 package, HMMs were created for each of the arranged segments as described in the HMMER2 user guide (http://hmmer.janelia.org). It should be noted that due to a distinctly different scoring algorithm, HMMER3 is less suited for the presented method. Applying the HMMSEARCH tool, a random sequence of 10^9 amino acids was scanned in order to validate the specificity of the individual HMM. Figure 10.1 shows sequence logo representations of the six HMMs generated that, collectively, cover the sequence of serpins. HMMs E and F each represent portions of the RSL sequences.

2.2. Processing of genomic sequences

Large contiguous genomic sequences were divided into segments no longer than 5.1×10^6 nucleotides (nt), each with an overlap of 10^5 nt to the adjacent segment, using the EMBOSS program SPLITTER. This enables a more efficient search for hits when applying HMMSEARCH, since the elapsed time increases subquadratically with the sequence length. The processed genomic fragments were translated into all six reading frames by applying TRANSEQ (EMBOSS package).

2.3. Identification of serpin genes with mRSL cassette exons

Translated genomic sequences were screened for matches to the six chosen HMMs (*E*-value cutoff: 0.01 for HMMs A-E and 10 for HMM F), and all matches arranged in the same direction were ordered according to their positions in the genome. The resulting list was screened according to the following criteria:

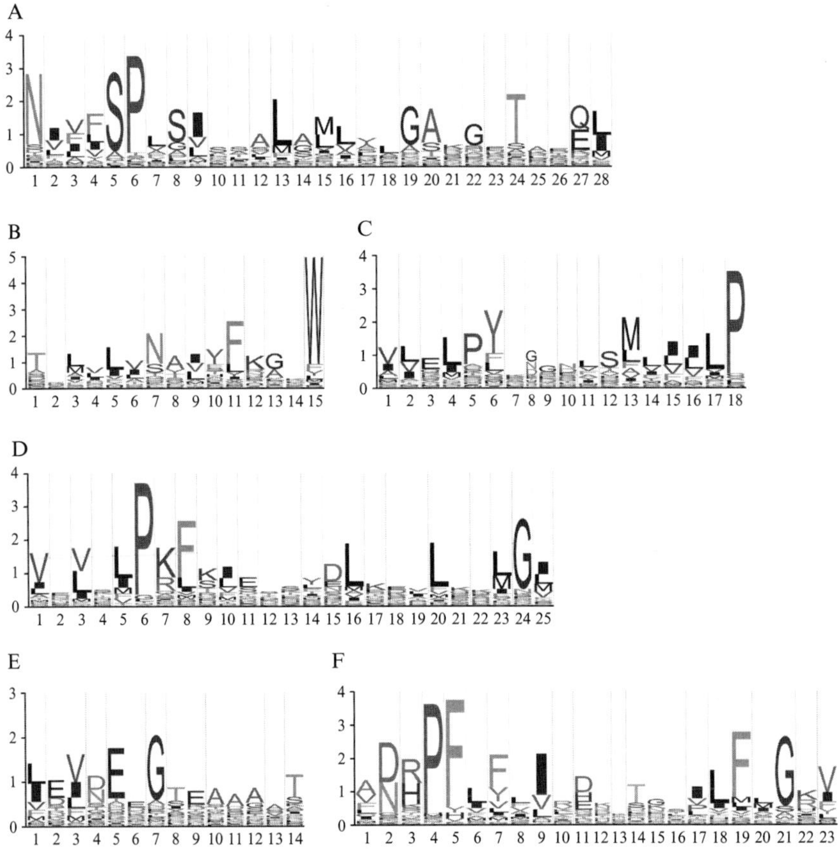

Figure 10.1 *Logos of profile HMMs used for extracting serpin scaffold sequences (A–D) and RSL cassettes (E, F) from translated genomic sequences.* The HMMs shown correspond to the following residues of the mature form (394 amino acids) of Hsap-A1AT: A, 49–76; B, 180–194; C, 239–255; D, 284–308; E, 338–351; F, 366–388. The relative sizes of the letters reflect the frequencies of amino acids in the serpin superfamily at a certain position of the six HMMs. Vertical lines indicate positions where indels (exceeding the background level) may occur. The logos were generated with a modified version of LOGOMAT-M (Schuster-Böckler *et al.*, 2004). (For color version of this figure, the reader is referred to the web version of this chapter.)

1. Discard matches with distances >20000 amino acids.
2. Search for hits matching to at least three HMMs that are arranged in the correct, serpin-specific order.
3. Discard matches to HMMs A–E that are arranged in the wrong order and that possess *E*-values >0.001.
4. Always include matches to HMM F in order to account for noncanonical RSLs, such as found in noninhibitory serpins or pseudoexons.

5. Evidently, several consecutive hits that match to HMM F are indicative of mRSL serpin genes. For further in-depth analysis, genomic sequences located upstream of HMM A (10,000 nt) and downstream of HMM F (15,000 nt) were included in the analyses.

2.4. Exon–intron structures of serpin genes

Exon–intron structures were established by comparing cDNA/EST sequences to genomic DNA whenever possible. For transcript-based mapping of gene structures, cDNA/EST sequences were transferred via FORMATDB (BLAST+ package, Camacho *et al.*, 2009) into a local, BLASTable database. Genomic serpin sequences identified as described above were searched against the transcript database, using the BLASTN tool (word size: 6; *E*-value cutoff: 10^{-5}). Gene and cDNA sequences of hits were aligned using the NEEDLE algorithm (match score: 1; mismatch score: -3; gap open score: -20; gap extend score: -0.5).

To deduce serpin gene structures in species with insufficient cDNA/EST data sets, the gene finding program AUGUSTUS (Stanke *et al.*, 2004) was consulted. The individual, computationally predicted amino acid sequences were aligned with three prototype serpins (Hsap-A1AT, srp-3 from *C. elegans*, and isoform F of Spn4 from *D. melanogaster*), applying NEEDLE (settings: EBLOSUM80 matrix; gap open penalty: 8; gap extend penalty: 2). In each case, the model with the highest cumulative score was chosen.

2.5. Searching for mRSL serpin genes in the genome of a model organism

The procedure described above was tested by analyzing the genomes of several model organisms, and as an example, the results obtained for the pea aphid *Acyrthosiphon pisum* (*A. pisum*), a plant pest, are presented. The 464-Mb genome of this insect (The International Aphid Genomics Consortium, 2010) contains about 17 serpin genes, 2 of which are endowed with several RSL encoding cassette exons (Fig. 10.2). One of these, named *Apis-Spn1*, is predicted to give rise to nine isoforms with different RSLs, while transcription of the other, *Apis-Spn2*, may produce four RSL variants. A search in the NCBI EST/cDNA database revealed only entries for a subset of these isoforms (Accession Numbers: FF329667.1, EX625808.1, FP917733.1, FF333497.1, FF315154.1, FF320908.1, CN584628.1, FF308176.1, and FF323630.1; date of database mining: September 21, 2010).

The Apis-Spn1 RSL variants are highly variable, depicting seven different residues at the presumed P1 position. One of these, Apis-Spn1-x1, terminates with an endoplasmic retrieval signal (KEEL) that is associated

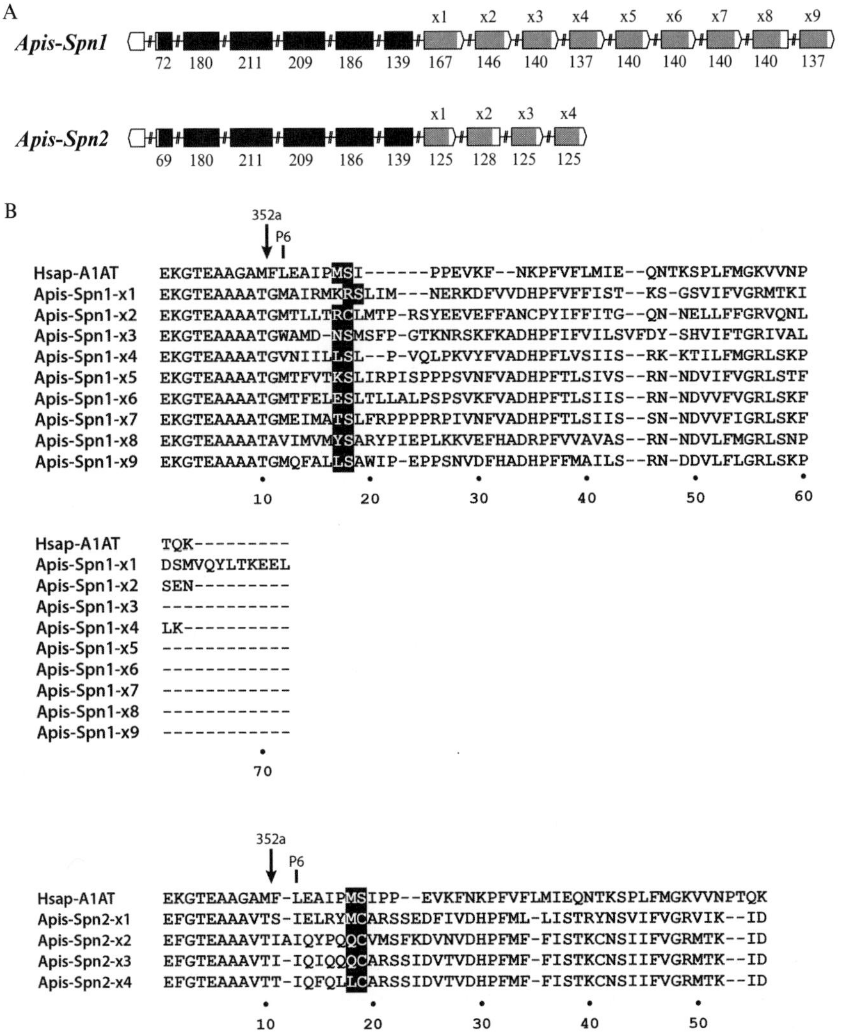

Figure 10.2 *Structure of two A. pisum serpin genes with multiple RSL encoding cassette exons and alignment of deduced RSL sequences.* (A) Constitutive exons and alternatively spliced RSL cassette exons are represented as black and gray boxes, respectively. Untranslated exon segments are shown in white. The exons are drawn to scale and their sizes (nt) are shown underneath. Introns (not drawn to scale) are displayed as interrupted lines. (B) RSL sequences were aligned with MAFFT using the E-INS-i strategy and adjusted manually. The RSL sequence of Hsap-A1AT is included for comparison. The intron at position 352a (not present in *Hsap-A1AT*) that separates the serpin core sequences from the exchangeable RSL cassettes is indicated by an arrow. Position P6 is marked by a vertical line. The scissile bond of Hsap-A1AT and the presumed P1–P1′ positions of the *A. pisum* serpin isoforms are marked by reverse white-on-black printing.

with an RSL sequence typically found in serpins that inhibit furin-like enzymes. Similar serpin isoforms have previously been detected in other insects (Krüger *et al.*, 2002) and have subsequently been shown to inhibit proprotein convertases (Oley *et al.*, 2004; Osterwalder *et al.*, 2004; Richer *et al.*, 2004). The currently known *Apis-Spn1* cDNAs, however, do not code for a signal peptide; it is therefore unclear, how, if at all, the KEEL variant may enter the secretory pathway, the cellular compartment, where proprotein convertases preferentially act. In any case, the Apis-Spn1 isoforms appear to cover a wide spectrum of biochemically different target enzymes. The *Apis-Spn2* gene, in contrast, encodes only four isoforms, and thus, its ability for antiproteolytic intervention is limited. The low divergence of Apis-Spn2 isoforms may either indicate a relatively young evolutionary age or ongoing selection through a recently emerging challenge by a group of similar proteases.

The RSL cassette exons of both *Apis-Spn1* and *Apis-Spn2* are each spliced at position 352a to the core of the serpin transcript (the intron position refers to the first base coding for amino acid 352 of the mature form of Hsap-A1AT; Ragg *et al.*, 2001). Apparently, this site is very appropriately positioned for generating functional variants, as it maps between the series of small N-terminal RSL residues enabling the conformational changes required for the inhibitory reaction and the downstream residues that determine target specificity. Whether the prevalent use of position 352a for linking RSL cassettes to the inhibitor scaffold (Börner and Ragg, 2008) is due to common ancestry or, alternatively, due to repeated independent invention remains to be investigated.

2.6. Step-by-step protocol

2.6.1. Construction of HMMs

1. Download serpin sequences from the NCBI RefSeq database. A considerable number of false predictions can be filtered out by restricting the protein size from 300 to 600 residues.

   Call: *serpin AND srcdb_refseq_known[prop]AND 300:600[SLEN]*
2. *Optional*: Align RefSeq serpin sequences with prototype serpin sequences (see Section 2.1) and discard sequences having less than 40% similarity to at least one of the reference serpins. Though not essential, this considerably reduces false positives and thus improves the sensitivity of the constructed HMMs. A filtering tool based on the EMBOSS program NEEDLE is available at our Web page (http://www.techfak.uni-bielefeld.de/ags/zellgen/tools/).

3. *Optional*: Split the leftover sequences into a training set and a test set. The training set is used to construct the HMMs. The test set serves to evaluate the performance of the HMMs.
4. Globally align all leftover sequences.
   Call: *mafft — globalpair [file containing sequence] > [output file]*
5. Identify all regions in the aligned sequences with a minimum length of 10 residues that contain at least six highly conserved amino acid positions. Load the alignment file into GENEDOC and set the threshold for shading to 60%.
6. Realign the identified regions using CLUSTALW2.
   Call: *clustalw2 – infile = [file containing sequences] – align*
7. Load the alignment into GENEDOC and remove all columns with more than 80% gaps. Export the alignment.
8. Create an HMM from the alignment.
   Call: *hmmbuild [hmm output file] [alignment file]*
9. Calibrate the HMM. The calibration parameters should be adapted to the sizes of the DNA sequences under scrutiny (see Section 2.6.2).
   Call: *hmmcalibrate — mean 500 – sd 300 [hmm file]*

2.6.2. Identification of serpin genes with mRSL cassette exons

1. Download a genome. A list of ongoing and finished sequencing projects may be obtained from the GOLD database (http://www.genomesonline.org/). Many genomes can be downloaded from the NCBI genome database (http://www.ncbi.nlm.nih.gov/sites/genome).
2. *Optional*: split the genome into overlapping chunks.
   Call: *splitter – sequence [genome file] – outseq [output file] – size 5000000 – overlap 200000*
3. Translate the DNA sequences into all six reading frames.
   Call: *transeq – sequence [splitted genome file] – outseq [output file] – frame 6*
4. Search the translated sequences for matches to the HMMs.
   Call: *hmmsearch – Z1 [hmm file] [sequence file] > [output file]*
5. A program parsing the HMMSEARCH output is available on our Web page (see link given above). The merged, sorted, and filtered HMM matches are assembled into a list comprising all identified serpin genes.

2.7. Limitations

In order to evaluate the performance of our HMM search strategy, the filtered serpin sequences were partitioned into a training set and a test set, encompassing 811 sequences each. All proteins of the test set were reverse translated, resulting in a set of DNA sequences (an artificial genome) to which our search strategy was applied. The recall rates for HMMs A–E,

Table 10.2 Performance of the hidden Markov models

	Recall rate in unmodified test set[a] (%)	Recall rate in modified test set[b] (%)	Random hits[c]
HMM A	92.4	80.9	34
HMM B	85.3	78.9	448
HMM C	74.1	63.0	398
HMM D	91.7	80.6	91
HMM E	71.9	64.5	848
HMM F	92.7	83.0	105
HMM F, unfiltered	96.8	96.1	5702
Gene predictions	96.9	94.1	0

[a] Randomly reverse-translated serpin sequences.
[b] Randomly reverse-translated serpin sequences, interspersed with three to six introns of 200–10,000 nt in length.
[c] Number of hits in a random sequence of 10^9 amino acids.

reflecting the ratio between correct, significant hits, and the total number of serpin genes encoded in the genome, ranged from 71.9% to 92.4% (Table 10.2). Since matches to HMM F, irrespective of their *E*-values, are never discarded, the recall rate for this model is even higher (96.8%). Thus, only a small proportion of mRSL serpin genes run the risk of being misclassified, in particular, those with only two RSL encoding cassette exons. The model, *per se*, does not consider whether the RSL sequences fit into the serpin reading frame; conceptual mRSL serpin genes thus always require visual scrutiny.

Introns that interrupt peptide segments represented by the HMMs chosen are another point of concern. In order to verify the stability of our approach against intron insertions, a modified test set was created by interspersing each synthetic genomic sequence with three to six introns between 200 and 10,000 nt in length. While the recall rates for HMMs A–E decreased notedly (63.0–83.0%), the percentage of identified RSL sequences and serpin genes did not change significantly.

3. Conclusion

This chapter describes a protocol for automated identification of multitasking serpin genes that appear to be widespread in the animal kingdom. Such serpin genes, via their repertoire of varying bait regions, provide a means of effective, adaptive evolution in the never-ending conflict with exogenous and/or endogenous proteases. The challenges now lying ahead include the identification of targets for individual isoforms under various physiological conditions and the unraveling of the rules that govern their expression. The procedures outlined here thus should help to reveal the role of mRSL serpin genes in innate immunity and other defense networks.

ACKNOWLEDGMENT

This work was supported by a grant from Deutsche Forschungsgemeinschaft, Graduate Program "Bioinformatics" at Bielefeld University.

REFERENCES

Börner, S., and Ragg, H. (2008). Functional diversification of a protease inhibitor gene in the genus *Drosophila* and its molecular basis. *Gene* **415,** 23–31.

Brandt, K. S., Silver, G. M., Becher, A. M., Gaines, P. J., Maddux, J. D., Jarvis, E. E., and Wisnewski, N. (2004). Isolation, characterization, and recombinant expression of multiple serpins from the cat flea, *Ctenocephalides felis*. *Arch. Insect Biochem. Physiol.* **55,** 200–214.

Brüning, M., Lummer, M., Bentele, C., Smolenaars, M. M. W., Rodenburg, K. W., and Ragg, H. (2007). The Spn4 gene from *Drosophila melanogaster* is a multipurpose defense tool directed against proteases from three different peptidase families. *Biochem. J.* **401,** 325–331.

Camacho, C., Coulouris, G., Avagyan, V., Ma, N., Papadopoulos, J., Bealer, K., and Madden, T. L. (2009). BLAST+: Architecture and applications. *BMC Bioinformatics* **10,** 421.

Danielli, A., Kafatos, F. C., and Loukeris, T. G. (2003). Cloning and characterization of four *Anopheles gambiae* serpin isoforms, differentially induced in the midgut by *Plasmodium berghei* invasion. *J. Biol. Chem.* **278,** 4184–4193.

Dufour, E. K., Denault, J. B., Bissonnette, L., Hopkins, P. C., Lavigne, P., and Leduc, R. (2001). The contribution of arginine residues within the P6-P1 region of α_1-antitrypsin to its reaction with furin. *J. Biol. Chem.* **276,** 38971–38979.

Forsyth, S., Horvath, A., and Coughlin, P. (2003). A review and comparison of the murine α_1-antitrypsin and α_1-antichymotrypsin multigene clusters with the human clade A serpins. *Genomics* **81,** 336–345.

Gettins, P. G. W. (2002). Serpin structure, mechanism, and function. *Chem. Rev.* **102,** 4751–4804.

Hegedus, D. D., Erlandson, M., Baldwin, D., Hou, X., and Chamankhah, M. (2008). Differential expansion and evolution of the exon family encoding the Serpin-1 reactive centre loop has resulted in divergent serpin repertoires among the Lepidoptera. *Gene* **418,** 15–21.

Hill, R. E., and Hastie, N. D. (1987). Accelerated evolution in the reactive centre regions of serine protease inhibitors. *Nature* **326,** 96–99.

Jiang, H., Wang, Y., and Kanost, M. R. (1994). Mutually exclusive exon use and reactive centre diversity in insect serpins. *J. Biol. Chem.* **269,** 55–58.

Jiang, H., Wang, Y., Huang, Y., Mulnix, A. B., Kadel, J., Cole, K., and Kanost, M. R. (1996). Organization of serpin gene-1 from *Manduca sexta*. Evolution of a family of alternate exons encoding the reactive site loop. *J. Biol. Chem.* **271,** 28017–28023.

Katoh, K., and Toh, H. (2008). Recent developments in the MAFFT multiple sequence alignment program. *Brief. Bioinform.* **9,** 286–298.

Krüger, O., Ladewig, J., Köster, K., and Ragg, H. (2002). Widespread occurrence of serpin genes with multiple reactive centre-containing exon cassettes in insects and nematodes. *Gene* **293,** 97–105.

Larkin, M. A., Blackshields, G., Brown, N. P., Chenna, R., McGettigan, P. A., McWilliam, H., Valentin, F., Wallace, I. M., Wilm, A., Lopez, R., Thompson, J. D., Gibson, T. J., and Higgins, D. G. (2007). Clustal W and Clustal X version 2.0. *Bioinformatics* **23,** 2947–2948.

Luke, C. J., Pak, S. C., Askew, D. J., Askew, Y. S., Smith, J. E., and Silverman, G. A. (2006). Selective conservation of the RSL-encoding, proteinase inhibitory-type, clade L serpins in *Caenorhabditis* species. *Front. Biosci.* **11,** 581–594.

Nicholas, K. B., Nicholas, H. B., Jr., and Deerfield, D. W., II (1997). GeneDoc: Analysis and visualization of genetic variation. *EMBNET News* **4,** 1–4.

Oley, M., Letzel, M. C., and Ragg, H. (2004). Inhibition of furin by serpin Spn4A from *Drosophila melanogaster*. *FEBS Lett.* **577,** 165–169.

Osterwalder, T., Kuhnen, A., Leiserson, W. M., Kim, Y. S., and Keshishian, H. (2004). *Drosophila* serpin 4 functions as a neuroserpin-like inhibitor of subtilisin-like proprotein convertases. *J. Neurosci.* **24,** 5482–5491.

Ragan, E. J., An, C., Yang, C. T., and Kanost, M. R. (2010). Analysis of mutually-exclusive alternative spliced serpin-1 isoforms and identification of serpin-1 proteinase complexes in *Manduca sexta* hemolymph. *J. Biol. Chem.* **285,** 29642–29650.

Ragg, H. (2007). The role of serpins in the surveillance of the secretory pathway. *Cell. Mol. Life Sci.* **64,** 2763–2770.

Ragg, H., Ulshöfer, T., and Gerewitz, J. (1990). On the activation of human leuserpin-2, a thrombin inhibitor, by glycosaminoglycans. *J. Biol. Chem.* **265,** 5211–5218.

Ragg, H., Lokot, T., Kamp, P. B., Atchley, W. R., and Dress, A. (2001). Vertebrate serpins: Construction of a conflict-free phylogeny by combining exon-intron and diagnostic site analyses. *Mol. Biol. Evol.* **18,** 577–584.

Rice, P., Longden, I., and Bleasby, A. (2000). EMBOSS: The European Molecular Biology Open Software Suite. *Trends Genet.* **16,** 276–277.

Richer, M. J., Keays, C. A., Waterhouse, J., Minhas, J., Hashimoto, C., and Jean, F. (2004). The Spn4 gene of *Drosophila* encodes a potent furin-directed secretory pathway serpin. *Proc. Natl. Acad. Sci. USA* **101,** 10560–10565.

Sasaki, T. (1991). Patchwork-structure serpins from silkworm (*Bombyx mori*) larval hemolymph. *Eur. J. Biochem.* **202,** 255–261.

Schechter, I., and Berger, A. (1967). On the size of the active site in proteases. *Biochem. Biophys. Res. Commun.* **27,** 157–162.

Schuster-Böckler, B., Schultz, J., and Rahmann, S. (2004). HMM Logos for visualization of protein families. *BMC Bioinformatics* **5,** 7–14.

Silverman, G. A., Bird, P. I., Carrell, R. W., Church, F. C., Coughlin, P. B., Gettins, P. G., Irving, J. A., Lomas, D. A., Luke, C. J., Moyer, R. W., Pemberton, P. A., Remold-O'Donnell, E., *et al.* (2001). The serpins are an expanding superfamily of structurally similar but functionally diverse proteins. Evolution, mechanism of inhibition, novel functions, and a revised nomenclature. *J. Biol. Chem.* **276,** 33293–33296.

Silverman, G. A., Whisstock, J. C., Bottomley, S. P., Huntington, J. A., Kaiserman, D., Luke, C. J., Pak, S. C., Reichhart, J. M., and Bird, P. I. (2010). Serpins flex their muscle: I. Putting the clamps on proteolysis in diverse biological systems. *J. Biol. Chem.* **285,** 24299–24305.

Stanke, M., Steinkamp, R., Waack, S., and Morgenstern, B. (2004). AUGUSTUS: A web server for gene finding in eukaryotes. *Nucleic Acids Res.* **32,** W309–W312.

The International Aphid Genomics Consortium (2010). Genome sequence of the pea aphid *Acyrthosiphon pisum. PLoS Biol.* **8,** e1000313.

Whisstock, J. C., Silverman, G. A., Bird, P. I., Bottomley, S. P., Kaiserman, D., Luke, C. J., Pak, S. C., Reichhart, J. M., and Huntington, J. A. (2010). Serpins flex their muscle: II. Structural insights into target peptidase recognition, polymerization, and transport functions. *J. Biol. Chem.* **285,** 24307–24312.

Zou, Z., Picheng, Z., Weng, H., Mita, K., and Jiang, H. (2009). A comparative analysis of serpin genes in the silkworm genome. *Genomics* **93,** 367–375.

CHAPTER ELEVEN

Methods to Measure the Kinetics of Protease Inhibition by Serpins

Anita J. Horvath,* Bernadine G. C. Lu,* Robert N. Pike,[†] *and* Stephen P. Bottomley[†]

Contents

Abstract

The serpin molecule has evolved an unusual mechanism of inhibition, involving an exposed reactive center loop (RCL) and conformational change to covalently trap a target protease. Successful inhibition of the protease is dependent on the rate of serpin–protease association and the efficiency with which the RCL inserts into β-sheet A, translocating the covalently bound protease and thereby completing the inhibition process. This chapter describes the kinetic methods used for determining the rate of protease inhibition (k_a) and the stoichiometry of inhibition. These kinetic variables provide a means to examine different serpin–protease pairings, assess the effects of mutations within a serpin on protease inhibition, and determine the physiologically cognate protease of a serpin.

1. Introduction

Serpins inhibit serine proteases by a unique process known as the "suicide substrate inhibition mechanism" (Potempa *et al.*, 1994). Irreversible protease inhibition is achieved through the formation of a covalent

* Australian Centre for Blood Diseases, Monash University, Melbourne, Victoria, Australia
† Department of Biochemistry and Molecular Biology, Monash University, Melbourne, Victoria, Australia

Methods in Enzymology, Volume 501
ISSN 0076-6879, DOI: 10.1016/B978-0-12-385950-1.00011-0

complex in which both serpin and protease are rendered inactive. This mechanism is distinct from other classes of protease inhibitors, such as Kunitz- and Kazal-type molecules, which inhibit the enzyme by using a reversible, tight, and noncovalent "lock and key" mechanism (Bode and Huber, 2000).

The exposed reactive center loop (RCL) acts as bait and is recognized as a substrate by target proteases. Cleavage of the RCL results in a dramatic conformational change in the serpin molecule and is characterized by insertion of the N-terminal portion of the cleaved RCL into β-sheet A (Carrell and Travis, 1985). The conformational change results in irreversible inhibition of the target protease as loop insertion translocates the protease and distorts its active site, thereby trapping it as an acyl-enzyme intermediate.

The initial interaction between a serpin (I) and protease (E) is a reversible Michaelis–Menten reaction, which proceeds to the formation of an acyl intermediate (Fig. 11.1). Cleavage of the peptide bond at P_1–P_1' occurs by nucleophilic attack by the catalytic serine of the protease on the carbonyl carbon of the P_1 residue. Thc attack proceeds through the standard covalent tetrahedral intermediate of a substrate cleavage reaction, with the subsequent formation of an ester bond between the hydroxyl group of the Ser residue and the carbonyl carbon of the P_1 residue. This forms the covalent acyl-enzyme intermediate (EI′) (Lawrence *et al.*, 1995; Wilczynska *et al.*, 1995). With the peptide bond between the serpin P_1–P_1' residues broken, the N-terminal strand of the RCL is released and the loop can insert into the β-sheet A.

The mechanism becomes branched at this point with the rate of loop insertion determining whether the serpin behaves as a substrate or an

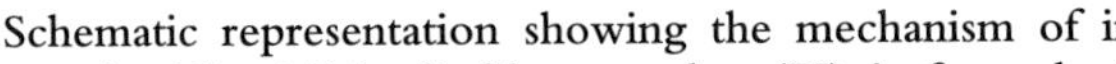

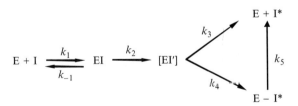

Figure 11.1 Schematic representation showing the mechanism of inhibition of a protease by a serpin. The Michaelis-like complex (EI) is formed after the initial interaction of the protease (E) and serpin (I), with the forward rate constant k_1 and the back constant of k_{-1}. This proceeds to formation of the covalent acylintermediate (EI′), with the rate constant of k_2. From this point, the reaction can proceed down two different pathways which primarily depends on how rapidly the cleaved serpin loop is able to insert into the A β-sheet. The formation of cleaved serpin (I*) and active protease, with the rate constant k_3, results from complete proteolysis. However, if the serpin kinetically traps the protease, the inhibitory complex (E – I*) is formed with the rate constant k_4.

inhibitor (Lawrence *et al.*, 2000). If the loop insertion process is rapid ($k_3 > k_4$), the inhibitory pathway proceeds toward full protease inhibition ($E - I^\star$) by trapping the acyl intermediate before the protease can complete the deacylation reaction. The rapid loop insertion, accompanied by conformational change in the serpin and translocation of the protease, distorts the protease active site and results in complete inhibition (Huntington *et al.*, 2000; Lawrence *et al.*, 2000). However, if loop insertion proceeds too slowly ($k_4 > k_3$), then the deacylation reaction continues, the serpin is cleaved without protease inhibition ($E + I$), and active protease is released. The inhibited serpin–protease complex is stable for days–weeks, depending on the serpin–protease pair (Calugaru *et al.*, 2001; Plotnick *et al.*, 2002; Zhou *et al.*, 2001). A typical inhibitory serpin forms a covalent complex with its cognate protease which is resistant to SDS and thermal denaturation, has a stoichiometry of inhibition (SI) close to 1 and an association constant (k_a) of $\geq 10^5\ M^{-1}\ s^{-1}$(Gettins, 2002).

Serpins are promiscuous inhibitors, able to inhibit multiple proteases. A number of serpins have been reported to possess inhibitory activity toward proteases outside the archetypal targets, that is, members of the chymotrypsin serine protease family. These include proteases of the subtilisin or cysteine protease family. α_1-Antitrypsin is a rapid inhibitor of elastase but also inhibits both subtilisin *Carlsberg* and proteinase K (Beatty *et al.*, 1980; Komiyama *et al.*, 1996). The intracellular serpin PI-9 is an inhibitor of both the serine protease granzyme B and subtilisin A (Dahlen *et al.*, 1997, 1998; Sun *et al.*, 1996).

Both the caspase and cathepsin families of cysteine proteases are targeted for inhibition by serpins. The viral serpin crmA inhibits caspase1, caspase3, and caspase8, while PI-9 is a weak inhibitor of caspase1 (Annand *et al.*, 1999; Young *et al.*, 2000). Cathepsin L has been reported to be targeted by human SCCA1, SCCA-2, hurpin/headpin, protein C inhibitor, and chicken MENT (Fortenberry *et al.*, 2010; Higgins *et al.*, 2010; Irving *et al.*, 2002a; Schick *et al.*, 1998; Welss *et al.*, 2003). In the presence of DNA, cathepsin V is inhibited 10- to 50-fold more rapidly by both SCCA-1 and MENT (Ong *et al.*, 2007). A similar mechanism of inhibition by serpins is proposed for the interaction with cysteine proteases (Irving *et al.*, 2002b).

The accurate determination of inhibition constants provides a means to compare the rate and efficiency at which a serpin inhibits different proteases and is used to determine its most physiologically relevant cognate protease. This review will describe methods to calculate the rate of protease inhibition (k_a) by conventional protease–substrate assays, termed as discontinuous and continuous assays and a method for determining the SI (Horvath *et al.*, 2005; Le Bonniec *et al.*, 1995; Olson *et al.*, 1993). Intrinsic fluorescence-based assays, such as stopped flow kinetics, have been reviewed extensively elsewhere and are as such not included in this report.

2. Determining the Rate of Protease Inhibition (k_a)

The type of assay used to determine the rate of inhibition is dependent on indications of how rapid the reaction is likely to be in preliminary experiments. The rate of protease inhibition by serpins is measured by utilizing either the discontinuous or the continuous/progress curve method. The discontinuous method is generally used for reactions when the rate of inhibition is $\leq 10^5\ M^{-1}\text{sec}^{-1}$. Faster reaction rates, at $\geq 10^6\ M^{-1}\text{sec}^{-1}$, are usually determined using progress curves (Horvath *et al.*, 2005; Olson *et al.*, 1993). If reactions proceed faster than can be measured by substrate-based assays, then stopped flow kinetics is commonly employed (Boudier *et al.*, 1999; Christensen *et al.*, 1996; Olson, 1988).

2.1. Initial assessment of the rate of inhibition

An initial assessment of the rate of inhibition can be made by incubating the protease with a more than five fold molar excess of serpin in the presence of substrate for the protease. The reaction is setup in a 96-well microtiter plate, thereby allowing continuous monitoring of the rate of substrate hydrolysis in a plate reader. The observed loss of proteolytic activity in the presence of the serpin, compared to protease alone, is then used to assess the rate of inhibition.

2.1.1. Example reaction 1: Assessing the rate of serpin inhibition

Assay

Into a well of a 96-well microtiter plate, add in the following order:
50 μl of 2–10 n*M* protease
50 μl of 5- and 10-fold molar excess of serpin
100 μl substrate solution
Place the plate into reader and follow the rate of substrate hydrolysis by taking readings every 10–30 s.

Figure 11.2 shows the expected profiles for slow and fast inhibitors of plasmin. The serpin α_2-antiplasmin is a fast inhibitor of plasmin with an association rate (k_a) of $1 \times 10^7\ M^{-1}\ \text{s}^{-1}$. As seen in Fig. 11.2, in the presence of α_2-antiplasmin, there is an initial rapid increase in substrate hydrolysis followed by a steady-state phase reflective of complete loss of plasmin activity due to complete inhibition by the serpin. Reaching this steady-state phase is required in order to accurately calculate the association rate (k_a) using the continuous method.

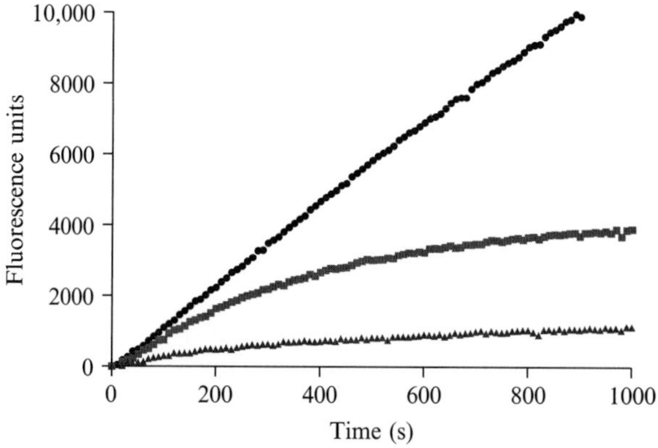

Figure 11.2 Example progress curves of the interaction of a fast and slow inhibitor of plasmin. Reactions were performed in the presence of plasmin (0.5 n*M*), wild-type α_2-antiplasmin (2.5 n*M*), and α_2-antiplasmin ΔC-term (40 n*M*) and the substrate H-Ala-Phe-Lys-AMC (1 m*M*). Change in fluorescence was measured by excitation/emission at 355 nm/460 nm. Reactions are plasmin + buffer (closed circle), plasmin + wild-type α_2-antiplasmin (closed triangle), and plasmin + α_2-antiplasmin ΔC-term (closed square). (For color version of this figure, the reader is referred to the Web version of this chapter.)

In the presence of a slow inhibitor, a C-terminally truncated form of α_2-antiplasmin, Lu *et al.* (2011) complete inhibition is not achieved and continued hydrolysis of the substrate is observed without reaching the steady-state phase. Therefore, the discontinuous assay is more suitable for measurement of the inhibition rate in this instance.

2.2. Discontinuous assay

The discontinuous assay is used when the rate of inhibition is low and cannot be determined through the use of progress curves, unless excessive amounts of protease and serpin are used; such high concentrations may not be feasible. In the discontinuous method, a constant concentration of protease is incubated with increasing amounts of serpin. The protease amount used should be low enough to achieve pseudo-first-order conditions with a more than five fold molar excess of serpin to protease. For each serpin–protease ratio, serpin and protease are preincubated together for various time lengths and a separate reaction mixture is prepared for each time point. If some initial estimate of the k_a for the reaction is available, the times and concentrations of serpin used for the assay may be estimated by calculating the predicted half-life ($t_{1/2}$) for the reaction using the following equation: $t_{1/2} = \ln 2/k_a \times [\text{Inhibitor}]$. The time for the full reaction may then be estimated by multiplying $t_{1/2}$ by 10. On completion of the incubation, substrate is added to the reaction and the rate of substrate hydrolysis is monitored. The rate of protease activity at each time point is then calculated by linear regression analysis. The pseudo-first-order constant, k_{obs}, is determined by the slope of a semilogarithmic plot of the residual protease activity against time of incubation (Fig. 11.3). Linear regression analysis of the points provide the k_{obs} value. k_{obs} values are then plotted against serpin concentration and the slope of the linear regression of the line of best fit produces the second-order rate constant k_a.

2.2.1. Example reaction 2: Measurement of the rate of inhibition of plasmin–antiplasmin (C-trunc) variant

Reagents	
A set concentration of protease	5 n*M* plasmin
Six different concentrations of serpin[a]	25–50 n*M* antiplasmin
Substrate	150 μ*M* H-Ala-Phe-Lys-AMC

[a]A separate reaction for eight time points for each serpin concentration (t = 0–5 min).

Assay: Reactions are set up in a 96-well plate. For each concentration of serpin, eight time points are required and a separate reaction mixture is prepared.Aliquot 50 μl of 5 n*M* protease into eight wells of 96-well microtiter plate. Assign each well as a time point for the reaction. Add 50 μl of serpin at desired concentration to the first well of the microtiter plate (this becomes t = 300 s, or the longest incubation time point). Add another 50 μl of serpin to the next well containing protease when the next time point is reached. This process is repeated until time = 0 s is reached, and after the final addition of serpin, 100 μl of substrate is immediately added to each reaction and followed directly by measurement of substrate hydrolysis.

Analysis: Figure 11.3 shows the semilog plots of the residual rate of plasmin activity over time. The pseudo-first-order constant, k_{obs}, is determined by the slope of the semilogarithmic plot of the residual protease activity against time. The k_{obs} values are then plotted against serpin concentration and linear regression analysis is used to fit the points to a line (Lu *et al.*, 2011). The slope of this line provides the second-order rate constant, k_a.

2.3. Continuous assay

The continuous assay is the process by which the inhibition of the protease is measured by progress curves which are generated from monitoring the rate of product formation over time (Horvath *et al.*, 2005). A constant amount of protease is mixed with increasing amounts of serpin and a fixed concentration of substrate.

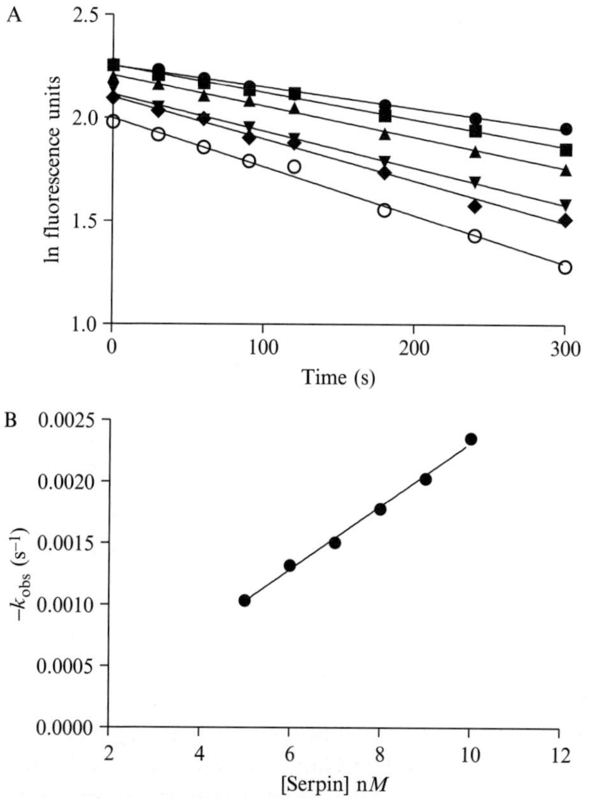

Figure 11.3 Discontinuous assay of the inhibition of plasmin by α_2-antiplasmin ΔC-term. (A) Semilogarithmic plots of residual plasmin activity versus time for reactions at various concentrations of antiplasmin ((5–10 n*M*), 5 n*M* closed circle, 6 n*M* closed square, 7 n*M* closed triangle, 8 n*M* closed inverted triangle, 9 n*M* closed diamond and 10 nM open circle). (B) Plot of k_{obs} as a function of antiplasmin concentration. Linear regression of the slope gave the second-order rate constant k_a for the inhibition of plasmin by antiplasmin ΔC-term.

2.3.1. Example reaction 3: Measurement of the rate of inhibition of plasmin by α_2-antiplasmin

Reagents	
A set concentration of protease	0.5 n*M* plasmin
Six different concentrations of serpin	1–10 n*M* antiplasmin
Substrate	1 m*M* H-Ala-Phe-Lys-AMC

Assay: A total of eight 200 μl reactions, one with protease and substrate alone and seven with increasing serpin concentration, along with constant protease and substrate concentrations, are set up in a BSA-coated 96-well microtiter plate. Into the microtiter plate, aliquot 100 μl of substrate into each well, followed by 50 μl of serpin at the desired concentration. Place microtiter plate into plate reader. Add 50 μl of protease to each well and immediately commence measurement of substrate hydrolysis by taking readings at 15–30 s time intervals.

Analysis: Rate constants are measured under pseudo-first-order conditions using progress curves for the interaction of plasmin (0.5 n*M*) and antiplasmin (1–10 n*M*) in the presence of 1 m*M* fluorogenic substrate, H-Ala-Phe-Lys-AMC (Bachem, Bubendorf, Switzerland) (Lu *et al.*, 2011). The rate of product formation is measured using an excitation and emission wavelengths of 355 and 460 nm, respectively (Fig. 11.4A).

Product formation is described by Eq. (11.1), where P is the concentration of product at time t, k_{obs} is the apparent first-order rate constant, and v_0 is the initial velocity.

$$P = \frac{v_0}{k_{obs}}\left[1 - e^{(1-k_{obs}t)}\right] \tag{11.1}$$

For each combination of protease and serpin, a k_{obs} value is calculated by nonlinear regression analysis of the data using Eq. (11.1) (Olson *et al.*, 1993; Rovelli *et al.*, 1992). The k_{obs} values are then plotted against the respective serpin concentration [serpin] and linear regression is used to provide a line to the points (Fig. 11.4B). The slope of this line provides the second-order rate constant, k'.

As the rate of inhibition is dependent on the SI and the inhibitor is in competition with the substrate [S], the second-order rate constant k' is corrected for substrate concentration, the K_M of the protease for substrate, and the SI (Eq. (11.2)) to calculate k_a, the rate of association.

$$k_a = k'\left(1 + \frac{[S]}{K_M}\right)SI \tag{11.2}$$

3. Efficiency of the Serpin Inhibitory Reaction

To ascertain the efficiency of the serpin inhibitory reaction, the SI is calculated. This is a measure of the balance between the inhibitory reaction and the substrate reaction and describes the number of moles of serpin required to inhibit 1 mol of protease. If the inhibitory pathway proceeds faster than the substrate pathway, then the SI approaches 1. However, if serpin proteolysis and the substrate pathway prevail over the inhibitory pathway, then the SI is greater than 1. Physiological serpin–protease pairs, such as thrombin–anthrombin, plasmin–antiplasmin have 1:1 serpin–protease molar relationships (Moroi and Aoki, 1976; Olson, 1985). The SI can influence the measured association rate, and therefore, it is an important parameter to calculate when examining any protease–serpin interaction.

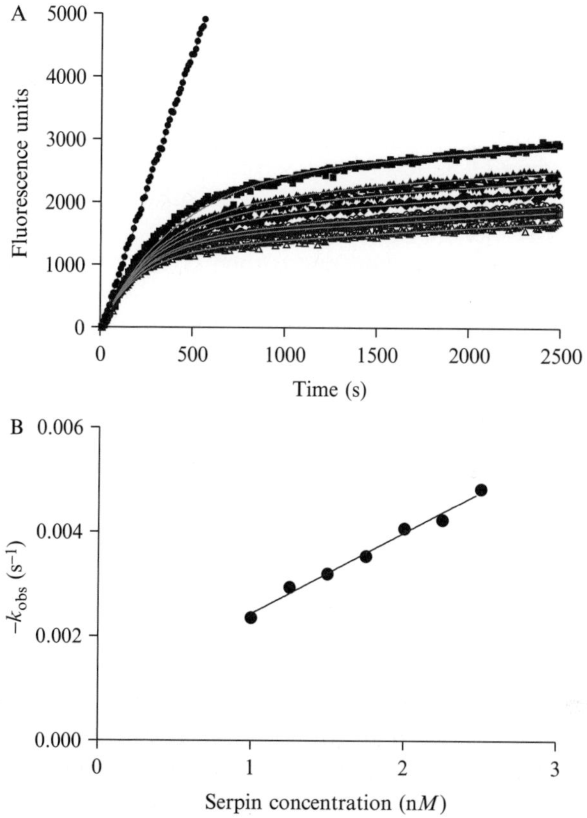

Figure 11.4 Continuous assay of the inhibition of plasmin by α_2-antiplasmin. (A) Progress curves of the interaction between plasmin (0.5 n*M*) and wild-type α_2-antiplasmin (1–2.5 n*M*) monitored by continuous measurement of the change in fluorescence excitation/emission at 355 nm/460 nm. k_{obs} at each serpin concentration was determined by nonlinear regression analysis of each curve using Eq. (11.1). (B) k_{obs} were plotted against α_2-antiplasmin concentration and linear regression analysis was used to determine the second-order rate constant (k). k_a was determined by accounting for K_M of the protease for substrate (Eq. (11.2)). (For color version of this figure, the reader is referred to the Web version of this chapter.)

To calculate the SI, a series of reactions are set up with a constant molar amount of protease and increasing molar amounts of serpin.

3.1. Example reaction 1: SI calculation using plasmin–antiplasmin

Reaction

30 μl of 16 n*M* plasmin
30 μl of serpin (0–24 n*M*)
Incubate at 37 °C for 1 h.

Assay: Aliquot duplicate 25 μl lots of each reaction into a 96-well microtiter plate. To assay for residual protease activity, add 75 μl of reaction buffer (20 m*M* Tris–HCl, pH 8.0, 150 m*M* NaCl, 0.01% Tween 20) and 100 μl of 150 μ*M* H-Ala-Phe-Lys-AMC substrate and measure the rate of substrate hydrolysis.

Analysis: Figure 11.5 shows that as serpin concentration increases the rate of substrate hydrolysis decreases. The rate of substrate hydrolysis which is used as a measure of residual protease activity is measured by linear regression (Fig. 11.5A). The resultant rate is then converted to a percentage of maximal protease activity (from reactions with protease and buffer alone) and is then plotted against the serpin:protease ratio (Fig. 11.5B). The SI is

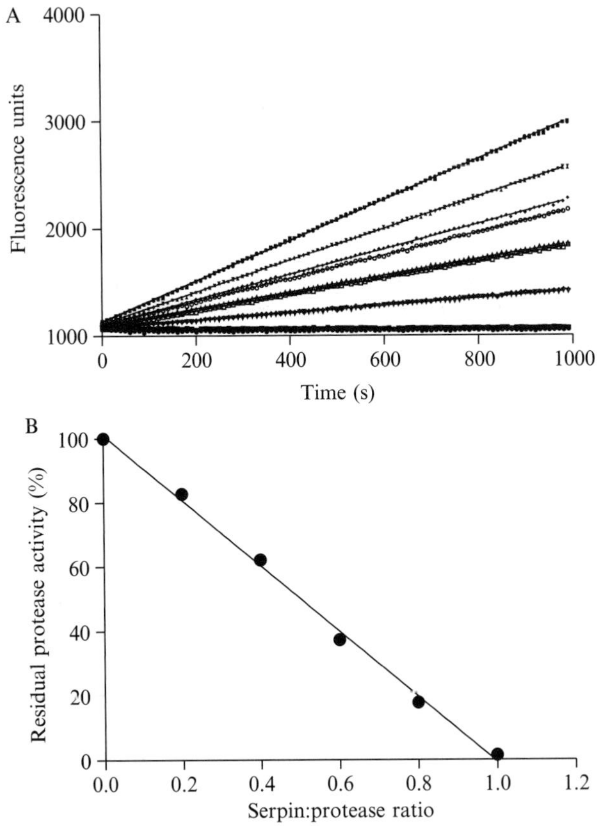

Figure 11.5 Stoichiometry of inhibition of plasmin by α_2-antiplasmin. (A) Plasmin (1 n*M*) was incubated with wild-type α_2-antiplasmin (0.2–1.5 n*M*) for 1 h at 37 °C. Residual protease activity was assayed in the presence of the substrate H-Ala-Phe-Lys-AMC (0.2 m*M*). Reactions were set up in duplicate. (B) The stoichiometry of inhibition was determined by plotting the percentage residual protease activity against serpin:protease ratio and extrapolation to the ratio which resulted in total loss of protease activity.

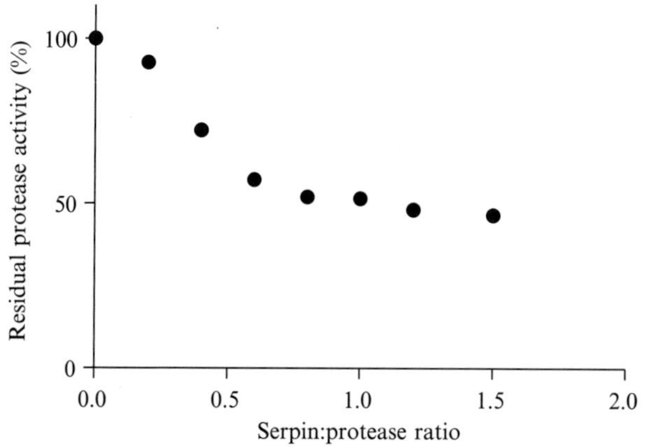

Figure 11.6 The effect of incubation time on the stoichiometry of inhibition. Plasmin (1 n*M*) was incubated with wild-type α_2-antiplasmin (0.2–1.5 n*M*) for 5 min at 37 °C. Residual protease activity was assayed in the presence of the substrate H-Ala-Phe-Lys-AMC (0.2 m*M*). The percentage residual protease activity was plotted against serpin: protease ratio.

determined by extrapolating to the serpin:protease ratio where protease activity is zero.

A linear relationship should be observed between serpin:protease ratio and residual protease activity. Alternatives to this relationship may be observed if the protease and serpin are not incubated for sufficient time to ensure the inhibitory reaction has occurred to completion. In this instance, as seen in Fig. 11.6, residual protease activity has plateaued at 50%. This requires either a longer incubation time or an increase in both serpin and protease concentration in the reaction. It is for this reason that it is important to incubate the serpin–protease reactions for at least $5 \times t_{1/2}$.

The SI obtained through the use of the protease/substrate assay as described can also be confirmed by following the formation of the covalent serpin–protease complex on SDS-PAGE (Horvath *et al.*, 2005).

ACKNOWLEDGMENTS

The work presented in this report is supported with grants from the National Health and Medical Research Council (NHMRC) of Australia.

REFERENCES

Annand, R. R., Dahlen, J. R., Sprecher, C. A., De Dreu, P., Foster, D. C., Mankovich, J. A., Talanian, R. V., Kisiel, W., and Giegel, D. A. (1999). Caspase-1 (interleukin-1beta-converting enzyme) is inhibited by the human serpin analogue proteinase inhibitor 9. *Biochem. J.* **342**(Pt. 3), 655–665.

Beatty, K., Bieth, J., and Travis, J. (1980). Kinetics of association of serine proteinases with native and oxidized alpha-1-proteinase inhibitor and alpha-1-antichymotrypsin. *J. Biol. Chem.* **255**(9), 3931–3934.

Bode, W., and Huber, R. (2000). Structural basis of the endoproteinase-protein inhibitor interaction. *Biochim. Biophys. Acta* **1477**(1–2), 241–252.

Boudier, C., CadÃ¨ne, M., and Bieth, J. G. (1999). Inhibition of neutrophil cathepsin G by oxidized mucus proteinase inhibitor. Effect of heparin. *Biochemistry* **38**(26), 8451.

Calugaru, S. V., Swanson, R., and Olson, S. T. (2001). The pH dependence of serpin-proteinase complex dissociation reveals a mechanism of complex stabilization involving inactive and active conformational states of the proteinase which are perturbable by calcium. *J. Biol. Chem.* **276**(35), 32446–32455.

Carrell, R. W., and Travis, J. (1985). α1-Antitrypsin and the serpins: Variation and counter variation.. *Trends Biochem. Sci.* **10,** 20–24.

Christensen, U., Bangert, K., and Thorsen, S. (1996). Reaction of human alpha2-antiplasmin and plasmin stopped-flow fluorescence kinetics. *FEBS Lett.* **387**(1), 58.

Dahlen, J. R., Foster, D. C., and Kisiel, W. (1997). Human proteinase inhibitor 9 (PI9) is a potent inhibitor of subtilisin A. *Biochem. Biophys. Res. Commun.* **238,** 329–333.

Dahlen, J. R., Jean, F., Thomas, G., Foster, D. C., and Kisiel, W. (1998). Inhibition of soluble recombinant furin by human proteinase inhibitor 8. *J. Biol. Chem.* **273,** 1851–1854.

Fortenberry, Y. M., Brandal, S., Bialas, R. C., and Church, F. C. (2010). Protein C inhibitor regulates both cathepsin L activity and cell-mediated tumor cell migration. *Biochim. Biophys. Acta* **1800**(6), 580–590.

Gettins, P. G. (2002). Serpin structure, mechanism, and function. *Chem. Rev.* **102**(12), 4751–4804.

Higgins, W. J., Fox, D. M., Kowalski, P. S., Nielsen, J. E., and Worrall, D. M. (2010). Heparin enhances serpin inhibition of the cysteine protease cathepsin L. *J. Biol. Chem.* **285**(6), 3722–3729.

Horvath, A. J., Irving, J. A., Rossjohn, J., Law, R. H., Bottomley, S. P., Quinsey, N. S., Pike, R. N., Coughlin, P. B., and Whisstock, J. C. (2005). The murine orthologue of human antichymotrypsin: A structural paradigm for clade A3 serpins. *J. Biol. Chem.* **280** (52), 43168–43178.

Huntington, J. A., Read, R. J., and Carrell, R. W. (2000). Structure of a serpin-protease complex shows inhibition by deformation. *Nature* **407**(6806), 923–926.

Irving, J. A., Shushanov, S. S., Pike, R. N., Popova, E. Y., Bromme, D., Coetzer, T. H., Bottomley, S. P., Boulynko, I. A., Grigoryev, S. A., and Whisstock, J. C. (2002a). Inhibitory activity of a heterochromatin-associated serpin (MENT) against papain-like cysteine proteinases affects chromatin structure and blocks cell proliferation. *J. Biol. Chem.* **277**(15), 13192–13201.

Irving, J. A., Pike, R. N., Dai, W., Bromme, D., Worrall, D. M., Silverman, G. A., Coetzer, T. H., Dennison, C., Bottomley, S. P., and Whisstock, J. C. (2002b). Evidence that serpin architecture intrinsically supports papain-like cysteine protease inhibition: Engineering alpha(1)-antitrypsin to inhibit cathepsin proteases. *Biochemistry* **41**(15), 4998–5004.

Komiyama, T., Gron, H., Pemberton, P. A., and Salvesen, G. S. (1996). Interaction of subtilisins with serpins. *Protein Sci.* **5**(5), 874–882.

Lawrence, D. A., Ginsburg, D., Day, D. E., Berkenpas, M. B., Verhamme, I. M., Kvassman, J. O., and Shore, J. D. (1995). Serpin-protease complexes are trapped as stable acyl-enzyme intermediates. *J. Biol. Chem.* **270**(43), 25309–25312.

Lawrence, D. A., Olson, S. T., Muhammad, S., Day, D. E., Kvassman, J. O., Ginsburg, D., and Shore, J. D. (2000). Partitioning of serpin-proteinase reactions between stable inhibition and substrate cleavage is regulated by the rate of serpin reactive center loop insertion into beta-sheet A. *J. Biol. Chem.* **275**(8), 5839–5844.

Le Bonniec, B. F., Guinto, E. R., and Stone, S. R. (1995). Identification of thrombin residues that modulate its interactions with antithrombin III and alpha 1-antitrypsin. *Biochemistry* **34**(38), 12241–12248.

Lu, B. C. G., Sofian, T., Law, R. H. P., Coughlin, P. B., and Horvath, A. J. (2011). Contribution of conserved lysine residues in the alpha 2 antiplasmin C terminus to plasmin binding and inhibition. *J.Biol. Chem.* **286**(28), 24544–24552.

Moroi, M., and Aoki, N. (1976). Isolation and characterization of alpha2-plasmin inhibitor from human plasma. A novel proteinase inhibitor which inhibits activator-induced clot lysis. *J. Biol. Chem.* **251**(19), 5956–5965.

Olson, S. T. (1985). Heparin and ionic strength-dependent conversion of antithrombin III from an inhibitor to a substrate of alpha-thrombin. *J. Biol. Chem.* **260**(18), 10153–10160.

Olson, S. T. (1988). Transient kinetics of heparin-catalyzed protease inactivation by antithrombin III. Linkage of protease-inhibitor-heparin interactions in the reaction with thrombin. *J. Biol. Chem.* **263**(4), 1698–1708.

Olson, S. T., Bjork, I., and Shore, J. D. (1993). Kinetic characterization of heparin-catalyzed and uncatalyzed inhibition of blood coagulation proteinases by antithrombin. *Methods Enzymol.* **222,** 525–559.

Ong, P. C., McGowan, S., Pearce, M. C., Irving, J. A., Kan, W. T., Grigoryev, S. A., Turk, B., Silverman, G. A., Brix, K., Bottomley, S. P., Whisstock, J. C., and Pike, R. N. (2007). DNA accelerates the inhibition of human cathepsin V by serpins. *J. Biol. Chem.* **282**(51), 36980–36986.

Plotnick, M. I., Rubin, H., and Schechter, N. M. (2002). The effects of reactive site location on the inhibitory properties of the serpin alpha(1)-antichymotrypsin. *J. Biol. Chem.* **277** (33), 29927–29935.

Potempa, J., Korzus, E., and Travis, J. (1994). The serpin superfamily of proteinase inhibitors: Structure, function, and regulation. *J. Biol. Chem.* **269**(23), 15957–15960.

Rovelli, G., Stone, S. R., Guidolin, A., Sommer, J., and Monard, D. (1992). Characterization of the heparin-binding site of glia-derived nexin/protease nexin-1. *Biochemistry* **31** (13), 3542–3549.

Schick, C., Pemberton, P. A., Shi, G. P., Kamachi, Y., Cataltepe, S., Bartuski, A. J., Gornstein, E. R., Bromme, D., Chapman, H. A., and Silverman, G. A. (1998). Cross-class inhibition of the cysteine proteinases cathepsins K, L, and S by the serpin squamous cell carcinoma antigen 1: A kinetic analysis. *Biochemistry* **37,** 5258–5266.

Sun, J. R., Bird, C. H., Sutton, V., Mcdonald, L., Coughlin, P. B., Dejong, T. A., Trapani, J. A., and Bird, P. I. (1996). A cytosolic granzyme B inhibitor related to the viral apoptotic regulator cytokine response modifier A is present in cytotoxic lymphocytes. *J. Biol. Chem.* **271,** 27802–27809.

Welss, T., Sun, J., Irving, J. A., Blum, R., Smith, A. I., Whisstock, J. C., Pike, R. N., von Mikecz, A., Ruzicka, T., Bird, P. I., and Abts, H. F. (2003). Hurpin is a selective inhibitor of lysosomal cathepsin L and protects keratinocytes from ultraviolet-induced apoptosis. *Biochemistry* **42**(24), 7381–7389.

Wilczynska, M., Fa, M., Ohlsson, P. I., and Ny, T. (1995). The inhibition mechanism of serpins. Evidence that the mobile reactive center loop is cleaved in the native protease-inhibitor complex. *J. Biol. Chem.* **270**(50), 29652–29655.

Young, J. L., Sukhova, G. K., Foster, D., Kisiel, W., Libby, P., and Schonbeck, U. (2000). The serpin proteinase inhibitor 9 is an endogenous inhibitor of interleukin 1beta-converting enzyme (caspase-1) activity in human vascular smooth muscle cells. *J. Exp. Med.* **191**(9), 1535–1544.

Zhou, A., Carrell, R. W., and Huntington, J. A. (2001). The serpin inhibitory mechanism is critically dependent on the length of the reactive center loop. *J. Biol. Chem.* **276**(29), 27541–27547.

CHAPTER TWELVE

Predicting Serpin/Protease Interactions

Jiangning Song,*,†,1 Antony Y. Matthews,*,1 Cyril F. Reboul,* Dion Kaiserman,* Robert N. Pike,* Phillip I. Bird,* *and* James C. Whisstock*,‡

Contents

* Department of Biochemistry and Molecular Biology, Monash University, Melbourne, Victoria, Australia
† State Engineering Laboratory for Industrial Enzymes and Key Laboratory of Systems Microbial Biotechnology, Tianjin Institute of Industrial Biotechnology, Chinese Academy of Sciences, Tianjin, China
‡ ARC Centre of Excellence in Structural and Functional Microbial Genomics, Monash University, Clayton, Victoria, Australia
1 These authors contributed equally to this work.

Methods in Enzymology, Volume 501
ISSN 0076-6879, DOI: 10.1016/B978-0-12-385950-1.00012-2

Abstract

Proteases are tightly regulated by specific inhibitors, such as serpins, which are able to undergo considerable and irreversible conformational changes in order to trap their targets. There has been a considerable effort to investigate serpin structure and functions in the past few decades; however, the specific interactions between proteases and serpins remain elusive. In this chapter, we describe detailed experimental protocols to determine and characterize the extended substrate specificity of proteases based on a substrate phage display technique. We also describe how to employ a bioinformatics system to analyze the substrate specificity data obtained from this technique and predict the potential inhibitory serpin partners of a protease (in this case, the immune protease, granzyme B) in a step-by-step manner. The method described here could also be applied to other proteases for more generalized substrate specificity analysis and substrate discovery.

1. Introduction

1.1. Structure and function of serpins: Unique suicide substrates of proteases

The Serpin superfamily is structurally characterized by a well-conserved fold consisting of 9 α-helices and 3 β-sheets as well as a 20–25 amino acid region named the reactive center loop (RCL, sometimes also referred to as reactive site loop) (Fig. 12.1A). A large body of biochemical and structural studies (with over a hundred serpin structures solved) has shown that this metastable fold is under considerable strain in the native state. This makes serpins sensitive to mutations promoting misfolding and resulting in serpin deficiency. The native, inhibitory complexes, cleaved, latent, and polymeric forms have been characterized to atomic resolution by protein crystallography techniques and differ primarily by the structure of the RCL (Whisstock *et al.*, 1998). We will here mostly focus on the native, cleaved, and inhibitory states of inhibitory serpins (see Law *et al.*, 2006; Silverman *et al.*, 2001, 2004; Whisstock and Bottomley, 2006; Whisstock *et al.*, 2010 for a full review).

Inhibitory serpins (Serine protease inhibitors) undergo a remarkable conformational change upon protease contact, switching from the native to the inhibitory complex (inhibitory pathway). The rearrangement forms an absolute requirement for the inhibitory function, where the serpin acts as a pseudosubstrate for the target protease. In the native state, the RCL adopts

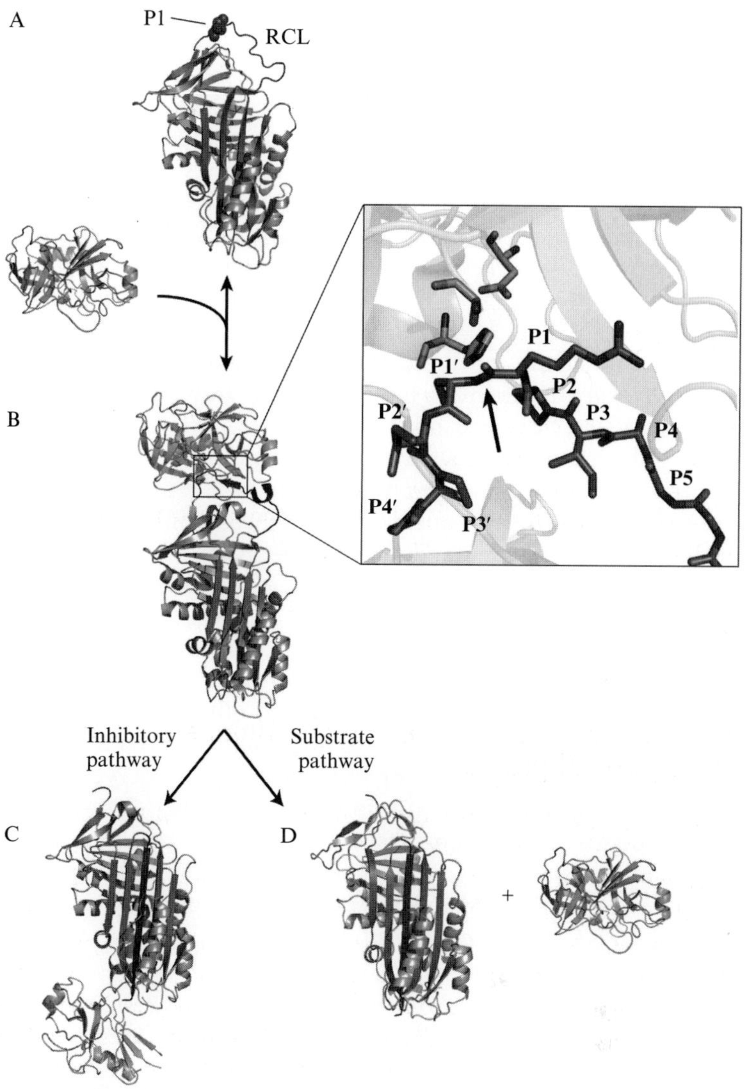

Figure 12.1 Serpin structural forms and inhibitory mechanism. The serpin β-sheet A is in red, the RCL in blue. The remaining of the serpin body is in gray; the target protease is in green. (A) Crystal structure of the serpin α-1-antitrypsin in the native state. Position of the P1 residue is indicated. (B) Reversible Michaelis complex. The boxed insert illustrates RCL/protease interactions prior to cleavage of the P1–P1′ bond (indicated by an arrow) and formation of the acyl-enzyme intermediate. Residues of the RCL are colored blue and numbered according to the nomenclature of Schechter and Berger (1967). Trypsin catalytic triad is displayed in green. (C) Inhibitory complex formed following the inhibitory pathway. (D) Serpin cleaved state and release of the protease, following the substrate pathway. PDB (Berman *et al.*, 2000) accession codes: (a) α-1-antitrypsin: 1ATU, trypsin: 1DP0; (b) cartoon and boxed representations: 1OPH; (c) 1EZX; (d) 2ACH. (See Color Insert.)

a disordered, solvent-exposed conformation accessible to the target protease. Situated on top of the serpin scaffold, it offers as bait a pseudosubstrate for the protease (Carrell *et al.*, 1991), which initially docks onto the RCL (Michaelis complex) (Fig. 12.1B). The RCL contains the P1–P1′ scissile bond cleaved by the target protease. Upon its cleavage, a reversible acyl-enzyme intermediate is formed, while the inhibitory serpin rapidly undergoes the stressed-to-relaxed transition: the RCL inserts into β-sheet A and adopts a β-strand conformation (Huntington *et al.*, 2000). The strain under which the serpin scaffold lies is thus lifted allowing the relocation of the protease to the opposite pole of the serpin, concomitant to the insertion of the amino-terminal part of the RCL into the β-sheet. This results in improved thermostability of the final serpin–protease complex, where the protease is covalently trapped by the serpin and partially disordered (Fig. 12.1C). Its active site is critically distorted, which prevents further hydrolysis of the intermediate and release of the protease. With the irreversible cleavage of the RCL, serpins achieve efficient removal of the target protease and are thus named “suicide” or “single use” inhibitors.

The RCL is key to serpin inhibitory specificity and extends from P15 to P3′ in most serpins. It is the most variable part of the molecule and bears the P1 residue, which exerts great influence over the target protease specificity. Mutation of this residue can alter the serpin inhibitory spectrum and promote disease. When the serpin α-1-antitrypsin is affected by the naturally occurring P1 Met to Arg mutation, it no longer inhibits elastase but targets coagulation cascade proteases, leading to hemophilia (Owen *et al.*, 1983).

While the first crystal structure of a Michaelis complex revealed residues P4–P3′ are implicated in direct protease/RCL contact (Ye *et al.*, 2001; Fig. 12.1B, boxed insert), more recent structures have highlighted the importance of residues outside the P4–P3′ positions in binding the target protease (Li *et al.*, 2004, 2008). Serpin exosites extending to the prime-side (P′-side) of the RCL can enhance protease interaction and seem to be a feature of serpins with more than one target (Whisstock *et al.*, 2010). In some other serpin/protease interactions, exosite contacts also play critical roles to improve the poor ability of target proteases to recognize certain RCL sequences and allow interactions with multiple target proteases (Börner and Ragg, 2008; Huntington, 2006; Johnson *et al.*, 2006; Li *et al.*, 2008; Whisstock *et al.*, 2010).

Macromolecular cofactors can also modulate serpin activity. Binding to specific sites of the protein, they can affect RCL flexibility and accessibility, thus altering serpin inhibitory function. In the case of antithrombin, the RCL is partially inserted into β-sheet A and is thus inefficient at binding target proteases. High-affinity binding of heparin provokes exposure of the RCL to the solvent and the protease, enabling antithrombin to efficiently target proteases (Li *et al.*, 2004), while concomitantly an exosite for factor Xa is exposed on the body of the serpin (Huntington, 2006). In another example, the active conformation of plasminogen activator inhibitor 1 is maintained by interaction with vitronectin, which slows the rate of insertion of the uncleaved RCL into β-sheet A. Spontaneous RCL insertion leads to the irreversible formation of the inactive latent conformation (Keijer *et al.*, 1991).

The kinetic nature of the inhibitory mechanism can lead serpins to follow the substrate pathway, where the serpin fails to capture the protease and is rendered unable to complete inhibition (Fig. 12.1D). If the RCL resembles the conformation of the protease's natural substrate too closely, cleavage of the intermediate will proceed before the protease can be relocated and trapped. Hence the substrate pathway leads to the insertion of the RCL in the cleaved, inactive state, while the active protease is released. The kinetic parameters that define the tendency to follow the inhibitory or the substrate pathways (k_{ass}, association constant; SI, stoichiometry of inhibition) can also be modulated by the addition of cofactors such as glycosaminoglycans to the serpin and/or protease (Olson *et al.*, 1992; Pike *et al.*, 2005).

1.2. The substrate specificity of proteases

Protease-controlled site-specific proteolysis is one of the most important posttranslational modifications. The key to understanding the physiological role of a protease is to identify the repertoire of its natural substrate(s) or to identify its substrate specificity (Song *et al.*, 2011). The specificity of proteases vary, primarily depending on their active sites, which display selectivity ranging from preferences for a number of specific amino acids at defined positions to more generic proteases with limited discrimination at one position. In addition to the primary amino acid sequence of the substrate, specificity is also influenced by the three-dimensional conformation of the substrate (secondary and tertiary structures). In particular, proteases preferentially cleave substrates within extended loop regions, while residues that are buried within the interior of the protein substrate are clearly inaccessible to the protease active site. Finally, cleavage is regulated by the temporal and physical colocation of the protease and substrate. In particular, some proteases are sequestered within specific compartments, with limited access to proteins, while others are able to cleave multiple substrates in different physiological compartments (Song *et al.*, 2011).

This chapter attempts to provide instruction in the analysis of the extended substrate specificity of proteases with a view to using the technique to mine for potential substrates. This technique has been previously used to examine the substrate specificities of a number of proteases, including the immune protease granzyme B (Kaiserman *et al.*, 2006a,b). A potent cytotoxin, granzyme B is the most widely studied member of the granzyme family of serine proteases. Granzymes are expressed by cytotoxic lymphocytes, which transfer the contents of specialized secretory lysosomes (including granzyme B) into the cytoplasm of target cells, whereupon granzyme B cleaves several specific substrates (including Bid and caspase-3), inducing the targeted cell to undergo apoptosis (for review, see Trapani and Sutton, 2003).

Human granzyme B has a murine counterpart which, upon superficial examination, appears to be very similar. They share significant sequence homology at the amino acid level and both hydrolyze substrates after an acidic residue (mainly aspartic acid) (Odake *et al.*, 1991). Indeed, many studies have assumed these two molecules to be functionally identical, and thus they are often used interchangeably in experimental systems. However, murine granzyme B cleaves the primary proapoptotic substrate Bid with a much lower efficiency than human granzyme B (Kaiserman and Bird, 2005; Kaiserman *et al.*, 2006a,b). This observation, along with data demonstrating differences in the cytotoxic potential between human and murine granzyme B, suggests that (at least on a functional level) these two molecules are not as similar as previously believed.

An examination of both human and mouse granzyme B molecules using the substrate phage display, and some of the bioinformatics techniques described in this chapter, reveals differences in the extended substrate specificities at two important subsites (Kaiserman *et al.*, 2006a,b) which could account for the inability of mouse granzyme B to efficiently cleave Bid. While these differences do not completely explain the comparatively reduced cytotoxicity of mouse granzyme B, it is convincing evidence that the origin of any granzyme B molecule should be matched with other reagents when designing *in vitro* or *in vivo* experimental systems in which the specific function of granzyme B may influence the outcome.

This chapter is intended to serve as an aid to researchers who are interested in employing phage display technology to characterize the extended substrate specificity of a protease (granzyme B is used as an example in this work), building the substrate specificity models using the PoPS (prediction of protease specificity) program (Boyd *et al.*, 2005) and further applying these models to predict specific serpin inhibitors containing potential substrate sequences in their RCL regions. We first outline the basic principles and requirements of phage display substrate technology. Next, we describe the methodology of substrate phage library design, preparation, biopanning, and analysis. Then we discuss bioinformatics approaches to analyze and determine the substrate specificity of granzyme B based on the data generated from the phage display selection. Finally, we illustrate how to build the specificity models and make predictions for novel putative substrates of granzyme B.

2. Phage Display Methods

In the following sections, we will describe how to generate and probe a substrate phage library to characterize the substrate specificity of a protease, how to use the data generated from this system to build the substrate

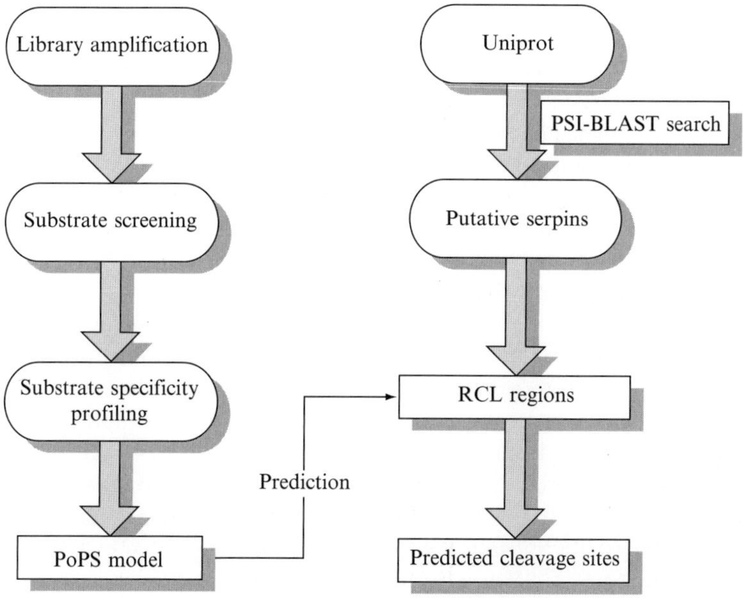

Figure 12.2 The execution steps of the entire process. There are two major sections. The first section describes how to perform the phage display experiments to characterize the substrate specificity of a protease and how to construct a valid substrate specificity model using PoPS engine (Boyd *et al.*, 2005) (using granzyme B as an example). The second section describes how to perform a PSI-BLAST search against the human and mouse proteomes in order to retrieve the putative serpins and how to make a prediction of cleavage sites in the RCL regions of the putative serpins using the built PoPS model in the first section.

specificity model, and how to make a prediction of candidate serpins with the optimal substrate specificity. We will examine a data set generated for the human immune protease, granzyme B, as an example to illustrate our methodology. An overview of the steps followed is presented in Fig. 12.2 and discussed in full detail below.

2.1. Application of phage display technology for determining protease substrate specificity

The ability to present a large number of potential substrates to a protease, combined with an inherent means of selecting only protease susceptible substrates, is the main advantage of a phage display substrate system for determining protease specificity (for review, see Deperthes, 2002; Diamond, 2007).

Peptide phage display has been in wide use since its development in 1985, when recombinant peptides were first expressed attached to the five copies of the coat protein (gIIIp) of the M13 filamentous phage (Smith, 1985). The phage display substrate technique described in this chapter utilizes the nonfilamentous T7 bacteriophage to take advantage of a commercially available kit (T7Select; Novagen) which streamlines the production of recombinant phage libraries. Peptide sequences are displayed on the surface of the 10B capsid protein (which is generated at low frequency by a frameshift of the 10A capsid protein). The T7 capsid is composed of a combination of 10A and 10B proteins, although 10B comprises less than 10% of the total capsid protein (Condron *et al.*, 1991) allowing easy construction of a monovalent substrate phage library.

Following the generation of the initial library (of phage displaying the pseudosubstrate sequence of interest), phage are immobilized on a solid support and exposed to protease. Phage which display protease-susceptible sequences are freed from the support, collected, and amplified to generate enriched sublibraries. Following several rounds of selection (see Fig. 12.3), analysis of the recombinant sequences in these enriched libraries is easily achieved by modern high-throughput DNA sequencing technology, yielding a pool of data for subsequent bioinformatics analysis.

2.2. General protocols and requirements

2.2.1. Substrate phage library design

Before employing a substrate phage technique to probe the substrate specificity of a protease, several factors must be considered. Of primary concern are conditions under which the protease is enzymatically active. The system which is described below was specifically chosen to analyze the substrate specificity of granzyme B, a serine protease which is active at near neutral pH (~7.4). This allowed the use of a hexahistidine affinity tag to immobilize the substrate phage on an IMAC (nickel ion) affinity gel medium. Thus a protease which is enzymatically active outside the pH range suitable for the binding of IMAC medium is not suitable for analysis using this particular system, unless another affinity tag is substituted (a number of options are reviewed in Deperthes, 2002).

When generating the substrate phage library, the user must consider whether the aim of the study is to examine the primary (P1) specificity of the protease or the extended substrate specificity. The library construction is flexible enough to allow the protease-susceptible linker region (pseudosubstrate) between the T7 phage capsid 10B protein and affinity tag to be varied in length and composition. The library described below utilizes a nonamer pseudosubstrate peptide with a fixed aspartic acid residue at position four to provide a target P1 for the granzyme B protease. The structure of the pseudosubstrate is thus: X-X-X-D-X-X-X-X-X. The intention of this anchored residue is twofold: first, to provide as many targets for the protease as possible (as each recombinant phage should display a pseudosubstrate

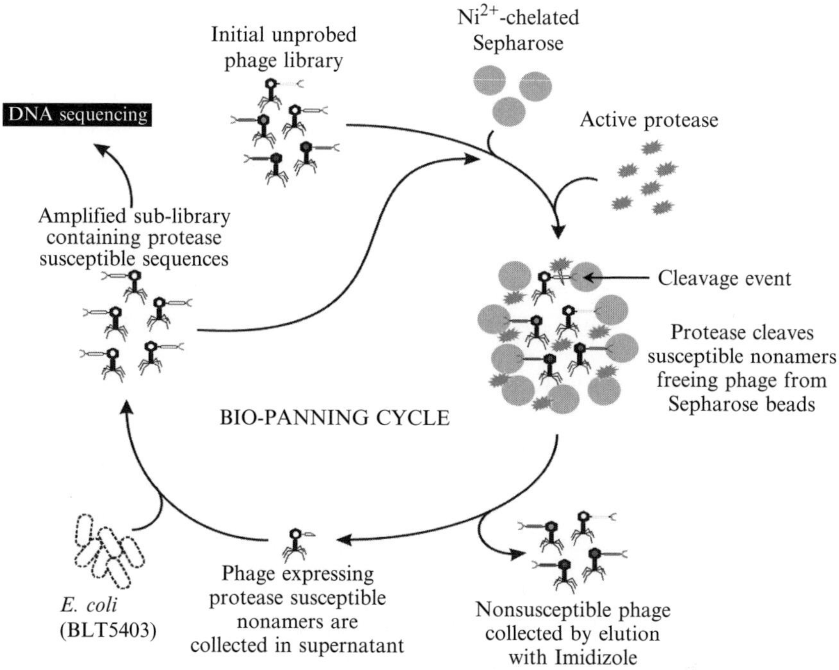

Figure 12.3 Schematic of the substrate phage biopanning cycle: an overview of the procedure of biopanning to select for enriched sublibraries of protease susceptible phage. A library of phage displaying varied nonamer pseudosubstrate sequences upstream of a hexahistidine affinity tag is immobilized on Ni^{2+}-chelating Sepharose® resin. The protease of interest is incubated with the immobilized phage, hydrolyzing susceptible pseudosubstrates. Hence the phage expressing these sequences are freed from the resin and can be collected and amplified in the host *E. coli* BLT5403 strain. This enriched sublibrary is then used for the subsequent cycle of biopanning, which can be repeated until a suitably enriched substrate phage population is obtained.

containing a residue compatible with the S1 subsite of the protease); second, to attempt to restrict the cleavage of the peptide to a position which will provide the most information about the extended substrate specificity of the protease in the P1′–P5′ subsites.

This approach yielded useful insights into the extended substrate specificity of both human and murine granzyme B in particular, highlighting a significant difference at both the P2 and P2′ positions (Kaiserman *et al.*, 2006a,b). To further elucidate the differences in extended substrate specificity, another library could be constructed in which other residues were anchored at positions P2 and P2′ as well as the aspartic acid at the P1 position. However, it may be preferable to present a completely randomized pseudosubstrate to a protease which is less stringent in the residues accommodated in the S1 pocket, to hunt for a substrate less reliant on the residue at P1.

2.2.2. Substrate phage library generation

The T7Select1-1b phage display system (Novagen, Merck KGaA, Damstadt, Germany) was specifically selected to restrict the number of recombinant phage coat proteins displayed per phage to one (or none). Thus during biopanning, only a single protease cleavage event is required to release a phage displaying a protease susceptible pseudosubstrate from the resin.

The T7Select Cloning Kit (Novagen, Merck KGaA, Damstadt, Germany) contains phage vector arms which are predigested at *Eco*RI and *Hin*dIII sites in the phagemid, to allow easy integration of double-stranded "insert" DNA predigested with these restriction enzymes (or which have compatible 5′ overhanging sequences). Three oligonucleotides are required to generate the pseudosubstrate "insert" with compatible 5′ overhanging sequences suitable for ligation with the phage vector arms (Karlson *et al.*, 2002). The randomized pseudosubstrate sequence is encoded by the nucleotides $(NNK)_9$ (where N = A, C, T, or G and K = T or G) which allow representation of all residues at any subsite, while excluding *ochre* (TAA) and *opal* (TGA) stop codons (see Table 12.1 for a full list of allowed and disallowed codons). Should an anchored residue be required at any position, NNK should be replaced with the desired codon at this position.

Table 12.1 The list of synonymous codons that are used to encode the nonamer substrate sequence

		Second base in codon								
		T		C		A		G		
First base in codon	T	TTT	Phe	TCT	Ser	TAT	Tyr	TGT	Cys	T
		TTC	Phe	TCC	Ser	TAC	Tyr	TCC	Cys	C
		TTA	Leu	TCA	Ser	TAA	STOP	TGA	STOP	A
		TTG	Leu	TCG	Ser	TAG	stop	TGG	Trp	G
	C	CTT	Leu	CCT	Pro	CAT	His	CGT	Arg	T
		CTC	Leu	CCC	Pro	CAC	His	CCC	Arg	C
		CTA	Leu	CCA	Pro	CAA	Gln	CCA	Arg	A
		CTG	Leu	CCG	Pro	CAG	Gln	CGG	Arg	G
	A	ATT	Ile	ACT	Thr	AAT	Asn	AGT	Ser	T
		ATC	Ile	ACC	Thr	AAC	Asn	AGC	Ser	C
		ATA	Ile	ACA	Thr	AAA	Lys	AGA	Arg	A
		ATG	Met	ACG	Thr	AAG	Lys	AGG	Arg	G
	G	GTT	Val	GCT	Ala	GAT	Asp	GGT	Gly	T
		GTC	Val	GCC	Ala	GAC	Asp	GGC	Gly	C
		GTA	Val	GCA	Ala	GAA	Glu	GGA	Gly	A
		GTG	Val	GCG	Ala	GAG	Glu	GGG	Gly	G
										Third base in codon

Allowed codons are shaded dark, and disallowed codons are shaded light. The two excluded STOP codons are highlighted.

```
            LeuThrProGlyGlyXaaXaaXaaXaaXaaXaaXaaXaaXaaHisHisHisHisHisHis *
5'-AATTCTCTCACTCCAGGCGGCNNKNNKNNKNNKNNKNNKNNKNNKNNKCATCACCATCACCATCACTA-3'
       |||||||||||||||||                           ||||||||||||||||||||
    3'-GAGAGTGAGGTCCGCCG                           GTAGAGGTAGAGGTAGAGATTCGA-5'
```

Figure 12.4 The structure of the nonamer insert cassette. The longer, sense (upper) and two shorter, antisense (lower) oligonucleotides are annealed, phosphorylated, and ligated to the phage vector arms to generate a recombinant phage vector (Karlson *et al.*, 2002). The 5′-AATT and 5′-AGTC overhangs allow direct ligation with the *Eco*RI and *Hin*dIII T7Select vector arms (T7Select manual; Novagen).

Three oligonucleotide sequences (as designed by Karlson *et al.*, 2002) are annealed to generate a randomized nonamer dsDNA insert (see Fig. 12.4):

Sense oligo

[5′-AATTCTCTCACTCCAGGCGGC(NNK)$_9$CATCACCATCAC-CATCACTA-3′]

Antisense oligos

[5′-GCCGCCTGGAGTGAGAG-3′] and [5′-AGCTTAGTGATGGT-GATGGTGATG-3′]

The substrate phage library is prepared as follows (adapted from Cwirla *et al.*, 1990):

1. Sense and antisense oligonucleotides are diluted in annealing buffer (100 m*M* potassium acetate, 2 m*M* magnesium acetate, 30 m*M* HEPES–KOH, pH 7.4; Sigma–Aldrich) to a final concentration of 5 n*M* each.
2. Diluted oligonucleotides are incubated at 95 °C for 4 min, then 70 °C for 10 min. The diluted oligonucleotides are then slowly cooled to 4 °C over a 30-min period to facilitate correct annealing (a PCR thermocycler is useful for this step).
3. The annealed oligonucleotides are then phosphorylated prior to ligation to the phage vector arms. This is done using T4 Polynucleotide Kinase, as per the manufacturer's instructions (New England Biolabs).
4. The annealed and phosphorylated "insert" is ligated to the T7Select1-1b vector arms (T7Select System Manual, Novagen) by T4 DNA ligase (New England Biolabs). This will generate a T7 phage vector suitable for *in vitro* packaging into new phage.
5. The T7 phage vector is mixed with T7 Packaging Extracts (Novagen) according to instructions to generate recombinant T7 phage. Note that these phage do not express the recombinant 10B capsid protein

displaying the pseudosubstrate. Expression will only occur after amplification of these phage in a suitable *Escherichia coli* host (see below).

6. Efficiency of the *in vitro* phage packaging is assessed by performing a plaque forming assay to estimate the titer of the library. A serial dilution of the packaging mixture is performed and plated with host bacteria (BLT5403 BL21) as per instructions (T7Select System Manual, Novagen). This allows an estimation of the sequence complexity of the library. Complexities in the order of 1.0×10^6–1.0×10^8 total phage are typically observed. Bear in mind that a theoretically "complete" nonamer library (i.e., every possible sequence is represented by at least one phage) would contain 5.12×10^{11} phage.
7. To generate recombinant phage which express the engineered 10B capsid protein, a plate-lysate amplification of the starting library is performed using 145 mm × 20 mm diameter sterile plastic Petri dishes (SARSTEDT AG & Co.), as per instructions (T7Select System Manual, Novagen). The plate-lysate amplification is preferred to a liquid-lysate amplification to ensure even amplification of all phage in the library. However, this may not be practical, depending on the number of phage in the library. In such a case, the library can be amplified using a liquid-lysate method (T7Select System Manual, Novagen).
8. Once the library has been amplified, clarified, and pooled (see T7Select System Manual, Novagen), 0.1 volume sterile 80% glycerol is added, the library is aliquoted into 50 ml aliquots, and stored frozen at −70 °C until required.

2.3. Screening the substrate phage display library

Before beginning the screening process, it is vital to assess the quality of the protease under examination. Thought should be given to the structure and purity of the protease, as well as the constituents of the buffer in which it is stored. Recombinant proteins are often purified by the use of a hexahistidine affinity tag. Should the protease contain this affinity tag, there is the possibility that it will bind directly to the IMAC medium, either displacing the immobilized phage or preventing the protease from easily accessing the pseudosubstrate. Removal of the hexahistidine tag (if possible) or purification using another affinity tag may abrogate this problem.

Regardless of the source of the protease (be that recombinant or native), steps should be taken to ensure that the protein is as free from contaminating proteins as possible. In particular, any contaminating proteases may also find compatible substrates in the library, and during the biopanning process the phage expressing these sequences will also be amplified, ultimately contaminating the final data set. Finally, ensure that no metal-chelating reagents (e.g., EDTA) are present in the protease buffer, as these may nonspecifically elute the phage from the IMAC support.

Prior to commencement of biopanning, it is also prudent to ensure the protease under examination does not target structural proteins of the bacteriophage. This is easily determined by overnight incubation of a sample of the amplified phage library with the protease at the concentration to be used for screening. Titration and comparison of this sample with a protease-free control should show no loss of titer (signifying disruption or destruction of the phage) due to the presence of the protease.

2.3.1. Biopanning procedure

Three to four rounds of biopanning are usually sufficient to produce a suitably enriched subpopulation of phage displaying protease-susceptible sequences. The biopanning procedure is performed as follows (for an overview of the procedure, see Fig. 12.3):

1. A sample (5–10 ml) of the amplified library is incubated with 500 μl of Ni Sepharose® 6 Fast Flow IMAC resin (GE Healthcare) on a rocking table at 4 °C for 1–2 h.
2. The resin is collected by centrifugation (2 min at 500 × *g*) and the supernatant discarded, taking care not to disturb the resin bed.
3. The resin is gently washed with 2 × 50 ml washes of wash buffer (1 *M* NaCl, 0.1% (v/v) Tween 20 (Amresco Inc.), 20 m*M* imidazole (Amresco Inc.) in PBS, pH 7.2). The imidazole is included in this buffer to prevent nonspecific binding to the resin (based upon the manufacturer's suggestion). If using another IMAC resin check the manufacturer's specifications before adding imidazole.
4. The resin is gently washed with 2 × 50 ml of incubating buffer (150 m*M* NaCl, 1 m*M* $MgSO_4$, 20 m*M* Tris, pH 7.2; all components from Amresco Inc.). Any cofactors required for full activity of the protease should be included in this buffer.
5. The resin is left suspended in 1 ml of incubating buffer, which is split into two equal 500 μl aliquots in 1.5 ml tubes. Protease is added to one aliquot to the required final concentration (200 n*M* is a suggested starting point—optimization may be required depending on the protease under investigation), and both aliquots are made up to 1 ml with incubating buffer.
6. Aliquots are incubated overnight at 37 °C with gentle shaking or rotation to prevent settling of the resin. During this incubation, the protease has the opportunity to cleave the randomized sequence between phage capsid and hexahistidine tag, releasing these phage into solution for subsequent collection.
7. Phage that have been released into solution are collected by transferring the total contents of the treatment tube into an empty 1.2-ml

disposable Bio-Spin chromatography column (Bio-Rad) and collecting the elution volume. The resin is gently washed with a further 500 μl of incubating buffer to collect remaining soluble phage.

8. Phage that are still bound to the resin are eluted and collected by washing with 1.5 ml of 0.5 *M* imidazole supplemented with 1 m*M* $MgSO_4$.
9. Both steps are repeated for the control sample. Again, soluble and bound phage are collected separately.
10. A 50 -μl sample of the phage eluted by protease treatment is set aside for later titration, while the remaining volume (1.45 ml) is mixed with 8.55 ml of host bacteria (BLT5403 BL21) which have been freshly grown to $OD_{600} = 1.0$ in M9LB (see T7Select System Manual).
11. After mixing phage and host bacteria, the 10 ml volume is split into 10 × 1 ml aliquots to minimize the chance that any phage will amplify faster than the general population and become artificially overrepresented in the sublibrary. All aliquots are incubated at 37 °C with shaking at 250 rpm for 3 h (or until lysis of the host strain is observed by partial clearing of the solution and the appearance of stringy debris).
12. Lysed aliquots are pooled and NaCl is added to a final concentration of 0.5 *M*. The lysate is clarified by centrifugation at 8000 × *g* for 10 min. Collect the supernatant into a clean and sterile container. This sublibrary can then be used for another round of biopanning by repeating steps 1 through 12, or DNA can be isolated for sequencing if a suitable enrichment is deemed to have been achieved.

2.3.2. Assessing sublibrary enrichment due to action of the protease

Our experience with this system suggests that some variability is observed in both the numbers of phage which are immobilized on the resin and the numbers of phage which spontaneously elute from the resin even when the protease is absent. It is essential then to have some way of assessing if enrichment has taken place and if this enrichment is due to the action of the protease.

This is achieved by observing the numbers of soluble phage collected after incubation with the protease as a fraction of those still bound to the resin (and subsequently removed with imidazole treatment). Titration of phage in the soluble and bound fractions is determined by the same plaque forming assay used to calculate the complexity of the initial library following *in vitro* packaging (see T7Select System Manual, Novagen). Hence,

$$\%\text{phage eluted} = \frac{\text{Titer(soluble phage)}}{\text{Titer(bound phage)} + \text{Titer(soluble phage)}} \times 100 \quad (12.1)$$

thus:

$$\text{Phage eluted by protease (fold difference)} = \frac{\%\text{phage eluted}(+\text{protease})}{\%\text{phage eluted}(\text{no protease})} \tag{12.2}$$

Over multiple rounds of enrichment for phage expressing protease-susceptible sequences, the fraction of phage eluted by the protease should increase exponentially. Once the fraction of phage eluted by the protease reaches a suitable level (we empirically determined this to be greater than 30-fold over control), individual phage can be isolated for DNA sequencing.

2.3.3. Sequence analysis of protease-selected recombinant phage

To determine the sequence of the peptide a particular phage is displaying, the phage must be isolated, disrupted, and its genomic DNA collected for PCR amplification and subsequent nucleotide sequencing.

It is suggested that this be done on a smaller scale (10–20 phage) first, to determine if (a) enrichment has indeed occurred (strong substrate specificity profiles can be observed in surprisingly few sequences); (b) the phage in the sublibrary are recombinant (some wild-type phage are generated during the initial library generation, and these may have a growth advantage under certain conditions); or (c) one or several phage have been so strongly enriched that they account for a majority of the sublibrary (rendering large-scale DNA sequencing redundant). Once the sublibrary has been deemed suitable for further analysis, large-scale sequencing (96-well format or larger) can be performed.

Phage DNA sequence determination is performed as follows:

1. A sample of the sublibrary is diluted to a suitable concentration to allow approximately 100 phage plaques to develop when a plaque forming assay is performed (see T7Select System Manual).
2. Individual plaques are "lifted" from the plate by pushing a sterile pipette tip into the top agar to remove a small plug and then pipetting this plug vigorously with 100 μl of sterile water. This solution now contains copies of a single phage and can be archived by mixing with 0.1 volume of sterile 80% glycerol and storing at −70 °C.
3. To extract genomic DNA, the phage coat must be disrupted. An aliquot of the plaque lift solution is mixed with an equal volume of 20 m*M* EDTA and heated to 65 °C for 10 min. This will generate a suitable template for PCR amplification of the section of the phage genome containing the randomized insert using T7SelectUP (5′-GGAGCTGTCGTATTCCAGTC-3′) and T7SelectDOWN

(5′-AACCCCTCAAGACCCGTTTA-3′) primers (Novagen, Merck KGaA, Damstadt, Germany) according to instructions (see T7Select System Manual, Novagen).

4. Following completion of the PCR, the size of the generated products (~200 bp) is confirmed by electrophoresis (80 V, 1 h) in a 2.5% (w/v) agarose/TBE gel (TBE: 45 m*M* Tris–HCl, pH 8.0, 45 m*M* boric acid, 1 m*M* EDTA).
5. PCR products are purified prior to DNA sequencing to remove unbound primer molecules. For small-scale sequencing reactions, Wizard SV Gel and PCR spin columns (Promega) are used. For large-scale analysis (96-well format), Multiscreen μ96 PCR Clean-Up Plates (Millipore) are used.
6. DNA Dye-Terminator sequencing is performed using the BigDye 3.1 reagent (Applied Biosystems) with the T7SelectUP primer.
7. For small-scale sequencing reactions, Dye-Terminator-labeled products are precipitated by diluting the 20 μl sequencing reaction to 100 μl with precipitation buffer (112.5 m*M* sodium acetate, pH 5.2, 75% (v/v) absolute ethanol) and incubating for 15 min at room temperature. Products are pelleted by centrifugation (16,000 × *g* for 30 min) and washed once with 70% (w/v) ethanol before air drying.
8. Large-scale (96-well) Dye-Terminator DNA sequencing reactions are cleaned using Montage μ96 Sequencing Clean-Up Plates (Millipore) and electropherograms generated by running the samples on a 3730S Genetic Analyser (Applied Biosystems).

2.4. Assessing the quality of the data set generated by substrate phage biopanning

Following PCR amplification and nucleotide sequencing of the phage genome sequences, a data set of the phage-susceptible sequences nonamers can be compiled and should be examined using software designed for handling multiple sequence alignments. Before a consensus pattern can be determined, the sequences must usually undergo alignment. The sequences in the data set should be aligned according to the following criteria: (i) known primary/secondary protease substrate specificity data, (ii) suspected protease substrate specificity data, and (iii) patterns that are present in the nonamer sequences.

It is our experience that a small number (usually 1–10%) of phage in each data set will express sequences containing one or more *amber* stop codons (the only allowed stop codon based on the NNK codon usage) in the nonamer sequence. These phage should not be able to anchor to the chelating resin, as they express a recombinant sequence which is truncated and hence will not contain a hexahistidine tag. It is currently unclear why these "background" phage are still present in the pool after four rounds of biopanning, but they are observed even in data sets which display very strong selection due to the protease. Regardless, these sequences must be identified and culled from the data set before alignment can proceed.

If the nonamer was designed with an "anchored" residue at a particular position (see 2.2.1), then alignment should be relatively straightforward. If the "anchored" residue was inserted to act as a P1 residue, it may be assumed that the protease in question has cleaved at this position. Thus initial alignment can take place based upon this anchored residue.

At this point, a consensus pattern in the sequences may be readily observable, at which point it may be possible to further align sequences based on this pattern to generate a better fit. This is especially relevant for sequences in which another occurrence of the "anchored" residue is observed, as in the following example: X-D-X-D-X-X-X-X-X. Here, the anchored aspartic acid (underlined) may not be the primary P1 residue, as another aspartic acid residue is present. A closer inspection of the consensus sequence generated from the full data set will be required to determine the alignment of this particular sequence.

In the case where a completely random nonamer library has been used, the alignment can be more problematic unless a very obvious pattern is present. In cases where the pattern is more subtle (and hence difficult to determine by eye), it may be useful to begin by using a multiple alignment algorithm (e.g., ClustalW, gaps prohibited (Larkin *et al.*, 2007)) to aid initial pattern discovery.

2.5. Analysis of phage display data and characterization of the substrate specificity of granzyme B

In this section, we describe in detail the method used to analyze the sequence data collected from phage display experiments and determine the substrate specificity of a protease.

Following alignment of the sequences in the data set (see Section 2.5), we generate a $\Delta\sigma$ value for each amino acid observed at each subsite position, based upon the following formula:

$$\Delta\sigma = \frac{\mathrm{Obs}(x) - nP(x)}{\sqrt{nP(x)[1 - P(x)]}} \tag{12.3}$$

where n is the total number of sequences being analyzed, Obs(x) is the occurrence of amino acid x at a subsite in the selected sequences, and $P(x)$ is the theoretical occurrence of amino acid x based on the coding of the nonamer sequence. Use of $\Delta\sigma$ to analyze the statistical distribution of amino acids in substrate sequences was originally proposed by Matthews

Table 12.2 The number of allowed synonymous codons

Amino acid	A	C	D	E	F	G	H
Synonymous codons	2	1	1	1	1	2	1
Amino acid	I	K	L	M	N	P	Q
Synonymous codons	1	1	3	1	1	2	1
Amino acid	R	S	T	V	W	Y	Total
Synonymous codons	3	3	2	2	1	1	31

et al. in their study of the substrate specificity of furin using substrate phage display (Matthews *et al.*, 1994).

The value of $\Delta\sigma$ shows the difference of the observed frequency from the expected frequency in terms of standard deviation, by assuming a binomial distribution of amino acids (Matthews *et al.*, 2004). The calculation of $\Delta\sigma$ takes into account the difference in the numbers of codons allowed for each residue (see Table 12.1), thus reducing background noise and highlighting amino acids at positions which are the most significant specificity determinants.

Therefore, the theoretical probability $P(x)$ can be calculated based on the number of allowed synonymous codons for amino acid x divided by the total number of allowed synonymous codons, as given in Table 12.2.

A $\Delta\sigma$ value of >1 suggests that a particular amino acid is overrepresented at that subsite. Once the $\Delta\sigma$ values for all subsites through P7 to P6′ positions are calculated according to the above procedures, they can be further normalized to generate the weight values for all the subsites based on which the PoPS model can be generated (an example of the $\Delta\sigma$ values calculated for all subsites through P7 to P6′ positions of human granzyme B substrates is provided in Fig. 12.5). Negative values for $\Delta\sigma$ are artifacts which appear due to strong positive selection and do not indicate selection against a residue at a particular subsite (for further discussion, see Section 2.6).

2.6. Generating a substrate specificity model for PoPS

The PoPS software engine (Boyd *et al.*, 2005) uses a protease specificity model to evaluate the probability of an input amino acid sequence containing a suitable protease cleavage site. The model can be derived directly from experimental data generated using the substrate phage biopanning procedure described above, or other bioinformatics approaches, for example, positional scanning substrate combinatorial peptide libraries, data from known biological substrates, etc.

Calculated Δσ values at each subsite									
Amino Acid	P4	P3	P2	P1	P1′	P2′	P3′	P4′	P5′
A		2.2	6.2		−0.6	−0.2	−0.6	−0.6	0.2
C		−0.7					−1.3	−1.3	−0.7
D	−1.3	1.0		55.3	1.5	1.0	−0.7	2.1	1.0
E	−1.3	4.9	−1.3		1.0	−1.3	1.5	2.6	1.0
F	−0.2	−0.2	1.0		2.6	−0.2	1.5	−0.2	3.8
G	−2.2	10.2	−1.4		−1.0	2.6	−0.2	−1.0	0.2
H						−1.3			−1.3
I	27.3		−1.3		−0.7	1.0	2.1	3.8	1.5
K		−1.3							−1.3
L	−2.0	−2.0	−1.0		0.4	7.7	0.4	0.7	−0.3
M	−0.2		−1.3		−1.3	−0.2	−0.7	−0.7	−0.7
N		−1.3			0.4	−1.3	−1.3	−0.2	
P	−1.8		3.4				−1.4	−1.8	−0.6
Q		−0.7	0.4		−1.3	−0.7		−1.3	−0.7
R							−1.0	−1.6	−1.6
S	−2.6	3.7	1.7		3.1	1.4	−2.3	−0.3	−1.0
T	−2.2	−1.4	0.6		−1.4	−2.2	1.8	−1.0	−2.2
V	9.8	−1.4	3.4		5.0	−1.4	6.6	5.4	2.2
W			−1.3		1.0	−1.3	−0.2	−0.7	1.0
Y	−1.3	−1.3	0.4		3.2	3.8	1.0	0.4	3.8

Figure 12.5 The $\Delta\sigma$ values for subsites through P4 to P5′. Values shown were obtained following phage display biopanning with human granzyme B on an Asp-anchored (position 4/9) phage library. The adjusted $\Delta\sigma$ values for all subsites through P4 to P5′ positions provides an overview of the substrate specificity of human granzyme B. Residues with $\Delta\sigma$ values greater than 5 are highlighted.

The protease specificity model contains information regarding the protease's preference for any given amino acid residue at each subsite using a system of positive and negative weightings. These weightings range from a value of -5 through $+5$ to indicate the preference of the protease for a particular residue at the subsite position. The values generated (positive or negative) for each subsite are added to compute a score indicating the likelihood of protease cleavage at a particular position in the input sequence.

While the values of $\Delta\sigma$ generated from the substrate phage display data set (for each residue at each subsite) are not constrained by specific limits, they can be normalized to conform to the weighting parameters under which PoPS operates in order to generate a protease specificity model. Subsites that display very strong positive selection (either through protease selection or through artificial anchoring) will have residues with large positive $\Delta\sigma$ values. Thus, to prevent these subsites from obscuring the value of other subsites, each subsite is considered independently, that is, $\Delta\sigma$ values are normalized to the largest $\Delta\sigma$ value in each subsite.

Substrate phage display biopanning is (by definition) a system of positive selection. This is useful for highlighting positive selection of sequences by the protease, that is, those residues which are overrepresented at any particular subsite. However, the reverse is not necessarily true, as the system is not suitable for showing negative selection, only the absence of positive selection. If a particular residue is under-represented at a subsite, one cannot infer that the protease "dislikes" the residue at this position, only that the protease does not prefer it. It is also not possible to gauge the magnitude or "degree" of "lack of preference" if $\Delta\sigma$ values are negative. Hence when constructing a PoPS model from the $\Delta\sigma$ values, we choose not to normalize negative values, instead replacing them with the absolute lowest value of -5. The remaining positive $\Delta\sigma$ values for each subsite are then normalized by dividing them by the largest $\Delta\sigma$ value at that subsite and multiplying the fraction by $+5$. Hence the largest $\Delta\sigma$ value at each subsite will be assigned a value of $+5$, the remaining positive values will scale between 0 and $+5$, and all negative $\Delta\sigma$ values will be assigned a value of -5.

In the special case of an anchored P1 residue, all other residues at this subsite are likely to have negative $\Delta\sigma$ values which will be normalized to values of -5 in the model. Some consideration should be given to the nature of the protease at this point. For example, human granzyme B is not only preferentially an Asp-ase but is also capable of hydrolyzing bonds after the acidic amino acid Glu. The substrate phage display data for human granzyme B was obtained using a phage library with an anchored Asp P1 residue. Thus as no Glu residues will be present at the P1 position (due to the anchoring of Asp), the value assigned to Glu will be -5. This will have the artificial effect of lowering the overall score of any potential cleavage site where Glu is present at the P1 position. In this case, the score assigned to Glu at the P1 position in the PoPS model was amended to a low positive number to prevent artificial exclusion of cleavage sites with an acidic (but not Asp) P1 residue.

3. Sequence Analysis Methods

3.1. Bioinformatics approaches for predicting protease substrates

Although proteases represent around 2% of the gene products across all the genomes, the complete repertoire of their target substrates and inhibitors remain to be fully characterized. In this context, bioinformatic modeling approaches that are capable of predicting putative cleavage sites of proteases and can provide experimentally testable information are invaluable tools. In the past few decades, a number of bioinformatics approaches have been developed to facilitate the *in silico* discovery of protease substrates and specificity profiling analysis.

Existing approaches can be generally categorized into two main classes: machine learning or empirical scoring function. The former is mainly based on the selection and subsequent representation of useful sequence features and further mapping of input features to the property of being cleaved (positive samples) or noncleaved (negative samples). Employed machine learning approaches include Bayesian biobasis function neural networks (Chen *et al.*, 2008; Yang, 2005), support vector machines (Barkan *et al.*, 2010; Chen *et al.*, 2008; Wee *et al.*, 2007, 2009), and support vector regression (Song *et al.*, 2010). The latter relies on learning the underlying rules using the distribution of positive and negative sequences and building empirical scoring functions to discriminate between the two classes. Recently developed modeling and prediction tools falling in this category include PoPS (Boyd *et al.*, 2005) and SitePrediction (Verspurten *et al.*, 2009). See Song *et al.* (2011) for a comprehensive review of these approaches.

Although machine learning-based approaches can generally accept amino acid sequences of substrates as the input to the built predictive models and, in some cases, allow more accurate prediction of potential cleavage sites of a protease of interest (Barkan *et al.*, 2010; Song *et al.*, 2010), information regarding the substrate specificity of the protease can only be derived from experimental approaches. In the absence of such data or given the insufficient availability of such data, it will not be feasible to curate a reasonable training set of cleaved (positive) and noncleaved (negative) substrate peptides. As a result, it is not viable to apply machine learning approaches to learn a model of the substrate specificity of a protease and make a prediction of cleavage sites in order to identify novel substrates.

However, empirical scoring tools, such as PoPS and SitePrediction, often have a user-friendly interface and facilitate the creation of specificity models of any protease from either experimental data obtained from phage display experiments or expert knowledge of substrate specificity. Further, PoPS can be used to augment and refine specificity models with dependency rules and validate hypotheses regarding the "subsite cooperativity," an important phenomenon which has been increasingly observed in multiple proteases (Boyd *et al.*, 2009; Ng *et al.*, 2009a,b). Therefore, in the following sections, we will describe how to perform phage display experiments, process experimental data, and build substrate specificity models using the PoPS engine.

3.2. Proteome-wide search of putative human serpins and identification of the RCL sequences as pseudosubstrates of proteases

Serpin structures contain a number of highly conserved regions, including the shutter, breach, and hinge (Law *et al.*, 2006; Whisstock and Bottomley, 2006). Locating these structures allows identification of the residues which encode the RCL. Once identified, the sequences of the RCL regions can be

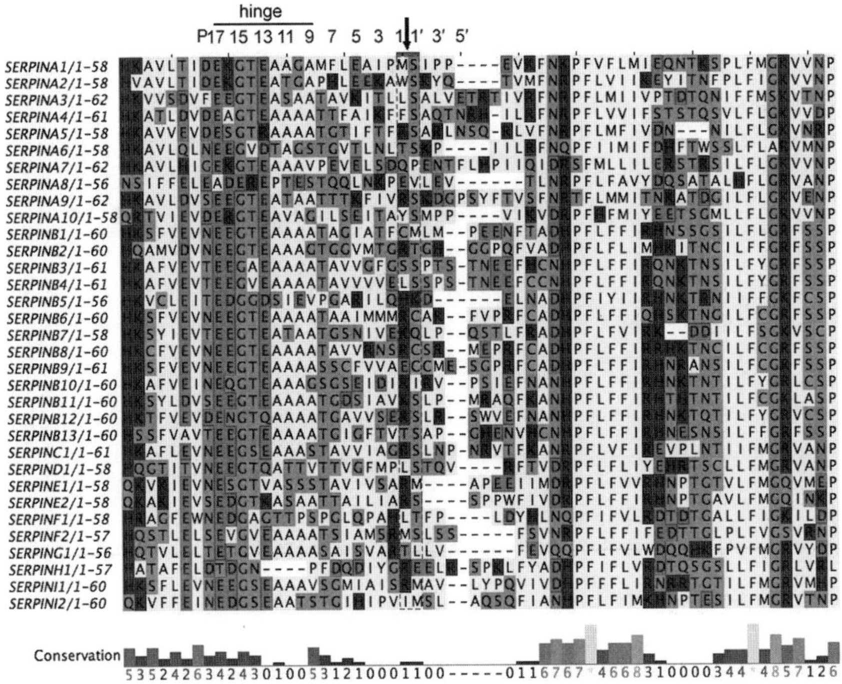

Figure 12.6 Amino acid sequence alignment of the RCL sequences of the 33 human serpins. Amino acid sequences were aligned using ClustalW 2.0 (Larkin *et al.*, 2007) and displayed using Jalview 2.4 (Waterhouse *et al.*, 2009). Using the nomenclature of Schechter and Berger (1967), the RCL sequences are numbered from P17 to P5′, which are considered to be protease specificity determining. The hinge region in the RCL is underlined. The scissile bond between P1 and P1′ sites is marked by a black arrow. Residues are colored according to their side chain chemical properties: Glu + Asp (negatively charged; red), Asn + Gln + Gly + Ser + Thr (polar uncharged; green), Ala + Leu + Ile + Val + Met + Phe + Pro + Trp + Tyr (nonpolar; yellow), Arg + Lys + His (positively charged; blue), Cys (light blue). The conservation, quality, and consensus scores, as three of the automatically quantitative alignment annotations, are also shown. (See Color Insert.)

extracted and aligned. An alignment of human serpin RCL sequences is shown in Fig. 12.6.

As can be seen from Fig. 12.6, the characteristic residues in the consensus hinge region of human serpins are P17(E), P16(E/K/R/D), P15(G), P14 (T/S), small aliphatic residues (A/G/S) at P12–P9 positions (Benarafa and Remold-O'Donnell, 2005) as well as at P8(T/S). Using this consensus signature motif, the RCL sequences of all the human serpins can be identified (Table 12.3).

Table 12.3 Human serpins: Uniprot ID, alternative name, RCL sequence, and possible protease targets

Name	Uniprot ID	Alternative name	Protein length	Sequence of hypervariable and distal regions of RCL	Protease targets
SERPINA1	P01009	Alpha-1-antitrypsin	418	GTEAAGAMFLEAIPMSIPPEVKFN	Extracellular; inhibition of neutrophil elastase
SERPINA2	P20848	Alpha-1-antitrypsin-related protein	420	GTEATGAPHLEEKAWSKYQTVMFN	Not characterized
SERPINA3	P01011	Alpha-1-antichymotrypsin	423	GTEASAATAVKITLLSALVETRTIVRFN	Extracellular; inhibition of cathepsin G
SERPINA4	P29622	Kallistatin	427	GTEAAAATSFAIKFFSAQTNRHILRFN	Extracellular; inhibition of kallikrein
SERPINA5	P05154	Plasma serine protease inhibitor	406	GTRAAAATGTIFTFRSARLNSQRLVFN	Extracellular; inhibition of active protein C
SERPINA6	P08185	Corticosteroid-binding globulin	405	GVDTAGSTGVTLNLTSKPIILRFNQPF	Extracellular; noninhibitory
SERPINA7	P05543	Thyroxine-binding globulin	415	GTEAAAVPEVELSDQPENTFL	Extracellular; noninhibitory
SERPINA8	P01019	Angiotensinogen	485	EREPTESTQQLNKPEVLEVTLN	Extracellular; noninhibitory
SERPINA9	Q86WD7	Centerin	417	GTEATAATTTKFIVRSKDGPSYFTVSFN	Extracellular
SERPINA10	Q9UK55	Protein Z-dependent protease inhibitor	444	GTEAVAGILSEITAYSMPPVIKVD	Extracellular; inhibition of activated factor Z and XI
SERPINB1	P30740	Leukocyte elastase inhibitor	379	GTEAAAATAGIATFCMLMPEENFTAD	Intracellular; inhibition of neutrophil elastase
SERPINB2	P05120	Plasminogen activator inhibitor 2	415	GTEAAAGTGGVMTGRTGHGGPQFVAD	Intracellular; inhibition of uPA
SERPINB3	P29508	Squamous cell carcinoma antigen 1	390	GAEAAAATAVVGFGSSPTSTNEEFHCN	Intracellular; inhibition of cathepsin L and V

(*Continued*)

Table 12.3 (*Continued*)

Name	Uniprot ID	Alternative name	Protein length	Sequence of hypervariable and distal regions of RCL	Protease targets
SERPINB4	P48594	Squamous cell carcinoma antigen 2	390	GVEAAAATAVVVVELSSPSTNEEFCCN	Intracellular; inhibition of cathepsin G and chymase
SERPINB5	P36952	Maspin	375	GGDSIEVPGARILQHKDELNAD	Intracellular; noninhibitory
SERPINB6	P35237	Placental thrombin inhibitor	376	GTEAAAATAAIMMMRCARFVPRFCAD	Intracellular; inhibition of cathepsin G
SERPINB7	O75635	Megsin	380	GTEATAATGSNIVEKQLPQSTLFRA	Intracellular
SERPINB8	P50452	Peptidase inhibitor 8	374	GTEAAAATAVVRNSRCSRMEPRFCAD	Intracellular; inhibition of furin
SERPINB9	P50453	Peptidase inhibitor 9	376	GTEAAAASSCFVVAECCMESGPRFCAD	Intracellular; inhibition of granzyme B
SERPINB10	P48595	Peptidase inhibitor 10	397	GTEAAAGSGSEIDIRIRVPSIEFNAN	Intracellular; inhibition of thrombin and trypsin
SERPINB11	Q96P15	Epipin	392	GTEAAAATGDSIAVKSLPMRAQFKAN	Intracellular
SERPINB12	Q96P63	Yukopin	405	GTQAAAATGAVVSERSLRSWVEFNAN	Intracellular; inhibition of trypsin
SERPINB13	Q9UIV8	Headpin	391	GTEAAAATGIGFTVTSAPGHENVHCN	Intracellular; inhibition of cathepsin L and K
SERPINC1	P01008	Antithrombin	464	GSEAAASTAVVIAGRSLNPNRVTFKAN	Extracellular; inhibition of thrombin and factor Xa

SERPIND1	P05546	Heparin cofactor II	499	GTQATTVTTVGFMPLSTQVRFTVD	Extracellular; inhibition of thrombin
SERPINE1	P05121	Plasminogen activator inhibitor 1	402	GTVASSSTAVIVSARMAPEEIIMD	Extracellular; inhibition of thrombin, uPA, tPA, and plasmin
SERPINE2	P07093	Protease nexin I	398	GTKASAATTAILIARSSPPWFIVD	Extracellular; inhibition of uPA and tPA
SERPINF1	P36955	Pigment epithelium-derived factor	418	GAGTTPSPGLQPAHLTFPLDYHLN	Noninhibitory
SERPINF2	P08697	Alpha-2-antiplasmin	491	GVEAAAATSIAMSRMSLSSFSVN	Extracellular; inhibition of plasmin
SERPING1	P05155	Plasma protease C1 inhibitor	500	GVEAAAASAISVARTLLVFEV QQPFLFVLWDQQH	Inhibition of C1 esterase
SERPINH1	P50454	Collagen-binding protein	418	GNPFDQDIYGREELRSPKLFYAD	Noninhibitory
SERPINI1	Q99574	Neuroserpin	410	GSEAAAVSGMIAISRMAVLYPQVIVD	Extracellular; inhibitor of uPA, tPA, and plasmin
SERPINI2	O75830	Myoepithelium-derived serine protease inhibitor	405	GSEAATSTGIHIPVIMSLAQSQFIAN	Extracellular

The P1 residue at the scissile bond is the critical determinant of the substrate specificity of a protease, and mutation of residues at this position leads to the alteration of the inhibitory spectrum of serpins, often resulting in disease (Kaiserman and Bird, 2010). However, different proteases exhibit different preferences for P1 residues. For example, granzyme B shows stringent specificity for aspartic acid (D) in the substrate P1 position (Cullen *et al.*, 2007; Kaiserman *et al.*, 2006a,b; Rotonda *et al.*, 2001; Ruggles *et al.*, 2004; Van Damme *et al.*, 2009; Waugh *et al.*, 2000), whereas thrombin preferentially cleaves at arginine in P1 positions in substrates (Cole *et al.*, 1967; Coughlin, 2000; Ng *et al.*, 2009a,b). In addition, adjacent residues near the scissile bond (P7–P2 and P1′–P6′ positions) can also influence the cleavage efficiency of a protease, thus conferring important additional determinants of specificity.

In the case of inhibitory serpins, however, the requirement for conventional specificity-determining residues at P1 positions is not absolute; many serpins may inhibit proteases by using other P1 residues that are not preferentially cleaved by a particular protease (Kaiserman and Bird, 2010). The reason for this may possibly be attributed to the underlying serpin mechanism. In these occasions, residues in the proximal regions to the P1 position provide additional important information regarding the inhibitory specificity for serpins. Here, our hypothesis is that the P7–P6′ sequences from the serpin RCL regions are likely to cover the most important residues that are determinants of the inhibitory specificity of serpins (Roberts and Hejgaard, 2008). Table 12.4 thus lists the P7–P6′ sequences of 33 human serpins. These P7–P6′ sequences will be used as the input to the PoPS program to predict potential cleavage sites of human granzyme B in the following section.

3.3. Construction of the substrate specificity models and prediction of cleavable human serpins by granzyme B

In this section, based on the normalized substrate specificity data obtained from Section 2.5, we describe how to build the substrate specificity models and predict cleavable human serpins of granzyme B using the PoPS engine, which is a comprehensive bioinformatic tool that allows users to model and profile protease substrate specificity and predict substrate cleavage sites for any protease.

Detailed procedures for building the PoPS model of the substrate specificity of human granzyme B are as follows:

Step 1: Create a specificity model

(a) Save (as an input file for PoPS) the normalized weight values for all the subsites from P7 to P6′ sites that were previously obtained (see Section 2.6). Name this input file as "PoPS_input.txt."

Table 12.4 P7–P6′ sequences of human serpins

Name	P7	P6	P5	P4	P3	P2	P1	P1′	P2′	P3′	P4′	P5′	P6′
SERPINA1	F	L	E	A	I	P	M	S	I	P	P	E	V
SERPINA2	H	L	E	E	K	A	W	S	K	Y	Q	T	V
SERPINA3	A	V	K	I	T	L	L	S	A	L	V	E	T
SERPINA4	T	F	A	I	K	F	F	S	A	Q	T	N	R
SERPINA5	G	T	I	F	T	F	R	S	A	R	L	N	S
SERPINA6	G	V	T	L	N	L	T	S	K	P	I	I	L
SERPINA7	E	V	E	L	S	D	Q	P	E	N	T	F	L
SERPINA8	Q	Q	L	N	K	P	E	V	L	E	V	T	L
SERPINA9	T	T	K	F	I	V	R	S	K	D	G	P	S
SERPINA10	L	S	E	I	T	A	Y	S	M	P	P	V	I
SERPINB1	A	G	I	A	T	F	C	M	L	M	P	E	E
SERPINB2	G	G	V	M	T	G	R	T	G	H	G	G	P
SERPINB3	A	V	V	G	F	G	S	S	P	T	S	T	N
SERPINB4	A	V	V	V	V	E	L	S	S	P	S	T	N
SERPINB5	G	A	R	I	L	Q	H	K	D	E	L	N	A
SERPINB6	A	A	I	M	M	M	R	C	A	R	F	V	P
SERPINB7	G	S	N	I	V	E	K	Q	L	P	Q	S	T
SERPINB8	A	V	V	R	N	S	R	C	S	R	M	E	P
SERPINB9	S	C	F	V	V	A	E	C	C	M	E	S	G
SERPINB10	G	S	E	I	D	I	R	I	R	V	P	S	I
SERPINB11	G	D	S	I	A	V	K	S	L	P	M	R	A
SERPINB12	G	A	V	V	S	E	R	S	L	R	S	W	V
SERPINB13	G	I	G	F	T	V	T	S	A	P	G	H	E
SERPINC1	A	V	V	I	A	G	R	S	L	N	P	N	R
SERPIND1	T	V	G	F	M	P	L	S	T	Q	V	R	F
SERPINE1	A	V	I	V	S	A	R	M	A	P	E	E	I
SERPINE2	T	A	I	L	I	A	R	S	S	P	P	W	F
SERPINF1	G	L	Q	P	A	H	L	T	F	P	L	D	Y
SERPINF2	S	I	A	M	S	R	M	S	L	S	S	F	S
SERPING1	A	I	S	V	A	R	T	L	L	V	F	E	V
SERPINH1	D	Q	D	I	Y	G	R	E	E	L	R	S	P
SERPINI1	G	M	I	A	I	S	R	M	A	V	L	Y	P
SERPINI2	G	I	H	I	P	V	I	M	S	L	A	Q	S

Cleavage presumably occurs between P1 and P1′ positions (highlighted in gray).

(b) Download the main PoPS program (JNLP/WebStart program) from http://pops.csse.monash.edu.au/pops-cgi/programs.php.

(c) Run the PoPS JNLP program. Once the program is downloaded and installed, you can simply run it by double-clicking the icon of the program. If it does not start, this probably means you have not installed Java. Go to http://www.java.com and click the "Free Java Download" link to initiate the download process. After Java is installed in your computer, you should be able to run the PoPS program locally.

(d) At the PoPS interface, click "load model from file" from the "model" menu and then select the "PoPS_input.txt" file. This will start to create a specificity model based on the input file. Repeatedly apply "Use these values" to all the P7–P6′ (corresponding to S7–S6′ site in PoPS) profiles. Finally, you can name the current model as "PoPS_granzyme_B," as shown in Fig. 12.7.

After the PoPS specificity model is created, the next two steps are to submit the substrate amino acids sequence and to predict substrate cleavage sites based on the built specificity model. Here, we use the RCL sequence of human SERPINB9 as the input and describe how to predict substrate cleavage using the PoPS program. SERPINB9 is the inhibitory serpin for human granzyme B and is known to contain a cleavage site for granzyme B in its RCL between E15 and C16 (Fig. 12.7).

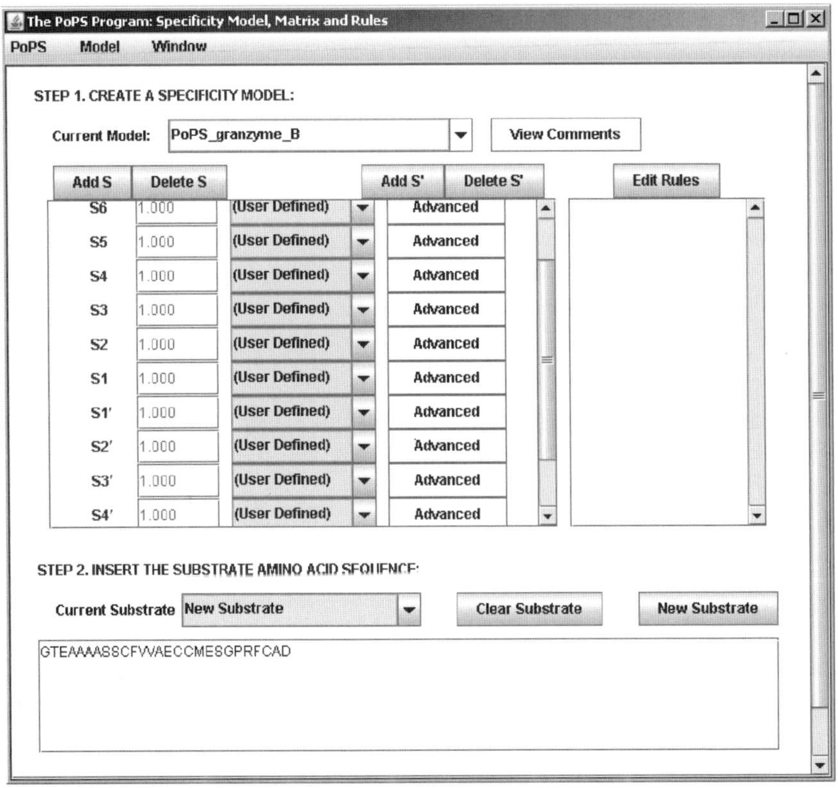

Figure 12.7 Creation of the PoPS substrate specificity model of human granzyme B. There are three steps for using PoPS to predict the substrate cleavage: (i) create a specificity model, (ii) insert the substrate amino acid sequence, and (iii) predict substrate cleavage.

Step 2: Insert the substrate amino acid sequence

(e) Input the amino acid sequence of the RCL region of human SERPINB9 in plain text in the substrate panel in PoPS, as shown in Fig. 12.7.

Step 3: Predict substrate cleavage

(f) Next, press the "Predict" button to submit the task. The prediction result will be displayed in both text and graphical formats in the results panel of PoPS, as shown in Fig. 12.8. In cases where there are many predicted cleavage sites generated, the threshold value can be increased so that some false positives with lower total scores are filtered out.

However, predicted cleavage sites using the preliminary PoPS model that was built based on the specificity data from phage display experiments may be biased or erroneous. Thus, it may be necessary to augment the PoPS

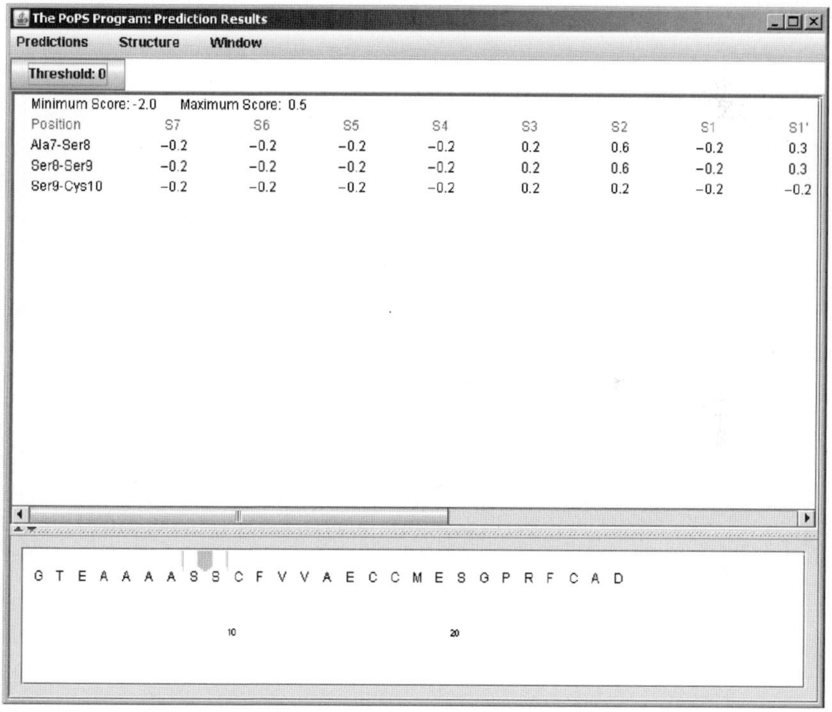
The PoPS Program: Prediction Results
Predictions Structure Window
Threshold: 0
Minimum Score: -2.0 Maximum Score: 0.5
Position S7 S6 S5 S4 S3 S2 S1 S1'
Ala7-Ser8 −0.2 −0.2 −0.2 −0.2 0.2 0.6 −0.2 0.3
Ser8-Ser9 −0.2 −0.2 −0.2 −0.2 0.2 0.6 −0.2 0.3
Ser9-Cys10 −0.2 −0.2 −0.2 −0.2 0.2 0.2 −0.2 −0.2
G T E A A A A S S C F V V A E C C M E S G P R F C A D
10
20

Figure 12.8 The prediction result of the substrate cleavage site of granzyme B for human SERPINB9 using the PoPS program. Thicker arrow indicates that the predicted cleavage site has a greater value for the score.

model using expert knowledge. For instance, it is known that granzyme B prefers to cleave after the acidic residue Asp but is also capable of cleaving after Glu. As the phage library was designed to utilize an anchored Asp residue at the P1 position, Glu is not represented at this subsite and will thus be assigned a negative value in the PoPS model generated from this data. Therefore, it is necessary to customize the P1 profile (S1 in PoPS) and augment the corresponding PoPS model by assigning a positive value (the weight) to P1 Glu (e.g., from −0.2 to 5.0) in order to take into account the relative importance of Glu at the P1 subsite. However, the three positively charged residues Arg, His, and Lys are given negative weights of −5.0, as they are very unlikely to be cleaved by human granzyme B. As a result, the modified PoPS model correctly predicts (i.e., the highest scoring event) the cleavage site of human granzyme B (VVAE|CC) in the RCL of SERPINB9 (Fig. 12.9).

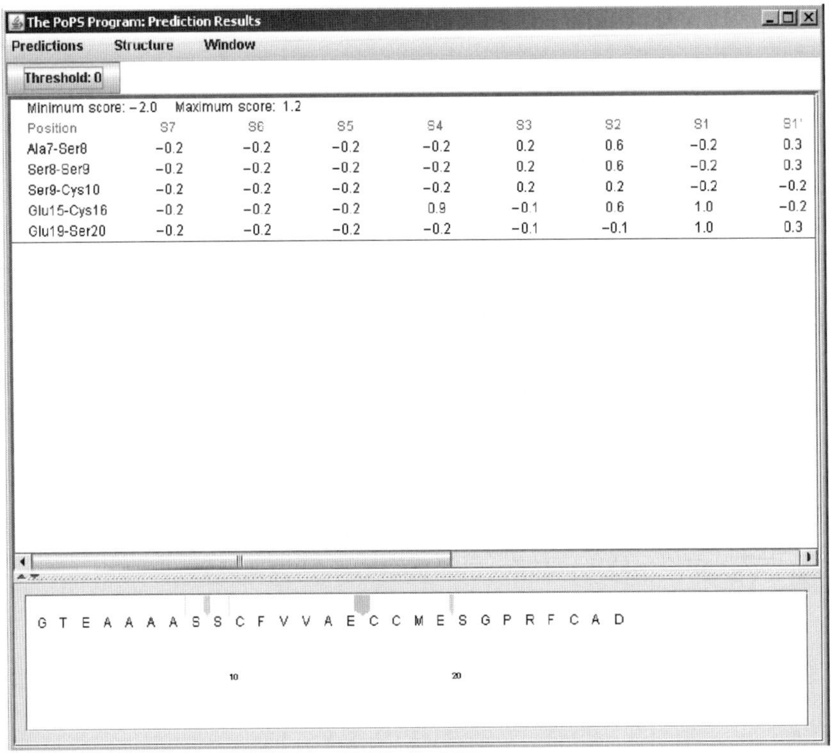
The PoPS Program: Prediction Results
Predictions Structure Window
Threshold: 0
Minimum score: −2.0 Maximum score: 1.2
Position S7 S6 S5 S4 S3 S2 S1 S1'
Ala7-Ser8 −0.2 −0.2 −0.2 −0.2 0.2 0.6 −0.2 0.3
Ser8-Ser9 −0.2 −0.2 −0.2 −0.2 0.2 0.6 −0.2 0.3
Ser9-Cys10 −0.2 −0.2 −0.2 −0.2 0.2 0.2 −0.2 −0.2
Glu15-Cys16 −0.2 −0.2 −0.2 0.9 −0.1 0.6 1.0 −0.2
Glu19-Ser20 −0.2 −0.2 −0.2 −0.2 −0.1 −0.1 1.0 0.3
G T E A A A A S S C F V V A E C C M E S G P R F C A D
10
20

Figure 12.9 The PoPS model can be further augmented using expert knowledge of the protease substrate specificity. The augmented PoPS model successfully predicted the cleavage site of human granzyme B in the RCL region of SERPINB9 between Glu 15 and Cys 16, as the top hit among five predicted cleavage sites.

Table 12.5 Top 5 ranked human serpins that are predicted to be human granzyme B substrates B by PoPS

Serpin	Location	Local sequence	Cleavage score	Ranking	Inhibitory likelihood
SERPINA10	E11-I12	VAGILSE|ITAYSM	6.2	1	Non
	Y15-S16	LSEITAY|SMPPVI	2.2	2	Likely
	I8-L9	TEAVAGI|LSEITA	0.3	3	Non
SERPINA8	E15-V16	QQLNKPE|VLEVTL	5.8	1	Likely
SERPINH1	E12-E13	QDIYGRE|ELRSPK	5.4	1	Unlikely
	D7-I8	GNPFDQD|IYGREE	4.5	2	Non
	E13-L14	DIYGREE|LRSPKL	3	3	Non
SERPINB10	E11-I12	AAGSGSE|IDIRIR	5.3	1	Unlikely
	D13-I14	GSGSEID|IRIRVP	4.3	2	Unlikely
	P19-S20	DIRIRVP|SIEFNA	1.9	3	Non
	G9-S10	EAAAGSG|SEIDIR	0.6	4	Non
	G7-S8	GTEAAAG|SGSEID	0.2	5	Non
SERPINB9	**E15-C16**	**SCFVVAE|CCMESG**	**5.2**	**1**	**Likely**
	E19-S20	VAECCME|SGPRFC	4.4	2	Non
	S8-S9	TEAAAAS|SCFVVA	0.5	3	Non
	A7-S8	GTEAAAA|SSCFVV	0.1	4	Non
	S9-C10	EAAAASS|CFVVAE	0.1	5	Non

Cleavage location between P1 and P1′ positions and the P7–P6′ sequences are given. The rankings are determined based on the cleavage score from PoPS. "|" indicates the cleavage site. The experimentally verified P1 residue in the cognate serpin of human granzyme B, SERPINB9, is shown in bold.

3.4. Inference of serpin inhibitory likelihood for a predicted cleavage site according to its PoPS cleavage score and relative distance to the theoretical P1 site within the serpin RCL region

Next, we apply the augmented PoPS model to predict the potential granzyme B cleavage sites of 33 human serpins (see Table 12.3 for the full list). Among them, 31 serpin sequences are predicted to contain cleavage sites by PoPS with cleavage scores above the threshold. Table 12.5 lists the top 5 ranked serpins that are predicted by PoPS to be cleaved by human granzyme B.

As the inhibition of a protease by serpin is critically dependent on the length of the RCL region of serpin (Zhou *et al.*, 2001), we can further infer the likelihood of a protease inhibition by serpin, for a predicted cleavage site, based on its PoPS cleavage probability score and its relative distance to the theoretical P1 site (Table 12.4). As can be seen in Fig. 12.10, in principal, there are four serpin inhibitory likelihoods: likely (predicted cleavage occurring in theoretical P1 site), possible (predicted cleavage

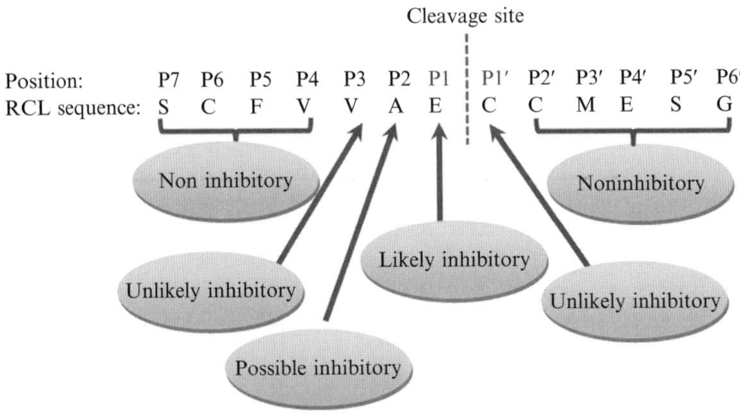

Figure 12.10 The serpin inhibitory likelihood can be inferred by taking into consideration the cleavage score from PoPS and the relative distance of a predicted cleavage site from the theoretical P1 position. The serpin inhibitory likelihood is categorized into four classes: likely inhibitory (predicted cleavage site is theoretical P1), possible inhibitory (predicted cleavage site is theoretical P2), unlikely inhibitory (predicted cleavage site is either theoretical P3 or P1′), and noninhibitory (predicted cleavage site is distant from the former three cases). See Table 12.5 for the list of theoretical or presumable cleavage sites of human serpins.

occurring in P2 site), unlikely (cleavage occurring in P3 or P1′ site), and noninhibitory (cleavage occurring in more distal regions). Note that the likelihood of protease inhibition by a particular serpin would generally increase if a predicted cleavage site is close to the theoretical P1 site (see Section 3.2). For the top ranking predicted serpin substrates of human granzyme B, their inhibitory likelihoods are provided in Table 12.5.

PoPS models that are refined based on the above procedures are expected to provide a more accurate prediction of the cleavage sites in the RCL regions of serpins. However, in cases where PoPS predicts a cleavage which is distant from the hypothetical P1 position in the RCL region (or elsewhere in the serpin molecule if the full primary sequence is used as input), three possible explanations exist.

First, when predicted sites fall outside the RCL, it must be remembered that PoPS determines its output primarily on the primary amino acid sequence input. Thus, while it is optional to use the PSIPRED algorithm (McGuffin *et al.*, 2000) to predict the secondary structure of an input sequence (and thus take into account the possibility that a particular predicted site may be buried), this is no substitute for expert user knowledge of three-dimensional serpin structure, which should be utilized to assess the relevance of any PoPS-predicted protease cleavage site. As the RCL is a surface exposed loop and therefore, more likely to be accessible to the protease than other more structured sites in the serpin molecule, predicted cleavage sites within the RCL region are potentially more significant than sites outside this region.

Second, although inhibitory serpin molecules function primarily to inhibit specific proteases, a cleavage within the RCL region might not result in inhibition of the participating protease due to kinetics of the interaction being unsuitable for capture of the protease by the serpin. In such a case, cleavage of the RCL by a particular protease might result instead in inactivation of the serpin, abrogating its inhibitory function. This seems more likely if the predicted cleavage site is not located very close to the hypothetical P1 position (see Fig. 12.10). For example, it has been shown that a metalloprotease from the Eastern diamond back rattlesnake (*Crotalus adamanteus*) is able to inactivate several human blood serpins by cleaving the RCL near the proximal hinge (Kress, 1986). Thus RCL cleavage that does not result in inhibition of the protease may represent a means for controlling the inhibitory action of the serpin.

Finally, although the P1 and surrounding residues are the primary determinants of serpin specificity, some serpins make use of exosite interactions to further narrow the specificity of their interactions with proteases (for review, see Gettins and Olson, 2009). Thus a predicted protease cleavage site in a particular serpin RCL which presents with a poor PoPS score may be overlooked as "background" without appreciating that an additional contact between protease and serpin exosite (or cofactor and serpin exosite) may render this interaction more favorable.

In summary, the predicted substrate cleavage using PoPS models and the cleavage score for each cleavage site can allow us to infer putative serpin partners of human granzyme B.

4. Concluding Remarks and Perspective

In-depth understanding of the substrate specificity of proteases is essential for obtaining deep insights into their function and is a prerequisite for the development of specific inhibitors to control their activities. Identification of the cognate serpin inhibitors of proteases is essential for the development of targeted therapies of protease-controlled pathogenesis. Substrate phage display is a powerful and invaluable tool for protease substrate specificity profiling and can aid in identification and selection of optimal substrates at the amino acid sequence level. In this chapter, we have described detailed methodology for characterizing the extended substrate specificity of human granzyme B using a phage display technique, building a substrate specificity model, and further applying this model to predict the cleavage sites of substrates (in this case, the RCL of human serpins). Our method provides a complementary approach to predict the human inhibitory serpins of granzyme B. This approach can be applied to other proteases to aid discovery of potential cognate serpins, or other more general substrates. It is our hope that the techniques described in our chapter can be extended to other problems in the field of protease systems biology.

ACKNOWLEDGMENTS

We thank the members of the Whisstock and Bird laboratories for support and discussion. This work was supported by grants from the National Health and Medical Research Council of Australia (NHMRC), the Australian Research Council (ARC), and the Chinese Academy of Sciences (CAS). J. S. is an NHMRC Peter Doherty Fellow and is also supported by the Hundred Talents Program of CAS. J. C. W. is an ARC Federation Fellow and an honorary NHMRC Principal Research Fellow.

REFERENCES

Barkan, D. T., Hostetter, D. R., Mahrus, S., Pieper, U., Wells, J. A., Craik, C. S., and Sali, A. (2010). Prediction of protease substrates using sequence and structure features. *Bioinformatics* **26,** 1714–1722.

Benarafa, C., and Remold-O'Donnell, E. (2005). The ovalbumin serpins revisited: Perspective from the chicken genome of clade B serpin evolution in vertebrates. *Proc. Natl. Acad. Sci. USA* **102,** 11367–11372.

Berman, H. M., Westbrook, J., Feng, Z., Gilliland, G., Bhat, T. N., Weissig, H., Shindyalov, I. N., and Bourne, P. E. (2000). The protein data bank. *Nucleic Acids Res.* **28,** 235–242.

Börner, S., and Ragg, H. (2008). Functional diversification of a protease inhibitor gene in the genus Drosophila and its molecular basis. *Gene* **415,** 23–31.

Boyd, S. E., Pike, R. N., Rudy, G. B., Whisstock, J. C., and Garcia de la Banda, M. (2005). PoPS: A computational tool for modeling and predicting protease specificity. *J. Bioinform. Comput. Biol.* **3,** 551–585.

Boyd, S. E., Kerr, F. K., Albrecht, D. W., de la Banda, M. G., Ng, N., and Pike, R. N. (2009). Cooperative effects in the substrate specificity of the complement protease C1s. *Biol. Chem.* **390,** 503–507.

Carrell, R. W., Evans, D. L. I., and Stein, P. E. (1991). Mobile reactive centre of serpins and the control of thrombosis. *Nature* **353,** 576–578.

Chen, C. T., Yang, E. W., Hsu, H. J., Sun, Y. K., Hsu, W. L., and Yang, A. S. (2008). Protease substrate site predictors derived from machine learning on multilevel substrate phage display data. *Bioinformatics* **24,** 2691–2697.

Cole, E. R., Koppel, J. L., and Olwin, J. H. (1967). Multiple specificity of thrombin for synthetic substrates. *Nature* **213,** 405–406.

Condron, B. G., Atkins, J. F., and Gesteland, R. F. (1991). Frameshifting in gene 10 of bacteriophage T7. *J. Bacteriol.* **173,** 6998–7003.

Coughlin, S. R. (2000). Thrombin signaling and protease-activated receptors. *Nature* **407,** 258–264.

Cullen, S. P., Adrain, C., Lüthi, A. U., Duriez, P. J., and Martin, S. J. (2007). Human and murine granzyme B exhibit divergent substrate preferences. *J. Cell Biol.* **176,** 435–444.

Cwirla, S. E., Peters, E. A., Barrett, R. W., and Dower, W. J. (1990). Peptides on phage: A vast library of peptides for identifying ligands. *Proc. Natl. Acad. Sci. USA* **87,** 6378–6382.

Deperthes, D. (2002). Phage display substrate: A blind method for determining protease specificity. *Biol. Chem.* **383,** 1107–1112.

Diamond, S. L. (2007). Methods for mapping protease specificity. *Curr. Opin. Chem. Biol.* **11,** 46–51.

Gettins, P. G., and Olson, S. T. (2009). Exosite determinants of serpin specificity. *J. Biol. Chem.* **284,** 20441–20445.

Huntington, J. A. (2006). Shape-shifting serpins—Advantages of a mobile mechanism. *Trends Biochem. Sci.* **31,** 427–435.

Huntington, J. A., Read, R. J., and Carrell, R. W. (2000). Structure of a serpin-protease complex shows inhibition by deformation. *Nature* **407,** 923–926.

Johnson, D. J., Li, W., Adams, T. E., and Huntington, J. A. (2006). Antithrombin-S195A factor Xa-heparin structure reveals the allosteric mechanism of antithrombin activation. *EMBO J.* **25,** 2029–2037.

Kaiserman, D., and Bird, P. I. (2005). Analysis of vertebrate genomes suggests a new model for clade B serpin evolution. *BMC Genomics* **6,** 167.

Kaiserman, D., and Bird, P. I. (2010). Control of granzymes by serpins. *Cell Death Differ.* **17,** 586–595.

Kaiserman, D., Bird, C. H., Sun, J., Matthews, A., Ung, K., Whisstock, J. C., Thompson, P. E., Trapani, J. A., and Bird, P. I. (2006a). The major human and mouse granzymes are structurally and functionally divergent. *J. Cell Biol.* **175,** 619–630.

Kaiserman, D., Whisstock, J. C., and Bird, P. I. (2006b). Mechanisms of serpin dysfunction in disease. *Expert Rev. Mol. Med.* **8,** 1–19.

Karlson, U., Pejler, G., Froman, G., and Hellman, L. (2002). Rat mast cell protease 4 is a beta-chymase with unusually stringent substrate recognition profile. *J. Biol. Chem.* **277,** 18579–18585.

Keijer, J., Ehrlich, H. J., Linders, M., Preissner, K. T., and Pannekoek, H. (1991). Vitronectin governs the interaction between plasminogen activator inhibitor 1 and tissue-type plasminogen activator. *J. Biol. Chem.* **266,** 10700–10707.

Kress, F. (1986). Inactivation of human plasma serine proteinase inhibitors (serpins) by limited proteolysis of the reactive site loop with snake venom and bacterial metalloproteinases. *J. Cell. Biochem.* **32,** 51–58.

Larkin, M. A., Blackshields, G., Brown, N. P., Chenna, R., McGettigan, P. A., McWilliam, H., Valentin, F., Wallace, I. M., Wilm, A., Lopez, R., Thompson, J. D., Gibson, T. J., *et al.* (2007). Clustal W and Clustal X version 2.0. *Bioinformatics* **23,** 2947–2948.

Law, R. H., Zhang, Q., McGowan, S., Buckle, A. M., Silverman, G. A., Wong, W., Rosado, C. J., Langendorf, C. G., Pike, R. N., Bird, P. I., and Whisstock, J. C. (2006). An overview of the serpin superfamily. *Genome Biol.* **7,** 216.

Li, W., Johnson, D. J., Esmon, C. T., and Huntington, J. A. (2004). Structure of the antithrombin-thrombin-heparin ternary complex reveals the antithrombotic mechanism of heparin. *Nat. Struct. Mol. Biol.* **11,** 857–862.

Li, W., Adams, T. E., Nangalia, J., Esmon, C. T., and Huntington, J. A. (2008). Molecular basis of thrombin recognition by protein C inhibitor revealed by the 1.6-A structure of the heparin-bridged complex. *Proc. Natl. Acad. Sci. USA* **105,** 4661–4666.

Matthews, D. J., Goodman, L. J., Gorman, C. M., and Wells, J. A. (1994). A survey of furin substrate specificity using substrate phage display. *Protein Sci.* **3,** 1197–1205.

McGuffin, L. J., Bryson, K., and Jones, D. T. (2000). The PSIPRED protein structure prediction server. *Bioinformatics* **16,** 404–405.

Ng, N. M., Quinsey, N. S., Matthews, A. Y., Kaiserman, D., Wijeyewickrema, L. C., Bird, P. I., Thompson, P. E., and Pike, R. N. (2009a). The effects of exosite occupancy on the substrate specificity of thrombin. *Arch. Biochem. Biophys.* **489,** 48–54.

Ng, N. M., Pike, R. N., and Boyd, S. E. (2009b). Subsite cooperativity in protease specificity. *Biol. Chem.* **390,** 401–407.

Odake, S., Kam, C. M., Narasimhan, L., Poe, M., Blake, J. T., Krahenbuhl, O., Tschopp, J., and Powers, J. C. (1991). Human and murine cytotoxic T lymphocyte serine proteases: Sub site mapping with peptide thioester substrates and inhibition of enzyme activity and cytolysis by isocoumarins. *Biochemistry* **30,** 2217–2227.

Olson, S. T., Björk, I., Sheffer, R., Craig, P. A., Shore, J. D., and Choay, J. (1992). Role of the antithrombin-binding pentasaccharide in heparin acceleration of antithrombin-proteinase reactions. Resolution of the antithrombin conformational change contribution to heparin rate enhancement. *J. Biol. Chem.* **267,** 12528–12538.

Owen, M. C., Brennan, S. O., Lewis, J. H., and Carrell, R. W. (1983). Mutation of antitrypsin to antithrombin. Alpha 1-antitrypsin Pittsburgh (358 Met leads to Arg), a fatal bleeding disorder. *N. Engl. J. Med.* **309,** 694–698.

Pike, R. N., Buckle, A. M., le Bonniec, B. F., and Church, F. C. (2005). Control of the coagulation system by serpins. Getting by with a little help from glycosaminoglycans. *FEBS J.* **272,** 4842–4851.

Roberts, T. H., and Hejgaard, J. (2008). Serpins in plants and green algae. *Funct. Integr. Genomics* **8,** 1–27.

Rotonda, J., Garcia-Calvo, M., Bull, H. G., Geissler, W. M., McKeever, B. M., Willoughby, C. A., Thornberry, N. A., and Becker, J. W. (2001). The three-dimensional structure of human granzyme B compared to caspase-3, key mediators of cell death with cleavage specificity for aspartic acid in P1. *Chem. Biol.* **8,** 357–368.

Ruggles, S. W., Fletterick, R. J., and Craik, C. S. (2004). Characterization of structural determinants of granzyme B reveals potent mediators of extended substrate specificity. *J. Biol. Chem.* **279,** 30751–30759.

Schechter, I., and Berger, A. (1967). On the size of the active site in proteases. I. Papain. *Biochem. Biophys. Res. Commun.* **27,** 157–162.

Silverman, G. A., Bird, P. I., Carrell, R. W., Church, F. C., Coughlin, P. B., Gettins, P. G., Irving, J. A., Lomas, D. A., Luke, C. J., Moyer, R. W., Pemberton, P. A., Remold-O'Donnell, E., *et al.* (2001). The serpins are an expanding superfamily of structurally similar but functionally diverse proteins. Evolution, mechanism of inhibition, novel functions, and a revised nomenclature. *J. Biol. Chem.* **276,** 33293–33296.

Silverman, G. A., Whisstock, J. C., Askew, D. J., Pak, S. C., Luke, C. J., Cataltepe, S., Irving, J. A., and Bird, P. I. (2004). Human clade B serpins (ov-serpins) belong to a cohort of evolutionarily dispersed intracellular proteinase inhibitor clades that protect cells from promiscuous proteolysis. *Cell. Mol. Life Sci.* **61,** 301–325.

Smith, G. P. (1985). Filamentous fusion phage: novel expression vectors that display cloned antigens on the virion surface. *Science* **228,** 1315–1317.

Song, J., Tan, H., Shen, H., Mahmood, K., Boyd, S. E., Webb, G. I., Akutsu, T., and Whisstock, J. C. (2010). Cascleave: Towards more accurate prediction of caspase substrate cleavage sites. *Bioinformatics* **26,** 752–760.

Song, J., Tan, H., Boyd, S. E., Shen, H., Mahmood, K., Webb, G. I., Akutsu, T., Whisstock, J. C., and Pike, R. N. (2011). Bioinformatic approaches for predicting substrates of proteases. *J. Bioinform. Comput. Biol.* **9,** 149–178.

Trapani, J. A., and Sutton, V. R. (2003). Granzyme B: Pro-apoptotic, antiviral and antitumor functions. *Curr. Opin. Immunol.* **15**(5), 533–543.

Van Damme, P., Maurer-Stroh, S., Plasman, K., Van Durme, J., Colaert, N., Timmerman, E., De Bock, P. J., Goethals, M., Rousseau, F., Schymkowitz, J., Vandekerckhove, J., and Gevaert, K. (2009). Analysis of protein processing by

N-terminal proteomics reveals novel species-specific substrate determinants of granzyme B orthologs. *Mol. Cell. Proteomics* **8,** 258–272.

Verspurten, J., Gevaert, K., Declercq, W., and Vandenabeele, P. (2009). SitePredicting the cleavage of proteinase substrates. *Trends Biochem. Sci.* **34,** 319–323.

Waterhouse, A. M., Procter, J. B., Martin, D. M. A., Clamp, M., and Barton, G. J. (2009). Jalview Version 2—A multiple sequence alignment editor and analysis workbench. *Bioinformatics* **25,** 1189–1191.

Waugh, S. M., Harris, J. L., Fletterick, R., and Craik, C. S. (2000). The structure of the pro-apoptotic protease granzyme B reveals the molecular determinants of its specificity. *Nat. Struct. Biol.* **7,** 762–765.

Wee, L. J., Tan, T. W., and Ranganathan, S. (2007). CASVM: Web server for SVM-based prediction of caspase substrates cleavage sites. *Bioinformatics* **23,** 3241–3243.

Wee, L. J., Tan, T. W., and Ranganathan, S. (2009). A multi-factor model for caspase degradome prediction. *BMC Genomics* **10**(Suppl. 3), S6.

Whisstock, J. C., and Bottomley, S. P. (2006). Molecular gymnastics: Serpin structure, folding and misfolding. *Curr. Opin. Struct. Biol.* **16,** 761–768.

Whisstock, J., Skinner, R., and Lesk, A. M. (1998). An atlas of serpin conformations. *Trends Biochem. Sci.* **23,** 63–67.

Whisstock, J. C., Silverman, G. A., Bird, P. I., Bottomley, S. P., Kaiserman, D., Luke, C. J., Pak, S. C., Reichhart, J. M., and Huntington, J. A. (2010). Serpins flex their muscle: Structural insights into target peptidase recognition, polymerization and transport functions. *J. Biol. Chem.* **285,** 24299–24305.

Yang, Z. R. (2005). Prediction of caspase cleavage sites using Bayesian bio-basis function neural networks. *Bioinformatics* **21,** 1831–1837.

Ye, S., Cech, A. L., Belmares, R., Bergstrom, R. C., Tong, Y., Corey, D. R., Kanost, M. R., and Goldsmith, E. J. (2001). The structure of a Michaelis serpin-protease complex. *Nat. Struct. Biol.* **8,** 979–983.

Zhou, A., Carrell, R. W., and Huntington, J. A. (2001). The serpin inhibitory mechanism is critically dependent on the length of the reactive center loop. *J. Biol. Chem.* **276,** 27541–27547.

CHAPTER THIRTEEN

Amino-Terminal Oriented Mass Spectrometry of Substrates (ATOMS): N-Terminal Sequencing of Proteins and Proteolytic Cleavage Sites by Quantitative Mass Spectrometry

Alain Doucet[1] *and* Christopher M. Overall[2]

Contents

1. Introduction 276
2. Overview of ATOMS 277
3. Control Experiment to Determine the Ratio Cutoff and Identify Natural N-Termini, Basal Proteolytic Products, and Outliers 281
4. Limited Proteolytic Processing of the Target Protein by the Test Protease *In Vitro* 282
5. Isotopic Labeling and Tryptic Digestion 283
6. Identification of Peptides by Liquid Chromatography-Tandem Mass Spectrometry 286
7. Mass Spectrometry Data Analysis 287
   7.1. Analysis of the control experiment and definition of the cutoff for elevated ratio data 287
   7.2. Identification of outliers, natural N-termini, and basal proteolysis products in the control experiment 289
   7.3. Identification of proteolytic cleavage sites 290
8. Discussion: Measuring the Effect of Protease Inhibitors on the Generation of Proteolytic Fragments 291
Acknowledgments 292
References 292

Centre for Blood Research, Life Sciences Institute, University of British Columbia, Vancouver, British Columbia, Canada
[1] Current address: Ottawa Institute of Systems Biology, University of Ottawa, Ottawa, Ontario, Canada
[2] Corresponding author.

Methods in Enzymology, Volume 501
ISSN 0076-6879, DOI: 10.1016/B978-0-12-385950-1.00013-4

Abstract

Edman degradation is a long-established technique for N-terminal sequencing of proteins and cleavage fragments. However, for accurate data analysis and amino acid assignments, Edman sequencing proceeds on samples of single proteins only and so lacks high-throughput capabilities. We describe a new method for the high-throughput determination of N-terminal sequences of multiple protein fragments in solution. Proteolytic processing can change the activity of bioactive proteins and also reveal cryptic binding sites and generate proteins with new functions (neoproteins) not found in the parent molecule. For example, extracellular matrix (ECM) protein processing often produces multiple proteolytic fragments with the generation of cryptic binding sites and neoproteins by ECM protein processing being well documented. The exact proteolytic cleavage sites need to be identified to fully understand the functions of the cleavage fragments and biological roles of proteases *in vivo*. However, the identification of cleavage sites in complex high molecular proteins such as those composing the ECM is not trivial. N-terminal microsequencing of proteolytic fragments is the usual method employed, but it suffers from poor resolution of sodium dodecylsulfate-polyacrylamide gel electrophoresis gels and is inefficient at identifying multiple cleavages, requiring preparation of numerous gels or membrane slices for analysis. We recently developed Amino-Terminal Oriented Mass spectrometry of Substrates (ATOMS) to overcome these limitations as a complement for N-terminal sequencing. ATOMS employs isotopic labeling and quantitative tandem mass spectrometry to identify cleavage sites in a fast and accurate manner. We successfully used ATOMS to identify nearly 100 cleavage sites in the ECM proteins laminin and fibronectin. Presented herein is the detailed step-by-step protocol for ATOMS.

1. Introduction

Identification of N-termini, whether natural or proteolytically generated, traditionally involves N-terminal sequencing of protein bands from SDS-PAGE (sodium dodecylsulfate-polyacrylamide gel electrophoresis) gels. This method suffers from the limited resolution of gels and its inability to analyze low molecular weight (< 4 Da) fragments. Moreover, N-terminal sequencing of proteolytic fragments must be performed separately for each fragment, which lowers throughput. To overcome these limitations and to increase the number of cleavage sites that can be identified in complex proteins, we developed and validated a new method named Amino-Terminal Oriented Mass spectrometry of Substrates (ATOMS) for the simultaneous identification of multiple N-terminal sequences of proteolytic sites in single complex proteins (Doucet and Overall, 2011).

Not only is the determination of the original mature protein desired but also the sequence of protease cleavage products is increasingly an area of importance. Processing of proteins can deeply modify their functions by switching activity (Cox *et al.*, 2008), exposing cryptic binding sites, and generating new proteins exhibiting functions not found in the intact protein, so called neoproteins (Giannelli *et al.*, 1997; Khan *et al.*, 2002; Morla *et al.*, 1994; Pozzi *et al.*, 2002; Rege *et al.*, 2005; Scapini *et al.*, 2002; Yi and Ruoslahti, 2001). Proteolytic processing of extracellular matrix (ECM) proteins such as laminins and fibronectins is known to reveal cryptic binding sites and neoproteins upon proteolysis (Cox *et al.*, 2008; Daudi *et al.*, 1991; Furie and Rifkin, 1980; Gold *et al.*, 1989; Goldfinger *et al.*, 1998; Horowitz *et al.*, 2008; Kulkarni *et al.*, 2008; Timpl *et al.*, 1979). The identification of the exact proteolytic cleavage sites is crucial to understand the biological functions of cryptic binding sites and neoproteins and their effects on cell behavior. However, the identification of the cleavage sites in these complex and high molecular weight proteins of the ECM has proven particularly difficult (Giannelli *et al.*, 1997; Horowitz *et al.*, 2008; Lambert Vidmar *et al.*, 1991; Rostagno *et al.*, 1989; Unger and Tschesche, 1999).

One application of ATOMS is for characterization of proteolytic fragments generated *in vitro* following incubation of the purified target protein with the protease of interest. ATOMS is then used to identify cleavage sites by liquid chromatography-tandem mass spectrometry. However, the method is also suitable for N-terminal sequencing of several intact proteins present in a sample. Using ATOMS, we recently identified 89 cleavage sites generated by serine and metalloproteinases in fibronectin-1 and laminin-1 (Doucet and Overall, 2011). We also identified membrane type-1 matrix metalloproteinase (MT1-MMP) cleavage sites in the versican G3 domain with ATOMS (Robert C. R. *et al.*, manuscript in preparation). In this chapter, we present a detailed protocol for performing ATOMS useful for the identification of N-termini of *in vitro*-generated proteolytic cleavage sites in complex high molecular weight proteins.

2. Overview of ATOMS

In ATOMS, N-terminal peptides of intact protein and proteolytic cleavage sites generated *in vitro* are identified by liquid chromatography-quantitative tandem mass spectrometry. *Bona fide* proteolytic sites produced by the protease under study are distinguished from natural N-termini and degradation products generated during the protein isolation/purification by comparing a protease-treated sample and a control sample (similar sample not treated with the protease) (Fig. 13.1A). Proteins and their proteolytic fragments are then denatured, reduced and alkylated, and their N-termini and lysine side chains are labeled by reductive dimethylation (Fig. 13.1B). The protease-treated and control samples are reacted with the heavy ($^{13}CD_2O$) (full circles) and light ($^{12}CH_2O$) (dashed circles) formaldehyde,

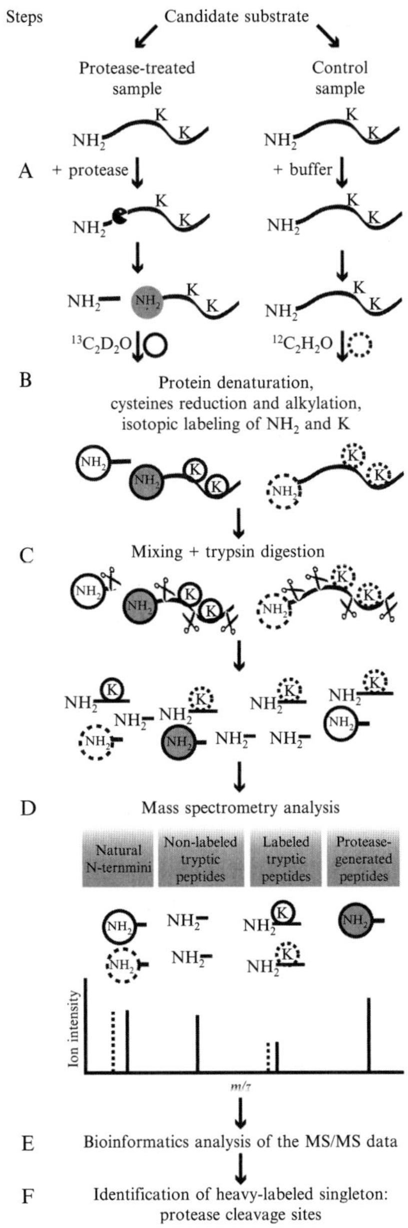

Figure 13.1 ATOMS workflow. (A) The target protein is divided in two identical samples. One sample is exposed to the protease under study (pacman) and the other sample is incubated with buffer only (control). The protease treatment generates proteolytic fragments with N-terminal peptides not found in the control aliquot (gray-shaded N-terminus). (B) The proteins from both samples are denatured, and

respectively (Fig. 13.1B). This introduces a 6-Da mass difference for each labeled residue between proteins from the two samples and allows for relative quantification of the labeled proteins. The two protein aliquots are combined and trypsinized (Fig. 13.1C). Dimethylation of lysines prevent trypsin from cleaving after this residue. Trypsin cleaves therefore only after arginines and presents the cleavage specificity of the endoprotease ArgC. The peptide solution is analyzed by quantitative liquid chromatography-tandem mass spectrometry (Fig. 13.1D). Natural N-termini, basal proteolytic peptides, and labeled internal tryptic peptides are represented by peak pairs of similar intensity in the MS1 spectra and separated by the mass difference of the labels. Protease-generated N-terminal peptides are only found in the protease-treated sample and are represented by heavy-labeled singletons (Fig. 13.1D). Bioinformatics analysis of the mass spectrometry data identifies and quantifies peptides (Fig. 13.1E), which results in the identification of the proteolytic cleavage sites (Fig. 13.1F). A detailed schematic of the bioinformatic analysis is depicted in Fig. 13.2.

ATOMS uses mass spectrometry, which is a very sensitive analytical method that requires some precautions to obtain optimal results. Any contamination of the sample by other proteins will be detected by the mass spectrometer. Wearing powder-free gloves and lab coat throughout the experiment will reduce keratin contamination, the most common protein contamination in mass spectrometry. Mass spectrometry samples are also plagued by chemical contaminations; in particular, plasticizers extracted from microcentrifuge tubes by organic solvents are very problematic. Always use microtubes resistant to organic solvents to avoid this problem (we suggest using Eppendorf brand microtubes). Also, expose

---

their cysteines are reduced and alkylated. The primary amine groups (N-termini and lysine side chains) of the protease-treated sample and the control samples are labeled with heavy (full circles) or light (dashed circles) formaldehyde, respectively. The differential labeling of the two samples results in a mass difference of 6 Da/label for similar peptides labeled with heavy versus light formaldehyde. (C) The two samples are combined and digested with trypsin (scissors), which cleaves only after arginine residues due to the lysine dimethylation. (D) This peptide mixture is analyzed by liquid chromatography-tandem mass spectrometry (LC-MS/MS). Four different types of peptides are observed: (i) protein natural N-termini and basal proteolysis products are represented by doublets of equal intensity separated by the mass difference induced by the isotope labeling in the MS1 spectra; (ii) tryptic peptides without lysine in their sequence are represented as unlabeled singletons; (iii) internal tryptic peptides with one or more lysines in their sequence and found in equal amounts in the protease- and buffer-treated sample are represented by doublets of the same intensity; (iv) protease-generated peptides are found only in the protease-treated samples and are represented by heavy-labeled singletons in the MS1 spectra, which is used as a signature to identify proteolytic cleavage sites. (E) The LC-MS/MS data are analyzed with the Mascot database search engine and ASAPRatio software tool. (F) The position of the heavy-labeled singletons in the target protein reveals the position of cleavage sites.

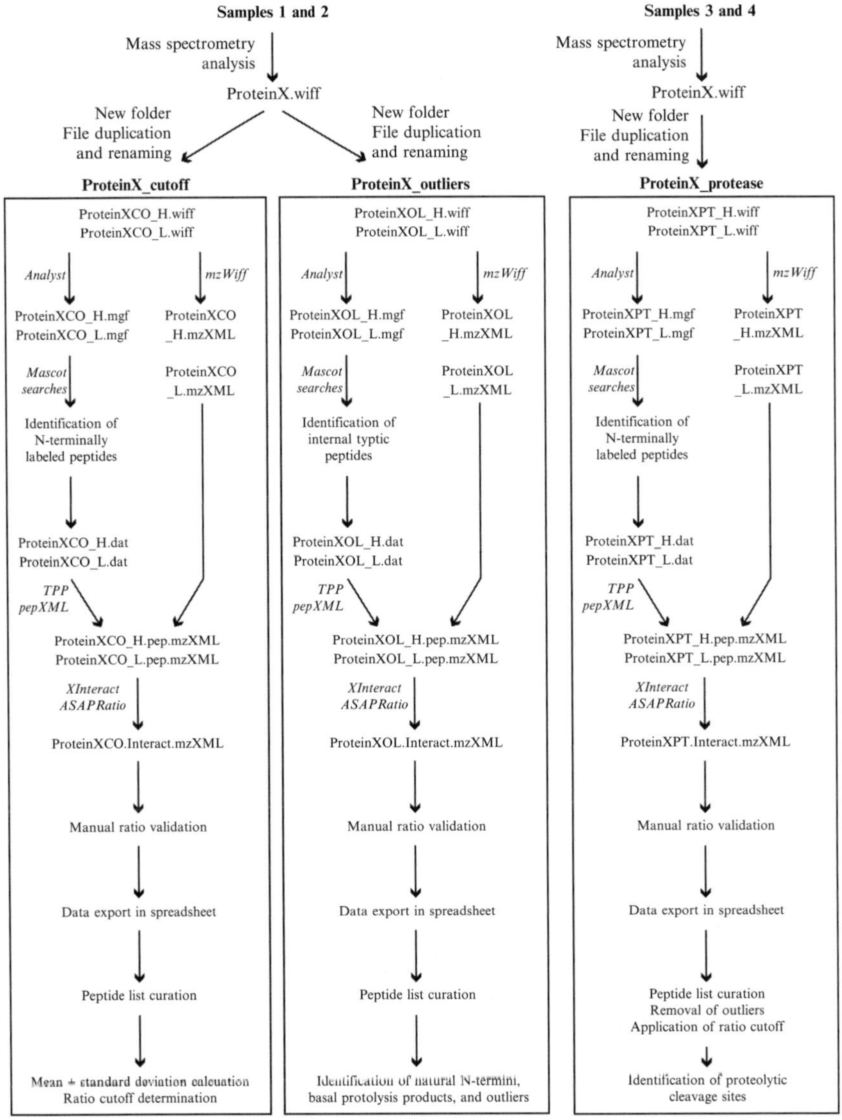

Figure 13.2 Bioinformatics analysis workflow. The peptide solution from combined Samples 1 and 2 (identical samples, none treated with the protease) is analyzed by LC-MS/MS on a QStar instrument, generating a .wiff file named ProteinX.wiff in the schematic (top left). This sample is used to determine the ratio cutoff (left column) and to identify the target protein natural N-termini, basal proteolysis products, and outliers (center column). The LC-MS/MS data from Samples 3 and 4 are used to identify proteolytic cleavage sites. The original .wiff files for the three analyses are copied, duplicated, renamed, and saved in three separate folder represented by three rectangles with their names in bold font at the top of each rectangle. For each analysis, the .wiff

the microcentrifuge tubes to organic solvent for the minimum time required and never heat microtubes with organic solvent. Always store organic solvent stocks in glass containers. Mass spectrometry is also incompatible with detergents, especially SDS, which prevent peptide ionization. Thus, remove any detergents from the target protein and protease stocks before performing ATOMS. Finally, polymeric molecules can also affect peptide ionization and detection by the mass spectrometer. Never concentrate protein solutions by dialysis against solid polyethylene glycol, as this results in a massive contamination of the samples with polyethylene glycol polymers and greatly decreases mass spectrometry data quality.

3. Control Experiment to Determine the Ratio Cutoff and Identify Natural N-Termini, Basal Proteolytic Products, and Outliers

As described in Fig. 13.1, proteolytic fragments will be found only in the protease-treated sample and not in the control sample. Thus, these fragments will show an elevated isotopic ratio in the quantitative mass spectrometry analyses. A ratio cutoff should be defined to discriminate between N-terminal peptides generated by the protease of interest from natural N-termini impurities, and basal proteolytic products. Basal proteolysis refers to any proteolytic processing occurring during the target protein expression and purification and before the ATOMS experiments. Also, in quantitative analyses, a few peptides not generated by the test protease will have a high ratio due to low ion intensity or misidentification by the database search engine. Such outliers should be identified and excluded; otherwise, they might be incorrectly interpreted as

---

files are converted to peak lists with the Analyst software (Applied Biosystems) (.mgf files) and to .mzXML files with the mzWiff software (software names are in italic next to the corresponding arrow). The .mgf files are then used to perform the database searches with Mascot, and the results are stored in .dat files. The .dat files and .mzXML files are used by the Trans-Proteomic Pipeline (TPP) to generate two peptide lists (.pep.mzXML files) for each analysis. The XInteract and ASAPRatio tools from the TPP are used to combine the two peptide lists and quantify peptide pairs for each analysis, resulting in one .XInteract.mzXML file for each analysis. The quantification data (ratios) are manually validated for the peptides associated to the target protein. The data are then exported in a spreadsheet and the peptides identified with $>95\%$ confidence and belonging to the target protein are selected. The mean ratio and standard deviation are calculated for the labeled internal peptides (left column) and are used to determine the peptides with a ratio significantly deviating from 1. Natural N-termini, basal proteolytic products, and N-terminal peptides with ratios significantly deviating from 1 are considered as outliers in this sample. Outliers are removed from the peptides in the spreadsheet from Samples 3 and 4 (right column), and the ratio cutoff (mean $\pm$ 3 SD) is used to identify protease-generated peptides and, thus, cleavage sites.

cleavage sites. The ratio cutoff, natural N-termini, impurities basal proteolytic products, and outliers are identified by using two identical samples that are analyzed as described in the workflow shown in Fig. 13.1, but no protease is added to either of the two samples. Note that many ECM proteins are isolated from cells in culture, and these cells are usually treated with trypsin when passaged. Not surprisingly, some tryptic degradation products may be found in the target protein preparation.

1. Prepare two 25-μg aliquots of the target protein (referred to as *Sample 1* and *Sample 2*) in 50 m*M* HEPES (4-(2-hydroxyethyl)-1-piperazineethanesulfonic acid) buffer or another suitable buffer (see Section 4, step 2 for details).
2. Incubate for 16 h at 37 °C.
3. Proceed to step 5 with these samples.

4. Limited Proteolytic Processing of the Target Protein by the Test Protease *In Vitro*

At this step, the target protein is incubated with the protease of interest, which will generate proteolytic fragments. An identical sample of the target protein is also incubated without the protease in the same conditions and this constitutes the control sample. The concentration of the target protein will vary greatly depending on its availability and source, but we recommend using target protein concentrations between 0.1 and 1 mg/ml. This concentration range minimizes sample loss associated with low protein concentration and aggregation/precipitation due to high protein concentration. The amount of protease to add will also vary greatly depending on efficiency of cleavage and its availability. Ideally, concentrations similar to *in vivo* situations should be used, but we are aware that the *in vivo* concentration of most proteases is unknown and also complicated by segregation at the cell surface or cellular compartments. A solution to define adequate protease concentration is presented in step 1. If protease concentration is already known, proceed directly to step 2.

1. This optional step will help define the protease concentration resulting in partial cleavage of the test protein, as is usually observed *in vivo*. The digestion should result in the generation of proteolytic fragments clearly visible on SDS-PAGE with a portion of the target protein still intact. We suggest performing small-scale assays using 1 μg of target protein and varying the molar ratio of protease-to-target protein from 1:10 to 1:1000 using the digestion conditions described in steps 2–5. Analyze the protein digests by SDS-PAGE, choose the appropriate protease concentration, and proceed to step 2.

2. Prepare two 25-μg aliquots of the target protein in 50 m*M* HEPES buffer with pH for optimal activity of the test protease. Other amine-free buffers can also be used. Absolutely avoid using chemicals with free amine groups such as Tris, ammonium bicarbonate buffer, and free amino acids (see Section 5, step 8). Avoid adding detergent unless it is absolutely required to maintain protein solubility. Sodium chloride and other amine-free salts can be used, but strive to maintain a low molarity.
3. Add your test protease to one aliquot. This sample will be referred to as *Sample 3*. The volume added should not dilute the test protein below 0.1 mg/ml. Note that some proteases such as MMPs need to be activated and their activity level verified prior to their addition to the test protein. In addition, removal of the activating agent is suggested to avoid activation of any contaminant proteases that might be present in the substrate sample.
4. Add the same volume of solution to the other target protein aliquot, referred to as *Sample 4*. This solution should have the same composition as the test protease, but without the protease.
5. Incubate both samples for 16 h at 37 °C.
6. Analyze 3 μg of protein from both samples by SDS-PAGE and reveal the target protein and proteolytic fragments with silver stain to confirm partial proteolysis of the test sample. Also include 3 μg of fresh target protein in this analysis. The electrophoretic pattern of the fresh target protein solution and Sample 4 should be similar. This will control for contamination of the target protein solution with unrelated active protease.
7. (Optional) Separate 10 μg of protein by SDS-PAGE and transfer on a PVDF membrane. Stain the membrane with freshly prepared Coomassie blue. Isolate and analyze bands from proteolytic fragments by N-terminal sequencing to complement the cleavage sites identified by ATOMS and to increase coverage.
8. Proceed to Section 5 with the remaining sample solutions.

5. Isotopic Labeling and Tryptic Digestion

This step will results in the denaturation, reduction of the disulfide bonds, alkylation of the cysteines, and the isotopic labeling of natural and protease-generated N-termini (Fig. 13.1B). Lysines side chains will also be labeled. The isotopic labeling is performed by reductive dimethylation of primary amine groups with "light" ($^{12}CH_2O$) and "heavy" ($^{13}CD_2O$) formaldehyde and catalyzed by the reductive agent sodium cyanoborohydride ($NaBH_3CN$) (Hsu *et al.*, 2003).

1. Add one sample volume of 8.0 *M* GuHCl to each sample (4 *M* GuHCl final concentration) to denature the proteins. Do not use urea as this will modify amino acids and prevents identification of the modified peptides by mass spectrometry.
2. Adjust the pH of the samples to 7.0 using 100 m*M* NaOH or HCl.
3. Add 5 m*M* DTT (final concentration) (1.0 *M* stock solution) to reduce disulfide bridges.
4. Incubate the samples at 65 °C for 1 h.
5. Cool the samples to room temperature.
6. Add 15 m*M* iodoacetamide (final concentration) (0.5 *M* stock solution freshly prepared) to alkylate the cysteine side chains.
7. Incubate the samples at room temperature, in the dark for 30 min.
8. Add DTT 30 m*M* (final concentration) to quench excess iodoacetamide.
9. Incubate the samples at room temperature for 30 min.
10. The following reaction should be performed in a chemical fume hood as it uses formaldehyde (a known carcinogen) and releases toxic cyanide gas from $NaBH_3CN$. Add heavy formaldehyde (formaldehyde containing the isotope ^{13}C and deuterium ($^{13}C^2D_2O$ from Cambridge Isotope Laboratories, Inc., 6.6 *M* stock solution)) to a final concentration of 40 m*M* to Samples 1 and 3. Add light formaldehyde (regular formaldehyde ($^{12}C^1H_2O$ from Sigma, 12.3 *M* stock solution)) to a final concentration of 40 m*M* to Samples 2 and 4 (formaldehyde concentration may vary depending on the sample, see below). NB: formaldehyde reacts with primary amines and many commercially available proteins are prepared in Tris buffer, which is a primary amine-containing molecule. In such samples, formaldehyde will preferentially react with the Tris molecules (m*M* concentration) and not with protein N-termini and lysines (n*M*-μ*M* concentration). This leads to incomplete isotopic labeling of the protein. Tris buffer from the protein solutions must be exchanged with an amine-free buffer such as HEPES prior to the experiment. If not possible, use a 40 m*M* excess of formaldehyde over the amine concentration. We observed labeling efficiencies of 96–100% with this strategy (Doucet and Overall, 2011).
11. Add $NaBH_3CN$ to a final concentration of 20 m*M* to all samples (or half of the total formaldehyde concentration if higher formaldehyde concentrations were used). $NaBH_3CN$ is toxic and we recommend purchasing this product already in solution (1 *M* solution is available from Sterogene Bioseparations (ALD reagent)).
12. Vortex the samples and adjust the pH to between 6 and 7.
13. Incubate the samples at 37 °C for 4 h. Overnight incubation is suggested.

14. Add 100 mM ammonium bicarbonate (final concentration) to quench the excess formaldehyde. This step is mandatory to avoid labeling of tryptic peptides in subsequent steps.
15. Adjust the pH between 6 and 7.
16. Incubate at 37 °C for 4 h.
17. Combine Samples 1 and 2.
18. Combine Samples 3 and 4.
19. Withdraw 1 μg of protein from each sample ("before precipitation" samples) and store at −20 °C.
20. Precipitate the remaining labeled proteins by addition of eight sample volumes of cold (−20 °C) acetone and one sample volume of cold (−20 °C) methanol. Acetone and methanol should be stored in glass containers.
21. Vortex and incubate the samples at −80 °C for 1 h.
22. Centrifuge the samples at 14,000×g at 4 °C for 10 min and discard the supernatant.
23. Add 1 ml of cold methanol to each tube, resuspend the pellet by vortexing vigorously, centrifuge, and discard the supernatants.
24. Repeat step 21 once.
25. Air-dry the samples.
26. Resuspend each dried protein pellet with 10 μl of 8.0 M GuHCl and vortex as necessary. If the pellets are not completely dissolved, add the minimum volume of 8.0 M GuHCl required to completely resolubilize the proteins.
27. Add 90 μl (or nine sample volumes) of 50 mM HEPES, pH 8.0 to each sample. This dilutes the GuHCl and permits trypsin activity in step 30.
28. Withdraw 1 μg of protein from each sample ("after precipitation" samples) and store at −20 °C.
29. Adjust the pH to 8.0 if required.
30. Add 1 μg of mass spectrometry grade trypsin.
31. Incubate overnight at 37 °C.
32. The next morning, withdraw 1 μg of protein from each sample ("after digestion" samples).
33. Analyze the "before precipitation," "after precipitation," and "after digestion" samples by SDS-PAGE. Stain the gel with silver nitrate and evaluate the precipitation and resolubilization efficiency by comparing the electrophoretic profiles of samples before and after acetone precipitation. Verify the completeness of the tryptic digestion. No band of molecular weight greater than 10 kDa should be visible after tryptic digestion (except the band of trypsin itself). If such bands persist, add 1 μg of trypsin to the sample and digest for 6 h. Verify the digestion on SDS-PAGE.

6. Identification of Peptides by Liquid Chromatography-Tandem Mass Spectrometry

We describe now the desalting of samples and identification of peptides by liquid chromatography-tandem mass spectrometry using a quadrupole-time-of-flight QStar instrument (Applied Biosystems; MDS-Sciex). Obviously, other types of tandem mass spectrometer can also be used for this task. Database search parameters such as precursor and fragment ion mass tolerance (Section 7) have to be modified in accordance with the instrument used. If the samples are to be shipped for analysis to a mass spectrometry core facility, you must contact the facility and enquire about their requirements for sample preparation prior to shipment.

1. Reduce the pH of the samples to 2.5 with formic acid.
2. Prepare a C_{18} Omix tip (Variant Inc.) according to the manufacturer recommendations.
3. Load the sample on the C_{18} resin of the Omix tip.
4. Wash the sample three times with 100 μl of 0.1% formic acid solution.
5. Elute the peptides with 100 μl of 80% acetonitrile, 0.1% formic acid solution in a new microtube.
6. Quickly evaporate the acetonitrile under vacuum. Do not dry the sample completely. Do not apply heat to the samples as this favors the release of plasticizers in the sample.
7. Resuspend the sample in 10 μl of 0.1% formic acid in water. Sample loss of around 30–50% is experienced during the sample desalting step.
8. Load 1 μg of peptide on a PepMap C_{18} reverse-phase chromatographic column (150 mm × 75 μm internal diameter) with buffer A (2% acetonitrile, 0.1% formic acid in water) using a nano-HPLC in-line with the mass spectrometer.
9. Elute the peptides from the reverse-phase column for 60 min at 200 nl/min with a linear gradient of 0–40% buffer B (80% acetonitrile, 0.1% formic acid in water) and inject the peptides directly into the mass spectrometer by electrospray ionization.
10. Operate the mass spectrometer in positive, data-dependant acquisition mode with a selected mass range of 400–1200. Peptides with +2 to +4 charge states are selected for tandem mass spectrometry, and the two most abundant peptides above a threshold of 30 counts are selected for MS/MS and dynamically excluded for 60 s with a 100 mmu mass tolerance.
11. The data of each mass spectrometry analysis are stored in a separate file (.wiff extension).

7. Mass Spectrometry Data Analysis

This section describes the bioinformatics analysis workflow for quantitative mass spectrometry data acquired with a QStar instrument as depicted in Fig. 13.2. The use of other mass spectrometers might require a different workflow. The Mascot software (Matrix Science) is used for the database search, and the Automated Statistical Analysis on Protein Ratio (ASAPRatio) tool (Li *et al.*, 2003) from the Trans-Proteomic Pipeline (TPP) developed at the Institute for Systems Biology in Seattle is used for peptide-relative quantification (Keller *et al.*, 2005). The TPP software requires the mass spectrometry files to be placed in specific folders. The details regarding the use of the TPP software is out of the scope of this chapter. Readers are referred to the TPP Wiki (http://tools.proteomecenter.org/wiki/index.php?title=Software:TPP) and the Institute for Systems Biology of Seattle website (http://www.proteomecenter.org/software.php) for details.

7.1. Analysis of the control experiment and definition of the cutoff for elevated ratio data

A ratio cutoff must be determined to distinguish protease-generated peptides with elevated abundance in the protease-treated sample from natural N-termini and N-termini generated by basal proteolysis. In this section, we calculate the mean quantification value and the standard deviation (SD) for internal tryptic peptides with dimethylated lysines not affected by proteolysis (Fig. 13.2, left column) for the mass spectrometry data from Samples 1 and 2. The values will be used to determine a cutoff value that defines the ratio at which a peptide is considered of elevated abundance in the experiments where the target protein is exposed to your protease of interest.

1. Copy the .wiff files from the experiments performed in Section 4 to a new folder (Fig. 13.2, left column) (keep the original .wiff files in a separate folder). This folder is named ProteinX_cutoff and is in bold in Fig. 13.2, and it is represented by the left rectangle. Duplicate the copied .wiff file and add "CO" (for cutoff) and H and L (for heavy and light label) at the end of their name. For example, the file named ProteinX.wiff becomes ProteinXCO_H.wiff and ProteinXCO_L.wiff in Fig. 13.2.
2. Generate the peak list files using the Mascot.dll script of the Analyst software (Applied Biosystems). These files have the extension ".mgf" (ProteinXCO_H.mgf and ProteinXCO_L.mgf). Save these files in the ProteinX_cutoff folder.

3. Perform database searches with Mascot (Matrix Science) using the .mgf files. Two independent searches are performed, one for heavy- and one for light-labeled peptides. For both searches, apply these parameters: fixed carboxymethylation of cysteines (+57.0214), variable modification for methionine oxidation (+15.9949), precursor and fragment ion mass tolerance of 0.4 Da, three miscleavages allowed, and enzyme specificity: ArgC and select the appropriate database (human, mouse, etc.) to perform the search. Search parameter specific for the heavy-labeled peptides is the modification of lysines with heavy formaldehyde (+34.0631) and for the light-labeled peptides, the modification of lysines with light formaldehyde (+28.0311). Verify the search results. The target protein should be identified and present the highest score. Also, note the peptide ion score resulting in 95% confidence identification at the peptide level. This score varies depending on the search parameters and the database used. These searches will generate two files (one for the heavy and one for the light label search) with the extension ".dat". Save these files (ProteinXCO_H.dat and ProteinXCO_L.dat) in the ProteinX_cutoff folder.
4. In parallel, convert the .wiff files to .mzXML files with the mzWiff software (available for free download at http://sourceforge.net/projects/sashimi/files/) using the default parameters and save the .mzXML files (ProteinXCO_H.mzXML and ProteinXCO_L.mzXML) in the ProteinX_cutoff folder.
5. In the Petunia interface of the TPP, use the .dat files as input and generate one pepXML file for each .dat file.
6. In the TPP, combine the two pepXML files with the XInteract tool. Include all peptides with a minimum of five amino acids and do not filter out peptides with low probability. Quantify the peptides with the ASAPRatio tool. Select the static modification quantification option, change the labeled residues to lysine (K), and quantify only the charge state where the collision-induced dissociation was acquired. Note that the peptide probability values calculated by the TPP cannot be used to measure the confidence at the identification level for this type of sample.
7. Open the resulting file, sort the list by protein identity, and find the peptides associated to your target protein and manually verify the quantification data for peptides with high, low, and nonavailable ratios. Correct the ratios if needed.
8. Include all useful columns and export the data in a spreadsheet. This file will be automatically saved in the ProteinX_cutoff folder.
9. Open the spreadsheet with Microsoft Excel (or equivalent software) and remove all the peptides not associated to your target protein and protease of interest.

10. Keep only the peptides with a Mascot ion score associated with >95% confidence in the peptide identification (as noted in Section 7, step 3).
11. For standardization purpose, our laboratory always expresses quantification ratios as protease/control, which is heavy-labeled/light-labeled (H/L) peptides in this protocol. ASAPRatio tool expresses the ratios as L/H. Thus, the ratios in the spreadsheet must be inverted (1/(L/H)).
12. Calculate the mean H/L ratio and the SD and note the values. A cutoff of the mean ± 3 SD will be used later to identify proteolytic cleavage sites.

7.2. Identification of outliers, natural N-termini, and basal proteolysis products in the control experiment

This analysis also uses Samples 1 and 3 of Section 3 and will identify outliers, natural N-termini, and basal proteolysis products found in the protein preparation and not related to the protease under study (Fig. 13.2, middle column). In this section, only the peptides labeled at their N-termini are identified; internal tryptic peptides are ignored by the database search engine due to the search parameters. Samples 3 and 4 are identical, and thus, all peptide pairs should present a heavy/light ratio centered to 1. This is the case for natural N-termini and the basal proteolysis products. However, peptides with high and low ratios are also identified and considered as outliers. These should be identified and eliminated from other datasets.

1. Copy the .wiff files from the experiments performed in Section 3 to a new folder. This folder is named ProteinX_outliers and is in bold in Fig. 13.2, and it is represented by the middle rectangle. Duplicate the copied .wiff file and add OL (for outliers) and H and L as previously (ProteinXOL_H.wiff and ProteinXOL_L.wiff) (Fig. 13.2).
2. Generate the peak list files (ProteinXOL_H.mgf and ProteinXOL_L.mgf) and save it in the ProteinX_outliers folder.
3. Two independent searches are performed, one for heavy- and one for light-labeled peptides. For both searches, apply the parameters described before, except for these: search parameters are the fixed modification of N-termini (n) and lysines (K) with heavy formaldehyde (+34.0631) or light formaldehyde (+28.0311) for the heavy- and light-labeled samples, respectively. Verify the search results and note the peptide ion score associated with >95% confidence identification. Save the .dat files (ProteinXOL_H.dat and ProteinXOL_L.dat) in the ProteinX_outliers folder. NB: for natural N-termini identification, other database searches can be performed by specifying N-terminal modifications expected for the target protein in the search parameters. This can be N-terminal

acetylation for intracellular proteins, conversion of glutamine to pyro-glutamate for proteins with N-terminal glutamine, etc.

4. In parallel, convert the .wiff files to .mzXML files (ProteinXOL_H.mzXML and ProteinXOL_L.mzXML and save it in the ProteinX_outliers folder).
5. Generate the pepXML files using the .dat files.
6. Combine the two pepXML files with the XInteract tool and quantify with the ASAPRatio tool as previously with the following modification: in the static modification quantification option, change labeled residues to lysine (K) and N-termini (n).
7. Process the data as before, export the spreadsheet, and keep the peptides from the target protein identified with >95% confidence.
8. The peptides with a heavy/light ratio close to 1 are natural N-terminus and basal proteolysis products not associated with the protease under study. Peptides with ratio equal to the mean ± 3 SD as determined in Section 7.1 are outliers.

7.3. Identification of proteolytic cleavage sites

This step allows the identification of proteolytic cleavage sites by using the information obtained in Sections 7.1 and 7.2 and the analysis of the mass spectrometry data of Samples 3 and 4 from Section 4.

1. Copy the .wiff files from the experiments performed in Section 4 to a new folder. This folder is named ProteinX_protease and is in bold in Fig. 13.2, and it is represented by the right rectangle. Duplicate the copied .wiff file and add PT (PT for protease) and H and L (for heavy and light) at the end of the file name (ProteinXPT_H.wiff and ProteinXPT_L.wiff) and save it in the ProteinX_protease folder (see Fig. 13.2).
2. Generate the peak list files (ProteinXPT_H.mgf and ProteinXPT_L.mgf saved in ProteinX_protease folder).
3. Perform Mascot database searches with the search parameters of Section 7.2. Save ProteinXPT_H.dat and ProteinXPT_L.dat files in the ProteinX_protease folder.
4. Convert the .wiff files to .mzXML files (ProteinXPT_H.mzXML and ProteinXPT_L.mzXML in the ProteinX_protease folder).
5. Generate the pepXML files using the .dat files.
6. Quantify the peptides as in Section 7.2, export the spreadsheet, and process the data as previously.
7. Discard the peptides identified as outliers in Section 7.2.
8. The peptides with H/L (protease-treated/control sample) ratio > (mean + 3 SD) (as calculated in Section 7.1) were generated by the protease under study, and the position of their N-terminus in the target protein indicates the cleavage sites.

9. Peptides with H/L ratios < (mean − 3 SD) are only in the control sample and have disappeared from the protease-treated sample due to proteolysis, which indicate a proteolytic cleavage site within this peptide. The exact position of the cleavage site can be determined for these peptides if the protease under study is very specific. However, the exact cleavage site cannot be determined if the protease has a broad cleavage specificity.
10. To increase the coverage and to identify rare cleavage sites, two or more replicates can be performed.
11. Identified peptides with high H/L ratios from these replicates are merged to form a single peptide list.
12. Peptides identified with >95% confidence or identified by at least two independent mass spectra are conserved.

8. Discussion: Measuring the Effect of Protease Inhibitors on the Generation of Proteolytic Fragments

Protease activity is in part controlled by endogenous inhibitors such as the family of irreversible inhibitors of serine proteases named serpins. The reactive loop of serpins interacts with the protease active site in a substrate-like manner, and the peptide bond between the P1 and P1′ residue is hydrolyzed. However, unlike substrates, the serpin structure is greatly modified upon hydrolysis, which distorts the protease active site and prevents the release of the proteolytic fragments, thus forming a covalently bound irreversible inhibitory complex (Bode and Huber, 1992; Huntington *et al.*, 2000). We can assume that serpins compete with the protease substrates for the protease active site. The binding affinity (K_m or K_i), the concentration of substrate and serpins, and their rate of association (k_{on}) with the protease will dictate the rate of inhibition and proteolysis (Bieth, 1984). This situation is more complicated for proteolysis of complex high molecular weight substrates presenting multiple cleavage sites. Each independent cleavage site will likely have different K_m and k_{on} values and will thus compete differently with the inhibitor for the protease active site. A number of MMPs also cleave and inactivate serpins (Li *et al.*, 2004; Lijnen *et al.*, 2001; Liu *et al.*, 2000; Mast *et al.*, 1991; Xu *et al.*, 2010), and the cleaved forms of serpins have been identified in several MMP degradomics screens (Dean and Overall, 2007; Morrison *et al.*, 2011). Performing ATOMS analyses with various concentrations of protease could help identifying substrate cleavage sites preferentially proteolyzed in the presence of endogenous inhibitors and would inform on the likelihood of a particular bond being cleaved *in vivo* where serpins are present.

ACKNOWLEDGMENTS

A. D. acknowledges the Fond Quebecois de la Recherche sur la Nature et les Technologies and the Michael Smith Foundation for Health Research for their financial support. C. M. O is supported by a Canada Research Chair in Metalloproteinase Proteomics and Systems Biology. This work was supported by research grants from the Cancer Research Society and from the Canadian Institutes of Health Research as well as with an Infrastructure Grant from the Michael Smith Foundation for Health Research.

REFERENCES

Bieth, J. G. (1984). In vivo significance of kinetic constants of protein proteinase inhibitors. *Biochem. Med.* **32,** 387–397.

Bode, W., and Huber, R. (1992). Natural protein proteinase inhibitors and their interaction with proteinases. *Eur. J. Biochem.* **204,** 433–451.

Cox, J. H., Dean, R. A., Roberts, C. R., and Overall, C. M. (2008). Matrix metalloproteinase processing of CXCL11/I-TAC results in loss of chemoattractant activity and altered glycosaminoglycan binding. *J. Biol. Chem.* **283,** 19389–19399.

Daudi, I., Gudewicz, P. W., Saba, T. M., Cho, E., and Vincent, P. (1991). Proteolysis of gelatin-bound fibronectin by activated leukocytes: A role for leukocyte elastase. *J. Leukoc. Biol.* **50,** 331–340.

Dean, R. A., and Overall, C. M. (2007). Proteomics discovery of metalloproteinase substrates in the cellular context by iTRAQTM labeling reveals a diverse MMP-2 substrate degradome. *Mol. Cell. Proteomics* **6,** 611–623.

Doucet, A., and Overall, C. M. (2011). Amino-Terminal Oriented Mass Spectrometry of Substrates (ATOMS): N-terminal sequencing of proteins and proteolytic cleavage sites by quantitative mass spectrometry. *Mol. Cell. Proteomics,* **10,** M110.003533.

Furie, M. B., and Rifkin, D. B. (1980). Proteolytically derived fragments of human plasma fibronectin and their localization within the intact molecule. *J. Biol. Chem.* **255,** 3134–3140.

Giannelli, G., Falk-Marzillier, J., Schiraldi, O., Stetler-Stevenson, W. G., and Quaranta, V. (1997). Induction of cell migration by matrix metalloprotease-2 cleavage of laminin-5. *Science* **277,** 225–228.

Gold, L. I., Schwimmer, R., and Quigley, J. P. (1989). Human plasma fibronectin as a substrate for human urokinase. *Biochem. J.* **262,** 529–534.

Goldfinger, L. E., Stack, M. S., and Jones, J. C. (1998). Processing of laminin-5 and its functional consequences: Role of plasmin and tissue-type plasminogen activator. *J. Cell Biol.* **141,** 255–265.

Horowitz, J. C., Rogers, D. S., Simon, R. H., Sisson, T. H., and Thannickal, V. J. (2008) Plasminogen activation induced pericellular fibronectin proteolysis promotes fibroblast apoptosis. *Am. J. Respir. Cell Mol. Biol.* **38,** 78–87.

Hsu, J. L., Huang, S. Y., Chow, N. H., and Chen, S. H. (2003). Stable-isotope dimethyl labeling for quantitative proteomics. *Anal. Chem.* **75,** 6843–6852.

Huntington, J. A., Read, R. J., and Carrell, R. W. (2000). Structure of a serpin-protease complex shows inhibition by deformation. *Nature* **407,** 923–926.

Keller, A., Eng, J., Zhang, N., Li, X. J., and Aebersold, R. (2005). A uniform proteomics MS/MS analysis platform utilizing open XML file formats. *Mol. Syst. Biol.* **1,** 2005.0017.

Khan, K. M. F., Laurie, G. W., McCaffrey, T. A., and Falcone, D. J. (2002). Exposure of cryptic domains in the alpha 1-chain of laminin-1 by elastase stimulates macrophages urokinase and matrix metalloproteinase-9 expression. *J. Biol. Chem.* **277,** 13778–13786.

Kulkarni, M. M., Jones, E. A., McMaster, W. R., and McGwire, B. S. (2008). Fibronectin binding and proteolytic degradation by Leishmania and effects on macrophage activation. *Infect. Immun.* **76,** 1738–1747.

Lambert Vidmar, S., Lottspeich, F., Emod, I., Planchenault, T., and Keil-Dlouha, V. (1991). Latent fibronectin-degrading serine proteinase activity in N-terminal heparin-binding domain of human plasma fibronectin. *Eur. J. Biochem.* **201,** 71–77.

Li, X. J., Zhang, H., Ranish, J. A., and Aebersold, R. (2003). Automated statistical analysis of protein abundance ratios from data generated by stable-isotope dilution and tandem mass spectrometry. *Anal. Chem.* **75,** 6648–6657.

Li, W., Savinov, A. Y., Rozanov, D. V., Golubkov, V. S., Hedayat, H., Postnova, T. I., Golubkova, N. V., Linli, Y., Krajewski, S., and Strongin, A. Y. (2004). Matrix metalloproteinase-26 is associated with estrogen-dependent malignancies and targets alpha1-antitrypsin serpin. *Cancer Res.* **64,** 8657–8665.

Lijnen, H. R., Van Hoef, B., and Collen, D. (2001). Inactivation of the serpin alpha(2)-antiplasmin by stromelysin-1. *Biochim. Biophys. Acta* **1547,** 206–213.

Liu, Z., Zhou, X., Shapiro, S. D., Shipley, J. M., Twining, S. S., Diaz, L. A., Senior, R. M., and Werb, Z. (2000). The serpin alpha1-proteinase inhibitor is a critical substrate for gelatinase B/MMP-9 in vivo. *Cell* **102,** 647–655.

Mast, A. E., Enghild, J. J., Nagase, H., Suzuki, K., Pizzo, S. V., and Salvesen, G. (1991). Kinetics and physiologic relevance of the inactivation of alpha 1-proteinase inhibitor, alpha 1-antichymotrypsin, and antithrombin III by matrix metalloproteinases-1 (tissue collagenase), -2 (72-kDa gelatinase/type IV collagenase), and -3 (stromelysin). *J. Biol. Chem.* **266,** 15810–15816.

Morla, A., Zhang, Z., and Ruoslahti, E. (1994). Superfibronectin is a functionally distinct form of fibronectin. *Nature* **367,** 193–196.

Morrison, C. J., Mancini, S., Kappelhoff, R., Cipollone, J., Roskelley C., and Overall, C. M. 2011. Microarray and Proteomic Analysis of Breast Cancer Cell and Osteoblast Co-cultures: The Role of Osteoblast Matrix Metalloproteinase (MMP)-13 in Bone Metastasis. *J. Biol. Chem,* (in press).

Pozzi, A., LeVine, W. F., and Gardner, H. A. (2002). Low plasma levels of matrix metalloproteinase 9 permit increased tumor angiogenesis. *Oncogene* **272,** 272–281.

Rege, T. A., Fears, C. Y., and Gladson, C. L. (2005). Endogenous inhibitors of angiogenesis in malignant gliomas: Nature's antiangiogenic therapy. *Neuro Oncol.* **7,** 106–121.

Robert, C. R., Maurice, C. B., Doucet, A., and Overall C. M. (in preparation). Cleavage of the proteoglycan versican by cell surface membrane type 1-matrix metalloproteinase (MT1-MMP).

Rostagno, A. A., Frangione, B., and Gold, L. I. (1989). Biochemical characterization of the fibronectin binding sites for IgG. *J. Immunol.* **143,** 3277–3282.

Scapini, P., Nesi, L., Morini, M., Tanghetti, E., Belleri, M., Noonan, D., Presta, M., Albini, A., and Cassatella, M. A. (2002). Generation of biologically active angiostatin kringle 1–3 by activated human neutrophils. *J. Immunol.* **168,** 5798–5804.

Timpl, R., Rohde, H., Robey, P. G., Rennard, S. I., Foidart, J. M., and Martin, G. R. (1979). Laminin—A glycoprotein from basement membranes. *J. Biol. Chem.* **254,** 9933–9937.

Unger, J., and Tschesche, H. (1999). The proteolytic activity and cleavage specificity of fibronectin-gelatinase and fibronectin-lamininase. *J. Protein Chem.* **18,** 403–411.

Xu, D., McKee, C. M., Cao, Y., Ding, Y., Kessler, B. M., and Muschel, R. J. (2010). Matrix metalloproteinase-9 regulates tumor cell invasion through cleavage of protease nexin-1. *Cancer Res.* **70,** 6988–6998.

Yi, M., and Ruoslahti, E. (2001). A fibronectin fragment inhibits tumor growth, angiogenesis, and metastasis. *Proc. Natl. Acad. Sci. USA* **98,** 620–624.

CHAPTER FOURTEEN

Computational Methods for Studying Serpin Conformational Change and Structural Plasticity

Itamar Kass,[1] Cyril F. Reboul,[1] *and* Ashley M. Buckle

Contents

1. Introduction 296
2. Local to Global Dynamics Simulations 299
   2.1. MD simulation 300
   2.2. Coarse-grained description—removing some trees so the forest can be seen 302
   2.3. Monte Carlo simulations 302
   2.4. Normal mode analysis—studying the energy landscape without simulating it 303
3. Pushing the Limits—Improving Conformational Sampling 304
   3.1. High-temperature simulations 305
   3.2. Improving the odds—replica exchange MD 306
   3.3. Do not go back—local elevation dynamics 306
4. I Know Where to Go—Directed Simulations 307
   4.1. Slow progress forward—umbrella sampling 307
   4.2. A jump forward—steered and targeted dynamics 308
5. Nondynamical Methods 308
6. Software 309
   6.1. Chemistry at HARvard Molecular Mechanics (CHARMM) 309
   6.2. GROningen MAchine for Chemical Simulations (GROMACS) 309
   6.3. Not (just) Another Molecular Dynamics program (NAMD) 310
   6.4. Assisted Model Building with Energy Refinement (AMBER) 310
   6.5. Visualization 310
7. Force Fields 311
8. Hardware 313
9. Case Study 314
10. Outlook 316
References 319

Department of Biochemistry and Molecular Biology, Monash University, Melbourne, Victoria, Australia
[1] These authors contributed equally.

Methods in Enzymology, Volume 501
ISSN 0076-6879, DOI: 10.1016/B978-0-12-385950-1.00014-6

Abstract

Currently, over a hundred high-resolution structures of serpins are available, exhibiting a wide range of conformations. However, our understanding of serpin dynamics and conformational change is still limited, mainly due to challenges of monitoring structural changes and characterizing transient conformations using experimental methods. Insight can be provided, however, by employing theoretical and computational approaches. In this chapter, we present an overview of such methods, focusing on molecular dynamics and simulation. As serpin conformational dynamics span a wide range of timescales, we discuss the relative merits of each method and suggest which method is suited to specific conformational phenomena.

1. Introduction

Among serine protease inhibitors, serpins are not only the largest family but also unique in their mechanism of action. While many serpins are inhibitory, many have also evolved to develop noninhibitory functions and display a large range of activity, including hormone transport (human corticosteroid binding globulin; Siiteri *et al.*, 1982), tumor suppression (maspin; Zou *et al.*, 1994), protein storage (ovalbumin; Hunt and Dayhoff, 1980), and heat-shock proteins (HSP47; Kurkinen *et al.*, 1984).

The serpin native fold is composed of eight or nine α-helices, three β-sheets, and an extended reactive center loop (RCL; Elliott *et al.*, 2000; Law *et al.*, 2008; Zhang *et al.*, 2007). The native serpin scaffold is a kinetically trapped metastable fold (Carrell and Owen, 1985)—following binding of a target serine protease and proteolysis of the RCL, the serpin undergoes a major irreversible conformational change to the Relaxed (or R) form, also known as the cleaved state. The resulting complex, in which the covalently bound acyl-protease is dragged from the top of β-sheet A to the bottom, a distance of >7 nm, is a stable form with a half-life in the order of years (Huntington *et al.*, 2000). A key event in this transition involves the insertion of the cleaved RCL into β-sheet A, forming an additional strand (labeled s4A). The conformational changes are not limited to the serpin—the bound protease also undergoes a large degree of structural change, leaving it vulnerable to proteolysis (Herve and Ghelis, 1991; Huntington *et al.*, 2000; Kaslik *et al.*, 1995; Stavridi *et al.*, 1996; Tew and Bottomley, 2001). Specifically, the active site is critically distorted, thus preventing further hydrolysis of the acyl-intermediate and leaving the bound protease trapped and inactive. A summary of the structural forms adopted by serpins is shown in Fig. 14.1.

Comparing this mechanism to the "lock and key" mechanism of other, smaller protease inhibitors (Bode and Huber, 1992; Otlewski *et al.*, 2005), one gets the notion of an overcomplicated system. However, the serpin

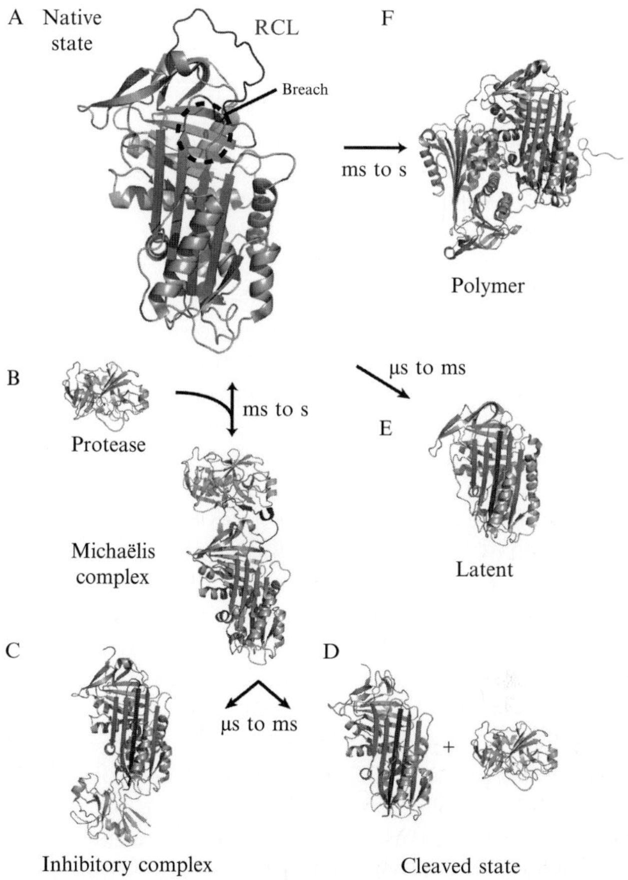

Figure 14.1 Different conformations accessible to serpins. Timescales for the structural changes shown are indicated. (A) The overall structure of native inhibitory α1-AT (Elliott *et al.*, 2000) is typical to all serpins and contains three β-sheets and nine α-helices. The breach, a functionally important region of the serpin, is circled in a dashed line and labeled. The central β-sheet A, into which the RCL is later incorporated, is colored in red. The RCL, colored in blue, can be seen at the top of the molecule. The first step of protease inhibition (here trypsin; Earnest *et al.*, 1991) is its binding to the serpin RCL and (B) the formation of a reversible Michaëlis complex (Dementiev *et al.*, 2003). (C) Following the proteolysis of the RCL, the serpin undergoes an S-to-R transition (inhibitory pathway). The transition results in the RCL inserting and forming a fourth strand in β-sheet A and distortion of the protease, covalently bound and inactive (Huntington *et al.*, 2000). (D) If the protease is able to complete the proteolysis cycle before its relocation and inactivation (substrate pathway), it is released active back to the medium (Baumann *et al.*, 1991). This pathway produces a cleaved inactive serpin after it has failed inhibiting the protease. The tight balance between both inhibitory and substrate pathways is of a kinetic nature (Kaiserman and Bird, 2010). Certain mutations, particularly within the RCL (Hopkins *et al.*, 1993), favor the nonproductive substrate pathway. (For interpretation of the references to color in this figure legend, the reader is referred to the Web version of this chapter.)

delivers major benefits such as a better control of the inhibitory function by cofactors, compared to small protease inhibitors. For example, the inhibition of thrombin and factor Xa (FXa) by human antithrombin is highly dependent on the binding of heparin to both the protease and the serpin (Rezaie, 1998; Rezaie and Olson, 2000). Structural studies of antithrombin in the native state (Carrell *et al.*, 1994), heparin-bound state (Jin *et al.*, 1997), and the antithrombin–thrombin–heparin ternary complex (Li *et al.*, 2004) have shed light on the allosteric regulation mechanism of antithrombin. Antithrombin circulates in the bloodstream, with its "bait" RCL partially inserted into the top of its A β-sheet (Carrell *et al.*, 1994). Upon binding of heparin, the RCL is fully released from the A β-sheet (Jin *et al.*, 1997) and adopts a protease-accessible conformation (Li *et al.*, 2004). Interestingly, the binding of long-chain heparin provides another advantage, as it can bind both thrombin and antithrombin and thus accelerates the formation of a Michaëlis complex (Lane *et al.*, 1984; Fig. 14.1B).

As well as providing advantages, the structural plasticity that is central to serpin function also renders them vulnerable to misfolding and aggregation. For example, several disease-causing variants of α1-AT have been identified. The most common of them is the "Z" variant (E342K) (Lomas *et al.*, 1992) located at the breach region, between the top of strand 5 of β-sheet A and the base of the RCL (Fig. 14.1A). Here, failure to properly maintain the metastable form leads to an increased propensity of α1-AT to polymerize in the endoplasmic reticulum of hepatocytes leading to a lack of secretion of α1-AT into the blood (Lomas *et al.*, 1992). The structural and the biochemical data available suggest that polymerization occurs via a domain-swapping event, where the RCL of one molecule inserts into β-sheet A of another to form an inactive serpin polymer (Dunstone *et al.*, 2000; Huntington *et al.*, 1999; Lomas *et al.*, 1992; Mast *et al.*, 1992; Yamasaki *et al.*, 2008; Fig. 14.1F).

In addition to polymerization, some serpins can spontaneously undergo a transition to a latent (denoted L) form. In this state, the uncleaved RCL is spontaneously inserted into β-sheet A, in a similar fashion similar to its conformation in the cleaved state (Beinrohr *et al.*, 2007; Stout *et al.*, 2000; Zhang *et al.*, 2007). Moreover, both states display high thermal stabilities (Lomas *et al.*, 1995). Interestingly, whereas the S-to-L transition is disease linked, such as in the T85M antithrombin variant (Beauchamp *et al.*, 1998), in some serpins (e.g., PAI-1), it is used as a built-in control mechanism (Mottonen *et al.*, 1992).

Currently, over a hundred high-resolution structures of serpins are available, exhibiting a wide range of conformations (Fig. 14.1). However, as serpin function is intimately coupled to their dynamic properties, a more complete understanding of these remarkable proteins requires insight into their conformational properties at the atomic level. Unfortunately, our understanding of serpin dynamics and conformational change is still limited, mainly due to challenges in monitoring structural changes and characterizing transient conformations using experimental methods. Specifically, dynamic processes cannot readily be investigated in atomic detail using experimental techniques such as nuclear magnetic resonance (NMR) spectroscopy or X-ray crystallography. While these methods provide details on the conformation in each state, they cannot provide information on key interactions and intermediate states along the pathway of conformational change. As a consequence, the use of molecular modeling techniques, in particular, molecular dynamics (MD) based methods, is required.

MD methods have proved to be useful for the study of many biological processes at the atomic level (for comprehensive reviews, see Hansson *et al.*, 2002; Karplus and McCammon, 2002; van Gunsteren *et al.*, 2006). For example, much insight into membrane structure, formation, and dynamics is due to the pioneering theoretical work of Berendsen and collaborators (Tieleman *et al.*, 1997). The behavior of membrane-bound peptides and pore formation in membranes is difficult to characterize using classic experimental methods and has also benefited from the use of MD (Bond and Sansom, 2006; Leontiadou *et al.*, 2006). Theoretical studies can also provide an insightful and atomistic view of complex biological processes such as allosteric regulation mechanisms, as in the case of hemoglobin (Cui and Karplus, 2008) and the chaperone GroEL (Keskin *et al.*, 2002; Ma and Karplus, 1998). Theoretical methods can also be used to study the binding and interaction of biomolecules. Theoretical studies of the interactions between the approved anti-HIV drug Amprenavir and wild-type and mutated protease have yielded free-energy binding profiles (Hou and Yu, 2007) as well as an atomic level understanding of the mechanism by which the protein mutates to become "immune" to the drug.

The combination of theoretical and experimental methods for the study of serpin dynamics and conformational change is in its infancy but holds much promise. MD studies can provide an atomic level insight into various dynamic processes central to serpin function, such as the S-to-R transition, cofactor activation, and polymerization. Accordingly, the main object of this chapter is to alert the reader to the array of theoretical methods available for the study of serpin dynamics. As such, we will discuss the different methods, in turn, highlighting their advantages and disadvantages. To get a better understanding of underlying algorithms and basic notions such as energy landscapes, readers are advised to consult references Allen and Tildesley (1987) or Leach (2001).

2. Local to Global Dynamics Simulations

The functional properties of serpins rely on their propensity to change conformation. Structural changes can be local, medium range, or global structural rearrangements. Examples for each case are, respectively, the effect of the Z variant mutation (E342K) on the stability, folding, and polymerization of α1-AT (Lomas *et al.*, 1992, 1993); shifts of secondary structure elements such as the F-helix during serpin folding, polymerization, and function (Cabrita *et al.*, 2004); and major structural rearrangements as seen in the S-to-R transition.

2.1. MD simulation

MD is based on classical mechanics in which the interactions between atoms are described using semiempirical sets of rules, known as force fields. Such force fields use simple chemical concepts to describe the potential energy of a system, usually composed of proteins, small molecules or nucleic acids, and solvent, in terms of the Cartesian coordinates of the atoms. Newton's equations of motion are then solved iteratively to generate the time evolution of the system. Applying concepts from statistical mechanics, the resulting trajectory can be used to evaluate various time-dependent structural, dynamic, and thermodynamic properties of the system. In recent years, MD has been successfully applied to the study of both global changes in complexes such as the human growth hormone receptor (Poger and Mark, 2010) or the local changes, such as protonation states in Influenza A M2 proton channel (Kass and Arkin, 2005).

In the past decade, relatively few MD studies of serpins dynamics have been performed. In 2001, the local dynamics of the Z variant of the modeled native fold α1-AT was studied by Jezierski and Pasenkiewicz-Gierula (2001). The simulations indicated that the Z mutation (E342K) affects the local hydrogen bonding and salt-bridging network, leading to conformational and dynamical deviations when compared to the wild type. Later on, the induced fit mechanism of antithrombin activation by heparin was studied by Verli and Guimaraes (2005). MD simulations of heparin-bound and unbound antithrombin were performed in order to understand the binding of heparin and its global effect on antithrombin conformation. The authors concluded that following binding of heparin, antithrombin undergoes global conformational changes, such as the elongation of the D-helix. In addition, the flexibility of the RCL was found to increase following the binding of heparin.

In a study by Zheng *et al.* (2008), radical-mediated footprinting and hydrogen/deuterium exchange were employed in conjugation with MD to study the dynamics of α1-AT, specifically the buried residues Met 374, Met 385, and Pro 255. The finding suggests that those residues, which are inaccessible to solvent in the crystal structure (Elliott *et al.*, 2000), are indeed solvent accessible due to α1-AT conformational dynamics. More recently, Chandrasekaran *et al.* (2009) studied the structure and dynamics of a modeled Michaëlis complex between Z-dependent protease inhibitor (ZPI) and FXa using MD simulations. The study results in a better understanding of

FXa binding to ZPI and emphasizes the role of the RCL and β-sheet C in binding specificity.

Although the above studies give a glimpse of serpins dynamics, it is important to notice that the simulation length was in the range of hundreds of picoseconds (ps) to less than 15 nanoseconds (ns). As relative motions of different globular regions take place at a microsecond (μs) to millisecond (ms) timescale (Fig. 14.2), and allosteric transitions can take up to seconds, it is clear that the previous studies cannot properly address global dynamics or physiologically relevant reactions, such as polymerization. This is due to too short a simulation length, defined as the number of iterations multiplied by the integration time step (Δt). The latter is the time difference between two consecutive evaluations of the system's potential energy, following the system's time evolution. This variable should be smaller than the fastest vibration in the system. A typical time step for classical MD is 2 fs (10^{-15} s), which means that 500,000 steps are needed in order to perform a 1-ns trajectory. This is currently a limitation in the usefulness of MD to study such processes. Nevertheless, algorithmic progress, mainly in scalability and parallelization (Hess *et al.*, 2008), in conjugation with hardware improvements and the use of graphical processing units (GPUs; Stone *et al.*, 2010)

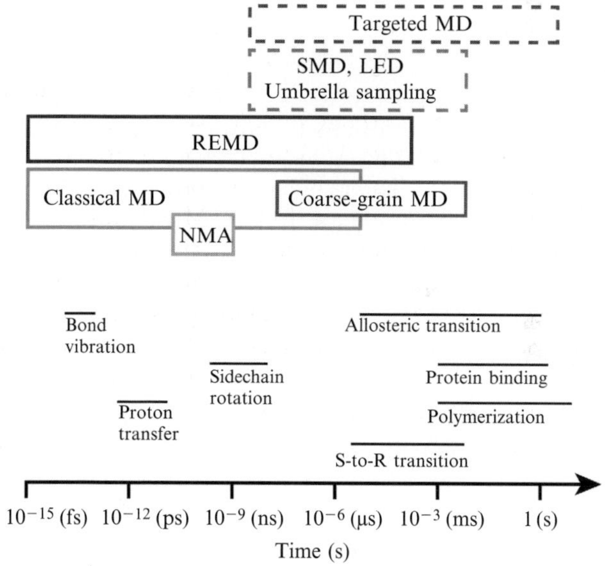

Figure 14.2 Timeline of general structural and biophysical processes experienced by serpins. The latter are displayed above the abscissa. The methods highlighted in this chapter (boxed) are represented according to the timescale they are able to reach during a simulation. Time-independent techniques are displayed in dashed-boxes. See text for abbreviations.

have recently allowed long-scale simulations, such as the 10-μs simulation of the WW domain by Freddolino *et al.* (2008). Therefore, it is now feasible to perform MD simulations of a serpin, hundreds of nanoseconds in length, providing insight into conformational changes in more physiologically relevant timescales.

2.2. Coarse-grained description—removing some trees so the forest can be seen

A simulation length of less than a microsecond is unfortunately still too short to study large-scale structural changes such as polymerization, and other approaches, or approximations, are necessary. In these cases, coarse grained (CG), rather than atomistic descriptions of the molecule, are employed. Such simplification, in which groups of atoms are replaced by beads with a distinctive physical nature (Marrink *et al.*, 2007), although sacrificing fine details, reduces the computational cost and allows longer time steps, thus allowing longer simulations (Sengupta and Marrink, 2010; Sengupta *et al.*, 2009). Such methods may allow the simulations of phenomena such as structural rearrangements following RCL cleavage, expected to take place in the μs to ms timescale, or even the polymerization of α1-AT. Although this method is inherently inferior to a full atomistic approach, the lack of atomic details can be overcome by reconstructing atomistic structures from CG representations of specific intermediates along the simulation pathway (Rzepiela *et al.*, 2010). Also, data resolution can be increased if one combines classical MD with CG simulations. For example, the simulation of a slow and large-scale transition can be performed at CG resolution, whereas the end states can be simulated using classical MD.

2.3. Monte Carlo simulations

In classical MD, new conformations are produced as the system evolves in time, based on Newtonian physics. An alternative way to collect conformations was suggested by Metropolis *et al.* (1953) based on the Monte Carlo (MC) simulation method: starting from a given configuration, perturbed configurations are generated randomly. Each new configuration is accepted or rejected based on its potential energy difference to the initial configuration. The resulting conformations are representative under given thermodynamic conditions such as temperature and volume.

As forces are not being calculated throughout the simulation, MC approaches are more computationally efficient than MD. Nevertheless, it is difficult to define simple structural perturbations that will give rise to major structural changes but will not generate unrealistic high-energy conformations. Moreover, MC simulations do not provide information about the time evolution of the system. Nevertheless, MC simulations were found to facilitate the refinement and study of protein loops (Hu *et al.*, 2007; Mandell *et al.*, 2009). In the case of serpins, it is therefore a tool that can be used to study different conformations accessible to the RCL, as well as the effect of mutations on RCL dynamics.

2.4. Normal mode analysis—studying the energy landscape without simulating it

Conformational changes that are characterized by low-frequency domain motions, or modes, such as the structural transition from the native to the latent state are arguably better studied using normal mode analysis (NMA; Levitt *et al.*, 1985). NMA of a protein will result in a set of collective motions, each describing an intrinsic molecular motion in which some parts of the protein move with the same frequency in a "harmonic" way. This is usually referred to as molecular movement along the least energy pathway, and hence NMA can describe or suggest a pathway between different conformations. NMA was found to be an important tool for the identification and characterization of low-frequency domain motions (Atilgan *et al.*, 2001; Hinsen, 1998). Simplified NMA (described below) has, for example, shown that the first step of the gating mechanism of the mechanosensitive channel can be explained in terms of low-frequency, collective motions of the integral membrane protein (Valadie *et al.*, 2003). It also identified such natural dynamics as being a factor defining the structural changes upon inhibitor binding in three different drug target enzymes (Bakan and Bahar, 2009).

However, NMA has some theoretical limitations, mainly the assumption of a harmonic energy "well" and the neglect of the effect of solvent. Nevertheless, NMA is a useful method for the study of intrinsic structural deformations and the identification of flexibility "hot spots."

Many software packages include the possibility of performing full-atom NMA (see Section 6). Following minimization of the full-atom conformation of the molecule, the mass-weighted Hessian matrix of second derivatives is calculated and diagonalized, yielding eigenvalues (inversely proportional to the frequencies) and associated eigenvectors. The latter represent the directions of motions identified. In Cartesian space, the Hessian matrix consists of a square matrix of dimension $(3N + 1)3N/2$, where N is the number of atoms in the molecule. This leads to computational difficulties for large systems. For example, in the case of a α1-AT dimer ($N \sim 3900$ atoms), this would produce a Hessian matrix with more than a quarter of a billion elements for the polymeric dimer. This size is hardly tractable with existing computer power, in addition to being slow to build and process. However, a range of alternatives has emerged, inspired by the elastic network model and the pioneering work of Tirion (1996). They make use of simplified protein models and/or force fields, where a residue can be reduced to its Cα and the protein potential energy solely influenced by the neighboring residues (Atilgan *et al.*, 2001; Hinsen, 1998). Full atom models can still be employed when associated with simplified degrees of freedom resulting in smaller and more manageable Hessians (Tama *et al.*, 2000). The different flavors of the reduced models and force fields achieve various accuracies and serve different purposes (Kondrashov *et al.*, 2007). They usually yield significant speed-ups, often larger than two orders of magnitude. These extreme simplifications have been consistently able to capture the concerted and functional motions arising from the topology of proteins and other biopolymers. Most of these techniques are available for free and via web servers (Eyal *et al.*, 2006; Hollup *et al.*, 2005; Suhre and Sanejouand, 2004), providing convenient analysis and visualization tools in a reasonable time scale.

To illustrate the ease and performance of NMA, we performed an analysis (Kass *et al.*, unpublished) of α1-AT using the ElNemo (Suhre and Sanejouand, 2004) web server. Within minutes, we were able to observe that the first few lower modes describe the rotation of the β-sheet B and C (top part of the serpin scaffold) with respect to the lower part of the protein (α-helices B to F and lower end of β-sheet A). Within these modes, the F-helix was identified as one of the most flexible parts of the protein (see Fig. 14.3), in agreement with the known importance of the F-helix for α1-AT initial conformation changes (Cabrita *et al.*, 2002).

3. Pushing the Limits—Improving Conformational Sampling

The structure and function of any molecular system cannot be accurately described by a single global minimum energy configuration, but only by a statistical–mechanical ensemble of (mainly) low-energy conformations. Therefore, an intensive sampling of the low-energy regions of the free-energy surface is a fundamental requirement in any simulation. Proteins are composed of tens of thousands of atoms, resulting in a very large number of degrees of freedom, and a very rugged energy surface, with numerous local maxima and minima. This also implies that an energy surface describing a complex conformational transition (e.g., the S-to-R transition) is composed of many energy barriers. The accurate sampling of such surfaces using MD methods is thus a very long and potentially unfeasible task and is thus widely known as the *search problem*. Therefore, the study of such energy surfaces demands the user to go beyond classical MD and use modified algorithms that allow better exploration of the conformation space.

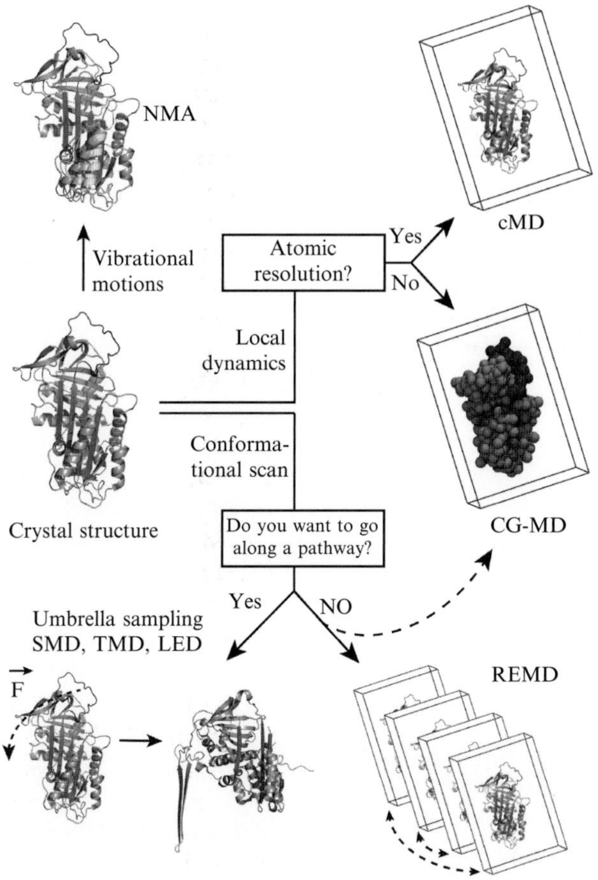

Figure 14.3 Schematic flowchart of the various methods presented in this chapter. See text for abbreviations. (See Color Insert.)

3.1. High-temperature simulations

Performing a simulation of a system at a temperature higher than the physiologic one (~300 K) increases the kinetic energy of the system and allows more conformations to be sampled, as the simulated system is less likely to spend time in local minima (Dinola *et al.*, 1984). Although very easy to implement, high-temperature simulations have three drawbacks. First, MD force fields are usually parameterized at physiological temperature and should be employed at this temperature. Second, simulating at high temperature leads to overestimation of the entropic contribution to the free energy. Third, although using such a method allows barrier crossing, conformational scanning shows little if no improvement compared to classical MD simulations. Therefore, we expect that these methods have little to offer for the study of serpin dynamics.

3.2. Improving the odds—replica exchange MD

Another method used to improve the scanning of conformational space is replica exchange molecular dynamics (REMD; Sugita and Okamoto, 1999; Swendsen and Wang, 1986), also referred to as parallel tempering. In REMD, one is simulating a series of noninteracting simulations of the same molecule each in the canonical ensemble, and each at a different temperature, termed *replicas*. At particular intervals, the conformations of these replicas are swapped (resulting in an exchange of states) with a transition probability *P*; such that each independent system maintains its canonical ensemble. Such exchanges occur through a simple swap of the simulation temperatures via velocity reassignment. Whereas the use of low-temperature replicas allows sufficient sampling of local minima, such replicas can be trapped in low-energy minima. Exchanges with high-temperature replicas, which can jump over energy barriers, allow better sampling of the energy surface compared to classical MD (Swendsen and Wang, 1986).

It is important to note that because a series of simulations are being done instead of a single simulation, REMD requires more computational resources. However, it is clear that REMD is more efficient than a single-temperature classical MD. This is achieved because the high-temperature replicas allow the lower temperature replicas to sample regions of phase space not accessible to even long trajectories using classical MD simulations. Despite the many advantages delivered by REMD, it is not widely used for the study of large proteins such as serpins, due mainly to the large computational resources needed and the theoretical and practical problems associated with replicas swap ping.

REMD enables the molecule under study to scale energy barriers. As a result, REMD can be used to study conformations that are difficult to access using classical MD, for example, the study of intermediate or activated states (Garcia and Onuchic, 2003; Zhou, 2004). As a result, this method is suited for investigating non-native serpins folds (James and Bottomley, 1998), which may provide insight into their pathological roles.

3.3. Do not go back—local elevation dynamics

Even with the use of high-temperature or REMD simulations, some energy surfaces are too rough and chaotic to be sufficiently simulated. It is not uncommon that a molecule is trapped in a small part of the energy surface. To overcome this limitation, Huber *et al.* have developed the local elevation dynamics (LED; Huber *et al.*, 1994) (also known as metadynamics; Laio and Parrinello, 2002) in which a penalty potential is added against any conformations already sampled. The addition of a "memory potential" prevents the system from returning to previously sampled conformations and results in exploration of new regions of the energy landscape. As the resulting energy surface is the sum of the true surface and the biased potential, one can easily recover the underlying free-energy surface.

The major problem facing the application of LED is the choice of the degree of freedom to which the potential should be applied. As discussed earlier, proteins have many degrees of freedom, making the price of visited configurations storage overwhelming. One way of overcoming this is to apply the modified potential to smaller subsets of degrees of freedom. This subset can be selected based on NMR data or preliminary NMA simulations.

Using LED simulations, one can "force" molecules to move out of energy minima, so conformations other than the most stable one can be studied. For serpins, LED can be used to accelerate dynamics over energy barriers. As such, a prudent selection of the degrees of freedom to vary may provide insight into the S-to-R conformational change.

4. I Know Where to Go—Directed Simulations

The dramatic structural changes that serpins undergo have relatively large energy barriers, which unfortunately make the simulation of such processes using classical MD impractical. Although the actual barrier crossing can be relatively rapid, the time spent in random fluctuations before a new conformation is explored can be long. To overcome this, the global potential energy of the system can be altered in order to improve sampling, as already discussed. Alternatively, given a prior knowledge of the reaction pathway (ideally provided by biochemical or biophysical evidence), the potential along a specific reaction coordinate can be changed. It is thus possible to enhance the sampling over a particular set of conformations, defining a reaction pathway, while keeping computational expense to a minimum.

4.1. Slow progress forward—umbrella sampling

The method of umbrella or biasing sampling, described by Torrie and Valleau (1977) and extended to protein simulations by Northrup *et al.* (1982), involves modifying the potential such that conformations along a reaction pathway become more favorable. One requirement is a prior knowledge of the reaction pathway. Moreover, the combination of umbrella sampling with thermodynamics integration, a free-energy calculation method, was found to improve the outcome (Ota and Brunger, 1997; Ota *et al.*, 1999). Due to its suitability for the study of pathways, it can be used to study structural transitions. In the case of serpins, for example, this method may be used to investigate not only the initial steps leading to the

S-to-L transition but also the free-energy profile of this state; specifically, this may be achieved by applying a biasing potential to effect translocation of the RCL, forcing it to move over strands 3 and 4 of β-sheet C.

4.2. A jump forward—steered and targeted dynamics

A different but related method, in which a time-dependent modified potential is employed, is termed steered MD (SMD; Grubmuller *et al.*, 1996; Izrailev *et al.*, 1997). In SMD, an external force is applied to the system to study its mechanical response, mimicking the principle of atomic force microscopy. SMD has proven useful in various applications, among them the study of ligand binding (Izrailev *et al.*, 1998), where external forces were applied to the ligands to dissociate them from the protein. The resulting accelerated dissociation enables one to gather information about enzyme flexibility and its response to ligand dissociation more efficiently than classical MD.

A related technique, in which a force is applied to one conformation in order to yield a different one, is called targeted MD (TMD; Engels *et al.*, 1992). This method is useful to predict a pathway between two known conformations. Using TMD, it would be possible to enforce *in silico* the transition between inactive and active antithrombin. Notably, both SMD and TMD apply an external force that is doing irreversible work on the system, hence driving the system from equilibrium. As a result, calculation of the free energy along the reaction pathway is not straightforward.

It is worth noting that when the forces applied by SMD or TMD are weak and the changes applied to the system are slow, it is equivalent to umbrella sampling. SMD and TMD both use a biasing potential to accelerate sampling along a pathway and are useful for the study of serpin interactions with different ligands. For example, both can be applied to the study of heparin binding to antithrombin and its influence on antithrombin RCL conformation.

5. Nondynamical Methods

The methods described above can all contribute toward our understanding of serpin conformational dynamics and therefore function. However, their use is usually limited by the available computational resources. A popular method for building a set of time-independent conformations that satisfy a set of distance constraints is called CONCOORD (de Groot *et al.*, 1997). The set of conformations derived is based on geometric considerations only, potentially yielding unrealistic structures. This may allow the identification of relevant conformations that might never be found using lengthy time-dependent MD-based techniques. It is therefore a complementary method.

6. Software

The wide use of theoretical methods is, in part, due to the availability of high-quality software packages such as CHARMM (Brooks *et al.*, 2009), GROMACS (Hess *et al.*, 2008), and NAMD (Phillips *et al.*, 2005). In addition, many biomolecular modeling packages (e.g., SYBYL) have some kind of MD capabilities. Although, in most cases, this is just an interface to one of the specialized MD software packages, it provides a simple way for nonexperts to take advantage of MD capabilities. There are inherent benefits to each available software packages, and choice comes down to personal preference. Taking the authors as an example, GROMACS and NAMD are routinely used for theoretical studies in our laboratory, where also more specific tasks use GROMOS (Christen *et al.*, 2005), Ghemical (Acton *et al.*, 2006), and others when needed.

6.1. Chemistry at HARvard Molecular Mechanics (CHARMM)

CHARMM is a macromolecular simulation software, which includes various computational tools, such as energy minimization, NMA, MD, REMD, and MC simulations. It is available for academic use for a small fee. Although originating from the group of Professor Karplus at Harvard University, it is being improved by the contributions of developers throughout the world. The CHARMM software is tightly developed with the CHARMM force field (MacKerell *et al.*, 1998). The CHARMM software consists of one application, which serves the user in all steps from system preparation through simulation to analysis. It also supports GPU acceleration.

6.2. GROningen MAchine for Chemical Simulations (GROMACS)

GROMACS, originally developed in the University of Groningen, is an open source software released under the GPL. GROMACS was initially a rewrite of the GROMOS package (van Gunsteren *et al.*, 1996). Unlike CHARMM, it is actually a set of applications; each designed for a specific task and enables a limited input. Being a versatile package, GROMACS enables the user to perform MD, MC, REMD, or COONCORD calculations and includes various analysis tools such as NMA. GROMACS enables the use of various force fields, such as GROMOS96 (van Gunsteren *et al.*, 1996), CHARMM (MacKerell *et al.*, 1998), and optimized potentials for liquid simulations (OPLS; Felts *et al.*, 2002). It supports GPU acceleration through the external OpenMM package (Eastman and Pande, 2010).

6.3. Not (just) Another Molecular Dynamics program (NAMD)

NAMD is a free of charge software, developed by the Theoretical and Computational Biophysics Group at the University of Illinois. Developed with parallelization in mind, NAMD can scale to hundreds of processors and is often used to simulate large systems (millions of atoms). Algorithms and protocols such as LED, SMD, and TMD are implemented; GPU acceleration is included. To improve efficiency, NAMD includes a scripting engine, which enables the automation of many tasks. Moreover, NAMD enables interactive simulations with the Visual Molecular Dynamics (VMD) molecular visualization software (Humphrey *et al.*, 1996).

NAMD enables the use of few force fields such as AMBER (Case *et al.*, 2005), CHARMM (MacKerell *et al.*, 1998), and OPLS (Felts *et al.*, 2002), but not GROMOS96 (van Gunsteren *et al.*, 1996).

6.4. Assisted Model Building with Energy Refinement (AMBER)

AMBER (Case *et al.*, 2005) is a collaborative effort from different groups around the world and is composed of a software suite and a force field (see below). Like most modern software, it includes a large range of techniques like the ones discussed above and is available for a small fee. GPU acceleration is supported or can optionally be added with the OpenMM package (Eastman and Pande, 2010).

6.5. Visualization

When dealing with thousands of snapshots of a molecule produced by the above-mentioned software, visualization and flexible analysis tools have become a necessity. While PyMOL (DeLano, W.L. The PyMOL Molecular Graphics System, Schrodinger, LLC; http://www.pymol.org) is a well-established standard for displaying and producing print-quality figures of a handful of conformations of a protein, it does not allow the user to analyze and visually study the kind of data generated by MD methods. Applications like YASARA ("Yet Another Scientific Artificial Reality Application"; http://www.yasara.com; free in its basic form) or UCSF Chimera (Pettersen *et al.*, 2004) (free of charge) provide a sophisticated graphical interface along with analysis tools and scripting languages adapted to dealing with MD trajectories. In addition, they are available on many platforms like OSX and LINUX. Another application, VMD (Humphrey *et al.*, 1996), offering complex 3D display capabilities, flexible analysis scripting (Tcl/Python) as well as supporting a large range of file formats, has become a *de facto* standard.

7. Force Fields

PDB files are the most common way of sharing biomolecular 3D structures (Berman *et al.*, 2007). The information within those files is essentially a list of atoms and their positions in space. The use of MD simulations requires the addition of physicochemical information to systems composed only of atomic coordinates. These data are incorporated in the force field and are added prior to any simulation.

As discussed earlier, force fields describe interactions between any bonded (e.g., bond length) and nonbonded atoms (e.g., electrostatic interactions). Most force fields in use in computer simulations of proteins are highly similar. Bond lengths and angles are described as harmonic springs, torsion angles are described as Fourier series, and the interactions between nonbonded atoms are described by a Lennard-Jones function and a Coulombic function. The main difference between the various force fields is due to different approaches taken when parameters are derived.

Although used for the study of macromolecules, force fields are usually parameterized based on experimental and/or quantum mechanical studies of small molecules or fragments. It is assumed that such parameters may be transferred to larger molecules. Different force fields have been developed from dissimilar types of experimental data, such as enthalpy of vaporization (OPLS) or free energy of solvation of small molecules in polar and apolar solvents (GROMACS), for different purposes. The most popular force fields are as follows:

1. AMBER: A family of force fields called AMBER (Cornell *et al.*, 1995) were parameterized with energies of solvation and experimental vibrational frequencies in mind. AMBER exists as a full-atom or as a united atom force field, meaning that only polar or aromatic hydrogens are explicitly represented, whereas nonpolar hydrogens are incorporated into the carbon to which they are bound. Recently added, the general AMBER force field (GAFF; Wang *et al.*, 2004) was built in order to allow automatic parameterization of most small pharmaceutically relevant compounds. It is therefore suitable for large-scale docking studies of small molecules.
2. CHARMM: Developed in conjugation to the CHARMM software; it has few variations, the most used among them is the all-atom CHARMM22 force field (MacKerell *et al.*, 1998) and its dihedral potential corrected variant CHARMM22/CMAP (Mackerell *et al.*, 2004). CHARMM22 parameters were refined to reproduce densities and heats of vaporization of liquids as well as unit cell parameters and heats of sublimation for small molecule crystals.

3. GROMOS: The GROMOS force field, developed at the University of Groningen and ETH Zurich, is aimed at the computer simulation of biomolecules, mainly proteins. Its newest parameter sets were optimized to reproduce the thermodynamic properties of pure liquids of a range of small polar molecules and the solvation free enthalpies of amino acid analogs in cyclohexane (53A5) and water (53A6) (Oostenbrink *et al.*, 2004). In contrast to CHARMM, which has more than 200 different atom types, GROMOS has only 53 different atoms. As a result, addition of new molecules to the force field is an easy task, which can be done in an automated way (http://compbio.biosci.uq.edu.au/atb/).
4. MARTINI: The MARTINI force field (Marrink *et al.*, 2007) is developed at the University of Groningen. It is a CG force field, in which groups of heavy atoms and bound hydrogens (e.g., four water molecules) are represented by one entity, such that the atomistic description is trade off for ease of calculation. The MARTINI force field, suited for biomolecular simulations, has been parameterized to reproduce the partitioning free energies between polar and apolar phases of a large number of chemical compounds.
5. OPLS: The OPLS force field was initially aimed at reproducing experimental properties, such as density and heat of vaporization, of small organic molecules liquids (Jorgensen and Tirado-Rives, 1988). Following versions were evolved to include amino acid parameters (Kaminski *et al.*, 2000).

Force fields are ever evolving in order to better reproduce experimental observation. Most force fields were initially developed during the 1980s and were parameterized and validated using simulations tens of ps in length, limiting the sampling of configurations and hence their accuracy. However, the recent availability of long simulations enables better parameterization of force fields. As mentioned above, the CHARMM22 was refined in order to better reproduce the ϕ and ψ torsion angles of the protein backbone in long simulations (Mackerell *et al.*, 2004). This is important for the simulations needed for the study of phenomena such as the S-to-R transition in serpins.

To improve performance, electrostatic interactions are reduced to simple point charges (SPCs) in most force fields, neglecting the effect of polarization. The need for two parameters sets, one for a polar and the other for an apolar environment (Oostenbrink *et al.*, 2004), indicates that this simplification does prevent the force field from being universal. Recent studies show an accuracy improvement for water models with polarization compared to simple point condensed phase models (Lamoureux *et al.*, 2003; Yu and van Gunsteren, 2004).

8. Hardware

While most MD applications are heavily demanding on available computational resources, gradual advances in software and particularly the speed and cost of hardware have brought MD simulation within the grasp of most biology research laboratories. Suitable multicore architectures range from eight-core workstations to high-performance clusters consisting of several hundred, or even thousands or cores connected by low-latency and high-bandwidth networks (e.g., InfiniBand). CPU time on the latter is typically accessible to most laboratories on a merit time-allocation basis.

System size (e.g., 130,000 atoms for systems composed of α1-AT and water), simulation length (tens of ns to study local dynamics, tens of ms to study conformational changes), and the number of replicas (in the case of REMD) are major factors that dictate the computational resources required. As an example of the traditional high-performance cluster and its ability to scale to a large number of nodes, we report here a simulation of the above-mentioned system employing classical MD. Each node consisted of two quad-core 2.66 GHz Intel Nehalem (8 CPUs) connected by InfiniBand and running GROMACS 4.0.7 (see case study below for details). In all, 1 node produced 3.9 ns/day, 2 nodes 7.6 ns/day, 4 nodes 14.5 ns/day, and 10 nodes (80 processors) 33.9 ns/day. Hence a classical MD approach on a typical high-performance cluster would require 10 months to reach 10 μs on 10 nodes.

Dedicated computing facilities have also appeared, such as the IBM BlueGene, where typically 64–256 CPUs are present per node, with the goal of efficiently suppressing requirements for internode communication. These PowerPC processors are, however, slower than the typical workstation processor. In our experience simulating a system of 160,000 atoms with NAMD2.7b2 (PME, NPT, 2 fs time step) produced 1.75 ns/day on 128 CPUs of a BlueGene/L at 0.7 GHz. This was outperformed for the same system and software by the use of 6 × 8 Xserve Twin Quad-Core Intel Xeon 2.8 GHz (48 processors) with a 10 GB Myrinet connection, which produced 3 ns/day. Nonetheless, supercomputers like the BlueGene provide an impressive number of processors that can scale up to the tens of thousands, a number out of reach for the typical multicore architecture cluster.

GPUs consist of graphical processing units stacked in a simple card (also known as a "video card" installed by default on most workstations). The number of processing units can scale up to 500. On a system of ∼91,000 atoms, AMBER 11 can produce 5.2 ns/day (PME, NVE) on an NVidia Tesla C2050 (448 GPUs at 1.3 GHz) (as reported on the AMBER (Cornell *et al.*, 1995) website), while it only generated 1.7 ns/day on a Dual × Quad-Core Intel E5462 at 2.80 GHz (8 CPUs). The threefold improvement is significant, but again, this would require a simulation time of ~5 years to reach 10 μs.

While computer power has become more widespread, faster, and specialized, it remains a strong limitation for classical MD. The use of algorithms accelerating the sampling and the exploration of the potential energy landscape is hence a necessity.

Another point one should consider is data storage. A classical MD simulation usually requires saving to hard drive a time series of snapshots, typically every 2–10 ps (every 1000–5000 time steps with a time step of 2 fs). Fine structural details such as side chain rotation can thus be captured with high sampling accuracy. As a single snapshot of a system consisting of 130,000 atoms is ~3 MB (CHARMM binary format), large files are being produced. One microsecond would require storage of 600 GB of data when saving conformations every 5 ps. Therefore, specialized data storage is a critical requirement.

9. Case Study

The study of the E342K variant of α1-AT is an example of the benefit arising from employing theoretical methods. E342K α1-AT, a mutation located at the breach between the top of strand 5 of β-sheet A and the base of the RCL (Fig. 14.1), is known to facilitate spontaneous polymerization (Lomas *et al.*, 1992). Despite the wide biochemical studies, the precise structural mechanism by which it affects α1-AT stability remains unknown. In our laboratory, we have studied the effect of this mutation on α1-AT in atomic detail. For both the wildtype and the E342K-mutated α1-AT, we have simulated three 80 ns trajectories using classical MD, each starting with a different assigned set of velocities. The results (unpublished) indicate that E342K α1-AT has a higher rate of β-sheet A opening compared to wild type (Fig. 14.4). Based on this study, a new mechanism for the E342K mutation effect is proposed, in which a higher tendency of β-sheet A opening leads to polymerization.

The atomic coordinates of the WT α1-AT were taken from the crystal structure of Elliott *et al.* (2000) (PDB entry 1QLP). A model of E342K α1-AT was obtained by performing an *in silico* mutation using PyMOL. Each protein[1] was placed in a simulation box[2] and solvated in water.[3]

[1] Structural data, in the PDB format, were translated into the GROMOS force field using *pdb2gmx* routine.
[2] Box size was set using *editconf* routine such that the minimum distance between any protein's atom and the box will be equal or more than 1.4 nm.
[3] Solvation was done using *genbox* routine.

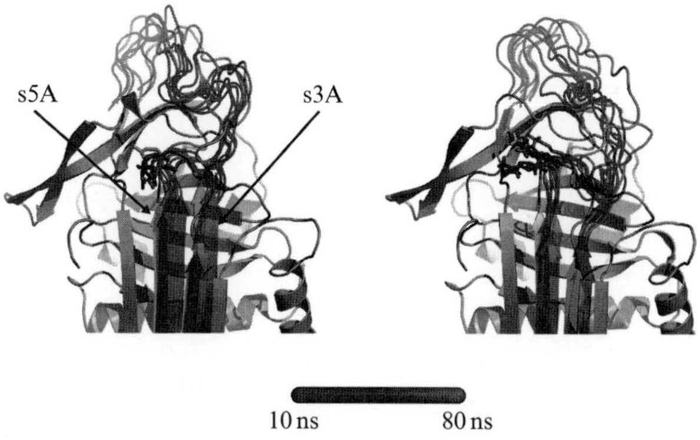

Figure 14.4 Superposition of snapshots taken every 10 ns from a 80-ns simulation of the initial structures of wildtype (left) and E342K-mutated α1-AT, shown in a cartoon representation. The color coding is according to the legend at the bottom, except the initial structure, which is colored in green. Residue 342 is shown in sticks representation. (For interpretation of the references to color in this figure legend, the reader is referred to the Web version of this chapter.)

Ions were added such that each box had a salt concentration of 0.1 *M* and a neutral total charge.[4] Each box was then energy minimized before commencing dynamics.[5] To avoid unnecessary distortion of the protein when the MD simulation is started, a gradual positional restraints procedure was used. Three consecutive equilibrations were performed in which all heavy protein atoms were restrained to their starting positions (using a force constant of 1000, 100, and 10 kJ/mol/nm^2, respectively), while the water was allowed to relax around the structure. Each system was then subjected to free simulation with configurations stored every 50 ps for analysis.

All simulations were performed using the GROMACS package version 4.0.7 (Hess *et al.*, 2008) in conjunction with the GROMOS 53A6 united atom force field (Oostenbrink *et al.*, 2004). Water was represented explicitly using the SPC model (Berendsen *et al.*, 1981). Each system was simulated under periodic conditions in a cubic box. The temperature throughout the simulation was maintained by independently coupling the protein and the solvent to an external temperature bath at 298 K (the temperature at which the force field was parameterized) with a coupling constant of $\tau_T = 0.1$ ps using a Berendsen thermostat (Berendsen *et al.*, 1984). The pressure was maintained at 1 bar by weakly coupling the system to an isotropic pressure bath (Berendsen *et al.*, 1984), using an isothermal compressibility of 4.6×10^{-5} bar^{-1} and a coupling constant of $\tau_P = 1$ ps. During the simulations, the length of all bonds

[4] Ions were added using the *genion* routine.
[5] The *grompp* routine was used in order to preprocess all data need for the *mdrun* program to commence simulations.

within the protein was constrained using the LINear Constraint Solver (LINCS) algorithm (Hess *et al.*, 1997). The time step for integrating the equations of motion was 2 fs. Nonbonded interactions were evaluated using a twin-range cutoff scheme: interactions falling within the 0.8 nm short-range cutoff were calculated every step, whereas interactions within the 1.4 nm long cutoff were updated every three steps, together with the pair list. A reaction-field correction was applied to the electrostatic interactions beyond the long-range cutoff (Tironi *et al.*, 1995), using a relative dielectric permittivity constant of $\varepsilon_{RF} = 62$ as appropriate for SPC water (Heinz *et al.*, 2001).

Numerical analysis of numerical properties such as Root Mean Square Deviation (RMSD) and Root Mean Square Fluctuation (RMSF)[6] was done using the analysis tools within the GROMACS package (Hess *et al.*, 2008) and were visualized using Grace (http://plasma-gate.weizmann.ac.il/Grace/). Trajectories were visualized using VMD molecular visualization software (Humphrey *et al.*, 1996).

10. Outlook

The serpin superfamily is vast and includes both inhibitory and noninhibitory proteins, all sharing a characteristic fold. Their inhibitory function depends on a metastable fold and its ability to rapidly undergo a dramatic structural change upon RCL cleavage by a target protease. The effects of naturally occurring mutations on the fold, activity, and polymerization of serpins have been studied intensely, yielding a large amount of structural and biochemical data. To complement the structural and biochemical data, one needs to study the dynamics of serpins at an atomic level. This proves to be a difficult task using conventional methods such as NMR but can be partially overcome by the use of theoretical methods.

In this chapter, we have described theoretical methods and algorithms that are available to study serpins and their characteristic structural plasticity. In addition, we have highlighted the questions that can be addressed by such methods. To help nonexpert readers to choose the right method, we briefly summarize the pros and cons of the discussed methods in Table 14.1.

In the past decade, the number of theoretical studies focusing on serpin dynamics has been modest, most likely due, in part at least, by the availability of computational resources. However, as the performance:cost ratio of CPU and disk storage is constantly improving, the past few years have brought significant CPU power within the reach of research laboratories. In addition, MD software continues to become both more efficient and easy to use by the nonspecialist. As a result, we anticipate that serpins will be increasingly studied using the MD methods outlined here.

Table 14.1 Summary of the methods and algorithms described in the text, including the major advantages and disadvantages associated with their use

Method	Pros	Cons	Examples
Classical MD	1. Gives a reliable atomic description of the system 2. Easy to set up and run	1. Can be slow for large systems 2. Can over-sample single energy well	Local dynamics, such as the effect of the E342K mutation on α1-AT
CG simulations	Speed up simulations	Loss of fine details, such as H-bonds	Large timescale phenomena, such as polymerization
MC	Faster than MD	Difficult to set up	Useful for loop conformations. Can be used to study the effect of mutation on RCL dynamics
NMA	Produce data on low-frequency domain motions	1. No dynamical data 2. Limited to a single conformation	Study natural deformation and flexibility of serpins
REMD	1. Better sampling of conformational space 2. Improved usage of computational resources	1. High computational demands 2. Can be hard to setup and interpretation	Study conformations that are difficult to access using classical MD or experimental techniques, such as structural intermediates of serpins
LED	Better sampling of conformational space	Caution in selecting degrees of freedom is affected by the modified potential	Enable the sampling of conformations differing from the most stable one. Can be used to study serpin S-to-L transition

(Continued)

Table 14.1 (*Continued*)

Method	Pros	Cons	Examples
Umbrella sampling	1. Better sampling of conformational space 2. Allow free-energy calculations along a pathway	Setup and analysis can be tricky	Umbrella sampling can be used for the study the initial steps of S-to-L or S-to-R transitions and polymerization
SMD	Allow study of molecules mechanical response	Careless settings can cause major artifacts	The study of ligand and/or protein binding to serpin, and their effect on serpin structure and dynamics. In addition, it can also be used for the study the initial steps of S-to-L or S-to-R transitions and polymerization
TMD	Predicts a pathway between two known conformations	Careless settings can cause major artifacts	

Using readily available computing clusters, single serpin molecule trajectories as long as tens of microseconds are accessible, providing significant insight into processes such as the incorporation of the RCL into the β-sheet A. Nevertheless, the calculation of millisecond simulations, as required for a full study of the S-to-R transition, is more challenging and will require the use of special methods.

REFERENCES

Acton, A., Banck, M., Bréfort, J., Cruz, M., Curtis, D., Hassinen, T., Heikkilä, V., Hutchison, G., Huuskonen, J., Jensen, J., Liboska, R., and Rowley, C. (2006). *Ghemical 2.0.* Department of Chemistry, University of Kuopio, Kuopio, Finland.

Allen, M. P., and Tildesley, D. J. (1987). Computer Simulation of Liquids. Oxford University Press, Oxford.

Atilgan, A. R., Durell, S. R., Jernigan, R. L., Demirel, M. C., Keskin, O., and Bahar, I. (2001). *Biophys. J.* **80,** 505–515.

Bakan, A., and Bahar, I. (2009). *Proc. Natl. Acad. Sci. USA* **106,** 14349–14354.

Baumann, U., Huber, R., Bode, W., Grosse, D., Lesjak, M., and Laurell, C. B. (1991). *J. Mol. Biol.* **218,** 595–606.

Beauchamp, N. J., Pike, R. N., Daly, M., Butler, L., Makris, M., Dafforn, T. R., Zhou, A., Fitton, H. L., Preston, F. E., Peake, I. R., and Carrell, R. W. (1998). *Blood* **92,** 2696–2706.

Beinrohr, L., Harmat, V., Dobo, J., Lorincz, Z., Gal, P., and Zavodszky, P. (2007). *J. Biol. Chem.* **282,** 21100–21109.

Berendsen, H. J. C., Postma, J. P. M., and van Gunsteren, W. F. (1981). *In* "Intermolecular Forces," (B. Pullman, ed.), pp. 331–342. Reidel, Dordrecht.

Berendsen, H. J. C., Postma, J. P. M., van Gunsteren, W. F., DiNola, A., and Haak, J. R. (1984). *J. Chem. Phys.* **81,** 3684–3690.

Berman, H., Henrick, K., Nakamura, H., and Markley, J. L. (2007). *Nucleic Acids Res.* **35,** D301–D303.

Bode, W., and Huber, R. (1992). *Eur. J. Biochem.* **204,** 433–451.

Bond, P. J., and Sansom, M. S. (2006). *J. Am. Chem. Soc.* **128,** 2697–2704.

Brooks, B. R., Brooks, C. L., 3rd, Mackerell, A. D., Jr., Nilsson, L., Petrella, R. J., Roux, B., Won, Y., Archontis, G., Bartels, C., Boresch, S., Caflisch, A., Caves, L., *et al.* (2009). *J. Comput. Chem.* **30,** 1545–1614.

Cabrita, L. D., Whisstock, J. C., and Bottomley, S. P. (2002). *Biochemistry* **41,** 4575–4581.

Cabrita, L. D., Dai, W., and Bottomley, S. P. (2004). *Biochemistry* **43,** 9834–9839.

Carrell, R. W., and Owen, M. C. (1985). *Nature* **317,** 730–732.

Carrell, R. W., Stein, P. E., Fermi, G., and Wardell, M. R. (1994). *Structure* **2,** 257–270.

Case, D. A., Cheatham, T. E., 3rd, Darden, T., Gohlke, H., Luo, R., Merz, K. M., Jr., Onufriev, A., Simmerling, C., Wang, B., and Woods, R. J. (2005). *J. Comput. Chem.* **26,** 1668–1688.

Chandrasekaran, V., Lee, C. J., Lin, P., Duke, R. E., and Pedersen, L. G. (2009). *J. Mol. Model.* **15,** 897–911.

Christen, M., Hunenberger, P. H., Bakowies, D., Baron, R., Burgi, R., Geerke, D. P., Heinz, T. N., Kastenholz, M. A., Krautler, V., Oostenbrink, C., Peter, C., Trzesniak, D., *et al.* (2005). *J. Comput. Chem.* **26,** 1719–1751.

Cornell, W. D., Cieplak, P., Bayly, C. I., Gould, I. R., Merz, K. M., Ferguson, D. M., Spellmeyer, D. C., Fox, T., Caldwell, J. W., and Kollman, P. A. (1995). *J. Am. Chem. Soc.* **117,** 5179–5197.

Cui, Q., and Karplus, M. (2008). *Protein Sci.* **17,** 1295–1307.

de Groot, B. L., van Aalten, D. M., Scheek, R. M., Amadei, A., Vriend, G., and Berendsen, H. J. (1997). *Proteins* **29,** 240–251.

Dementiev, A., Simonovic, M., Volz, K., and Gettins, P. G. (2003). *J. Biol. Chem.* **278,** 37881–37887.

Dinola, A., Berendsen, H. J. C., and Edholm, O. (1984). *Macromolecules* **17,** 2044–2050.

Dunstone, M. A., Dai, W., Whisstock, J. C., Rossjohn, J., Pike, R. N., Feil, S. C., Le Bonniec, B. F., Parker, M. W., and Bottomley, S. P. (2000). *Protein Sci.* **9,** 417–420.

Earnest, T., Fauman, E., Craik, C. S., and Stroud, R. (1991). *Proteins* **10,** 171–187.

Eastman, P., and Pande, V. (2010). *Comput. Sci. Eng.* **12,** 34–39.

Elliott, P. R., Pei, X. Y., Dafforn, T. R., and Lomas, D. A. (2000). *Protein Sci.* **9,** 1274–1281.

Engels, M., Jacoby, E., Kruger, P., Schlitter, J., and Wollmer, A. (1992). *Protein. Eng.* **5,** 669–677.

Eyal, E., Yang, L. W., and Bahar, I. (2006). *Bioinformatics* **22,** 2619–2627.

Felts, A. K., Gallicchio, E., Wallqvist, A., and Levy, R. M. (2002). *Proteins* **48,** 404–422.

Freddolino, P. L., Liu, F., Gruebele, M., and Schulten, K. (2008). *Biophys. J.* **94,** L75–L77.

Garcia, A. E., and Onuchic, J. N. (2003). *Proc. Natl. Acad. Sci. USA* **100,** 13898–13903.

Grubmuller, H., Heymann, B., and Tavan, P. (1996). *Science* **271,** 997–999.

Hansson, T., Oostenbrink, C., and van Gunsteren, W. (2002). *Curr. Opin. Struct. Biol.* **12,** 190–196.

Heinz, T. N., van Gunsteren, W. F., and Hünenberger, P. H. (2001). *J. Chem. Phys.* **115,** 1125–1136.

Herve, M., and Ghelis, C. (1991). *Arch. Biochem. Biophys.* **285,** 142–146.

Hess, B., Bekker, H., Berendsen, H. J. C., and Fraaije, J. G. E. M. (1997). *J. Comput. Chem.* **18,** 1463–1472.

Hess, B., Kutzner, C., van der Spoel, D., and Lindahl, E. (2008). *J. Chem. Theory Comput.* **4,** 435–447.

Hinsen, K. (1998). *Proteins* **33,** 417–429.

Hollup, S. M., Salensminde, G., and Reuter, N. (2005). *BMC Bioinformatics* **6,** 52.

Hopkins, P. C., Carrell, R. W., and Stone, S. R. (1993). *Biochemistry* **32,** 7650–7657.

Hou, T., and Yu, R. (2007). *J. Med. Chem.* **50,** 1177–1188.

Hu, X., Wang, H., Ke, H., and Kuhlman, B. (2007). *Proc. Natl. Acad. Sci. USA* **104,** 17668–17673.

Huber, T., Torda, A. E., and Vangunsteren, W. F. (1994). *J. Comput. Aided Mol. Des.* **8,** 695–708.

Humphrey, W., Dalke, A., and Schulten, K. (1996). *J. Mol. Graph.* **14**(1), 33–38.

Hunt, L. T., and Dayhoff, M. O. (1980). *Biochem. Biophys. Res. Commun.* **95,** 864–871.

Huntington, J. A., Pannu, N. S., Hazes, B., Read, R. J., Lomas, D. A., and Carrell, R. W. (1999). *J. Mol. Biol.* **293,** 449–455.

Huntington, J. A., Read, R. J., and Carrell, R. W. (2000). *Nature* **407,** 923–926.

Izrailev, S., Stepaniants, S., Balsera, M., Oono, Y., and Schulten, K. (1997). *Biophys. J.* **72,** 1568–1581.

Izrailev, S., Stepaniants, S., Isralewitz, B., Kosztin, D., Lu, H., Molnar, F., Wrigers, W., and Schulten, K. (1998). *In* "Computational Molecular Dynamics: Challenges, Methods, Ideas," (P. Denflhard, J. Hermans, B. Leimkuhler, A. Mark, R. Skeel, and S. Reich, eds.), pp. 39–65. Springer-Verlag, Berlin.

James, E. L., and Bottomley, S. P. (1998). *Arch. Biochem. Biophys.* **356,** 296–300.

Jezierski, G., and Pasenkiewicz-Gierula, M. (2001). *Acta Biochim. Pol.* **48,** 65–75.

Jin, L., Abrahams, J. P., Skinner, R., Petitou, M., Pike, R. N., and Carrell, R. W. (1997). *Proc. Natl. Acad. Sci. USA* **94,** 14683–14688.
Jorgensen, W. L., and Tirado-Rives, J. (1988). *J. Am. Chem. Soc.* **110,** 1657–1666.
Kaiserman, D., and Bird, P. I. (2010). *Cell Death Differ.* **17,** 586–595.
Kaminski, G. A., Friesner, R. A., Tirado-Rives, J., and Jorgensen, W. L. (2000). *Abstr. Pap. Am. Chem. Soc.* **220,** U279–U279.
Karplus, M., and McCammon, J. A. (2002). *Nat. Struct. Biol.* **9,** 646–652.
Kaslik, G., Patthy, A., Balint, M., and Graf, L. (1995). *FEBS Lett.* **370,** 179–183.
Kass, I., and Arkin, I. T. (2005). *Structure* **13**(12), 1789–1798.
Keskin, O., Bahar, I., Flatow, D., Covell, D. G., and Jernigan, R. L. (2002). *Biochemistry* **41,** 491–501.
Kondrashov, D. A., Van Wynsberghe, A. W., Bannen, R. M., Cui, Q., and Phillips, G. N., Jr. (2007). *Structure* **15,** 169–177.
Kurkinen, M., Taylor, A., Garrels, J. I., and Hogan, B. L. M. (1984). *J. Biol. Chem.* **259,** 5915–5922.
Laio, A., and Parrinello, M. (2002). *Proc. Natl. Acad. Sci. USA* **99,** 12562–12566.
Lamoureux, G., MacKerell, J. A. D., and Roux, B. (2003). *J. Chem. Phys.* **119,** 5185–5197.
Lane, D. A., Denton, J., Flynn, A. M., Thunberg, L., and Lindahl, U. (1984). *Biochem. J.* **218,** 725–732.
Law, R. H., Sofian, T., Kan, W. T., Horvath, A. J., Hitchen, C. R., Langendorf, C. G., Buckle, A. M., Whisstock, J. C., and Coughlin, P. B. (2008). *Blood* **111,** 2049–2052.
Leach, A. R. (2001). Molecular Modelling: Principles and Applications. Prentice Hall, Harlow.
Leontiadou, H., Mark, A. E., and Marrink, S. J. (2006). *J. Am. Chem. Soc.* **128,** 12156–12161.
Levitt, M., Sander, C., and Stern, P. S. (1985). *J. Mol. Biol.* **181,** 423–447.
Li, W., Johnson, D. J., Esmon, C. T., and Huntington, J. A. (2004). *Nat. Struct. Mol. Biol.* **11,** 857–862.
Lomas, D. A., Evans, D. L., Finch, J. T., and Carrell, R. W. (1992). *Nature* **357,** 605–607.
Lomas, D. A., Evans, D. L., Stone, S. R., Chang, W. S., and Carrell, R. W. (1993). *Biochemistry* **32,** 500–508.
Lomas, D. A., Elliott, P. R., Chang, W. S., Wardell, M. R., and Carrell, R. W. (1995). *J. Biol. Chem.* **270,** 5282–5288.
Ma, J., and Karplus, M. (1998). *Proc. Natl. Acad. Sci. USA* **95,** 8502–8507.
MacKerell, A. D., Bashford, D., Bellott, M., Dunbrack, R. L., Evanseck, J. D., Field, M. J., Fischer, S., Gao, J., Guo, H., Ha, S., Joseph-McCarthy, D., Kuchnir, L., *et al.* (1998). *J. Phys. Chem. B* **102,** 3586–3616.
Mackerell, A. D., Jr., Feig, M., and Brooks, C. L., 3rd (2004). *J. Comput. Chem.* **25,** 1400–1415.
Mandell, D. J., Coutsias, E. A., and Kortemme, T. (2009). *Nat. Methods* **6,** 551–552.
Marrink, S. J., Risselada, H. J., Yefimov, S., Tieleman, D. P., and de Vries, A. H. (2007). *J. Phys. Chem. B* **111,** 7812–7824.
Mast, A. E., Enghild, J. J., and Salvesen, G. (1992). *Biochemistry* **31,** 2720–2728.
Metropolis, N., Rosenbluth, A. W., Rosenbluth, M. N., Teller, A. H., and Teller, E. (1953). *J. Chem. Phys.* **21,** 1087–1092.
Mottonen, J., Strand, A., Symersky, J., Sweet, R. M., Danley, D. E., Geoghegan, K. F., Gerard, R. D., and Goldsmith, E. J. (1992). *Nature* **355,** 270–273.
Northrup, S. H., Pear, M. R., Lee, C. Y., Mccammon, J. A., and Karplus, M. (1982). *Proc. Natl. Acad. Sci.* **79,** 4035–4039.
Oostenbrink, C., Villa, A., Mark, A. E., and van Gunsteren, W. F. (2004). *J. Comput. Chem.* **25**(13), 1656–1676.
Ota, N., and Brunger, A. T. (1997). *Theor. Chem. Acc.* **98,** 171–181.

Ota, N., Stroupe, C., Ferreira-da-Silva, J. M. S., Shah, S. A., Mares-Guia, M., and Brunger, A. T. (1999). *Proteins Struct. Funct. Genet.* **37,** 641–653.
Otlewski, J., Jelen, F., Zakrzewska, M., and Oleksy, A. (2005). *EMBO J.* **24,** 1303–1310.
Pettersen, E. F., Goddard, T. D., Huang, C. C., Couch, G. S., Greenblatt, D. M., Meng, E. C., and Ferrin, T. E. (2004). *J. Comput. Chem.* **25,** 1605–1612.
Phillips, J. C., Braun, R., Wang, W., Gumbart, J., Tajkhorshid, E., Villa, E., Chipot, C., Skeel, R. D., Kale, L., and Schulten, K. (2005). *J. Comput. Chem.* **26,** 1781–1802.
Poger, D., and Mark, A. E. (2010). *Proteins* **78,** 1163–1174.
Rezaie, A. R. (1998). *J. Biol. Chem.* **273,** 16824–16827.
Rezaie, A. R., and Olson, S. T. (2000). *Biochemistry* **39,** 12083–12090.
Rzepiela, A. J., Schafer, L. V., Goga, N., Risselada, H. J., De Vries, A. H., and Marrink, S. J. (2010). *J. Comput. Chem.* **31,** 1333–1343.
Sengupta, D., and Marrink, S. J. (2010). *Phys. Chem. Chem. Phys* **12,** 12987–12996.
Sengupta, D., Rampioni, A., and Marrink, S. J. (2009). *Mol. Membr. Biol.* **26,** 422–434.
Siiteri, P. K., Murai, J. T., Hammond, G. L., Nisker, J. A., Raymoure, W. J., and Kuhn, R. W. (1982). *Rec. Prog. Horm. Res.* **38,** 457–510.
Stavridi, E. S., O'Malley, K., Lukacs, C. M., Moore, W. T., Lambris, J. D., Christianson, D. W., Rubin, H., and Cooperman, B. S. (1996). *Biochemistry* **35,** 10608–10615.
Stone, J. E., Hardy, D. J., Ufimtsev, I. S., and Schulten, K. (2010). *J. Mol. Graph. Model.* **29,** 116–125.
Stout, T. J., Graham, H., Buckley, D. I., and Matthews, D. J. (2000). *Biochemistry* **39,** 8460–8469.
Sugita, Y., and Okamoto, Y. (1999). *Chem. Phys. Lett.* **314,** 141–151.
Suhre, K., and Sanejouand, Y. H. (2004). *Nucleic Acids Res.* **32,** W610–W614.
Swendsen, R. H., and Wang, J. S. (1986). *Phys. Rev. Lett.* **57,** 2607–2609.
SYBYL, Tripos International, 1699 South Hanley Rd., St. Louis, Missouri, 63144, USA.
Tama, F., Gadea, F. X., Marques, O., and Sanejouand, Y. H. (2000). *Proteins* **41,** 1–7.
Tew, D. J., and Bottomley, S. P. (2001). *FEBS Lett.* **494,** 30–33.
Tieleman, D. P., Marrink, S. J., and Berendsen, H. J. (1997). *Biochim. Biophys. Acta* **1331,** 235–270.
Tirion, M. M. (1996). *Phys. Rev. Lett.* **77,** 1905–1908.
Tironi, I. G., Sperb, R., Smith, P. E., and Vangunsteren, W. F. (1995). *J. Chem. Phys.* **102,** 5451–5459.
Torrie, G. M., and Valleau, J. P. (1977). *J. Comput. Phys.* **23,** 187–199.
Valadie, H., Lacapcre, J. J., Sanejouand, Y. H., and Etchebest, C. (2003). *J. Mol. Biol.* **332,** 657–674.
van Gunsteren, W. F., Billeter, S. R., Eising, A. A., Hunenberger, P. H., Kruger, P., Mark, A. E., Scott, W. R. P., and Tironi, I. G. (1996). Biomolecular Simulation: The GROMOS96 manual and user guide. Vdf Hochschulverlag AG an der ETH Zuerich, Zuerich.
van Gunsteren, W. F., Bakowies, D., Baron, R., Chandrasekhar, I., Christen, M., Daura, X., Gee, P., Geerke, D. P., Glattli, A., Hunenberger, P. H., Kastenholz, M. A., Oostenbrink, C., *et al.* (2006). *Angew. Chem. Int. Ed. Engl.* **45,** 4064–4092.
Verli, H., and Guimaraes, J. A. (2005). *J. Mol. Graph. Model.* **24,** 203–212.
Wang, J., Wolf, R. M., Caldwell, J. W., Kollman, P. A., and Case, D. A. (2004). *J. Comput. Chem.* **25,** 1157–1174.
Yamasaki, M., Li, W., Johnson, D. J., and Huntington, J. A. (2008). *Nature* **455,** 1255–1258.
Yu, H., and van Gunsteren, W. F. (2004). *J. Chem. Phys.* **121,** 9549–9564.

Zhang, Q., Buckle, A. M., Law, R. H., Pearce, M. C., Cabrita, L. D., Lloyd, G. J., Irving, J. A., Smith, A. I., Ruzyla, K., Rossjohn, J., Bottomley, S. P., and Whisstock, J. C. (2007). *EMBO Rep.* **8,** 658–663.
Zheng, X. J., Wintrode, P. L., and Chance, M. R. (2008). *Structure* **16,** 38–51.
Zhou, R. H. (2004). *J. Mol. Graph. Model.* **22,** 451–463.
Zou, Z., Anisowicz, A., Hendrix, M. J., Thor, A., Neveu, M., Sheng, S., Rafidi, K., Seftor, E., and Sager, R. (1994). *Science* **263,** 526–529.

CHAPTER FIFTEEN

Probing Serpin Conformational Change Using Mass Spectrometry and Related Methods

Yuko Tsutsui,* Anindya Sarkar,† *and* Patrick L. Wintrode†

Contents

1. Introduction 326
2. Applications of HXMS 331
   2.1. Native state dynamics 331
3. Determination of Thermodynamic Stability Using Hydrogen–Deuterium Exchange Combined with Mass Spectrometry 336
   3.1. Equilibrium unfolding study of WT α_1-AT using the pulse-labeling technique 337
   3.2. Equilibrium unfolding of G117F mutant studied by HXMS 340
   3.3. Effects of glycosylation on the equilibrium unfolding of α_1-AT observed using HXMS 341
4. "Functional Unfolding" During the Native → Cleaved Transition 343
5. Investigating the Polymerization Pathway and Polymer Structure of α_1-AT by HXMS and Ion Mobility MS 344
6. Future Prospects 348
References 348

Abstract

The folding, misfolding, and inhibitory mechanisms of serpins are linked to both thermodynamic metastability and conformational flexibility. Characterizing the structural distribution of stability and flexibility in serpins in solution is challenging due to their large size and propensity for aggregation. Structural mass spectrometry techniques offer powerful tools for probing the mechanisms of serpin function and disfunction. In this chapter, we review the principles of the two most commonly employed structural mass spectrometry techniques—hydrogen/deuterium exchange and chemical footprinting—and describe their application to studying serpin flexibility, stability, and conformational change in

* Division of Chemistry and Chemical Engineering, California Institute of Technology, Pasadena, California, USA
† Department of Physiology & Biophysics, Case Western Reserve University, Cleveland, Ohio, USA

Methods in Enzymology, Volume 501
ISSN 0076-6879, DOI: 10.1016/B978-0-12-385950-1.00015-8

solution. We also review the application of both hydrogen/deuterium exchange and ion mobility mass spectrometry to probe the mechanism of serpin polymerization and the structure of serpin polymers.

1. Introduction

Conformational mobility is key to serpin function, disfunction, and regulation (Gettins, 2002). In order to inhibit their target proteases, serpins must undergo a massive conformational change in which the solvent-exposed reactive center loop (RCL) inserts into the five stranded β-sheet A, becoming a sixth strand. A similar rearrangement occurs when serpins such as PAI-1 spontaneously transition from the active to the latent form. Serpins also undergo subtler conformational changes, such as those responsible for the allosteric activation of antithrombin by heparin. Pathological serpin disfunction is also mediated by conformational change: all of the currently proposed models for serpin polymerization require considerable remodeling of sheet A (Sivasothy *et al.*, 2000; Yamasaki *et al.*, 2008).

While nuclear magnetic resonance (NMR) is the premier technique for characterizing the conformational dynamics of proteins in solution, the serpins' relatively large size (~45 kDa) and propensity for aggregation make them poor candidates for high-resolution NMR studies. Techniques such as fluorescence resonance energy transfer and site-directed spin labeling can provide information on both the structure and dynamics of large proteins in solution. However, these techniques require engineering cysteins into the protein of interest and then attaching optical or paramagnetic probes that can potentially perturb native structure. Additionally, the structural information provided by these methods, while valuable, is limited to the probe site(s).

Relative to other techniques for probing solution structure and dynamics, structural mass spectrometry methods offer a number of attractive features. They provide the ability to simultaneously monitor a large number of intrinsic, nonperturbing probes (amide hydrogens or reactive side chains) that are distributed throughout the structure. Mass spectrometry-based methods require small amounts of sample (as little as nanomoles of protein) and, because the analysis step is separate from the labeling step, interrogation of protein structure and dynamics can be performed under a wide range of conditions. In this chapter, we review the principles and procedures of the most common structural mass spectrometry methods—hydrogen/deuterium exchange and oxidative footprinting—and describe their application to probing serpin structure, dynamics, and misfolding.

Protein hydrogen/deuterium exchange exploits the fact that backbone amide hydrogen atoms will readily exchange with deuterium when a protein is incubated in D_2O (Bai *et al.*, 1993; Fig. 15.1). Exchange is catalyzed by both acid and base and is thus strongly pH dependent: a fact of critical importance for hydrogen exchange mass spectrometry (HXMS). The rate of exchange is at a minimum at pH 2.5 and increases by approximately a factor of 10 with each pH unit increase or decrease. The intrinsic exchange rate is also sequence dependent, and the exchange rates in unstructured peptides can vary by a factor of ~30 depending on the amino acid sequence. Intrinsic exchange rates can be predicted using data from extensive NMR studies of model peptides (Bai *et al.*, 1993).

In folded proteins, exchange rates depend on the local structural environment. Amide hydrogen atoms that are sequestered from solvent or involved in hydrogen bonding are protected from exchange. In order for exchange at these positions to occur, local and/or global fluctuations must disrupt the structure and expose amide hydrogens to solvent (Wales and Engen, 2006). The frequency with which such fluctuations occur determine the exchange rate for a given amide hydrogen. Rates of hydrogen exchange in folded proteins span eight orders of magnitude, making them extremely sensitive probes of local stability and dynamics. Interpretation of measured H/D exchange rates depends on the relative rates of the underlying processes: the "breaking" of local structure that exposes an amide hydrogen for exchange, k_1; the reformation of this structure, k_{-1}; and the intrinsic chemical rate of exchange for the exposed hydrogen, k_{int}. The process is described schematically as

$$\text{Closed(H)} \underset{k_{-1}}{\overset{k_1}{\rightleftharpoons}} \text{Open(H)} \overset{k_{\text{int}}}{\rightarrow} \text{Open(D)} \underset{k_1}{\overset{k_{-1}}{\rightleftharpoons}} \text{Closed(D)}$$

When the dynamics of protein fluctuations are fast compared with the chemical rate of exchange, then $k_{-1} >> k_{\text{int}}$ and the observed rate of exchange can be expressed as

$$k_{\text{obs}} = \frac{k_1}{k_{-1}} k_{\text{int}} = K_{\text{unf}} k_{\text{int}} \tag{15.1}$$

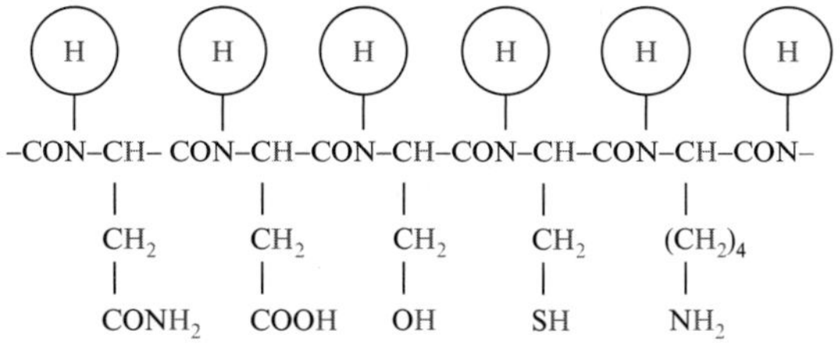

Figure 15.1 Hydrogens in proteins. The amide hydrogens monitored by H/D exchange are circled.

where K_{unf} is the equilibrium constant for the unfolding/refolding process(es) that expose the amide hydrogen in question for exchange. The case when $k_{-1} >> k_{int}$ is referred to as the EX2 regime of exchange, and Eq. (15.1) reflects the fact that under these conditions, unfolding/refolding will typically occur many times before exchange occurs, and thus the observed rate of exchange is proportional to the amount of time that the amide hydrogen spends in an exposed conformation. Hydrogen atoms in less stable regions that unfold frequently will exchange quickly while those in stable regions that rarely unfold will exchange slowly.

The conditions for EX2 exchange will typically hold for most proteins under physiological conditions. Denaturants such as urea or guanidine hydrochloride (GuHCl) will slow the rate of refolding, while alkaline pH will dramatically increase the chemical rate of exchange. As a result, in the presence of denaturants or at high pH, $k_{-1} \ll k_{int}$, and

$$k_{obs} = k_1 \tag{15.2}$$

This is referred to as the EX1 regime of exchange, and the observed exchange rate is a direct read out of the rate of unfolding. EX1 exchange is especially significant for H/D exchange studies carried out using mass spectrometry, as it allows the direct quantification of different structural populations in solution. This will be described in more detail below.

The basic HXMS experiment is illustrated schematically in Fig. 15.2. Protein is diluted into D_2O under the desired conditions and incubated for a fixed time period. The exchange reaction is then "quenched" by dropping

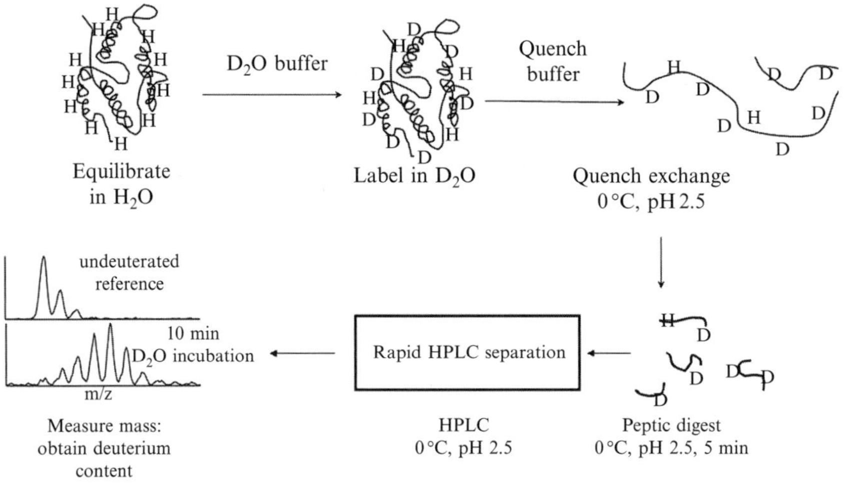

Figure 15.2 Overview of the experimental procedure for measuring local H/D exchange rates using mass spectrometry.

the pH to 2.5 and the temperature to 0 °C. This slows the chemical rate of H/D exchange by approximately five orders of magnitude. Protein is then digested with the acid protease pepsin. This can be accomplished either by adding pepsin to the quenched solution and incubating at 0 °C for 5 min or by passing the quenched protein over a column of immobilized pepsin. The latter method has generally been reported to give more efficient digestion. The digest is then loaded on to a reverse phase HPLC column and eluted directly into an electrospray ionization mass spectrometer. As deuterium is one mass unit heavier than hydrogen, the extent of deuterium uptake in each peptic fragment can be determined from the mass shift relative to an undeuterated reference sample.

Even under quenched conditions, back exchange of deuterium with hydrogen occurs, albeit slowly, and it is therefore desirable to correct for this deuterium loss due to back exchange. This is accomplished by preparing a 100% deuterated reference sample and subjecting it to the identical procedures of digestion and analysis that are used on the experimental samples. A 100% deuterated sample can be prepared by incubation the protein in a solution of D_2O containing high concentrations of a denaturant such as guanidine deuterochloride. By comparing the measured level of deuteration for this reference sample with the theoretical maximum deuterium level, the degree of back exchange for each peptide can be determined. When conditions are optimized, it is generally possible to achieve average back exchange levels of 15–20%. The reference sample can be used to correct for back exchange using the following equation:

$$D = \left(\frac{m - m_{0\%}}{m_{100\%} - m_{0\%}} \right) N$$

where D is the corrected level of deuterium incorporation, $m_{0\%}$ is the measured mass of an undeuterated reference sample, $m_{100\%}$ is the measured mass of the fully deuterated reference sample, m is the measured mass of the experimental sample, and N is the total number of exchangeable amide hydrogens on the peptide of interest. Plotting this corrected deuterium level versus time for each peptic fragment reveals the local rates of H/D exchange in different regions of a protein's structure. It should be noted that fully deuterated reference samples can be difficult to prepare because denaturants result in noisy spectra with the result that some peptides of interest may not be detectable. If the same protein is being compared under different conditions (e.g., in the presence/absence of a ligand), then relative deuterium levels may be compared directly without correcting for the extent of back exchange, which should be identical within error provided that experimental conditions such as temperature, pH, and ionic strength are held constant. The above description applies when exchange occurs in the EX2 regime.

EX1 exchange results in characteristic "double isotopic envelopes" which will be discussed in more detail below when equilibrium unfolding studies of serpins are described.

Hydroxyl-radical-mediated (•OH) protein footprinting combined with mass spectrometric detection has recently been developed to define protein structure, assembly, and conformational changes in solution based on measurements of the reactivity of amino acid side chains (Takamoto and Chance, 2006). This method probes protein structure by monitoring solvent accessibility using •OH modification (Xu and Chance, 2007). Solvent-exposed reactive side chains are readily oxidized by •OH, while buried side chains resist oxidation. Oxidized side chains can subsequently be identified by characteristic mass shifts using mass spectrometry (Fig. 15.3). •OH radicals can be generated by several methods including Fenton reagent, photooxidation of peroxide, and radiolysis of water (Guan and Chance, 2005). The intrinsic reactivity of side chains with •OH radicals has been extensively studied, and their relative susceptibility to oxidation is Cys > Met > Trp > Tyr > Phe > His > Leu, Ile > Arg, Lys, Val > Pro, Ser, Thr > Gln, Glu > Asp, Asn > Ala > Gly (Takamoto and Chance, 2006). Due to either low reactivity or low detectability of the resulting modified species, Gly, Ala, Asp, Asn, Arg, and Glu make for poor probes in footprinting experiments, leaving the remaining 14 standard amino acids as potential probes.

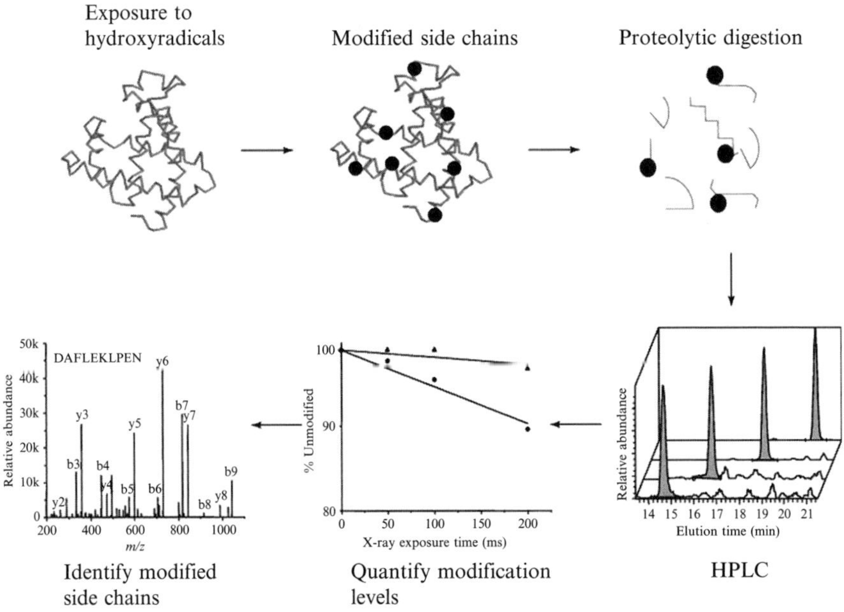

Figure 15.3 Overview of the experimental procedure for monitoring side-chain solvent accessibility using radiolytic footprinting.

Hydroxyradical footprinting has both advantages and disadvantages compared with HXMS. Footprinting does not provide information on entire polypeptide chain as HXMS potentially does (excluding prolines). However, the 14 usable side chains are usually well distributed throughout the sequence, and for these side chains footprinting provides single amino acid resolution. Perhaps more importantly, side-chain oxidation is a stable covalent modification that is not subject to rapid back exchange like H/D exchange. This allows for larger and more complex protein systems to be studied because time-consuming postlabeling purification procedures can be employed.

2. Applications of HXMS

2.1. Native state dynamics

2.1.1. Dynamics of a canonical serine protease inhibitor, α_1-antitrypsin studied by H/D exchange and radiolytic footprinting

Because the serpins require extensive conformational mobility when carrying out their functions, the dynamical properties of the native state are of interest. Tsutsui *et al.* (2006) applied HXMS to probe the native state dynamics of the canonical serpin human wild-type α_1-antitrypsin (WT α_1-AT) in the metastable form.

This study employed a version of the standard protocol for HXMS of native proteins:

1. Purified protein (5 μg, 565 n*M*) in H_2O, 10 m*M* phosphate buffer, 50 m*M* NaCl at pH 7.8 was diluted 24-fold into D_2O, 10 m*M* phosphate buffer pD 7.8 and incubated for different time periods at room temperature. For preparing D_2O buffers recall that pD is the reading on a standard pH meter +0.4.
2. At time intervals ranging 5–3000 s, the reaction was quenched by the addition of an equal volume of cold 100 m*M* phosphate, pH 2.4 and immediately frozen by immersion in a dry ice ethanol bath. Samples were stored at −80 °C.
3. Pepsin was dissolved at 1 mg/ml in H_2O + 0.05% TFA, pH 2.5. Frozen samples were thawed and pepsin was added to give a final mass/mass ratio of 1:1. Digestion was allowed to proceed on ice for 5 min.
4. The digest was loaded onto a reverse phase C18 column immersed in ice and eluted directly into an electrospray ionization mass spectrometer with a water–acetonitrile gradient from 10% to 45% acetonitrile in 12 min. The acetonitrile contained 0.05% TFA.

This basic protocol can be varied in a number of ways. To reduce the final volume to be injected onto the HPLC column, concentrated acid may be added to lower the pH rather than 100 m*M* phosphate buffer. The elution gradient can also be varied in order to optimize the separation of peptic fragments, although it is recommended to keep the total elution time < 20 min.

In Fig. 15.4, the flexibility of native α_1-AT is depicted as follows: for each peptic fragment, the ratio of the number of slowly exchanging amide hydrogens (measured by HXMS) to the number of amides participating in hydrogen bonds (determined from the crystal structure) is calculated. A ratio close to 1 indicates stable hydrogen bonding and therefore rigid structure. As expected from the crystal structure of WT α_1-AT in the metastable state (Elliott *et al.*, 2000), amide hydrogen atoms in the RCL underwent rapid exchange with deuterium, reaching 80% deuterium uptake within 10 s of labeling time due to solvent-exposed amide hydrogen atoms (Fig. 15.3). This lack of protection in the RCL also indicates the absence of protein aggregation during the experiment, as the formation of the loop-sheet polymers requires burial of amide hydrogen atoms in the RCL region between β-strands 3A and 5A. HXMS data further indicate that not only the formation of the loop-sheet polymers but also a spontaneous conformational change to the latent form does not occur under the experimental conditions. For example, structural rigidity in β-strand 1C hinders "peeling-off" of the strand from the adjacent β-strand 4C, thereby restricting the translational mobility of the RCL. Further, slow deuterium uptake of amide hydrogen atoms in β-strands 3A and 5A indicates stable

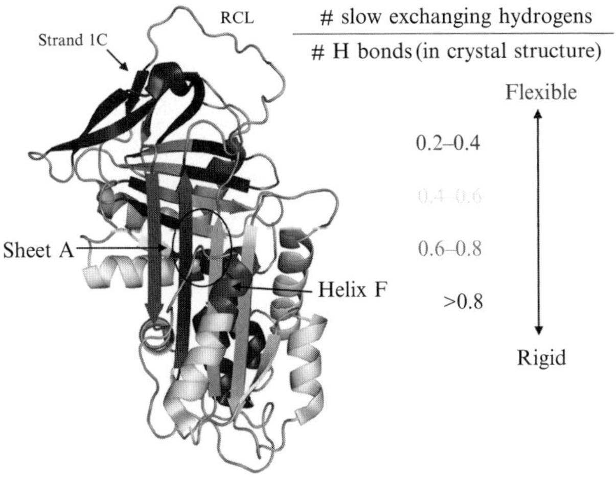

Figure 15.4 Distribution of conformational flexibility in the metastable form of α_1-AT determined by HXMS. (See Color Insert.)

hydrogen bonds between these strands. These observations suggest that a conformer with a partial unfolding of β-strands 3A and 5A is not significantly populated in the native state ensemble.

Comparable deuterium uptake to the RCL region was observed in the N-terminal half of helix A and the C-terminal half (the top half) of F-helix. Moderate flexibility was observed in the C-terminal half of helix A, the N-terminal half of F-helix, and helices D and E. Although these regions appear to form stable helices in the crystal structure (Elliott *et al.*, 2000), HXMS data indicate structural flexibility in these regions. These findings highlight the strength of hydrogen–deuterium exchange, allowing one to characterize transiently populated conformers that cannot be detected by many other techniques. Flexibility in these regions appears to be important for the inhibitory function. Since F-helix is situated nearby the RCL insertion site, a conformational change to form a serpin–protease covalent complex requires either rigid-body movement of F-helix away from β-sheet A or unfolding of F-helix to avoid a steric clash between the protease and the F-helix. Rapid deuterium uptake in the top of F-helix suggests a partial unfolding of this region; therefore, the conformation of F-helix in the metastable state is amenable to undergo the conformational change prior to the protease binding. Flexibility in helices A and D is suggested to play an important role in functional regulation of other serpins such as antithrombin and plasminogen activator inhibitor-I through allostery (Gettins, 2002). Interestingly, viral and thermophilic bacterial serpins completely lack helix D or have their N-terminal half of helix A or D truncated (Gettins, 2002). Since evolution tends to exploit flexible regions to optimize protein function, structural fluctuations in helices A and D hint an intimate relationship between evolution of allostery and function.

The metastable form of α_1-AT was also studied using radiolytic footprinting. Footprinting indicates that, as expected, solvent-exposed side chains are readily modified while most buried side chains are protected (Zheng *et al.*, 2008). There are however, several exceptions. Tyr 160, which is located near the C terminus of the F-helix and is largely buried, is readily oxidized, indicating that this region is highly labile in solution. This finding is consistent with results from H/D exchange. Met 374 and Met 385 are both nearly totally buried, and their sulfur atoms are completely inaccessible. Nevertheless, both are oxidized during footprinting. These residues are located near cavities that have been identified as contributing to inhibitory efficiency through cavity-filling mutagenesis studies (Lee *et al.*, 2001). Facile oxidation of these putatively buried side chains suggests that the structure near these cavities is highly dynamic, and this is supported by molecular dynamics simulations, which indicate that structural fluctuations transiently expose the sulfur atoms of both Met 374 and Met 385. The structural environment around these cavities is thus highly dynamic, and such structural lability may contribute to inhibition.

2.1.2. The effects of a stabilizing cavity-filling mutation on the native state dynamics of α_1-antitrypsin

Numerous mutagenesis studies have been performed to investigate a relationship between the inhibitory function and the thermodynamic stability of α_1-AT. It has been proposed that the metastability of α_1-AT arises from inefficient structural packing, and such structural defects in the metastable state facilitate the conformational change to form a thermodynamically more stable protease–serpin complex during inhibition (Lee *et al.*, 2000). Therefore, mutations that improve side-chain packing are expected to increase the thermodynamic stability of α_1-AT relative to wild type and result in decreased inhibitory activity. In particular, G117F represents such a mutation that fills a cavity between β-sheet 2A and F-helix (Lee *et al.*, 2000). In the crystal structure of G117F, no obvious structural difference was observed compared to wild type except a shift in the position of the F-helix (Gooptu *et al.*, 2009). The downward shift of the F-helix in G117F introduces repacking of residues at an interface between F-helix and β-sheet 2A, and these local interactions at the interface were proposed to contribute to the stabilization of the mutant in the metastable state.

Sengupta *et al.* (2009) investigated the dynamics of G117F mutant by HXMS. Several regions that are involved in the conformational change during the protease inhibition showed reduced exchange with deuterium compared to that of the same regions in wild type (Fig. 15.5). As expected

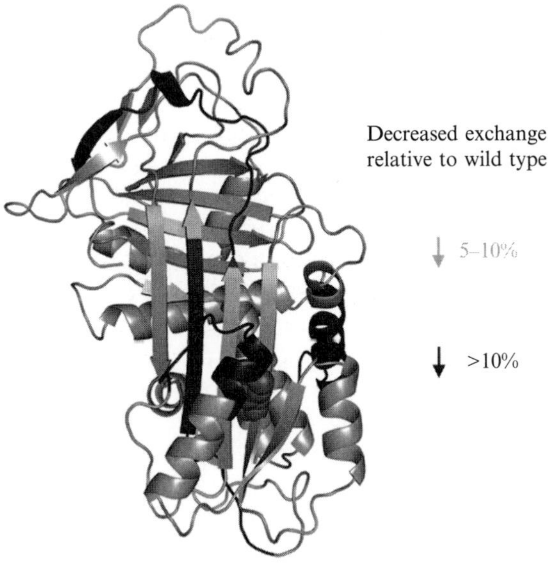

Figure 15.5 Allosteric rigidification resulting from the cavity-filling G117F mutation in α_1-AT. F117 is shown as magenta spheres. (For interpretation of the references to color in this figure legend, the reader is referred to the web version of this chapter.)

from the crystal structure of G117F, significant changes in deuterium uptake were observed in the F-helix. In wild type, the C-terminal region of the F-helix underwent rapid exchange with deuterium, with rates comparable to the RCL region; in contrast, the same region in G117F was found to be structurally more rigid as judged by reduced deuterium uptake. However, the effects of the mutation were not confined locally. Instead, the effects of the mutation were propagated to distant key regions involved in inhibition. A key region for the RCL insertion, β-strand 5A, also showed slower deuterium exchange than wild type. Decreased structural fluctuations in these regions indicate a difficulty to access to a conformer that can readily form the protease–serpin complex, as structural flexibility reflects frequency of structural deviations from the ensemble averaged structure represented by the crystal structure. Therefore, the underlying mechanism of reduced inhibitory activity seen in G117F is due to an inability to access to a conformer that readily undergoes a global conformational change during the inhibition. Reduced exchange was also seen in β-sheet C and helix D. When introduced into the pathological Glu342 → Lys (Z) mutant of α_1-AT, the G117F mutation leads to decreased polymerization and increased secretion from cultured cells. It has been shown that removal of strand 1C from the rest of sheet C is required for polymer formation, and one potential mechanism for resisting polymerization is strengthened interactions between strands 1C and 2C, as evidenced by decreased H/D exchange rates in this region in G117F (Meagher *et al.*, 2000).

2.1.3. Effects of glycosylation on the native state dynamics of human α_1-antitrypsin

Glycosylation, a common posttranslational modification, plays a number of crucial roles, including modulating intermolecular interactions, helping prevent aggregation, increasing the lifetime of circulating proteins by conferring resistance to proteolytic degradation, and intrinsic stabilization of protein structure (Shental-Bechor and Levy, 2009; Varki, 1993). The majority of previous biophysical studies of α_1-AT have employed a recombinant form that is produced in *Escherichia coli* and is therefore unglycosylated. Glycosylation usually increases the global stability of proteins, and further, global stabilization is often accompanied by reduced flexibility in the native state (Shental-Bechor and Levy, 2009). It has been shown previously that glycosylation increases the stability of α_1-AT against both thermal and chemical denaturation (Kwon and Yu, 1997).

Human plasma α_1-AT (HPα_1-AT) is normally fully glycosylated at three different Asn residues (Asn46, Asn83, and Asn247) with a mixture of bi- and triantennary complex glycans (Mills *et al.*, 2001). HXMS revealed that, as with recombinant α_1-AT (RCα_1-AT), the regions containing significant α-helical and β-sheet content exchanged more slowly than the regions consisting largely of turns or loops (Sarkar and Wintrode, 2011). Residues

62–77 (helix C), 100–118 (helix D and strand 2A), 227–240 (strands 1B, 2B, and part of 3B), 318–329, 325–338 (strand 5A), 374–384 (strands 4B and 5B), and 266–275 (helix H) all showed relatively slow exchange. Also as expected, the RCL and other surface loops showed rapid exchange. While HXMS patterns in recombinant and plasma α_1-AT were similar overall there were several differences. Peptides 188–205, containing a loop and a portion of strand 4C, and 318–329, containing a loop between helix I and strand 5A, both showed increased exchange in HPα_1-AT. Peptides 208–227, 227–240, and 353–372 showed reduced exchange in HPα_1-AT relative to RCα_1-AT. Peptide 208–227 contains residues that are spatially adjacent to the glycan at position 46 and might be stabilized by direct interactions with the sugar moieties. Such stabilization could propagate to neighboring residues in β-sheets B and C. Stabilization of residues in peptide 227–240 by interactions with the glycan moieties at position 83 is also possible. There is no obvious structural explanation for the destabilization seen at the bottom of strand 5A, as it is not close to either a site of glycosylation or a region that is affected by glycosylation.

The H/D exchange rates in the key regions which play important roles in the metastable → stable transition were not affected by glycosylation. Flexibility was retained in the F-helix, while rigidity in the "breach" and "shutter" regions was likewise unaltered. This study did identify one difference in the native state dynamics of recombinant and glycosylated α_1-AT that has potential functional significance. It demonstrated that the release of β-strand 1C from the rest of β-sheet C is required for the formation of pathological serpin polymers (Chang *et al.*, 1997). In RCα_1-AT, strand C showed substantial protection from H/D exchange indicating a stable interaction with the rest of sheet C. This interaction would clearly deter polymer formation. Strand 1C showed increased protection from H/D exchange in HPα_1-AT. Previous work has shown that HPα_1-AT is more resistant to aggregation/polymerization than RCα_1-AT (Kwon and Yu, 1997). These results suggest that increased stability of strand 1C is one factor contributing to increased resistance to the formation of pathological polymers.

3. Determination of Thermodynamic Stability Using Hydrogen–Deuterium Exchange Combined with Mass Spectrometry

Protein flexibility and stability are not interchangeable. Flexibility in folded proteins reflects the structure of the free energy landscape near the free energy minimum that defines the native state. Observed flexibility will generally be dominated by nonnative, possibly partially unfolded, structures that are close in energy to the native structure. In contrast, thermodynamic stability reflects the free energy difference between the native state ensemble and the denatured state ensemble. Both quantities—flexibility and thermodynamic stability—are clearly relevant to understanding the behavior of inhibitory serpins, which rely on both metastability and conformational mobility for their function.

The addition of denaturants such as guanidine hydrochloride (GuHCl) or urea will slow the rate of protein folding and thus give rise to EX1 exchange. In the case of EX1 exchange, all amide hydrogens in a given region exchange when that region unfolds. If unfolding is slow ($t_{1/2} > 1$ s), this leads to characteristic bimodal isotopic envelopes in the mass spectra of peptides from that region, with the relative areas of the upper and lower mass-to-charge ratio (m/z) peaks corresponding to the populations in solution that have sampled the unfolded state (upper m/z peak) or remained folded (lower m/z peak). After a short ($\sim$10 s) pulse with D_2O, the relative areas under the low and high m/z envelopes are often directly proportional to the equilibrium populations of the folded and unfolded states. It is not guaranteed that the relative areas under the peaks give an accurate estimate of the relative equilibrium populations. However, in the presence of low to moderate concentrations of denaturant, the rates of both unfolding and refolding are frequently slow enough that the peak areas do accurately reflect the equilibrium. Excellent agreement between folding/unfolding equilibrium constants estimated from pulse-labeling HXMS and more traditional methods has been found in many cases (Deng and Smith, 1999). While small proteins typically unfold in a concerted two-state manner, larger proteins often contain distinct cooperatively folding domains. In proteins that unfold in a multistate manner, such cooperative domains are readily identified by HXMS.

3.1. Equilibrium unfolding study of WT α_1-AT using the pulse-labeling technique

WT α_1-AT was incubated in different concentration of GuHCl–H_2O at pH 7.8 for 1 h at room temperature to equilibrate followed by 10-fold dilution with pD 7.8 D_2O buffer containing the same concentration of guanidine as in the preincubated sample (Tsutsui and Wintrode, 2007). In GuHCl concentrations ranging from 0.5 to 1.2 *M*, double isotopic envelopes were clearly evident for all peptides, indicating the coexistence of folded and unfolded populations. For example, two mass envelops observed in a peptic fragment covering residue 64–77 correspond to a population with the region 64–77 folded and another population with the same region unfolded (Fig. 15.6A). Thereby, populations with a particular region folded and unfolded at different guanidine concentrations can be quantified by assessing the area under each of the two mass envelops using a Gaussian equation, and

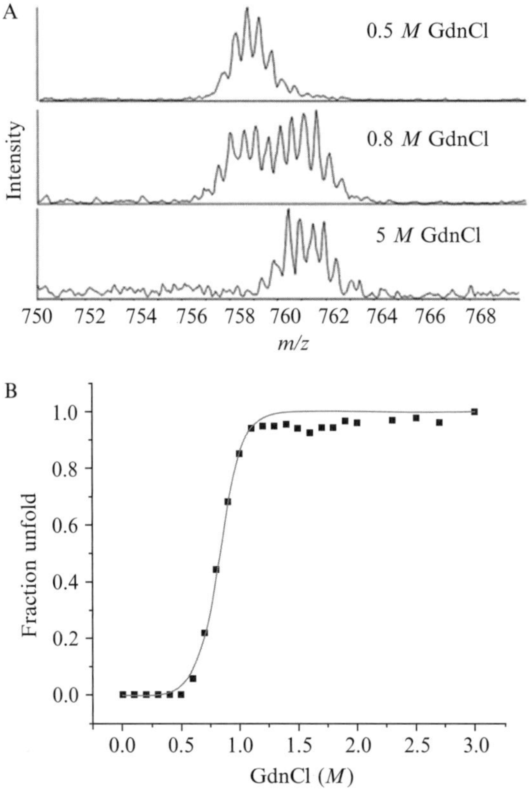

Figure 15.6 (A) Example of a double isotopic envelope observed during GuHCl denaturation of α_1-AT. The peptide shown corresponds to residues 64–77. (B) Equilibrium unfolding curve for peptide 64–77 constructed from the relative areas under the low (native) and high (denatured) mass peaks in the mass spectra at different GuHCl concentrations.

the unfolded fraction for each peptic fragment can be plotted against guanidine concentration to obtain $\Delta G_{\mathrm{U-N}}$ (Fig. 15.6B).

$\Delta G_{\mathrm{U\text{-}N}}$ values for different regions within WT α_1-AT structure were found to be in a range of 2.6–5.6 kcal/mol (Tsutsui and Wintrode, 2007). Although the C-terminal region of F-helix was found to be as flexible as the RCL region in the continuous-labeling experiment, this region was not the least stable region. Therefore, structural flexibility does not always correlate with the local thermodynamic stability of a protein. For all regions analyzed, the denaturation midpoint ($[\mathrm{GuHCl}]_{1/2}$) was found to be 0.8–1 *M* GuHCl. Denaturation curves for several representative peptic fragments from different structural regions are shown in Fig. 15.7. Chemical denaturation studies using circular dichroism (CD) spectroscopy found two unfolding transitions in α_1-AT: the first with a midpoint $[\mathrm{GuHCl}]_{1/2} = 0.8$ *M* resulted in the loss

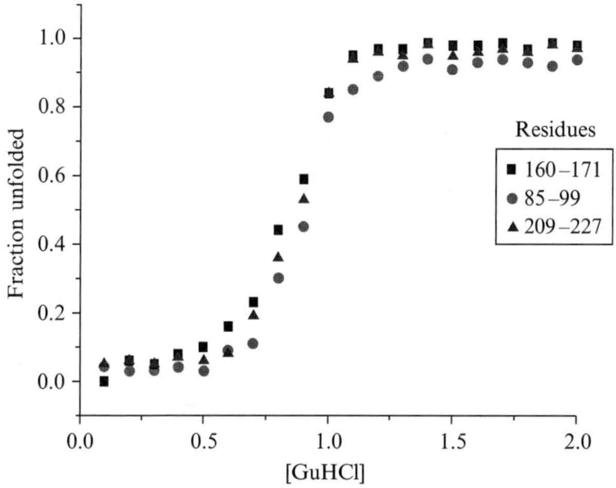

Figure 15.7 HXMS derived denaturation curves for peptic fragments derived from different regions of α_1-AT.

of ~30% of the native signal, while the remaining ~70% was lost during a broad transition with a midpoint $[GuHCl]_{1/2} = 2.8\ M$ (James *et al.*, 1999). The transition observed by HXMS corresponds very well with the first transition observed by CD. The presence of clear double isotopic envelopes indicates that unfolding is cooperative, while the fact that the unfolding curves are near superimposable for all but one peptide indicates, surprisingly, that the whole α_1-AT structure acts as a single cooperative unit, despite the appearance of multiple transitions in the CD unfolding curve. The fact that this cooperative transition coincides very well with the first transition observed by CD spectroscopy indicates that the transition is not well described as "unfolding" *per se*, since nearly 70% of secondary structure is retained after the transition. Further, gel chromatography shows that α_1-AT at 1.5 *M* GuHCl is somewhat expanded relative to the native state but still significantly more compact than the fully denatured state. Finally, Trp194 has a significant fluorescence signal above 1.2 *M* GuHCl (Cabrita *et al.*, 2002), while the nearby peptide 241–251 shows persistent protection against H/D exchange even above 2 *M* GuHCl (the only peptide in this study to do so). This suggests that, while there is a cooperative loss of stable hydrogen bonding throughout most of the molecule, there is nonetheless considerable persistent structure. Taken together with differential scanning calorimetry data, these observations indicate that the first denaturation transition at 0.8–1 *M* guanidine is cooperative and involves an accumulation of a molten globule form that has solvent accessible amide hydrogen atoms while retaining considerable secondary structure and perhaps residual side-chain packing. HXMS detected only the first transition from the metastable to the molten globule form. The second transition detected by CD and florescent spectroscopy involves unfolding of the molten globule form. Protease inhibition requires partial disruption of β-sheet A (Lee *et al.*, 2001; Seo *et al.*, 2000, 2002). The fact that the entire α_1-AT molecule unfolds as a single cooperative unit suggests that local unfolding in β-sheet A may induce unfolding in other regions of the molecule. Therefore, an alternative molecular mechanism of the protease inhibition involves a formation of the molten globule form to allow RCL insertion between β-strands 3A and 5A. In support of this hypothesis, extensive transient unfolding of α_1-AT structure during the inhibition was observed by another HXMS study, described below (Baek *et al.*, 2009). As described in the next section, an inability to accumulate the molten globule form appears to be correlated with the loss of the protease inhibitory activity.

3.2. Equilibrium unfolding of G117F mutant studied by HXMS

An equilibrium unfolding study of a cavity-filling mutant G117F by Sengupta *et al.* (2009) provided an important insight into the mechanism of the mutant stabilization and decreased inhibitory activity. The denaturation midpoint ($[GuHCl]_{1/2}$) determined by CD spectroscopy and HXMS agreed at 1.4 *M* GuHCl, 0.6 *M* higher than that found in wild-type unfolding. In HXMS data, coexistence of two populations was observed at 1.4 *M* GuHCl, one corresponding to the folded and the other to the unfolded form. Subsequently, the mass envelop corresponding to the folded form disappeared above 3 *M* GuHCl indicating the two-state cooperative transition. The degree of stabilization relative to wild type ($\Delta\Delta G$) for peptic peptides analyzed in the HXMS study was found to be clustered around 4.2 kcal/mol, in a good agreement with $\Delta\Delta G$ of 5 kcal/mol found in a previous fluorescent spectroscopic study (Lee *et al.*, 2000). However, unlike wild type, the second transition was absent in the equilibrium unfolding curve monitored by CD spectroscopy. A loss of tertiary and secondary structures was detected above 1.5 *M* guanidine by fluorescent spectroscopy and above 3 *M* GuHCl by CD spectroscopy, respectively (Lee *et al.*, 2000; Sengupta *et al.*, 2009). Although noncoincidence of two denaturation curves obtained by spectroscopic methods implies the existence of an intermediate, it is not a direct evidence for the three-state protein unfolding (Dill and Shortle, 1991). Nonoverlapping denaturation curves are observed in the variable two-state model, where properties of one state, such as the radius of the molecule change with external variables such as denaturant concentration and temperature (Dill and Shortle, 1991). Equilibrium unfolding studies by HXMS provide a direct evidence of a two-state transition through an observation of populations. Thus, the unfolding of G117F involves only a single transition from the metastable to the unfolded state without accumulating the molten globule form observed in the wild-type unfolding.

Based on the crystal structure of the mutant, favorable aromatic side-chain interactions around the mutation site were thought to contribute to the stabilization of the mutant (Gooptu *et al.*, 2009). However, there was no significant difference in local $[\mathrm{GuHCl}]_{1/2}$ determined by HXMS between F-helix region and other regions, indicating similar local stability in all regions analyzed. Instead, HXMS along with spectroscopic data suggested that the mutant stabilization is achieved by changing the unfolding mechanism through destabilization of the molten globule form. According to the lattice model for protein denaturation, a single mutation can have a large effect on the stability of a protein by changing the distribution of conformations in the denatured state (Shortle *et al.*, 1992). Therefore, the effect of a mutation is not always evident in crystal structures, if mutations affect the thermodynamic stability of proteins by altering sampling in the denatured state ensemble.

Based on the equilibrium unfolding behavior of WT α_1-AT, it was proposed that the molten globule form is a possible intermediate for the RCL insertion during the inhibition (Tsutsui and Wintrode, 2007). This leads to the suggestion that the decreased inhibitory activity of G117F could be due to its inability to access and accumulate the molten globule form during the inhibition. That partial denaturation is required during the inhibitory conformational change is supported by the detection of "functional unfolding" during inhibition, described below. Together, these results reveal that the folding and functional energy landscapes of inhibitory serpins are intimately related.

3.3. Effects of glycosylation on the equilibrium unfolding of α_1-AT observed using HXMS

Glycosylation generally thermodynamically stabilizes the native states of proteins (Shental-Bechor and Levy, 2009), and spectroscopic studies have shown that glycosylated α_1-AT is more stable than unglycosylated RCα_1-AT (Kwon and Yu, 1997). It has also been found that significant stabilization of α_1-AT can lead to impaired activity, and yet glycosylated α_1-AT is fully active. To gain further insight into the relationship between stability and activity in glycosylated α_1-AT, pulse-labeling HXMS was used to monitor the equilibrium unfolding of glycosylated HPα_1-AT (Sarkar and Wintrode, 2011).

The most surprising result of this study was the finding that glycosylation appears to stabilize the molten globule state of α_1-AT rather than the native state. CD spectroscopy shows that the midpoint of the GuHCl-induced unfolding transition is shifted from $\sim$0.8 *M* (for RCα_1-AT) to 2.3 *M*. Further, the transition appears to be two-state for glycosylated α_1-AT rather than three-state as observed for recombinant. Pulse-labeling HXMS, however, reveals a very different picture of the unfolding transition. All peptides in glycosylated α_1-AT displayed cooperative unfolding with a transition midpoint at ~1 *M* GuHCl and were fully unfolded by 1.2 *M* GuHCl, similar to recombinant (Fig. 15.8).

It therefore appears that glycosylation stabilizes the compact denatured form of AT, but not the native state. Further, glycosylation appears to alter the properties of the compact denatured state of α_1-AT. Unlike recombinant, which loses ~30% of its native ellipticity above 1 *M* GuHCl, plasma AT retains near native levels of ellipticity. It has been argued that compaction of the polypeptide chain induces secondary structure (Yee *et al.*, 1994), and our results suggest that HPα_1-AT at 1.2 *M* GuHCl is more compact than RCα_1-AT under the same conditions. Despite the increased secondary structure content of plasma AT at low GuHCl concentrations, it retains virtually no protection against H/D exchange under these conditions, indicating that whatever secondary structure is present must be marginally stable and subject to frequent and significant fluctuations. α_1-AT is thus highly unusual in that glycosylation does not appear to stabilize the native state. In the case of inhibitory serpins, this peculiar behavior may reflect functional requirements since significant stabilization of the native metastable form of AT could led to compromised inhibitory activity. Previous studies on the effects of glycosylation on protein folding and stability have found that the primary effect of glycosylation is to destabilize the denatured state (Shental-Bechor and Levy, 2008). In the case of α_1-AT, the effects of glycosylation on the unfolded state appear to be decoupled from the stability of the native state. In a protein with a simple two-state folding mechanism, such destabilization would naturally drive the protein to the native state. The equilibrium folding/unfolding mechanism of α_1-AT, however, is not two-state. Instead, AT unfolds through a complex mechanism involving a molten globule intermediate. These results suggest that glycosylation does destabilize the unfolded state of α_1-AT at mild GuHCl concentrations, but this drives α_1-AT into the compact denatured state rather than to the native state.

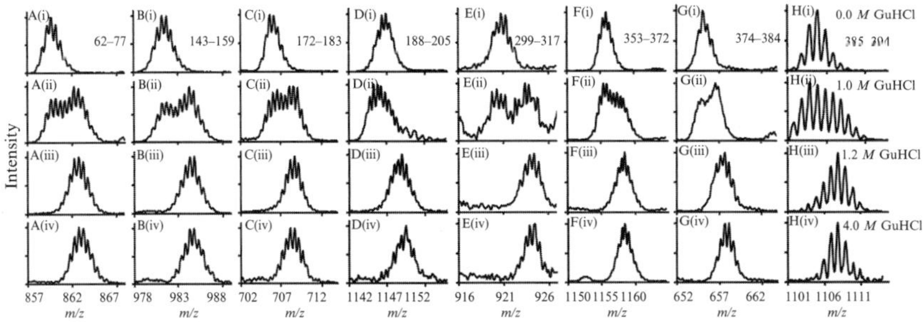

Figure 15.8 Mass spectra for peptides derived from different regions of glycosylated α_1-AT at 0, 1.0, 1.2, and 4 *M* GuHCl.

4. "Functional Unfolding" During the Native → Cleaved Transition

It is clear from crystal structures that large-scale structural rearrangements are required for the translocation of trapped protease and the insertion of the RCL into β-sheet A to proceed. However, crystal structures of the native and cleaved states provide limited information on the extent of the structural rearrangements that occur during the transition. The apparent correlation between inhibitory efficiency and the ability to populate a compact denatured state upon mild perturbations to solvent conditions, discussed above, suggest that it may be necessary to visit partially unfolded conformations during the inhibitory conformational change.

Baek *et al.* (2009) tested this hypothesis using HXMS. They incubated α_1-antitrypsin in D_2O under three conditions: (1) native α_1-antitrypsin alone, (2) the covalent complex between cleaved α_1-antitrypsin and inhibited trypsin, and (3) α_1-antitrypsin diluted into D_2O containing an equimolar amount of trypsin. By comparing local H/D exchange rates under these three conditions, Baek *et al.* were able to identify regions that transiently unfold during the native → cleaved transition as well as regions that are stabilized in the inhibitory complex.

Baek *et al.* found multiple regions that showed increased deuterium uptake under condition (3), indicating transient local unfolding during the native → cleaved transition (Fig. 15.9). Some of these transient unfolding regions are readily predicted from the crystal structures of the native and cleaved forms. The very top of β-strands 3A and 5A shows evidence of unfolding (although they are both part of larger peptides that include additional structural elements), as does the C-terminal portion of the F-helix. However, transient unfolding also occurs in regions that would not be easily predicted from crystal structures, including regions quite distant from the RCL insertion site. β-Strands 2B, 3B and 6B, 4C, and 6A, as well as most of α-helices A and G, all appear to undergo at least partial transient unfolding during the conformational change. These results support the contention that extensive structural disruptions throughout the molecule are required during the native → cleaved transition. Interestingly, there is not evidence for unfolding in most of β-strands 3A and 5A, despite the fact that these two must clearly separate in order to accommodate the RCL. This indicates that excepting the tops of these strands (forming part of the "breach" region where RCL insertion begins) the "opening" must be very transient and lead quickly to either reclosing or RCL insertion. Baek *et al* suggest that rapid displacement of strands 3A and 5A interactions by the RCL in a "zipping"-type mechanism could accomplish RCL insertion without significant solvent exposure in strands 3A and 5A.

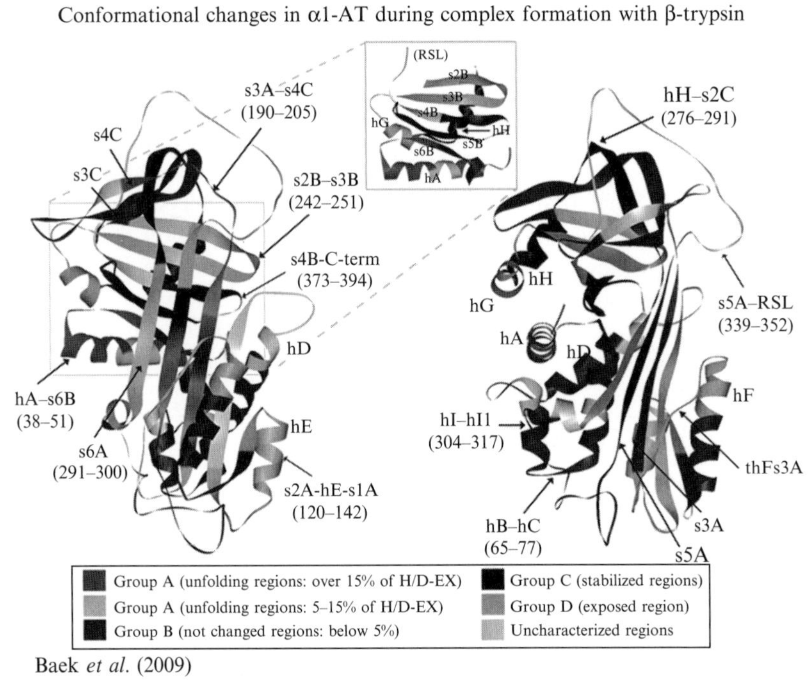

Figure 15.9 Patterns of deuterium uptake in different regions of α_1-AT during the inhibitory conformational change. Red and orange indicate regions that undergo "functional unfolding" during inhibition. (For interpretation of the references to color in this figure legend, the reader is referred to the web version of this chapter.)

Additionally, HXMS identified several regions that were stabilized in the inhibitory complex relative to the metastable state. Notably, β-strand 5A is stabilized in the complex. This is consistent with the observation that stabilizing mutations in strand 5A lead to impaired inhibitory activity, which suggested that instability of strand 5A was important for inhibitory function. β-Strand 2C was also stabilized in the complex, perhaps due to improved interactions with strand 1C after the "strain" imposed by the RCL is removed.

5. Investigating the Polymerization Pathway and Polymer Structure of α_1-AT by HXMS and Ion Mobility MS

The formation of serpin polymers *in vivo* is associated with a variety of diseases including dementia, emphysema, and liver cirrhosis collectively known as serpinopathies (Lomas and Carrell, 2002). Due to its medical significance, the mechanism of serpin polymerization has been vigorously investigated primarily by spectroscopic techniques (Dafforn *et al.*, 1999; Dunstone *et al.*, 2000; Lomas *et al.*, 1995). *In vitro*, the formation of polymers is facilitated at acidic pH, mild denaturant concentration, or elevated temperature above 45 °C (Dafforn *et al.*, 1999; Devlin *et al.*, 2002). The presence of an aggregation-prone intermediate, often denoted as M^*, has been inferred from the kinetics of polymerization, and possible structural features of M^* may be gleaned from a crystal structure of α_1-antichymotrypsin in which β-strands 3A and 5A are separated and both the RCL and part of helix F are inserted to stabilize the separation (Gooptu *et al.*, 2000). The classical serpin polymerization pathway proposes an accumulation of an intermediate with separated β-strands 3A and 5A prior to the polymer formation (Ekeowa *et al.*, 2010). Subsequent annealing of the RCL of one serpin molecule to the opening between β-strands 3A and 5A of another serpin molecule leads to the formation of polymers (Lomas and Carrell, 2002). Recently, an alternative polymerization pathway was proposed based on a crystal structure of antithrombin polymers (Yamasaki *et al.*, 2008). This β-hairpin polymerization pathway requires complete expulsion of β-strand 5A from β-sheet A of one serpin and subsequent insertion of both strand 5A and the RCL into sheet A of another serpin. Both proposed polymerization mechanisms require the formation of an intermediate in which β-sheet A is substantially disrupted and therefore increased H/D exchange should be observed in this region if the intermediate(s) accumulate to a significant degree prior to polymerization.

Incubation at slightly elevated temperatures such as 45 °C leads to polymerization of wild-type α_1-AT, and HXMS studies were performed under these conditions (Tsutsui *et al.*, 2008). At 45 °C, polymerization is very slow compared with the time scale of the hydrogen–deuterium exchange reaction so that dynamics of the intermediate can be investigated. WT α_1-AT was incubated at 45 °C for different time periods up to 5000 s in prewarmed D_2O buffer, and deuterium uptake of different regions at 45 °C in the monomeric α_1-AT intermediate structure was compared to that of the monomeric metastable state at 25 °C. The deuterium-labeling time at 25 °C was corrected for temperature dependence of the hydrogen–deuterium exchange reaction by incubating the sample in D_2O buffer for six times longer than corresponding labeling time at 45 °C. Surprisingly, deuterium uptake in β-strands 3A and 5A at 25 and 45 °C was comparable. Because the separation of β-strands 3A and 5A would certainly lead to increased deuterium uptake, this result suggests that either the strand separated M^* conformation is not populated to a significant extent or it is a transient species that quickly proceeds to the polymeric form. A substantial increase in deuterium uptake was observed in β-sheet C. These observations led to the suggestion that mild heating initially weakens interactions in sheet C, and that subsequently, an encounter with a second serpin molecule leads to the displacement of strand 1C and transient formation of M*, followed quickly by the formation of an intermolecular loop-sheet linkage.

In contrast, a recent ion mobility mass spectrometry (IM-MS) result argues against the notion that the M* intermediate is not populated to a significant degree. IM-MS measures the drift time of protein ions in a tube filled with a carrier buffer gas. Drift time can be converted into a collision cross section, which depends on the protein ions' shape and size. IM-MS can thus separate and characterize proteins with different oligomerization states and/or conformations. Performing IM-MS on α_1-AT polymers formed by heating, Ekeowa *et al.* detected native monomers, polymers, and a monomeric intermediate with an enlarged collision cross section, indicating an expanded structure (Fig. 15.10; Ekeowa *et al.*, 2010). This indicates that a nonnative monomeric species does in fact accumulate during polymerization. There are several possible explanations for the apparent discrepancy between the HXMS and IM-MS results. Polymers for IM-MS were formed by heating at 60 °C, while polymers for HXMS were formed by heating at 45 °C. The increased temperature may have increased the

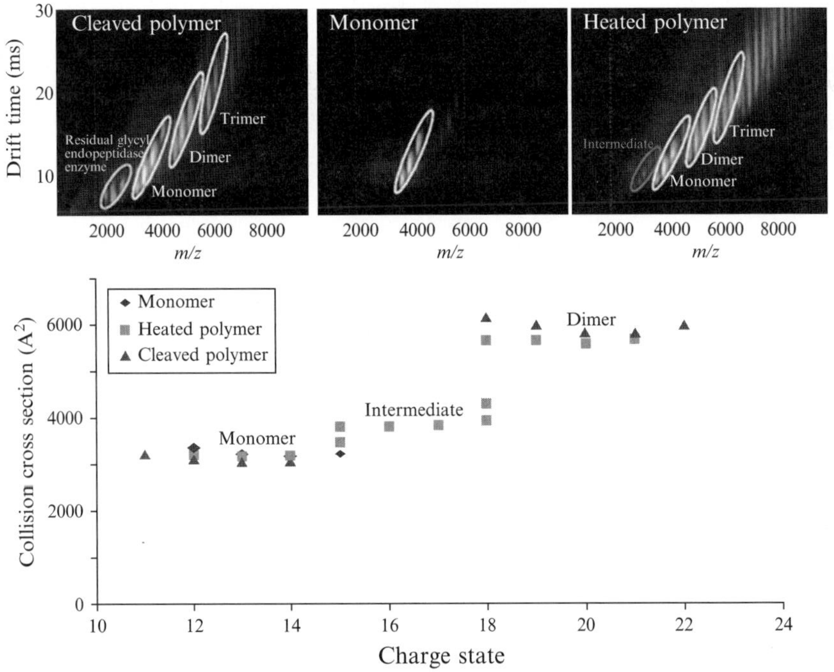

Figure 15.10 Ion mobility MS of α_1-AT monomer and polymers. After heating, an additional monomeric species with a slightly increased collision cross section (indicating a somewhat expanded structure) is evident. (For color version of this figure, the reader is referred to the web version of this chapter.)

population of M* to detectable levels. Additionally, the length of the incubation prior to measurement was different in each case.

The structure of polymers formed by extended incubation at 45 °C was probed using HXMS (Fig. 15.11; Tsutsui *et al.*, 2008). Compared with monomeric α_1-AT, dramatically reduced exchange was seen in the RCL, the C-terminal portion of helix F and helix E. Smaller but still significant decreases were seen in strands 3A and 5A, strand 3C and helix D. There are currently two major structural models for serpin polymers: a long-standing "loop-sheet" model, in which the RCL of one serpin is inserted between strands 3A and 5A of another serpin, and a more recent "domain swap" model in which strand 5A is extracted from sheet A of one serpin, and both it and the RCL insert into sheet A of another serpin becoming strands 4A and 5A (expelling the other serpins' strand 5A in the process). The pattern of decreased H/D exchange in the polymer relative to the monomer is generally consistent with both models. Decreased exchange in either model would be expected in the RCL, due to its sequestration from solvent, and in strands 3A and 5A, where the conversion from parallel interactions with each other to antiparallel interactions with the newly introduced strand 4A is expected to result in increased stability. Decreased exchange in helix F could result from burial in the serpin–serpin interface, as might be expected from some of the loop-sheet models, but it could also result from improved packing against the six stranded β-sheet A.

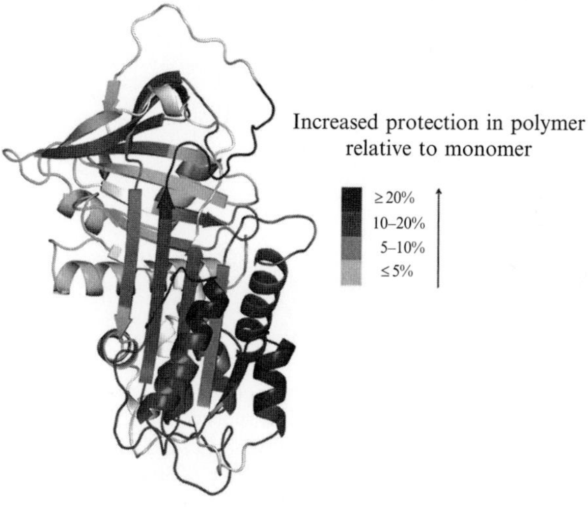

Figure 15.11 Changes in deuterium uptake in different regions of α_1-AT upon polymerization. (For color version of this figure, the reader is referred to the web version of this chapter.)

6. Future Prospects

There are several areas in which structural mass spectrometry techniques can shed light on serpin structure and function. As mentioned above, the structure of pathological serpin polymers remains a subject of debate, and it is possible that polymers formed under different conditions (temperature, pH, denaturant) may adopt distinct structures. The application of HXMS to different polymer preparations and to polymers of serpins other than α_1-AT may clarify this. Additionally, the application of radiolytic footprinting to serpin polymers would provide single residue resolution information on which side chains become sequestered from solvent upon polymerization. Such information would provide valuable constrains for attempts to construct plausible structural models of serpin polymers.

In addition to the mechanism of polymerization, the mechanism of serpin folding remains to be elucidated. A detailed view of the serpin folding pathway would help explain the remarkable fact that serpins do not fold to the global free energy minimum but are trapped in a metastable conformation. The relatively slow rates of serpin folding (Kim and Yu, 1996) combined with the ability of HXMS to directly quantify the populations of folded and unfolded species in a structurally resolved manner suggests that HXMS is a promising technique for resolving the kinetic folding pathways of serpins in detail. Finally, serpins *in vivo* interact with and are regulated by a variety of proteins and other molecules. Examples include the allosteric activation of antithrombin by heparin and the regulation of PAI-1 by vitronectin (Gettins, 2002). Structural mass spectrometry provides an array of powerful methods for defining the mechanisms of serpin regulation by proteins, lipids, and small molecules.

References

Baek, J. H., Yang, W. S., Lee, C., and Yu, M. H. (2009). Functional unfolding of alpha1-antitrypsin probed by hydrogen-deuterium exchange coupled with mass spectrometry. *Mol. Cell. Proteomics* **8,** 1072–1081.

Bai, Y., Milne, J. S., Mayne, L., and Englander, S. W. (1993). Primary structure effects on peptide group hydrogen exchange. *Proteins* **17,** 75–86.

Cabrita, L. D., Whisstock, J. C., and Bottomley, S. P. (2002). Probing the role of the F-helix in serpin stability through a single tryptophan substitution. *Biochemistry* **41,** 4575–4581.

Chang, W. S. W., Whisstock, J. C., Hopkins, P. C. R., Lesk, A. M., Carrell, R. W., and Wardell, M. R. (1997). Importance of the release of strand 1C to the polymerization mechanism of inhibitory serpins. *Protein Sci.* **6,** 89–98.

Dafforn, T. R., Mahadeva, R., Elliott, P. R., Sivasothy, P., and Lomas, D. A. (1999). A kinetic mechanism for the polymerization of alpha1-antitrypsin. *J. Biol. Chem.* **274,** 9548–9555.

Deng, Y. Z., and Smith, D. L. (1999). Rate and equilibrium constants for protein unfolding and refolding determined by hydrogen exchange-mass spectrometry. *Anal. Biochem.* **276,** 150–160.

Devlin, G. L., Chow, M. K., Howlett, G. J., and Bottomley, S. P. (2002). Acid Denaturation of alpha1-antitrypsin: Characterization of a novel mechanism of serpin polymerization. *J. Mol. Biol.* **324,** 859–870.

Dill, K. A., and Shortle, D. (1991). Denatured states of proteins. *Annu. Rev. Biochem.* **60,** 795–825.

Dunstone, M. A., Dai, W., Whisstock, J. C., Rossjohn, J., Pike, R. N., Feil, S. C., Le Bonniec, B. F., Parker, M. W., and Bottomley, S. P. (2000). Cleaved antitrypsin polymers at atomic resolution. *Protein Sci.* **9,** 417–420.

Ekeowa, U. I., Freeke, J., Miranda, E., Gooptu, B., Bush, M. F., Perez, J., Teckman, J., Robinson, C. V., and Lomas, D. A. (2010). Defining the mechanism of polymerization in the serpinopathies. *Proc. Natl. Acad. Sci. USA* 17146–17151.

Elliott, P. R., Pei, X. Y., Dafforn, T. R., and Lomas, D. A. (2000). Topography of a 2.0 A structure of alpha1-antitrypsin reveals targets for rational drug design to prevent conformational disease. *Protein Sci.* **9,** 1274–1281.

Gettins, P. G. (2002). Serpin structure, mechanism, and function. *Chem. Rev.* **102,** 4751–4804.

Gooptu, B., Hazes, B., Chang, W. S., Dafforn, T. R., Carrell, R. W., Read, R. J., and Lomas, D. A. (2000). Inactive conformation of the serpin alpha(1)-antichymotrypsin indicates two-stage insertion of the reactive loop: Implications for inhibitory function and conformational disease. *Proc. Natl. Acad. Sci. USA* **97,** 67–72.

Gooptu, B., Miranda, E., Nobeli, I., Mallya, M., Purkiss, A., Brown, S. C., Summers, C., Phillips, R. L., Lomas, D. A., and Barrett, T. E. (2009). Crystallographic and cellular characterisation of two mechanisms stabilising the native fold of alpha1-antitrypsin: Implications for disease and drug design. *J. Mol. Biol.* **387,** 857–868.

Guan, J. Q., and Chance, M. R. (2005). Structural proteomics of macromolecular assemblies using oxidative footprinting and mass spectrometry. *Trends Biochem. Sci.* **30,** 583–592.

James, E. L., Whisstock, J. C., Gore, M. G., and Bottomley, S. P. (1999). Probing the unfolding pathway of alpha1-antitrypsin. *J. Biol. Chem.* **274,** 9482–9488.

Kim, D., and Yu, M. H. (1996). Folding pathway of human alpha 1-antitrypsin: Characterization of an intermediate that is active but prone to aggregation. *Biochem. Biophys. Res. Commun.* **226,** 378–384.

Kwon, K. S., and Yu, M. H. (1997). Effect of glycosylation on the stability of alpha1-antitrypsin toward urea denaturation and thermal deactivation. *Biochim. Biophys. Acta Gen. Subj.* **1335,** 265–272.

Lee, C., Park, S. H., Lee, M. Y., and Yu, M. H. (2000). Regulation of protein function by native metastability. *Proc. Natl. Acad. Sci. USA* **97,** 7727–7731.

Lee, C., Maeng, J. S., Kocher, J. P., Lee, B., and Yu, M. H. (2001). Cavities of alpha(1)-antitrypsin that play structural and functional roles. *Protein Sci.* **10,** 1446–1453.

Lomas, D. A., and Carrell, R. W. (2002). Serpinopathies and the conformational dementias. *Nat. Rev. Genet.* **3,** 759–768.

Lomas, D. A., Elliott, P. R., Sidhar, S. K., Foreman, R. C., Finch, J. T., Cox, D. W., Whisstock, J. C., and Carrell, R. W. (1995). alpha 1-Antitrypsin Mmalton (Phe52-deleted) forms loop-sheet polymers in vivo. Evidence for the C sheet mechanism of polymerization. *J. Biol. Chem.* **270,** 16864–16870.

Meagher, J. L., Olson, S. T., and Gettins, P. G. (2000). Critical role of the linker region between helix D and strand 2A in heparin activation of antithrombin. *J. Biol. Chem.* **275,** 2698–2704.

Mills, K., Mills, P. B., Clayton, P. T., Johnson, A. W., Whitehouse, D. B., and Winchester, B. G. (2001). Identification of alpha(1)-antitrypsin variants in plasma with the use of proteomic technology. *Clin. Chem.* **47,** 2012–2022.

Sarkar, A., and Wintrode, P. L. (2011). Effects of glycosylation on the stability and flexibility of a metastable protein: The human serpin alpha-1 antitrypsin. *Int. J. Mass Spectrom.* **302,** 69–75.

Sengupta, T., Tsutsui, Y., and Wintrode, P. L. (2009). Local and global effects of a cavity filling mutation in a metastable serpin. *Biochemistry* **48,** 8233–8240.

Seo, E. J., Im, H., Maeng, J. S., Kim, K. E., and Yu, M. H. (2000). Distribution of the native strain in human alpha 1-antitrypsin and its association with protease inhibitor function. *J. Biol. Chem.* **275,** 16904–16909.

Seo, E. J., Lee, C., and Yu, M. H. (2002). Concerted regulation of inhibitory activity of alpha 1-antitrypsin by the native strain distributed throughout the molecule. *J. Biol. Chem.* **277,** 14216–14220.

Shental-Bechor, D., and Levy, Y. (2008). Effect of glycosylation on protein folding: A close look at thermodynamic stabilization. *Proc. Natl. Acad. Sci. USA* **105,** 8256–8261.

Shental-Bechor, D., and Levy, Y. (2009). Folding of glycoproteins: Toward understanding the biophysics of the glycosylation code. *Curr. Opin. Struct. Biol.* **19,** 524–533.

Shortle, D., Chan, H. S., and Dill, K. A. (1992). Modeling the effects of mutations on the denatured states of proteins. *Protein Sci.* **1,** 201–215.

Sivasothy, P., Dafforn, T. R., Gettins, P. G. W., and Lomas, D. A. (2000). Pathogenic a1-antitrypsin polymers are formed by reactive loop-b-sheet A linkage. *J. Biol. Chem.* **275,** 33663–33668.

Takamoto, K., and Chance, M. R. (2006). Radiolytic protein footprinting with mass spectrometry to probe the structure of macromolecular complexes. *Annu. Rev. Biophys. Biomol. Struct.* **35,** 251–276.

Tsutsui, Y., and Wintrode, P. L. (2007). Cooperative unfolding of a metastable serpin to a molten globule suggests a link between functional and folding energy landscapes. *J. Mol. Biol.* **371,** 245–255.

Tsutsui, Y., Liu, L., Gershenson, A., and Wintrode, P. L. (2006). The conformational dynamics of a metastable serpin studied by hydrogen exchange and mass spectrometry. *Biochemistry* **45,** 6561–6569.

Tsutsui, Y., Kuri, B., Sengupta, T., and Wintrode, P. L. (2008). The structural basis of serpin polymerization studied by hydrogen/deuterium exchange and mass spectrometry. *J. Biol. Chem.* **283,** 30804–30811.

Varki, A. (1993). Biological roles of oligosaccharides: All of the theories are correct. *Glycobiology* **3,** 97–130.

Wales, T. E., and Engen, J. R. (2006). Hydrogen exchange mass spectrometry for the analysis of protein dynamics. *Mass Spectrom. Rev.* **25,** 158–170.

Xu, G. H., and Chance, M. R. (2007). Hydroxyl radical-mediated modification of proteins as probes for structural proteomics. *Chem. Rev.* **107,** 3514–3543.

Yamasaki, M., Li, W., Johnson, D. J., and Huntington, J. A. (2008). Crystal structure of a stable dimer reveals the molecular basis of serpin polymerization. *Nature* **455,** 1255–1258.

Yee, D. P., Chan, H. S., Havel, T. F., and Dill, K. A. (1994). Does compactness induce secondary structure in proteins? A study of poly-alanine chains computed by distance geometry. *J. Mol. Biol.* **241,** 557–573.

Zheng, X. J., Wintrode, P. L., and Chance, M. R. (2008). Complementary structural mass spectrometry techniques reveal local dynamics in functionally important regions of a metastable serpin. *Structure* **16,** 38–51.

CHAPTER SIXTEEN

Determining Serpin Conformational Distributions with Single Molecule Fluorescence

Nicole Mushero* *and* Anne Gershenson†

Contents

1. Introduction 353
2. Labeling Serpins and Proteases with Fluorophores 354
   2.1. A general protocol for labeling proteins with maleimide derivatized fluorophores 354
3. Overview of Single Molecule Fluorescence Techniques 355
   3.1. SMF instrumentation 357
4. Serpin Polymerization 358
   4.1. α_1AT polymerization and depolymerization monitored by fluorescence correlation spectroscopy 359
   4.2. Early events in neuroserpin polymerization measured by two-color coincidence detection 364
   4.3. The utility of FCS and TCCD for studying polymerization 367
5. Conformational Distributions of Protease–Serpin Complexes 368
   5.1. Introduction to single pair Förster resonance energy transfer 368
   5.2. Conformational distributions of trypsin–α_1AT complexes 370
   5.3. A protocol for spFRET on protease–serpin complexes 372
6. Conclusions and Future Directions 373
Acknowledgments 373
References 374

Abstract

Conformational plasticity is key to inhibitory serpin function, and this plasticity gives serpins relatively easy access to alternative, dysfunctional conformations. Thus, a given serpin population may contain both functional and dysfunctional proteins. Single molecule fluorescence (SMF), with its ability to

* University of Massachusetts, School of Medicine, Worcester, Massachusetts, USA
† Department of Biochemistry and Molecular Biology, University of Massachusetts, Amherst, Massachusetts, USA

Methods in Enzymology, Volume 501
ISSN 0076-6879, DOI: 10.1016/B978-0-12-385950-1.00016-X

interrogate one fluorescently labeled protein at a time, is a powerful method for elucidating conformational distributions and monitoring how these distributions change over time. SMF and related methods have been particularly valuable for characterizing serpin polymerization. Fluorescence correlation spectroscopy experiments have revealed a second lag phase during *in vitro* α_1-antitrypsin polymerization associated with the formation of smaller oligomers that then condense to form longer polymers [Purkayastha, P., Klemke, J. W., Lavender, S., Oyola, R., Cooperman, B. S., and Gai, F. (2005). Alpha 1-antitrypsin polymerization: A fluorescence correlation spectroscopic study. *Biochemistry* **44**, 2642–2649.]. SMF studies of *in vitro* neuroserpin polymerization have confirmed that a monomeric intermediate is required for polymer formation while providing a test of proposed polymerization mechanisms [Chiou, A., Hägglöf, P., Orte, A., Chen, A. Y., Dunne, P. D., Belorgey, D., Karlsson-Li, S., Lomas, D., and Klenerman, D. (2009). Probing neuroserpin polymerization and interaction with amyloid-beta peptides using single molecule fluorescence. *Biophys. J.* **97**, 2306–2315.]. SMF has also been used to monitor protease–serpin interactions. Single pair Förster resonance energy transfer studies of covalent protease–serpin complexes suggest that the extent of protease structural disruption in the complex is protease dependent [Liu, L., Mushero, N., Hedstrom, L., and Gershenson, A. (2006). Conformational distributions of protease-serpin complexes: A partially translocated complex. *Biochemistry* **45**, 10865–10872.]. SMF techniques are still evolving and the combination of SMF with encapsulation methods has the potential to provide more detailed information on the conformational changes associated with serpin polymerization, protease–serpin complex formation, and serpin folding.

Abbreviations

α_1AT	α_1-antitrypsin
AF488	Alexa Fluor 488
AF647	Alexa Fluor 647
APD	avalanche photodiode
BSA	bovine serum albumin
BTryp	bovine trypsin
FRET	Förster resonance energy transfer
FCS	fluorescence correlation spectroscopy
RTryp	rat trypsin
SMF	single molecule fluorescence
spFRET	single pair FRET
TCCD	two-color coincidence detection

1. Introduction

Serpin structural remodeling is required for inhibition of target proteases. This functionally necessary structural plasticity allows serpins relatively easy access to multiple significantly different conformations only one of which is active resulting in conformational heterogeneity ranging from productive mixtures of active serpins and protease–serpin covalent complexes to dysfunctional mixtures of serpin oligomers. The difficulties inherent in differentiating between conformations were recently illustrated for the inhibitory serpin α_1-antichymotrypsin where an inactive state assumed to be the latent state, with full insertion of the intact reactive center loop into β sheet A, turned out to be the δ conformation, with only partial loop insertion (Pearce *et al.*, 2010). Thus, methods that can monitor serpin conformational distributions and how these distributions change over time would aid our understanding of serpin function, dysfunction, and folding.

While most experimental methods report on the average conformation of a heterogeneous protein sample, single molecule fluorescence (SMF) microscopy collects data from individual molecules allowing direct measurement of conformational distributions (Dahan *et al.*, 1999; Li *et al.*, 2003; Orte *et al.*, 2008a). SMF can also reveal how conformational distributions change over time during processes such as polymer formation (Chiou *et al.*, 2009; Purkayastha *et al.*, 2005) and protein folding (Chattopadhyay *et al.*, 2005; Chen *et al.*, 2007; Schuler and Eaton, 2008; Sherman *et al.*, 2008). Thus, along with other methods such as mass spectrometry and NMR that can also report on conformational heterogeneity, SMF and related fluorescence techniques provide a window unto the complicated structural landscape populated by serpins.

In this review, we begin with general technical aspects of SMF experiments from how to label proteins to the necessary instrumentation. Three different SMF methods have been applied to serpins. Gai and colleagues have used fluorescence correlation spectroscopy (FCS) to monitor α_1-antitrypsin (α_1AT) polymerization (Chowdhury *et al.*, 2007; Purkayastha *et al.*, 2005) while Klenerman and coworkers studied neuroserpin polymerization using two-color coincidence detection (TCCD) (Chiou *et al.*, 2009). We have used single pair Förster resonance energy transfer (spFRET) to determine conformational distributions of trypsin-α_1AT covalent complexes (Liu *et al.*, 2006, 2007). For all of these studies, we introduce the relevant SMF technique, provide protocols, and review the experimental findings. SMF is a growing field in which technical advances, particularly for imaging proteins in living cells (Patterson *et al.*, 2010), are continuing at a fast pace and we hope to inspire scientists working on serpins to interrogate their systems using SMF.

2. Labeling Serpins and Proteases with Fluorophores

SMF experiments require bright, photostable fluorophores that emit in the visible or near-IR. Commonly used fluorophores include the Alexa Fluor dyes (Invitrogen), Atto dyes (Atto-Tec, Germany), cyanine dyes such as Cy3 and Cy5 (GE Healthcare), and HiLyte dyes (Anaspec). These are photostable fluorophores with high-absorption cross-sections and high fluorescence quantum yields (Buschmann *et al.*, 2003; Roy *et al.*, 2008). Serpins have been covalently labeled with such fluorophores using Cys-maleimide linkages because of the stability and specificity this reaction affords. Native or unfolded proteins may be labeled in solution, and efficient labeling for ammonium sulfate precipitated proteins was recently reported (Kim *et al.*, 2008).

If a single, solvent-accessible Cys is present in wildtype protein it can be specifically labeled for FCS and/or TCCD experiments. For proteins that lack solvent-accessible Cys residues and for spFRET experiments where two fluorescent labels are required, solvent-accessible Cys residues may be introduced at various positions and, when necessary, native Cys residues may be mutated, usually to Ser or Ala. For proteins such as trypsin and related serine proteases in which all of the Cys residues are disulfide bonded, an additional solvent-accessible Cys may be introduced without mutating existing Cys residues (Mellet *et al.*, 2002). The resulting Cys variants, both unlabeled and labeled, should be checked for function, stability and, for serpins, polymerization propensity.

2.1. A general protocol for labeling proteins with maleimide derivatized fluorophores

1. Maleimide reactions are sulfhydryl specific between pH 6.5 and 7.5 (Hermanson, 1996), and labeling reactions are generally performed between pH 7.0 and 7.5 in a variety of buffers.
2. Protein concentrations should be at least 10 μM for effective labeling and higher concentrations, 25–100 μM, are generally recommended by the dye manufacturers.
3. Prior to labeling, the protein should be reduced using at least a 10-fold molar excess of DTT or TCEP usually for 10–15 m at room temperature. Note that TCEP can irreversibly inhibit Ser proteases. DTT must be removed prior to labeling by gel filtration. PD-10 or NAP-10 columns (GE Healthcare) containing Sephadex G-25, as well as Toyopearl HW-40C or HW-50F (Tosoh Biosciences) resins are all commonly used. TCEP need not be removed prior to labeling.

4. Prepare a 1–10 m*M* fluorophore stock solution immediately before use and protect this solution from light. To attain millimolar concentrations, hydrophobic fluorophores may need to be dissolved in an organic solvent such as dimethyl sulfoxide. The fluorophore is added to the protein solution with final fluorophore to protein molar ratios usually between 10:1 and 20:1. The final protein solution should contain 5% or less organic solvent to avoid protein denaturation. Samples may be layered with argon gas to avoid oxidation, and tubes should be sealed with parafilm and protected from light. Reactions are generally allowed to proceed for at least 2 h at room temperature and/or at 4 °C overnight.
5. The reaction is stopped by adding β-mercaptoethanol or glutathione, and free dye is removed by gel filtration (see Section 2 for a list of commonly used resins). Assuming that the protein environment does not alter the spectral properties of the fluorophore, labeling efficiency may be determined from an absorption spectrum using the extinction coefficient of the free fluorophore at or near its peak absorption (where the protein does not absorb), $\varepsilon_{\mathrm{max,fl}}$, the fluorophore's relative extinction coefficient at 280 nm, $\varepsilon_{280,\mathrm{fl}}/\varepsilon_{\mathrm{max,fl}}$, and the protein's extinction coefficient at 280 nm, $\varepsilon_{280,\mathrm{prot}}$:

$$\begin{aligned}\mathrm{DOL} &= \frac{[\mathrm{fluorophore}]}{[\mathrm{protein}]} = \frac{A_{\mathrm{max,fl}}/(\varepsilon_{\mathrm{max,fl}} l)}{A_{280,\mathrm{prot}}/(\varepsilon_{280,\mathrm{prot}} l)} \\ &= \frac{A_{\mathrm{max,fl}}\varepsilon_{280,\mathrm{prot}}}{(A_{280} - A_{\mathrm{max,fl}}(\varepsilon_{280,\mathrm{fl}}/\varepsilon_{\mathrm{max,fl}}))\varepsilon_{\mathrm{max,fl}}}\end{aligned} \quad (16.1)$$

where A is the measured absorbance, l is the pathlength of the cuvette, and DOL is the degree of labeling or the labeling efficiency. The DOL may also be measured using mass spectrometry, and this method is often more accurate.

3. Overview of Single Molecule Fluorescence Techniques

To our knowledge, all serpin SMF experiments have been conducted on serpin monomers, polymers, and molecular complexes *freely diffusing* in solution (Chiou *et al.*, 2009; Chowdhury *et al.*, 2007; Liu *et al.*, 2006; Purkayastha *et al.*, 2005) or, in one case, when microinjected into cells (Speil and Kubitscheck, 2010). While many of the most familiar SMF experiments on other biomolecules have been performed using total internal reflection microscopy and surface immobilized biomolecules (for reviews see Moerner and Fromm, 2003; Myong and Ha, 2010; Peterman *et al.*, 2004), our efforts to immobilize serpins have resulted in dysfunctional molecules. We therefore focus on SMF techniques that interrogate freely diffusing molecules (Table 16.1) using microscope-based instrumentation as depicted in Fig. 16.1.

In order to interrogate an individual molecule, there must be a very low probability that two or more molecules are in the observation volume at the same time. This places two restrictions on the experimental conditions: (i) low (10–100 p*M*) concentrations of labeled molecules must be used and (ii) observation volumes must be small, sub-femtoliter. To put the concentration and volume limitations in context, for a 1 n*M* solution a 1 fl volume will, on average, contain 1 molecule. Thus, 1 n*M* is too high a concentration for true single molecule experiments because the probability that 2 molecules will be detected is not negligible. The concentration limitations apply only to the fluorescently labeled protein of interest allowing unlabeled molecules to be in solution at arbitrary concentrations. Concentration restrictions are also relaxed for experiments on oligomers where the relevant concentration is that of the oligomer and not of the monomer as well as for FCS experiments that rely on small numbers of molecules rather than single molecules allowing a broader concentration range, from p*M* to 100 s of n*M*. The size of the observation volume is determined by the microscope optics and is diffraction limited so that shorter excitation wavelengths, λ_{ex}, and higher microscope objective numerical apertures, NA, result in smaller observation volumes.

Table 16.1 Applications of SMF techniques to biological problems

Technique	Applications
FCS	Serpin polymerization (Chowdhury *et al.*, 2007; Purkayastha *et al.*, 2005)
	Protein folding (Chattopadhyay *et al.*, 2005; Sherman *et al.*, 2008)
	Biomolecular interactions (Thompson, 1991)
	Protein conformational changes (Chen *et al.*, 2007; Neuweiler *et al.*, 2007)
TCCD	Serpin polymerization (Chiou *et al.*, 2009) and amyloid formation (Orte *et al.*, 2008a)
	Biomolecular interactions (Orte *et al.*, 2006)
spFRET	Conformational distributions during protein folding (Nienhaus, 2009; Schuler and Eaton, 2008)
	Protein conformational changes (Hoffmann *et al.*, 2007; Sharma *et al.*, 2008)
	Biomolecular interactions (Hoffman *et al.*, 2010; Liu *et al.*, 2006; Sharma *et al.*, 2008)

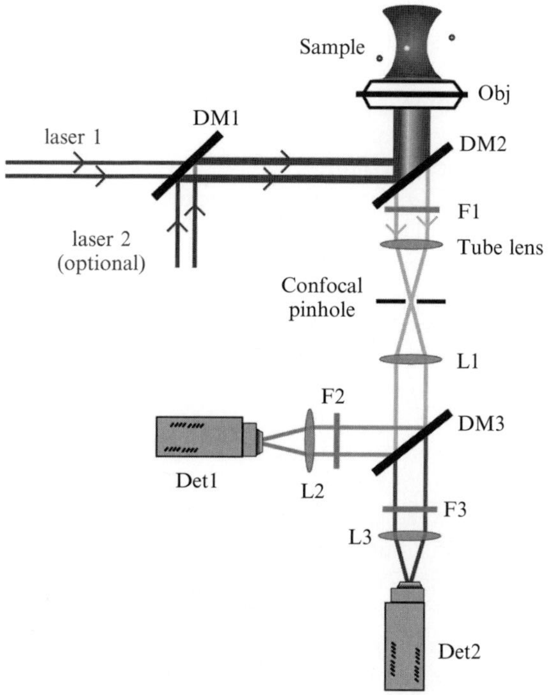

Figure 16.1 Schematic of a one-photon SMF microscope designed to interrogate fluorescently labeled molecules freely diffusing in solution. The excitation wavelength(s) is reflected into the microscope objective, Obj, by a dichroic mirror, DM2, and is focused into the sample. The emitted fluorescence is transmitted by DM2, any residual laser light is blocked by a longpass or notch filter, F1, and the tube lens focuses the fluorescence onto the confocal pinhole which blocks fluorescence from out-of-focus molecules, spatially restricting the observation volume. The fluorescence is then collimated by lens L1. For spFRET or TCCD, the fluorescence is split by color using another dichroic mirror, DM3, and focused onto high quantum efficiency detectors, Det1 and Det2, by lenses L2 and L3, respectively. The bandpass filters in front of the detectors, F2 and F3, block Raman scattering from water and minimize cross talk between the detectors. DM3 may be replaced by a polarizing beamsplitter for anisotropy experiments or by a 50–50 beamsplitter. The first dichroic mirror, DM1, allows the use of two different wavelength excitation beams. (For interpretation of the references to color in this figure legend, the reader is referred to the Web version of this chapter.)

3.1. SMF instrumentation

SMF instrumentation is often based on an inverted microscope where a dichroic mirror both reflects the laser light into the microscope objective and transmits the lower energy, higher wavelength fluorescence emitted by the sample (Fig. 16.1; Dahan *et al.*, 1999; Petersen and Elson, 1986; Sisamakis *et al.*, 2010). For conventional one-photon excitation, fluorescence arises from the entire illuminated volume, and a confocal pinhole is used to restrict the observation volume to a small volume centered at the focus (Dahan *et al.*, 1999; Magde *et al.*, 1974; Petersen and Elson, 1986). The pinhole is placed in the conjugate image plane and acts as a spatial filter, rejecting fluorescence from out-of-focus molecules thereby restricting the observation volume.

For spFRET, TCCD, and other experiments involving fluorophores with different colors, a dichroic mirror in the emission path is used to split the fluorescence by color and the fluorescence is collected by high-efficiency detectors such as single photon counting avalanche photodiodes (APDs). Some common alternative configurations include anisotropy experiments where the fluorescence may be split by polarization using a polarizing beamsplitter in place of or in addition to the dichroic mirror (Liu *et al.*, 2006; Sisamakis *et al.*, 2010) and single color experiments requiring high time resolution where the dichroic mirror may be replaced by a nonpolarizing 50–50 beamsplitter. Bandpass filters are often positioned in front of the detectors to restrict cross talk between detectors for multicolor experiments and to filter out the Raman scattering from water (Fig. 16.1).

In many two-color experiments, such as the TCCD work by Klenerman and coworkers (Chiou *et al.*, 2009; Orte *et al.*, 2008a), fluorophores with very different excitation spectra must be simultaneously excited. This requires colinear alignment of two different wavelength laser lines that are then focused in the sample (Fig. 16.1). For spFRET experiments, the acceptor fluorophore absorbs at redder wavelengths than does the donor fluorophore. While spFRET experiments may be performed by exciting only the donor fluorophore, alternately exciting the donor and the acceptor with two different wavelengths allows discrimination between molecules with low FRET efficiencies and those in which the acceptor is missing or has photobleached (Kapanidis *et al.*, 2005; Müller *et al.*, 2005). This again requires proper alignment of two different wavelength laser lines.

More comprehensive discussions of FCS instrumentation may be found in a review by Krichevsky and Bonnet (2002). TCCD experimental setups have been thoroughly described by Klenerman and coworkers (Li *et al.*, 2003; Orte *et al.*, 2008b). More detailed descriptions of SMF instrumentation for FRET may be found in recent reviews by Seidel and coworkers (Sisamakis *et al.*, 2010) as well as by Ha and coworkers (Roy *et al.*, 2008).

4. Serpin Polymerization

Mutations in metastable serpins can lead to polymerization resulting in disease (Davis *et al.*, 1999; Lomas *et al.*, 1992). While it is relatively easy to monitor polymers once they are formed, interrogating intermediates in polymer formation can be quite difficult. FCS and TCCD studies of serpin polymers allow researchers to probe early events in polymer formation including both changes in monomer structure that significantly alter translational diffusion and the formation of small oligomers (Chiou *et al.*, 2009; Chowdhury *et al.*, 2007; Purkayastha *et al.*, 2005).

4.1. α_1AT polymerization and depolymerization monitored by fluorescence correlation spectroscopy

Increases in molecular size due to oligomerization slow translational diffusion and increase the diffusion time, τ_D, of fluorescent molecules through the microscopic observation volume as measured by FCS (Fig. 16.2) (Krichevsky and Bonnet, 2002; Thompson, 1991). Protein conformational changes that significantly alter the overall protein shape can also change τ_D (Sherman *et al.*, 2008), making FCS sensitive to both the monomer conformation and oligomerization. Feng Gai, Barry Cooperman, and coworkers have used FCS to study both the formation of α_1AT polymers by heating (Purkayastha *et al.*, 2005) and the effects of peptide inhibitors on polymerization and depolymerization (Chowdhury *et al.*, 2007). FCS revealed that for α_1AT polymerization at 45 °C, τ_D increases in less than 30 m from ~2 ms for monomeric α_1AT to 4–6 ms indicating the formation of small oligomers and/or monomer conformational changes (Purkayastha *et al.*, 2005). After a lag of approximately 200 m, τ_D increases to 8–16 ms, but long polymers appear only after 400 m. As expected, the lag times in the production of oligomers and long polymers is concentration and temperature dependent, and higher concentrations result in longer polymers. The FCS data also suggest that large polymers form not by monomer addition but rather through the association of smaller polymers. Peptide driven depolymerization occurs in the opposite manner, large polymers dissociate into smaller polymers that then further dissociate (Chowdhury *et al.*, 2007). Below we describe how to analyze FCS data, further describe the findings of Gai and colleagues, and provide a protocol for FCS polymerization and depolymerization experiments based on their work (Chowdhury *et al.*, 2007; Purkayastha *et al.*, 2005). It should be noted that the choice of fluorescent dye, excitation wavelength, and excitation power can easily be optimized for the system of interest.

4.1.1. Introduction to FCS

Diffusion of molecules in and out of the observation volume results in fluorescence fluctuations, and the time scale of these fluctuations is related to the diffusion time, τ_D, which can be determined using FCS (Elson and Magde, 1974; Magde *et al.*, 1974; Thompson, 1991). The concentrations of fluorescently labeled molecules needed for FCS experiments range from <100 n*M* to p*M*. If the concentrations are too high, $\gtrsim 1000$ molecules in

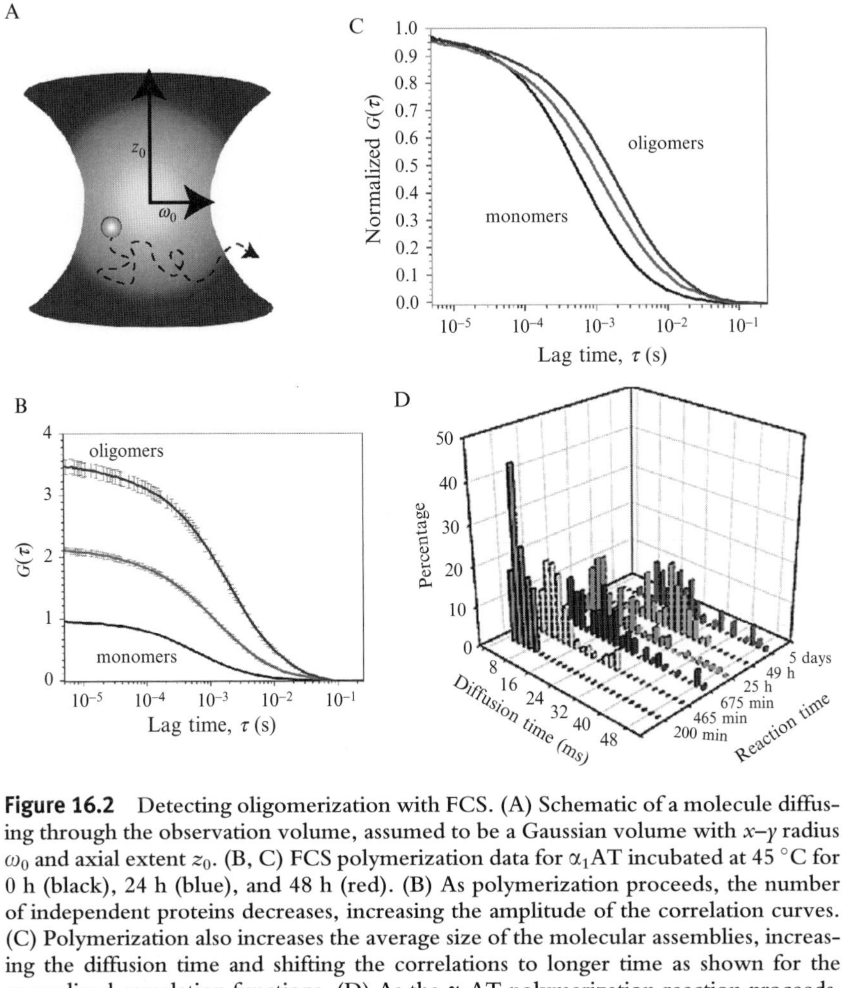

Figure 16.2 Detecting oligomerization with FCS. (A) Schematic of a molecule diffusing through the observation volume, assumed to be a Gaussian volume with x–y radius ω_0 and axial extent z_0. (B, C) FCS polymerization data for α_1AT incubated at 45 °C for 0 h (black), 24 h (blue), and 48 h (red). (B) As polymerization proceeds, the number of independent proteins decreases, increasing the amplitude of the correlation curves. (C) Polymerization also increases the average size of the molecular assemblies, increasing the diffusion time and shifting the correlations to longer time as shown for the normalized correlation functions. (D) As the α_1AT polymerization reaction proceeds, the distribution of diffusion times, τ_D, shifts to longer times. These data, adapted with permission from Purkayastha *et al.* (2005) copyright 2005 American Chemical Society, are for fits to Eq. (16.6) rather than Eq. (16.7) so the percentage of large polymers may be overestimated (see Section 4.1.1). (See Color Insert.)

the observation volume (Thompson, 1991), for every molecule that leaves the observation volume another one enters damping the fluctuations and degrading the FCS signal. At lower concentrations, diffusion of molecules in and out of the observation volume results in large changes in the number of molecules in the observation volume at a given time, large fluorescence fluctuations, and larger amplitude FCS signals (Fig. 16.2). The diffusion time, τ_D, may be determined by correlating the time-dependent fluorescence intensity, *I(t)*, and fitting the resulting autocorrelation, $G(\tau)$ (Elson and Magde, 1974; Krichevsky and Bonnet, 2002; Magde *et al.*, 1974; Thompson, 1991) (Fig. 16.2):

$$G(\tau) = \frac{\langle \delta I(t) \delta I(t+\tau) \rangle}{\langle I(t) \rangle^2} \tag{16.2}$$

$$\delta I(t) = I(t) - \langle I(t) \rangle \tag{16.3}$$

where $\langle\ \rangle$ indicates a time average, τ is the time lag between the time-resolved fluorescence intensities, and $\delta I(t)$ are the fluorescence fluctuations. Correlation functions may be calculated using a hardware correlator (e.g., Correlator.com) or by software correlaters using either commercial software (e.g., ISS, Zeiss, or Picoquant) or by programs written in various laboratories (e.g., Purkayastha *et al.*, 2005). Assuming that the excitation intensity in the observation volume is essentially a Gaussian ellipsoid that is brightest at the center and less bright toward the edges (Fig. 16.2A), the radius and axial extent of the observation volume are given by ω_0 and z_0, respectively. For a single fluorescent species, the calculated correlations may then be fit to

$$G(\tau) = \frac{1}{\langle N \rangle} \left(1 + \frac{\tau}{\tau_D}\right)^{-1} \left(1 + \frac{\tau}{S^2 \tau_D}\right)^{-1/2} \tag{16.4}$$

$$\tau_D = \frac{\omega_0^2}{4D} \tag{16.5}$$

where $\langle N \rangle$ is the average number of molecules in the observation volume, τ_D is related to the translational diffusion coefficient, *D*, by Eq. (16.5), and $S = z_0/\omega_0$ is the axial to radial ratio of the observation volume. Oligomerization (Chowdhury *et al.*, 2007; Purkayastha *et al.*, 2005) or protein unfolding (Sherman *et al.*, 2008) will slow translational diffusion, increasing the time it takes for the molecule to diffuse through the observation volume and shifting the correlation function to longer times (Fig. 16.2C).

The number of fitting parameters for FCS is often large compared to the number of data points making the fits underdetermined and much care must be taken to ensure that the results from the fits are experimentally meaningful. To avoid this problem, the radius and extent, ω_0 and z_0, of the observation volume are calibrated using a fluorescent dye with a known diffusion coefficient (Sherman *et al.*, 2008), and these values may be fixed when fitting protein data. The fit given in Eq. (16.4) is for a single molecular species. Many FCS experiments are performed on mixtures of species. If the fluorescence intensity, also called the molecular brightness, of all of the species is the same then $G(\tau)$ may be fit to (Chowdhury *et al.*, 2007; Purkayastha *et al.*, 2005)

$$G(\tau) = \frac{1}{\langle N \rangle} \sum_k f_k \left(1 + \frac{\tau}{\tau_D^k}\right)^{-1} \left(1 + \frac{\tau}{S^2 \tau_D^k}\right)^{-1/2} = \frac{1}{\langle N \rangle} \sum_k f_k g_k(\tau) \quad (16.6)$$

where $\langle N \rangle$ is the time-averaged total number of molecules in the observation volume, f_k is the fraction of species k in solution with diffusion time τ_D^k, and $g_k(\tau)$ is the autocorrelation for species k.

When fluorescently labeled proteins oligomerize the brightness of the oligomers increases with size, so a dimer is twice as bright as a monomer, a trimer is three times as bright as a monomer, etc. Note that this simple description assumes that changes in the fluorophore environment due to oligomerization do not affect the brightness. FCS is sensitive to brightness and the autocorrelations dependent on both the number of each species in the observation volume, N_k, and the brightness of each species, B_k (Krichevsky and Bonnet, 2002; Thompson, 1991):

$$G(\tau) = \frac{\sum_k N_k B_k^2 g_k(\tau)}{\left(\sum_j N_j B_j\right)^2} = \frac{\sum_k N_k \alpha_k^2 g_k(\tau)}{\left(\sum_j N_j \alpha_j\right)^2} \quad (16.7)$$

where $\alpha_k = B_k/B_1$ is the brightness of species k relative to species 1, the monomer for oligomerization experiments. Thus, $\alpha = 1$ for monomers, 2 for dimers, 3 for trimers, etc., in an ideal polymerization experiment. For polymerization experiments, Eq. (16.7) effectively means that $G(\tau)$ is biased toward higher-order oligomers.

4.1.2. Applying FCS to α_1AT polymerization

Heat-induced polymerization results in heterogeneous samples containing monomers, dimers, and higher-order oligomers. As shown in Eq. (16.7), brighter species such as long polymers containing many labeled monomers will make a larger contribution to the autocorrelation than will less bright species such as monomers. To account for this heterogeneity, multiple short datasets were collected for each sample and each individual dataset was analyzed using Eq. (16.6) (Purkayastha *et al.*, 2005). This allowed the distribution of diffusion times to be determined at the various time points providing a time-resolved profile of polymerization or depolymerization (Fig. 16.2D) (Chowdhury *et al.*, 2007; Purkayastha *et al.*, 2005). While the diffusion times determined from Eq. (16.6) are correct, because the molecular brightness terms are omitted from this equation (see Eq. (16.7)) the fraction of the population corresponding to the brightest species, that is, the largest oligomers at each time point, may have been overestimated by Gai and colleagues. Nonetheless, the FCS results show that heat-induced polymerization of α_1AT has two, concentration- and temperature-dependent lag phases (Purkayastha *et al.*, 2005), that the average polymer length increases with increasing temperature (Purkayastha *et al.*, 2005), and that small peptides such as WMDF can both prevent polymerization at 39 °C and lead to depolymerization of polymers formed at 45 °C. FCS can easily be applied to other polymer-prone serpins.

4.1.3. A protocol for monitoring α_1AT polymerization with FCS (based on the work by Gai, Cooperman, and collaborators; Chowdhury *et al.*, 2007; Purkayastha *et al.*, 2005)

1. Monomers of the α_1AT variant Cys232Ser/Ser359Cys (P1′C α_1AT) were labeled with the Cys reactive dye tetramethyl-rhodamine at a fluorophore to protein ratio of 20:1 in 50 m*M* Tris, 50 m*M* KCl, pH 8.3.
2. FCS experiments were performed using ~270 μW of 514.5 nm excitation from an argon ion laser and a 50 μm confocal pinhole. The setup was calibrated with the dye rhodamine 6G ($D = 280\ \mu m^2/s$ at 22 °C (Magde *et al.*, 1974)). All FCS experiments were performed at 23 °C.
3. Monomer experiments: P1′C α_1AT was diluted to 1 n*M* in 50 m*M* Tris, 50 m*M* KCl, pH 8.3, and FCS data were collected. Due to the presence of free dye, the resulting correlations were fit to two species using Eq. (16.6) where the first species corresponds to <10% free, tetramethyl-rhodamine and the second species, with a much longer diffusion time, to P1′C α_1AT.
4. Polymer formation: 5 μ*M* labeled P1′C α_1AT in 50 m*M* Tris, 50 m*M* KCl, pH 8.3, was incubated at 45, 50, or 55 °C.
5. FCS measurements: At various times, aliquots were removed from the heated samples and quickly diluted with 50 m*M* Tris, 50 m*M* KCl, pH 8.3, at 23 °C to 1 n*M* P1′C α_1AT (where 1 n*M* is the monomer concentration). FCS data were collected for a total of 100 s with an integration of 100 μs per time point. 10–100 datasets were collected for each diluted sample.
6. Polymerization blocking experiments (Chowdhury *et al.*, 2007): Polymers were formed by incubating 7 μ*M* of labeled P1′C α_1AT in 50 m*M* Tris, 50 m*M* KCl, pH 8.3, at 39 or 52 °C in the presence or absence of 1.4 μ*M* peptide, and FCS experiments were performed as described above.
7. Depolymerization experiments (Chowdhury *et al.*, 2007): Polymers were preformed at 45 °C for ~36 h using 7 μ*M* of labeled P1′C α_1AT in 50 m*M* Tris, 50 m*M* KCl, pH 8.3, the temperature was decreased to 39 °C and a 200-fold molar excess of peptide was added. FCS experiments were performed as described above.

4.2. Early events in neuroserpin polymerization measured by two-color coincidence detection

Amyloid formation by Aβ peptides is associated with Alzheimer's disease and interactions between neuroserpin and Aβ peptides have been shown to reduce neuroserpin polymerization both *in vitro* and in model systems (Kinghorn *et al.*, 2006). Conversely, neuroserpin reduces the toxicity of $A\beta_{1-42}$ oligomers in cells by accelerating aggregation (Kinghorn *et al.*, 2006). David Klenerman, David A. Lomas, and coworkers have taken advantage of SMF-based coincidence methods to monitor polymerization of neuroserpin and how interactions between neuroserpin and $A\beta_{1-40}$ peptides affect polymer formation (Chiou *et al.*, 2009).

Similarly to the α_1AT FCS experiments, polymers for TCCD experiments were formed at μ*M* neuroserpin concentrations and then diluted to ~50 p*M* (monomer concentration) for SMF microscopy (Chiou *et al.*, 2009). Dilutions of this magnitude can lead to dissociation of loosely bound species, and while serpin polymers are still intact at low concentrations (Chiou *et al.*, 2009; Purkayastha *et al.*, 2005), very few neuroserpin/$A\beta_{1-40}$ complexes were detected (Chiou *et al.*, 2009). Nonetheless, a dissociation constant of 10 ± 5 n*M* for neuroserpin/$A\beta_{1-40}$ complexes could be estimated from the TCCD data. While overnight incubation with $A\beta_{1-40}$ inhibited neuroserpin polymerization, addition of $A\beta_{1-40}$ immediately before or during polymerization accelerated neuroserpin polymerization (Chiou *et al.*, 2009). TCCD experiments on neuroserpin polymerization in the absence of $A\beta_{1-40}$ show that a conformational change in monomeric neuroserpin is the rate-limiting step for polymerization. In the following sections, the principles behind TCCD are introduced, the TCCD results are further described and a protocol for TCCD serpin experiments is given based on the work of Klenerman, Lomas, and colleagues (Chiou *et al.*, 2009).

4.2.1. Introduction to TCCD

Oligomerization, particularly early events in polymerization, may be monitored using two pools of protein labeled with spectrally distinct fluorophores, for example, a blue and a red fluorophore (Li *et al.*, 2003; Orte *et al.*, 2008a). At single molecule concentrations, association of blue and red labeled monomers results in diffusion of a blue and red labeled oligomer through the observation volume and coincident signals will be observed on the two detectors (Figs. 16.1 and 16.3). The intensity of the signal from the blue (red) detector is proportional to the number of monomers in the oligomer that are labeled with the blue (red) fluorophore. To discriminate signal from noise, the number of photons in a burst must be above a minimum threshold (number of photons/unit time) and the thresholds are individually optimized for the red and blue channels, respectively (Clarke *et al.*, 2007). The association quotient, Q, which increases as

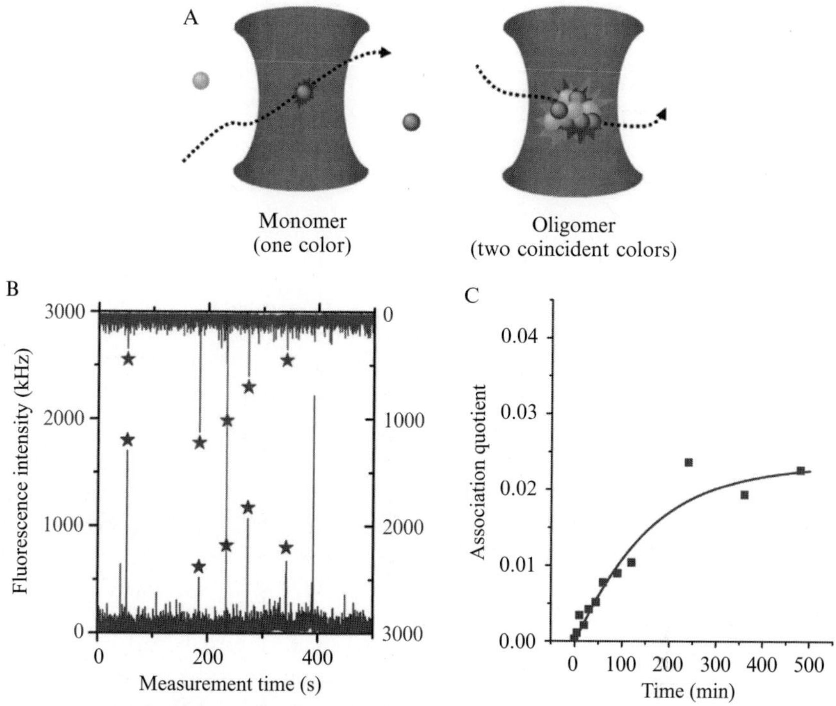

Figure 16.3 Detecting oligomerization with TCCD. (A) For monomers (left), photons with a single color are detected on one of the detectors. Oligomers emit photons with two colors (right) resulting in coincident photon bursts. (B) Fluorescence intensity in the blue (bottom) and red (top) channels as a function of time with coincident photon bursts indicated by stars. The intensity of the signal in each channel is proportional to the number of labeled monomers in an oligomer. (A) and (B) were adapted from Orte and colleagues, copyright (2008) National Academy of Sciences, USA (Orte *et al.*, 2008a). (C) As neuroserpin polymerizes the probability of coincident photon bursts increases leading to a time-dependent increase in the association quotient, Q (Eq. (16.8)), that can be fit using a kinetic model for neuroserpin polymerization (Chiou *et al.*, 2009). Reprinted from Chiou *et al.* (2009) with permission from Elsevier. (For interpretation of the references to color in this figure legend, the reader is referred to the Web version of this chapter.)

oligomers/polymers get larger, may then be determined as (Chiou *et al.*, 2009; Orte *et al.*, 2006)

$$Q = \frac{r_s}{r_B + r_R - r_S} \tag{16.8}$$

$$r_S = r_C - r_E \tag{16.9}$$

where the photon burst rates in the blue and red detection channels are given by r_B and r_R, respectively. The rate of bursts that are coincident on both channels is given by r_C which must be corrected for the rate of random coincidences, r_E, to reveal the rate of coincident events, r_s, arising from association (Eq. (16.9)). As polymers form and grow, Q increases with the time of incubation, t, according to Chiou *et al.* (2009):

$$Q(t) = y_0 + A\exp(-t/\tau) \tag{16.10}$$

where y_0 is the value of Q when polymerization has reached a steady state, A ($A < 0$) is the normalization constant, and τ is the rise time for polymerization.

The simplest application of TCCD requires two spectrally distinct fluorophores so that the signal in the blue (red) channel comes almost exclusively from the blue (red) fluorophore and FRET between the fluorophores is unlikely (although a more recent application combines TCCD and spFRET; Orte *et al.*, 2008b). The most important controls include experiments on monomers labeled with only the red or blue fluorophore to determine the signal intensity expected for each fluorophore and to make sure that the two detection channels are specific for each color as well as experiments in which red and blue monomers are mixed in order to determine the probability that simultaneous signals on both detectors occur by chance, r_E. TCCD was developed by Klenerman and colleagues, and more detailed descriptions of experimental considerations and data analysis are described in their work (Chiou *et al.*, 2009; Clarke *et al.*, 2007; Orte *et al.*, 2006, 2008a,b).

4.2.2. Applying TCCD to neuroserpin polymerization

At the earliest stages of polymerization TCCD reveals that the percentage of polymers is less than 1%. The initial change in Q, as determined from $(\partial Q/\partial t)_{t=0}$, is approximately $2.2 \pm 0.8 \times 10^{-6}\ \mathrm{s}^{-1}$ at all neuroserpin concentrations indicating that the initial rate-limiting step in polymerization is unimolecular, presumably a conformational change in the neuroserpin monomer. As observed in the FCS α_1AT polymerization experiments (Purkayastha *et al.*, 2005), the final neuroserpin polymer length is concentration dependent with higher concentrations resulting in larger polymers (Chiou *et al.*, 2009). In addition, as shown by a combination of modeling and fits to the TCCD data, TCCD may be used to test kinetic models of polymerization (Fig. 16.3C) (Chiou *et al.*, 2009).

4.2.3. A protocol for monitoring neuroserpin polymerization with TCCD (based on the work of Klenerman, Lomas, and collaborators; Chiou *et al.*, 2009)

1. Protein labeling: The Ser229Cys (S229C) mutation was introduced into neuroserpin (Chiou *et al.*, 2009) and the S229C variant with an

N-terminal 6× His tag was purified (Belorgey *et al.*, 2002; Chiou *et al.*, 2009). S229C neuroserpin was labeled with 10- to 20-fold molar excess of Alexa Fluor 488 (AF488, blue) or Alexa Fluor 647 (AF647, red).
2. Polymer formation: Equimolar concentrations of blue and red labeled neuroserpin were mixed in phosphate buffered saline at total neuroserpin concentrations ranging from 2.2 to 6.6 μ*M* and incubated at 45 °C for up to 480 m. Aliquots were removed at various times and snap-frozen in liquid nitrogen. More detailed protocols for neuroserpin polymerization may be found in a recent paper by Lomas and colleagues (Belorgey *et al.*, 2011).
3. Brief description of the TCCD setup: AF488 was excited with 220 μW of 488 nm laser light while AF647 was excited with 60 μW of 633 nm laser light from two overlapping Gaussian laser beams in a confocal microscope with a 60× NA 1.40 oil objective and a 50 μm pinhole (Fig. 16.1). The fluorescence was separated using a dichroic mirror (585DRLP, Omega Optical) and sent to two APDs, one for each color (Chiou *et al.*, 2009).
4. Glass coverslips used for TCCD experiments were incubated with 1 mg/ml bovine serum albumin (BSA) for at least 1 h at room temperature to prevent nonspecific protein adsorption. The BSA solutions were removed, snap-frozen neuroserpin samples were thawed, quickly diluted to 50 p*M* for TCCD experiments and applied to the coverslips.
5. The association quotient, Q, was calculated from the time-resolved photon counts from the blue and red detectors according to Eq. (16.8) (Chiou *et al.*, 2009; Orte *et al.*, 2006, 2008a) and the time dependence of Q was fit to Eq. (16.10). These experiments were performed for neuroserpin alone or in the presence of $A\beta_{1-40}$ (1:1 neuroserpin: $A\beta_{1-40}$ ratio).
6. Neuroserpin–Aβ interactions: $A\beta_{1-40}$ labeled at the N-terminus with HiLyte Fluor 488 was purchased from Anaspec. A 1:1 ratio of AF647 labeled neuroserpin (red) and HiLyte Fluor 488 labeled $A\beta_{1-40}$ (red) were incubated at μ*M* concentrations and 45 °C to promote neuroserpin polymerization, and TCCD experiments were performed as described above.

4.3. The utility of FCS and TCCD for studying polymerization

Both FCS and TCCD can easily be applied to any serpin that can be fluorescently labeled. Because both methods can measure the size of oligomers, they can be used to obtain detailed information on early events in polymerization and how protein–protein as well as protein–peptide interactions affect early and late events in polymerization. Finally, both methods may be combined with FRET to provide even more information on polymer structures (Li *et al.*, 2005; Nath *et al.*, 2010; Orte *et al.*, 2008b). Unlike FCS, which is often used to interrogate small numbers of molecules, TCCD is a real single molecule technique collecting data from one molecule at a time. However, TCCD is still being developed, and for the moment FCS is easier to apply with commercially available hardware and software.

5. Conformational Distributions of Protease–Serpin Complexes

Serpins inhibit target proteases by translocating the covalently attached protease more than 70 Å from one pole of the serpin to the other, concomitantly mechanically deforming the protease active site (Dementiev *et al.*, 2006; Huntington *et al.*, 2000; Shin and Yu, 2006). Ensemble FRET and other fluorescence techniques have been used to monitor protease interactions with inhibitory serpins providing information on the average conformation of the population (Lawrence *et al.*, 1994; Mellet *et al.*, 2002; Shin and Yu, 2002, 2006; Shore *et al.*, 1995; Swanson *et al.*, 2007; Tew and Bottomley, 2001). SpFRET experiments provide additional information by revealing the conformational distributions of protease–serpin complexes.

SpFRET experiments have been conducted on covalent complexes between α_1AT and bovine trypsin (BTryp) as well as its more stable cousin, rat trypsin (RTryp) (Liu *et al.*, 2006, 2007). SpFRET results suggest that BTryp is more disrupted in the covalent complex than is the more stable RTryp (Liu *et al.*, 2006). The orientation of the protease relative to α_1AT is also different in BTryp and RTryp complexes. X-ray crystal structures of covalent complexes between α_1AT and two noncognate serine proteases, BTryp (Huntington *et al.*, 2000) and porcine pancreatic elastase (Dementiev *et al.*, 2006), also revealed differences in protease structural disruption and orientation. While the extent of protease disruption may be affected by protease self-cleavage in crystals (Dementiev *et al.*, 2006), self-cleavage is unlikely in spFRET experiments due to the limited, 15 m, incubation time required for complex formation at μ*M* concentrations and the p*M* concentrations used when collecting spFRET data. Thus, both the crystal structures and the spFRET results suggest that the protease conformation and orientation in protease–serpin covalent complexes may be affected by protease stability. In addition, spFRET has revealed a fluorophore-trapped covalent complex in which trypsin is only partially disrupted and partially translocated by α_1AT. Below, we give a brief introduction to spFRET and describe how it has been applied to trypsin–α_1AT complexes.

5.1. Introduction to single pair Förster resonance energy transfer

FRET takes advantage of distance dependent dipole–dipole interactions between a donor fluorophore in the excited state and a ground-state acceptor. These interactions can result in energy transfer, with the probability of energy transfer increasing as, r, the distance between the donor and acceptor dipole moments decreases (Lakowicz, 2006; Sisamakis *et al.*, 2010; Steinberg, 1971; Stryer, 1978):

$$E = \frac{R_0^6}{(r^6 + R_0^6)} \quad (16.11)$$

where E is the energy transfer efficiency ($0 \le E \le 1$) and R_0 is the inter-fluorophore distance at which the probability of energy transfer is 50%. The efficiency of energy transfer, and thus R_0, depends on the fluorescence quantum yield of the donor, Q_D, the overlap between the emission spectrum of the donor and the absorption spectrum of the acceptor calculated using the overlap integral, $J(\lambda)$, the relative orientation of the donor and acceptor dipole moments given by the orientation factor, κ^2, and the refractive index, n, generally assumed to be that of water, 1.33 (Lakowicz, 2006; Sisamakis *et al.*, 2010; Steinberg, 1971; Stryer, 1978):

$$R_0(\text{Å}) = 9.78 \times 10^3 \left(\kappa^2 n^{-4} Q_D J(\lambda)\right)^{1/6} \quad (16.12)$$

If the donor and acceptor are freely rotating, $\kappa^2 = 2/3$, and the validity of this assumption may be tested by measuring the fluorescence anisotropy of the donor and acceptor. Detailed discussions of how to calculate R_0 may be found in work by Lakowicz (2006) and for single molecule measurements by Seidel and coworkers (Sisamakis *et al.*, 2010). For most donor–acceptor pairs R_0 is between 30 and 70 Å, and typical R_0 values for a number of donor–acceptor pairs used for spFRET experiments are given in Table 16.2. FRET is most sensitive in the distance range between $\sim 0.5R_0$ and $\sim 1.5R_0$ and donor–acceptor pair should be carefully chosen so that expected separations fall within this range. In general, FRET is very good at determining relative changes in distance, for example, changes in the donor–acceptor separation upon protein unfolding or protein–protein interactions, but absolute distance determinations should only be made with great care.

For spFRET experiments, fluorescence emission is split by color and donor fluorescence is collected by one detector (Det1 in Fig. 16.1) while acceptor fluorescence is collected by a second detector (Det2 in Fig. 16.1). The energy transfer efficiency, E, is then calculated from these fluorescence intensities (Dahan *et al.*, 1999):

$$E = I_A/(I_A + \gamma I_D) \quad (16.13)$$

where I_A and I_D are the intensities of a photon burst from the donor, D, and acceptor, A, respectively, corrected for background and cross talk between

Table 16.2 Reported R_0 values for donor–acceptor pairs commonly used for spFRET experiments[a]

Donor	Acceptor	R_0 (Å)
Alexa Fluor 488	Alexa Fluor 594	54 (Mukhopadhyay *et al.*, 2007; Müller-Späth *et al.*, 2010)
Alexa Fluor 488	Texas Red	48[b] (Liu *et al.*, 2006)
Alexa Fluor 555	Atto 610	59[b] (Liu *et al.*, 2006)
Alexa Fluor 555	Cy5	59[b] (Liu *et al.*, 2006)
Cy3	Cy5	53 (Ishii *et al.*, 1999) ~60 (Murphy *et al.*, 2004)

[a] R_0 depends on the donor quantum yield and emission spectrum, the acceptor absorption spectrum, and the rotational freedom of the fluorophores (Eq. (16.12)) all of which may be influenced by the local fluorophore environment. R_0 should therefore be calculated based on experimental data for the biomolecule(s) of interest.

[b] Used in serpin spFRET experiments.

the donor and acceptor detectors. In Eq. (16.13), γ corrects for differences in quantum yield and detection efficiency between the donor and acceptor fluorophores. To discriminate between signal and noise, E is only calculated for photon bursts in which the total number of photons, $I_A + I_D$, are above a minimum threshold, and the resulting spFRET efficiencies from each photon burst (Fig. 16.4A) are compiled into a histogram where the number of peaks and the peak widths detail the conformational distribution (Fig. 16.4B and C) (Dahan *et al.*, 1999). SpFRET efficiency histograms are often fit to Gaussian distributions and detailed discussions of data analysis for spFRET may be found in work by Seidel and coworkers (Antonik *et al.*, 2006; Kalinin *et al.*, 2007; Sisamakis *et al.*, 2010) and by Gopich and Szabo (2007, 2010).

5.2. Conformational distributions of trypsin–α_1AT complexes

Peaks in spFRET efficiency histograms have an inherent width associated with noise, and conformational heterogeneity can further increase this width (Antonik *et al.*, 2006; Gopich and Szabo, 2007). In agreement with differences in protease structural disruption between X-ray crystal structures for BTryp–α_1AT (Huntington *et al.*, 2000) and elastase–α_1AT (Dementiev *et al.*, 2006), BTryp–α_1AT covalent complexes have wider spFRET peaks than do corresponding RTryp–α_1AT complexes indicating that the structure around trypsin residue 113 is more disrupted in BTryp complexes. Despite the conformational differences between the BTryp and RTryp covalent complexes, time-dependent spFRET experiments show that both complexes dissociate with similar rate constants (6–7 × 10^{-6} s^{-1}) at pH 7.4 when the protease is fully translocated. Interestingly, the

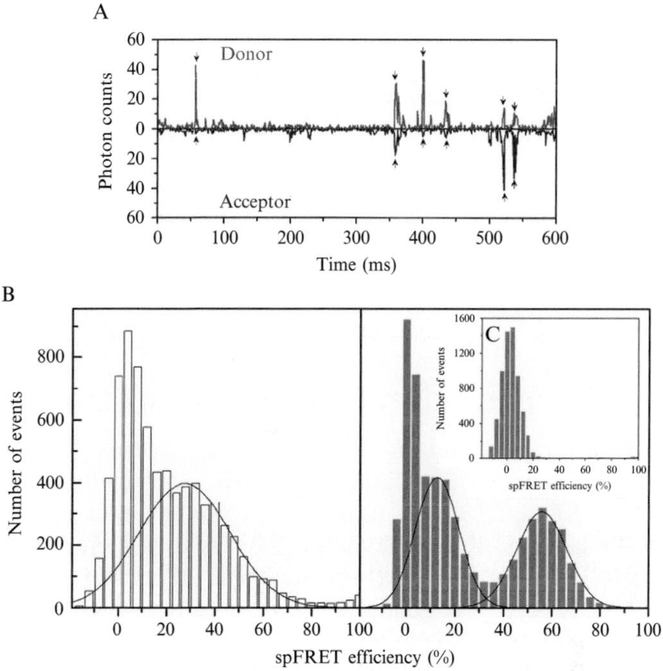

Figure 16.4 Detecting conformational distributions with spFRET (adapted from Liu *et al.* (2006)). (A) Detection of photons from single protease–serpin complexes, indicated by arrows, traversing the observation volume. These complexes are heterogeneous with some showing mostly donor fluorescence (low spFRET efficiency) and some showing mostly acceptor fluorescence (high spFRET efficiency). (B, C) SpFRET efficiencies from the individual complexes are compiled into a histogram to reveal the conformational distribution for protease–serpin complexes. The large "zero peak" centered around 0% efficiency arises from complexes containing only donor fluorophores or complexes in which the acceptor has photobleached. (B) Aside from the zero peak, BTryp–α_1AT complexes show a single, broad conformational distribution with a Gaussian fit to the distribution given by the black line. (C) When cationic fluorophores are attached to Cys232 on α_1AT, interactions between anionic RTryp and the cationic fluorophore trap partially translocated RTryp resulting in a second peak at high FRET efficiency. The inset shows that this conformation is not observed for the RTryp–α_1AT encounter complex where the serpin and protease are not covalently attached. (For color version of this figure, the reader is referred to the Web version of this book.)

fluorophores used in this study mediated trapping of covalent complexes in which RTryp is not fully translocated 70 Å across α_1AT (Liu *et al.*, 2006, 2007). This species was identified from an additional peak in RTryp–α_1AT spFRET histograms (Fig. 16.4C) and only occurs when α_1AT is labeled with a cationic fluorophore at Cys232. Interactions between cationic fluorophores and anionic RTryp stabilize the partially translocated RTryp resulting in RTryp–α_1AT complexes with a shorter lifetime, ~3 h at pH

7.4, and dissociation of the partially translocated complexes is accelerated at both low and high pH indicating that the RTryp active site is not fully disrupted in these complexes (Liu *et al.*, 2007).

These experiments demonstrate the utility of spFRET for monitoring protease–serpin covalent complexes both for its ability to delineate the conformational distribution and to monitor changes in this distribution over time. The serendipitous trapping of shorter-lived covalent complexes suggest that protease–serpin covalent complex stability may be tuned not only by altering the length of the reactive center loop (Plotnick *et al.*, 2002; Zhou *et al.*, 2001) but also by modulating electrostatic interactions between proteases and serpins.

5.3. A protocol for spFRET on protease–serpin complexes

1. Trypsin purification and labeling: Both trypsins have 12 native Cys residues all of which are disulfide bonded. For labeling, the surface accessible residue 113 was mutated to Cys (Liu *et al.*, 2006; Mellet *et al.*, 2002). Ser113Cys BTryp was expressed as inclusion bodies in *Escherichia coli* cells, purified and activated (Peterson *et al.*, 2001). Lys113Cys RTryp was expressed in yeast cells as a secreted protein, purified and activated (Hedstrom *et al.*, 1994). The trypsin variants were generally labeled with the donor fluorophore Alexa Fluor 555 maleimide (Invitrogen) and a soybean trypsin inhibitor column (Sigma Chemical) was used to separate trypsin from unreacted fluorophores (Liu *et al.*, 2006).
2. α_1AT purification and labeling: α_1AT containing the single native Cys, Cys232, or single Cys variants in the Cys232Ser background, was expressed as inclusion bodies in *E. coli*, purified and checked for activity (Liu *et al.*, 2006). α_1AT was generally labeled with the acceptor fluorophore Atto610 maleimide.
3. Covalent complex formation: Trypsin and α_1AT (0.5–2 μ*M*) were incubated for 15 m at room temperature at a 1:2 molar ratio and complex formation was confirmed by SDS-PAGE. Samples were diluted to 100–50 p*M* in buffer containing 10 m*M* of the trypsin inhibitor benzamidine to inhibit free trypsin. SpFRET experiments were performed on 300 μl samples on coverglass (Nalgene/Nunc) that had been incubated with 1 mg/ml BSA to prevent nonspecific protein adsorption.
4. Brief description of the spFRET setup: Samples were excited with the 488 nm laser light from an Ar–Kr laser in a confocal microscope based on an IX-70 inverted microscope (Olympus) with a 60× NA 1.2 water objective and a 100 μm pinhole (Fig. 16.1). The donor and acceptor fluorescence were separated using a dichroic mirror (Q605LP, Chroma Technologies) and sent to two APDs, one for each color (Liu *et al.*, 2006).

5. SpFRET analysis: The photon counts were collected using hardware and software from ISS, photon bursts were identified, and *E* was calculated according to Eq. (16.13) using programs written in Matlab (Mathworks). Peaks in the spFRET histograms were fit using Origin (OriginLab).

6. Conclusions and Future Directions

The work by Gai and colleagues (Chowdhury *et al.*, 2007; Purkayastha *et al.*, 2005) and by Klenerman and coworkers (Chiou *et al.*, 2009) demonstrate the utility of SMF methods for monitoring early events in polymer formation. More detailed structural information on serpin conformational distributions during the lag phases in polymerization may arise from studies that combine FCS or TCCD with spFRET (Li *et al.*, 2005; Orte *et al.*, 2008b). SpFRET studies of serpin folding, which can help elucidate the structure of folding intermediates (Schuler and Eaton, 2008), could also contribute to an understanding of serpin polymerization by identifying conformational distributions associated with serpin folding intermediates.

Ideally, one would like to follow serpin folding, protease–serpin complex formation or serpin polymer formation by following the same molecule over time. While serpin immobilization has been difficult, a number of different approaches have been developed for monitoring one or more molecules encapsulated in a small volume (Huebner *et al.*, 2009; Reiner *et al.*, 2006; Srisa-Art *et al.*, 2010). Encapsulation of a labeled serpin and a labeled protease in the same droplet should allow protease–serpin complex formation to be followed from the initial encounter complex to the final covalent complex. Similarly, early events in polymer formation might be followed by encapsulating just a few serpins in a single droplet. Finally, newly developed super-resolution methods for imaging single molecules in cells (Patterson *et al.*, 2010) may allow for detailed studies of the motions and localization of intracellular serpins as was recently demonstrated for ovalbumin injected into cells (Speil and Kubitscheck, 2010).

ACKNOWLEDGMENTS

We thank Dr. Lu Liu for the FCS data shown in Fig. 16.2B and C. Funding was provided by NIH grant GM094848 and NSF grant MCB-0446220.

REFERENCES

Antonik, M., Felekyan, S., Gaiduk, A., and Seidel, C. A. (2006). Separating structural heterogeneities from stochastic variations in fluorescence resonance energy transfer distributions via photon distribution analysis. *J. Phys. Chem. B* **110,** 6970–6978.

Belorgey, D., Crowther, D., Mahadeva, R., and Lomas, D. (2002). Mutant Neuroserpin (S49P) that causes familial encephalopathy with neuroserpin inclusion bodies is a poor proteinase inhibitor and readily forms polymers in vitro. *J. Biol. Chem.* **277,** 17367–17373.

Belorgey, D., Irving, J. A., Ekeowa, U. I., Freeke, J., Roussel, B. D., Miranda, E., Pérez, J., Marciniak, S. J., Crowther, D. C., Michel, C. H., and Lomas, D. A. (2011). Characterisation of serpin polymers in vitro and in vivo. *Methods* **53,** 255–266.

Buschmann, V., Weston, K. D., and Sauer, M. (2003). Spectroscopic study and evaluation of red-absorbing fluorescent dyes. *Bioconjugate Chem.* **14,** 195–204.

Chattopadhyay, K., Saffarian, S., Elson, E. L., and Frieden, C. (2005). Measuring unfolding of proteins in the presence of denaturant using fluorescence correlation spectroscopy. *Biophys. J.* **88,** 1413–1422.

Chen, H., Rhoades, E., Butler, J. S., Loh, S. N., and Webb, W. W. (2007). Dynamics of equilibrium structural fluctuations of apomyoglobin measured by fluorescence correlation spectroscopy. *Proc. Natl. Acad. Sci. USA* **104,** 10459–10464.

Chiou, A., Hägglöf, P., Orte, A., Chen, A. Y., Dunne, P. D., Belorgey, D., Karlsson-Li, S., Lomas, D., and Klenerman, D. (2009). Probing neuroserpin polymerization and interaction with amyloid-beta peptides using single molecule fluorescence. *Biophys. J.* **97,** 2306–2315.

Chowdhury, P., Wang, W., Lavender, S., Bunagan, M. R., Klemke, J. W., Tang, J., Saven, J. G., Cooperman, B. S., and Gai, F. (2007). Fluorescence correlation spectroscopic study of serpin depolymerization by computationally designed peptides. *J. Mol. Biol.* **369,** 462–473.

Clarke, R. W., Orte, A., and Klenerman, D. (2007). Optimized threshold selection for single-molecule two-color fluorescence coincidence spectroscopy. *Anal. Chem.* **79,** 2771–2777.

Dahan, M., Deniz, A. A., Ha, T., Chemla, D. S., Schultz, P. G., and Weiss, S. (1999). Ratiometric measurement and identification of single diffusing molecules. *Chem. Phys.* **247,** 85–106.

Davis, R. L., Shrimpton, A. E., Holohan, P. D., Bradshaw, C., Feiglin, D., Collins, G. H., Sonderegger, P., Kinter, J., Becker, L. M., Lacbawan, F., Krasnewich, D., Muenke, M., Lawrence, D. A., Yerby, M. S., Shaw, C. M., Gooptu, B., Elliott, P. R., Finch, J. T., Carrell, R. W., and Lomas, D. A. (1999). Familial dementia caused by polymerization of mutant neuroserpin. *Nature* **401,** 376–379.

Dementiev, A., Dobo, J., and Gettins, P. G. W. (2006). Active site distortion is sufficient for proteinase inhibition by serpins: Structure of the covalent complex of α1 proteinase inhibitor with porcine pancreatic elastase. *J. Biol. Chem.* **281,** 3452–3457.

Elson, E. L., and Magde, D. (1974). Fluorescence correlation spectroscopy. I. Conceptual basis and theory. *Biopolymers* **13,** 1–27.

Gopich, I. V., and Szabo, A. (2007). Single-molecule FRET with diffusion and conformational dynamics. *J. Phys. Chem. B* **111,** 12925–12932.

Gopich, I. V., and Szabo, A. (2010). FRET efficiency distributions of multistate single molecules. *J. Phys. Chem. B* **114,** 15221–15226.

Hedstrom, L., Perona, J. J., and Rutter, W. J. (1994). Converting trypsin to chymotrypsin: Residue 172 is a substrate specificity determinant. *Biochemistry* **33,** 8757–8763.

Hermanson, G. T. (1996). Bioconjugate Techniques. Academic Press, San Diego, CA.

Hoffman, H., Hilger, F., Pfeil, S. H., Hoffmann, A., Streich, D., Haenni, D., Nettels, D., Lipman, E. A., and Schuler, B. (2010). Single-molecule spectroscopy of protein folding in a chaperonin cage. *Proc. Natl. Acad. Sci. USA* **107,** 11793–11798.

Hoffmann, A., Kane, A., Nettels, D., Hertzog, D. E., Baumgartel, P., Lengefeld, J., Reichardt, G., Horsley, D. A., Seckler, R., Bakajin, O., and Schuler, B. (2007). Mapping protein collapse with single-molecule fluorescence and kinetic synchrotron radiation circular dichroism spectroscopy. *Proc. Natl. Acad. Sci. USA* **104,** 105–110.

Huebner, A., Bratton, D., Whyte, G., Yang, M., deMello, A. J., Abell, C., and Hollfelder, F. (2009). Static microdroplet arrays: A microfluidic device for droplet trapping, incubation and release for enzymatic and cell-based assays. *Lab Chip* **9,** 692–698.

Huntington, J. A., Read, R. J., and Carrell, R. W. (2000). Structure of a serpin-protease complex shows inhibition by deformation. *Nature* **407,** 923–926.

Ishii, Y., Yoshida, T., Funatsu, T., Wazawa, T., and Yanagida, T. (1999). Fluorescence resonance energy transfer between single fluorophores attached to a coiled-coil protein in aqueous solution. *Chem. Phys.* **247,** 163–173.

Kalinin, S., Felekyan, S., Antonik, M., and Seidel, C. A. (2007). Probability distribution analysis of single-molecule fluorescence anisotropy and resonance energy transfer. *J. Phys. Chem. B* **111,** 10253–10262.

Kapanidis, A. N., Laurence, T. A., Lee, N. K., Margeat, E., Kong, X., and Weiss, S. (2005). Alternating-laser excitation of single molecules. *Acc. Chem. Res.* **38,** 523–533.

Kim, Y., Ho, S. O., Gassman, N. R., Korlann, Y., Landorf, E. V., Collart, F. R., and Weiss, S. (2008). Efficient site-specific labeling of proteins via cysteines. *Bioconj. Chem.* **19,** 786–791.

Kinghorn, K. J., Crowther, D. C., Sharp, L. K., Nerelius, C., Davis, R. L., Chang, H. T., Green, C., Gubb, D. C., Johansson, J., and Lomas, D. A. (2006). Neuroserpin binds Aβ and is a neuroprotective component of amyloid plaques in Alzheimer disease. *J. Biol. Chem.* **281,** 29268–29277.

Krichevsky, O., and Bonnet, G. (2002). Fluorescence correlation spectroscopy: The technique and its applications. *Rep. Prog. Phys.* **65,** 251–297.

Lakowicz, J. R. (2006). Principles of Fluorescence Spectroscopy. 3rd edn. Springer, New York, NY.

Lawrence, D. A., Olson, S. T., Palaniappan, S., and Ginsburg, D. (1994). Serpin reactive center loop mobility is required for inhibitor function but not for enzyme recognition. *J. Biol. Chem.* **269,** 27657–27662.

Li, H., Ying, L., Green, J. J., Balasubramanian, S., and Klenerman, D. (2003). Ultrasensitive coincidence fluorescence detection of single DNA molecules. *Anal. Chem.* **75,** 1664–1670.

Li, G., Levitus, M., Bustamante, C., and Widom, J. (2005). Rapid spontaneous accessibility of nucleosomal DNA. *Nat. Struct. Mol. Biol.* **12,** 46–53.

Liu, L., Mushero, N., Hedstrom, L., and Gershenson, A. (2006). Conformational distributions of protease-serpin complexes: A partially translocated complex. *Biochemistry* **45,** 10865–10872.

Liu, L., Mushero, N., Hedstrom, L., and Gershenson, A. (2007). Short-lived protease serpin complexes: Partial disruption of the rat trypsin active site. *Protein Sci.* **16,** 2403–2411.

Lomas, D. A., Evans, D. L., Finch, J. T., and Carrell, R. W. (1992). The mechanism of Z α1-antitrypsin accumulation in the liver. *Nature* **357,** 605–607.

Magde, D., Elson, E. L., and Webb, W. W. (1974). Fluorescence correlation spectroscopy. II. An experimental realization. *Biopolymers* **13,** 29–61.

Mellet, P., Mély, Y., Hedstrom, L., Cahoon, M., Belorgey, D., Srividya, N., Rubin, H., and Bieth, J. G. (2002). Comparative trajectories of active and S195A inactive trypsin upon binding to serpins. *J. Biol. Chem.* **277,** 38901–38914.

Moerner, W. E., and Fromm, D. P. (2003). Methods of single-molecule fluorescence spectroscopy and microscopy. *Rev. Sci. Instrum.* **74,** 3597–3619.

Mukhopadhyay, S., Krishnan, R., Lemke, E. A., Lindquist, S., and Deniz, A. A. (2007). A natively unfolded yeast prion monomer adopts an ensemble of collapsed and rapidly fluctuating structures. *Proc. Natl. Acad. Sci. USA* **104,** 2649–2654.

Müller, B. K., Zaychikov, E., Bräuchle, C., and Lamb, D. C. (2005). Pulsed interleaved excitation. *Biophys. J.* **89,** 3508–3522.

Müller-Späth, S., Soranno, A., Hirschfeld, V., Hofmann, H., Rüegger, S., Reymond, L., Nettels, D., and Schuler, B. (2010). Charge interactions can dominate the dimensions of intrinsically disordered proteins. *Proc. Natl. Acad. Sci. USA* **107,** 14609–14614.

Murphy, M., Rasnik, I., Cheng, W., Lohman, T. M., and Ha, T. (2004). Probing single-stranded DNA conformational flexibility using fluorescence spectroscopy. *Biophys. J.* **86,** 2530–2537.

Myong, S., and Ha, T. (2010). Stepwise translocation of nucleic acid motors. *Curr. Opin. Struct. Biol.* **20,** 121–127.

Nath, S., Meuvis, J., Hendrix, J., Carl, S. A., and Engelborghs, Y. (2010). Early aggregation steps in alpha-synuclein as measured by FCS and FRET: Evidence for a contagious conformational change. *Biophys. J.* **98,** 1302–1311.

Neuweiler, H., Löllmann, M., Doose, S., and Sauer, M. (2007). Dynamics of unfolded polypeptide chains in crowded environment studied by fluorescence correlation spectroscopy. *J. Mol. Biol.* **365,** 856–869.

Nienhaus, G. U. (2009). Single-molecule fluorescence studies of protein folding. *Methods Mol. Biol.* **490,** 311–337.

Orte, A., Clarke, R., Balasubramanian, S., and Klenerman, D. (2006). Determination of the fraction and stoichiometry of femtomolar levels of biomolecular complexes in an excess of monomer using single-molecule, two-color coincidence detection. *Anal. Chem.* **78,** 7707–7715.

Orte, A., Birkett, N. R., Clarke, R. W., Devlin, G. L., Dobson, C. M., and Klenerman, D. (2008a). Direct characterization of amyloidogenic oligomers by single-molecule fluorescence. *Proc. Natl. Acad. Sci. USA* **105,** 14424–14429.

Orte, A., Clarke, R. W., and Klenerman, D. (2008b). Fluorescence coincidence spectroscopy for single-molecule fluorescence resonance energy-transfer measurements. *Anal. Chem.* **80,** 8389–8397.

Patterson, G., Davidson, M., Manley, S., and Lippincott-Schwartz, J. (2010). Superresolution imaging using single-molecule localization. *Annu. Rev. Phys. Chem.* **61,** 345–367.

Pearce, M., Powers, G., Feil, S., Hansen, G., Parker, M., and Bottomley, S. (2010). Identification and characterization of a misfolded monomeric serpin formed at physiological temperature. *J. Mol. Biol.* **403,** 459–467.

Peterman, E. J., Sosa, H., and Moerner, W. E. (2004). Single-molecule fluorescence spectroscopy and microscopy of biomolecular motors. *Annu. Rev. Phys. Chem.* **55,** 79–96.

Petersen, N. O., and Elson, E. L. (1986). Measurements of diffusion and chemical kinetics by fluorescence photobleaching recovery and fluorescence correlation spectroscopy. *Methods Enzymol.* **130,** 454–484.

Peterson, F. C., Gordon, N. C., and Gettins, P. G. (2001). High-level bacterial expression and 15N-alanine-labeling of bovine trypsin. Application to the study of trypsin-inhibitor complexes and trypsinogen activation by NMR spectroscopy. *Biochemistry* **40,** 6275–6283.

Plotnick, M. I., Rubin, H., and Schechter, N. M. (2002). The effects of reactive site location on the inhibitory properties of the serpin alpha(1)-antichymotrypsin. *J. Biol. Chem.* **277,** 29927–29935.

Purkayastha, P., Klemke, J. W., Lavender, S., Oyola, R., Cooperman, B. S., and Gai, F. (2005). Alpha 1-antitrypsin polymerization: A fluorescence correlation spectroscopic study. *Biochemistry* **44,** 2642–2649.

Reiner, J. E., Crawford, A. M., Kishore, R. B., Goldner, L. S., Helmerson, K., and Gilson, M. (2006). Optically trapped aqueous droplets for single molecule studies. *Appl. Phys. Lett.* **89,** 013904.

Roy, R., Hohng, S., and Ha, T. (2008). A practical guide to single-molecule FRET. *Nat. Methods* **5,** 507–516.

Schuler, B., and Eaton, W. A. (2008). Protein folding studied by single-molecule FRET. *Curr. Opin. Struct. Biol.* **18,** 16–26.

Sharma, S., Chakraborty, K., Müller, B. K., Astola, N., Tang, Y. C., Lamb, D. C., Hayer-Hartl, M., and Hartl, F. U. (2008). Monitoring protein conformation along the pathway of chaperonin-assisted folding. *Cell* **133,** 142–153.

Sherman, E., Itkin, A., Kuttner, Y. Y., Rhoades, E., Amir, D., Haas, E., and Haran, G. (2008). Using fluorescence correlation spectroscopy to study conformational changes in denatured proteins. *Biophys. J.* **94,** 4819–4827.

Shin, J.-S., and Yu, M.-H. (2002). Kinetic dissection of α1-antitrypsin inhibition mechanism. *J. Biol. Chem.* **277,** 11629–11635.

Shin, J.-S., and Yu, M.-H. (2006). Viscous drag as the source of active site perturbation during protease translocation: Insights into how inhibitory processes are controlled by serpin metastability. *J. Mol. Biol.* **359,** 378–389.

Shore, J. D., Day, D. E., Francis-Chmura, A. M., Verhamme, I., Kvassman, J., Lawrence, D. A., and Ginsburg, D. (1995). A fluorescent probe study of plasminogen activator inhibitor-1. Evidence for reactive center loop insertion and its role in the inhibitory mechanism. *J. Biol. Chem.* **270,** 5395–5398.

Sisamakis, E., Valeri, A., Kalinin, S., Rothwell, P. J., and Seidel, C. A. (2010). Accurate single-molecule FRET studies using multiparameter fluorescence detection. *Methods Enzymol.* **475,** 455–514.

Speil, J., and Kubitscheck, U. (2010). Single ovalbumin molecules exploring nucleoplasm and nucleoli of living cell nuclei. *Biochim. Biophys. Acta* **1803,** 396–404.

Srisa-Art, M., Demello, A. J., and Edel, J. B. (2010). High-efficiency single-molecule detection within trapped aqueous microdroplets. *J. Phys. Chem. B* **114,** 15766–15772.

Steinberg, I. Z. (1971). Long-range nonradiative transfer of electronic excitation energy in proteins and polypeptides. *Annu. Rev. Biochem.* **40,** 83–114.

Stryer, L. (1978). Fluorescence energy transfer as a spectroscopic ruler. *Annu. Rev. Biochem.* **47,** 819–846.

Swanson, R., Raghavendra, M. P., Zhang, W., Froelich, C., Gettins, P. G., and Olson, S. T. (2007). Serine and cysteine proteases are translocated to similar extents upon formation of covalent complexes with serpins. Fluorescence perturbation and fluorescence resonance energy transfer mapping of the protease binding site in CrmA complexes with granzyme B and caspase-1. *J. Biol. Chem.* **282,** 2305–2313.

Tew, D. J., and Bottomley, S. P. (2001). Intrinsic fluorescence changes and rapid kinetics of proteinase deformation during serpin inhibition. *FEBS Lett.* **494,** 30–33.

Thompson, N. L. (1991). Fluorescence correlation spectroscopy. *In* "Topics in Fluorescence Spectroscopy," (J. R. Lakowicz, ed.), pp. 337–378. Plenum Press, New York.

Zhou, A., Carrell, R. W., and Huntington, J. A. (2001). The serpin inhibitory mechanism is critically dependent on the length of the reactive center loop. *J. Biol. Chem.* **276,** 27541–27547.

CHAPTER SEVENTEEN

Serpin Polymerization *In Vitro*

James A. Huntington[1] *and* Masayuki Yamasaki

Contents

1. Introduction 380
2. Methods of Inducing Polymerization 383
   2.1. Introduction 383
   2.2. Heat 383
   2.3. Denaturants 391
   2.4. Low pH 392
   2.5. Chemical modification 393
   2.6. Proteolytic cleavage and short peptides 393
   2.7. Refolding 394
   2.8. *In vitro* translation 395
3. Kinetics of Polymerization 395
   3.1. General features and kinetic scheme 395
   3.2. Loss of monomer 396
   3.3. Spectroscopic methods 399
4. Effect of Mutations and "Drugs" on Polymerization 402
   4.1. Early studies on the effect of mutations 402
   4.2. Destabilizing and stabilizing mutations 403
   4.3. Cavity-filling mutations and drugs 404
   4.4. Inhibition by small molecules 404
   4.5. Inhibition by peptides 405
   4.6. Reversal by peptides 406
5. Mechanisms of Polymerization 408
   5.1. Introduction 408
   5.2. Structures and models 408
   5.3. Biochemical methods for distinguishing between models 411
6. Conclusions 415

Acknowledgments 416

References 417

Department of Haematology, Cambridge Institute for Medical Research, University of Cambridge, Cambridge, United Kingdom

[1] Corresponding author.

Methods in Enzymology, Volume 501

ISSN 0076-6879, DOI: 10.1016/B978-0-12-385950-1.00017-1

Abstract

Serpin polymerization is an event which generally occurs within living tissue as a consequence of a folding defect caused by point mutations. Major advances in cell biology and imaging have allowed detailed studies into subcellular localization, processing, and clearance of serpin polymers, but to understand the molecular basis of the misfolded state and polymeric linkage, it has been and continues to be necessary to generate polymers *in vitro*. The goal of this chapter is to outline the principal techniques that have been developed over the past 20 years to produce and characterize serpin polymerization *in vitro*. For the majority of this time, all data were interpreted in accordance with the so-called "loop-sheet" hypothesis, where polymers form through the intermolecular incorporation of the reactive center loop (RCL) of one serpin monomer into the β-sheet A of another. This hypothesis is supported by the ability of serpins to incorporate exogenous peptides into sheet A in an identical manner to the insertion of its own RCL upon cleavage by protease or conversion to the latent state. However, a recent crystal structure of an intact serpin dimer showed that much larger "domain swaps" are possible that would also lead to hyperstable linkage between serpin monomers. This chapter is therefore not limited to a description of experimental technique, but discusses the findings in light of the two current models of serpin polymerization. We would encourage readers to reevaluate the literature on serpin polymerization and to expand on the experiments outlined here in order to differentiate between possible domain-swapping mechanisms.

1. Introduction

Mutations in serpins (Fig. 17.1A; balls for mutation sites) commonly cause misfolding and accumulation in the endoplasmic reticulum of secretory cells. Deficiency of active serpin can lead to disease (loss-of-function), but surprisingly, the accumulation of aggregates within cells is also thought to lead to cell death by an unknown gain-of-function mechanism. While it is not unusual for mutated proteins to misfold, it is unusual for stable aggregates of the misfolded protein to accumulate in cells. This is believed to be due to the special structural feature of the serpins, namely, their ability to adopt a hyperstable state through β-sheet expansion (Fig. 17.1B–D). There is little known about the properties of serpin polymers from patient tissue. The inclusions appear to be stable, requiring sonication to liberate soluble fragments, and when run on nondenaturing (native) polyacrylamide gel electrophoresis (PAGE) to be composed of high molecular mass species (Fig. 17.2A), often referred to as "polymers." Soluble polymers of α1-antitrypsin (α1AT) have also been found in the circulation, and this material appears on electron micrographs as "beads-on-a-string," with each protomer separated from the other by a flexible linker (Fig. 17.2B). The two

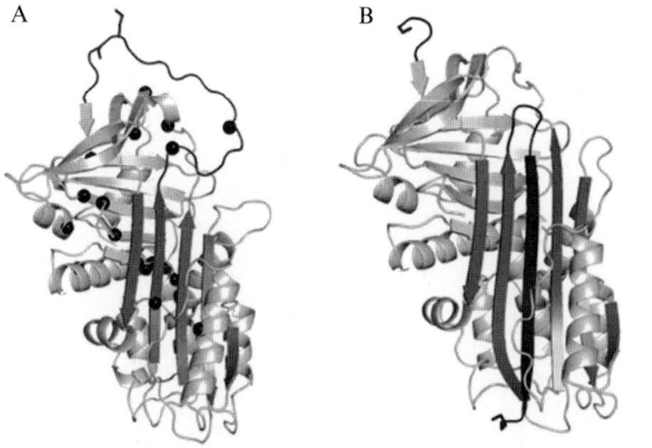

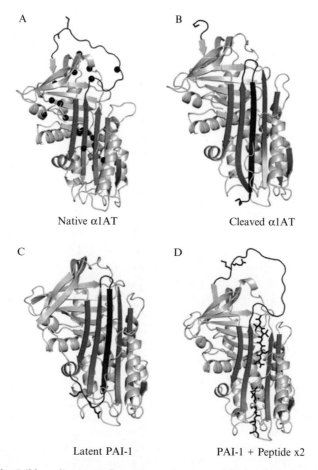

Figure 17.1 Ribbon diagrams of structures of native and β-sheet A-expanded forms of serpins. (A) Native α1AT (1QLP; Elliott *et al.*, 2000) is shown in the standard orientation with β-sheet A facing (dark gray) and the RCL on top (black). The scissile bond is illustrated by the side chains of the P1 and P1′ residues shown as sticks. The positions of mutations that have been reported to lead to polymerization of various serpins are shown as black balls. (B) The structure of α1AT cleaved at the scissile bond is shown (7API; Loebermann *et al.*, 1984), colored as before. The P15–P3 portion of the RCL has become strand 4 of β-sheet A. (C) PAI-1 and other serpins are capable of incorporating the RCL in the absence of cleavage to adopt the "latent" conformation (1LJ5; Mottonen *et al.*, 1992). (D) Most serpins are able to incorporate peptides into β-sheet A in a manner similar to RCL incorporation. Here, PAI-1 has incorporated two copies of the same peptide (black rods) into sheet A (1A7C; Xue *et al.*, 1998).

ends of a polymer are complementary and occasionally interact to form a "necklace" (Fig. 17.2B, inset).

The nature of the stable intermolecular polymer linkage was enigmatic until two key facts about serpins came to light: (1) serpins fold into a

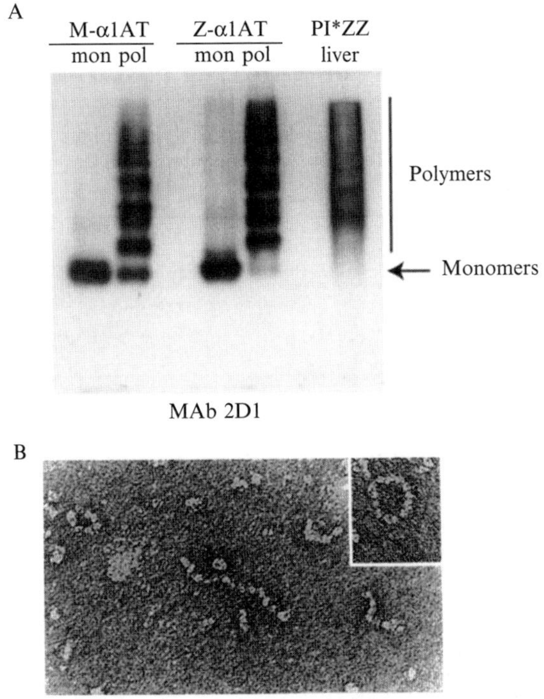

Figure 17.2 Laddering on native PAGE and beads-on-a-string morphology are hallmarks of serpin polymers. (A) Monomeric (mon) and polymeric (pol) forms of α1AT are visualized by Western blot of a native gel. Polymers were formed from native purified wild type (M) and Z-α1AT obtained from plasma by incubation of 0.2 mg ml^{-1} at 60 °C for 1 h. The polymers run as ladders, similar to what is obtained from the liver of a ZZ homozygous patient after sonication in detergent. The reader should note the difference between polymers made *in vitro* and those obtained from inclusions, and the treatment required to permit inclusion-derived "polymers" to enter the gel (Miranda *et al.*, 2010). Reprinted with permission. (B) Electron micrograph image of Siiyama α1AT polymers purified from patient plasma reveals clumps, linear polymers and circlets resembling beads-on-a-string (Lomas *et al.*, 1993b). This morphology and the apparent flexibility between protomers are thought to be typical of serpin polymers but are difficult to reproduce from inclusion-derived samples. Reproduced with permission.

metastable native state, with an exposed reactive center loop (RCL; Fig. 17.1A); and (2) serpins become hyperstable through the incorporation of the RCL as strand 4 of β-sheet A (Fig. 17.1B and C) or by incorporating peptides of similar sequence (Fig. 17.1D). The ability of serpins to incorporate exogenous peptides suggested that the basis of polymerization might be RCL incorporation from one molecule into the β-sheet A of another (in *trans*). This has become known as the "loop-sheet" hypothesis, and it has prevailed since the early 1990s. To study the molecular basis of serpin polymerization, it was necessary to develop techniques to "induce" serpins to polymerize. Several conditions have been found that result in the stable association of once native serpins into polymers, with laddering on native PAGE and beads-on-a-string morphology by EM. It has been widely assumed that the resulting polymers are essentially identical regardless of the method of induction, and that all *in vitro*-made polymers are the same as those deposited in the ER of secretory cells, regardless of the identity of the serpin or the position of the mutation. These assumptions have been called into question by recent work. In this chapter, we outline the methods used to polymerize serpins *in vitro*, describe the analysis of kinetic data, and discuss models of serpin polymers and how they might be distinguished experimentally.

2. Methods of Inducing Polymerization

2.1. Introduction

Mast *et al.* were the first to demonstrate that α1AT could polymerize *in vitro* at elevated temperatures and to speculate that this polymerization may be the cause for the liver inclusions or high molecular forms in plasma commonly observed among the patients for Z-deficiency variant (Mast *et al.*, 1992). In this study, plasma α1AT was incubated at 48 °C for 15 or 2 h with the assistance of 0.025% of a detergent, NP-40 (Fig. 17.3A). The heat-induced polymers showed discrete ladders in native PAGE and on transverse urea gradient gels (Fig. 17.3B) but ran as a single monomeric band in SDS-PAGE, suggesting that the polymer linkage was stable but noncovalent. These results demonstrated that serpin polymers could be formed *in vitro* through partial denaturation of the native state, and it was subsequently confirmed that other denaturants such as chaotropic agents, low pH, and organics could polymerize serpins. Work has focused on α1AT because of the prevalence of the Z mutation and the abundance of M-α1AT in plasma, although a handful of polymerization studies have been conducted on α1-antichymotrypsin (α1AC), antithrombin (AT), neuroserpin, plasminogen activator inhibitor 1 (PAI-1), and heparin cofactor II (HCII). A large selection of the published techniques used to produce serpin polymers *in vitro* is given as Table 17.1.

2.2. Heat

Among a variety of methods for *in vitro* polymerization, heat treatment is the most widely used. At physiological pH and salt concentrations, α1AT has been polymerized at a range of temperatures (37–80 °C) and protein concentrations (0.1–2.0 mg ml^{-1}) but showed some differences in kinetics

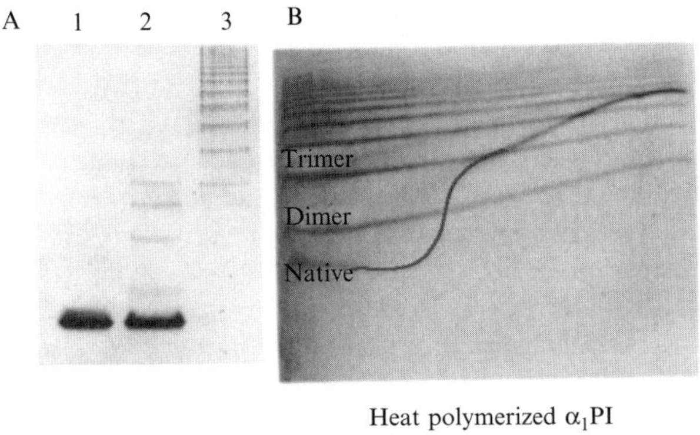

Figure 17.3 Typical native and urea gradient gels of α1AT polymers. (A) A native gel of virgin α1AT (lane 1) and α1AT treated at 48 °C for 2 h in the absence (lane 2) and presence of 0.025% NP-40 detergent shows the typical laddering of polymers and the effect of increasing stringency (Mast *et al.*, 1992). (B) Polymers formed in this manner are stable up to 8 *M* urea, as illustrated by a 0–8 *M* transverse urea gradient gel. The monomer band undergoes a normal unfolding transition (emphasized here by a shaded line), while the polymeric species do not (Mast *et al.*, 1992). Reproduced with permission.

depending on the conditions. For example, plasma M-α1AT (traditional nomenclature for wild-type native α1AT derived from plasma) polymerizes very slowly at temperatures just above physiological (e.g., 41 °C) but polymerizes rapidly when heated to just below its melting temperature (T_m) (50–55 °C). Several studies have been conducted where α1AT was incubated above its T_m (61–65 °C), as this dramatically decreases the incubation time for polymer formation. Indeed, a 150-h incubation of 2 mg ml^{-1} α1AT yielded no polymers at 37 °C, 18% polymers at 41 °C, 80% at 50 °C, while polymerization was complete within 1 h at 65 °C (Lomas *et al.*, 1993a). For plasma-derived M- and Z-α1AT, it is now common to incubate 0.2 mg ml^{-1} in PBS at 60 °C for 1 h to give polymers such as those shown in Fig. 17.2A (Miranda *et al.*, 2010). The addition of denaturants, detergents, or other compounds to heat treatment effectively accelerates polymerization, indicating that partial unfolding must precede intermolecular linkage (see Table 17.1). It is, however, worth mentioning that conditions of varying conditions and/or stringency may not produce the same results or the same types of polymers. An example of this is the so-called citrate polymer, made by incubating α1AT for 2 h at 70 °C in 0.7 *M* sodium citrate at pH 8 (Devlin *et al.*, 2002). Conditions that lead to partial unfolding, such as heat, establish an equilibrium between native, intermediate(s), and unfolded states, with aggregation, latency, and polymerization as alternative end states. Therefore, enhancing polymerization rates using harsh conditions will increase the possibility of off-pathway nonspecific aggregation rather than the formation of ordered polymer, and

Table 17.1 A selection of published methods for forming serpin polymers *in vitro*

Protein	Concentration (mg ml^{-1})	Buffer, pH	Temperature (°C)	Time	Reference
Heat method					
Plasma M-α1AT	0.2–1.0	50 m*M* Tris–HCl, 100 m*M* NaCl, pH 8	48	15 h	Mast *et al.* (1992)
	2	50 m*M* Tris–HCl, 50 m*M* KCl, 0.1 *M*% (v/v) β-mercaptoethanol, pH 7.5	37–65	~150 h	Lomas *et al.* (1993a)
	0.03–1.0	50 m*M* Tris–HCl, 50 m*M* KCl, 0.1 *M*% (v/v) β-mercaptoethanol, pH 7.5	65	~25 h	
	0.2	50 m*M* Tris–HCl, 50 m*M* KCl, 0.1 *M*% (v/v) β-mercaptoethanol, pH 7.5	30–100	3	
	0.2	50 m*M* Tris–HCl, 50 m*M* KCl, pH 7.4	60	3	Lomas *et al.* (1995)
	0.6	50 m*M* Tris–HCl, 50 m*M* KCl, pH 7.4	40	~12 days	
	0.1	50 m*M* Tris–HCl, 50 m*M* KCl, pH 7.4	45	~120 h	Dafforn *et al.* (1999)
	0.25	50 m*M* Tris–HCl, 50 m*M* KCl, pH 7.4	50	12 h	
	0.25	50 m*M* Tris–HCl, 50 m*M* KCl, pH 7.4	60	3 h	
	1	15 m*M* Tris–HCl, 150 m*M* NaCl, pH 7.4 or in the presence of 0.7 *M* citrate, pH 8.0	60	3 h	Janciauskiene *et al.* (2004)
	0.25	40 m*M* Tris–HCl, pH 8.0 or 20 m*M* ammonium acetate, pH 8.0	60		Ekeowa *et al.* (2010)
(Latent α1AT)	0.2	50 m*M* Tris–HCl, 50 m*M* KCl, 0.1 *M*% (v/v) β-mercaptoethanol, pH 7.4	30–100	2 h	Lomas *et al.* (1995)

(*Continued*)

Table 17.1 *(Continued)*

Protein	Concentration (mg ml^{-1})	Buffer, pH	Temperature (°C)	Time	Reference
Plasma Z-α1AT	2	50 m*M* Tris–HCl, 50 m*M* KCl, 0.1 *M*% (v/v) β-mercaptoethanol, pH 7.5	37, 41	~150 h	Lomas *et al.* (1992)
	2	50 m*M* Tris–HCl, 50 m*M* KCl, 0.1 *M*% (v/v) β-mercaptoethanol, pH 7.5	37	~400 h	Lomas *et al.* (1993a)
	0.6	50 m*M* Tris–HCl, 50 m*M* KCl, pH 7.4	40	~12 days	Lomas *et al.* (1995)
rt α1AT wild type and variants (from *E. coli*)	0.1	10 m*M* NaPi, 50 m*M* NaCl, 1 m*M* EDTA, 0.1% (v/v) β-mercaptoethanol, pH 7.5	55	~3 h	Kwon *et al.* (1994)
	0.4	20 m*M* Tris–HCl, pH 7.4	56	~2 h	Chang *et al.* (1997)
	1.0	20 m*M* Tris–HCl, 50 m*M* NaCl, 1 m*M* EDTA, pH 7.4	37–65	1 h	Zhou *et al.* (2003)
	0.8	50 m*M* Tris–HCl, 50 m*M* KCl, pH 8.3	45	>350 min	Purkayastha *et al.* (2005)
			50	350 min	
			55	275 min	
			60	100 min	
	0.1	10 m*M* NaPi, 50 m*M* NaCl, pH 7.8	45	4 days	Tsutsui *et al.* (2008)
	0.1	10 m*M* NaPi, 50 m*M* NaCl deuterium buffer, pD 7.8	45	~5000 s	
	0.025	50 m*M* NaPi, pH 7.5	53	3 h	Yamasaki *et al.* (2008)
	0.1	PBS, pH 7.4	50	o/n	Yamasaki *et al.* (2010)

(from Yeast)	1.0	50 m*M* Tris–HCl, 100 m*M* NaCl, 5 m*M* β-mercaptoethanol, 1 m*M* EDTA, pH 7.4	40	~114 h	Levina *et al.* (2009)
Plasma α1AC	0.5	10 m*M* KPi, 50 m*M* KCl, 5 m*M* EDTA, pH 6.8	50–100	2 h	Chang *et al.* (1998)
	0.1–0.4	50 m*M* Tris–HCl, 50 m*M* KCl, pH 7.4	42–48	~24 h	Crowther *et al.* (2003)
Plasma AT	0.5	10 m*M* NaPi, pH 6.5	53	~2 h	Yamasaki *et al.* (2008)
rt NS wild type and variants	0.7	50 m*M* Tris–HCl, 50 m*M* KCl, pH 7.4	37–45	17 h	Belorgey *et al.* (2002)
	0.25	50 m*M* Tris–HCl, 50 m*M* KCl, pH 7.4	37	~8 h	
	0.5	50 m*M* Tris–HCl, 50 m*M* KCl, pH 7.4	0–22	24 h	
	1.0	PBS, pH 7.4	45	~24 h	Onda *et al.* (2005)
	0.25	PBS, pH 7.4	37	~24 h	
	1.0	PBS ± 0.5 *M* sodium citrate, pH 7.4	37–55	24 h	
	5.0	20 m*M* NaPi, pH 7.4	42	1–3 h	Takehara *et al.* (2010)
(Latent NS)	1.0–3.0	PBS, pH 7.4	45, 55	~24 h	Onda *et al.* (2005)
	1.0	PBS ± 0.5 *M* sodium citrate, pH 7.4	37–55	24 h	
Denturant method					
Plasma M-α1AT	1.0	0–2 *M* GndHCl, 50 m*M* Tris–HCl, 50 m*M* KCl, 0.1 *M* β-mercaptoethanol, pH 7.5	4	12 h	Lomas *et al.* (1993a)
	0.25	1–3 *M* GndHCl, 40 m*M* Tris–HCl, pH 8.0	25	24 h	Ekeowa *et al.* (2010)
	0.25	1–4 *M* Urea, 40 m*M* Tris–HCl, pH 8.0	25	24 h	

(Continued)

Table 17.1 (*Continued*)

Protein	Concentration (mg ml^{-1})	Buffer, pH	Temperature (°C)	Time	Reference
Plasma Z-α1AT	2.0	1 *M* GndHCl, 50 m*M* Tris–HCl, 50 m*M* KCl, 0.1 *M* β-mercaptoethanol, pH 7.5	37	~24 h	Lomas *et al.* (1992)
	1.0	0–2 *M* GndHCl, 50 m*M* Tris–HCl, 50 m*M* KCl, 0.1 *M* β-mercaptoethanol, pH 7.5	4	12 h	Lomas *et al.* (1993a)
rt α1AT wild type and variants	0.25	0.75 *M* GndHCl, 50 m*M* NaPi, pH 7.0	37	8 h	Yamasaki *et al.* (2008)
	0.1	0.5–1.0 *M* GndHCl, PBS, pH 7.4	37	3 h	Yamasaki *et al.* (2010)
pH method					
Plasma M-α1AT	0.5	20 m*M* Na acetate, pH 4.0	25	~2 h	Devlin *et al.* (2002)
	0.25	0.1 *M* Na acetate, pH 4.5	25	24 h	Ekeowa *et al.* (2010)
Plasma AT	0.5	0.1 *M* Na acetate (pH 4–6), NaPi (pH 6–8), Tris–HCl (pH 8–9), glycine (pH 9–10)	50	16 h	Zhou *et al.* (2003)
	0.5	10 m*M* Na acetate, pH 5.7	37	16 h	Yamasaki *et al.* (2008)
rt NS wild type and variants	0.4	pH 5.0	45	5 h	Belorgey *et al.* (2010)
	0.4	pH 6.0–8.0	45	~3 h	
rt PAI-1 wild type and variants	0.1	0.1 *M* Na acetate, pH 4.0	37	12 h	Zhou *et al.* (2001)

Chemical method					
Plasma M-α1AT	0.2–1.0	0.025% NP-40, 50 m*M* Tris–HCl, 100 m*M* NaCl, pH 8	48	2 h	Mast *et al.* (1992)
(and plasma Z-α1AT)	1.25	5 m*M* lithocholic acid, 15 m*M* Tris–HCl, 150 m*M* NaCl, pH 7.4	RT, 48	~48 h	Gerbod *et al.* (1998)
Plasma AT	1.0	0.4 μ*M* peptidylarginine deiminase, 100 m*M* Tris–HCl, 5 m*M* $CaCl_2$, pH 7.4	37	~24 h	Ordonez *et al.* (2009)
rt PAI-1 wild type and variants	~0.2, 0.3	100 μ*M* 1-dodecyl sulfuric acid, 10 m*M* Hepes, 140 m*M* NaCl, pH 7.4	37	30 min	Pedersen *et al.* (2003)
	0.3	10 μ*M* bis-ANS or 500 μ*M* ANS in 10 m*M* Hepes, 140 m*M* NaCl, pH 7.4	37	30 min	
HCII	0.3	100 μ*M* 1-dodecyl sulfuric acid, 10 m*M* Hepes, 140 m*M* NaCl, pH 7.4	37	30 min	Pedersen *et al.* (2003)
	0.3	10 μ*M* bis-ANS or 500 μ*M* ANS in 10 m*M* Hepes, 140 m*M* NaCl, pH 7.4	37	30 min	
Protein	**Concentration (mg ml^{-1})**	**Protease, buffer, pH**	**Temperature (°C)**	**Time**	**Reference**
Proteolytic method					
Plasma M-α1AT	0.2–1.0	CA2 (cleaves at P10–P9), 50 m*M* Tris–HCl, 150 m*M* NaCl, pH 8.0	37	30 min	Mast *et al.* (1992)
	0.2–1.0	PP4 (cleaves at P11–P10), 50 m*M* Tris–HCl, 150 m*M* NaCl, pH 8.0	37	30 min	
	1.0	0.01 mg ml^{-1} PP4, 50 m*M* Tris–HCl, 150 m*M* NaCl, pH 7.4	37	24 h	Zhou *et al.* (2004)
	0.25	Glycyl endopeptidase (cleaves at P10–P9), 20 m*M* ammonium acetate buffer	37	4 h	Ekeowa *et al.* (2010)

(*Continued*)

Table 17.1 (*Continued*)

Protein	Concentration (mg ml^{-1})	Protease, buffer, pH	Temperature (°C)	Time	Reference
rt α1AT variant	0.5	0.01 mg ml^{-1} bovine trypsin, PBS, pH 7.4	25	1 h	Yamasaki *et al.* (2010)
Peptide method					
Plasma M-α1AT	0.25	50 μ*M* P14–P9 from AT, 40 m*M* Tris–HCl, pH 8.0	37	o/n	Ekeowa *et al.* (2010)
rt α1AT wild type and variants	0.5	~25 μ*M* P14–P9 from AT, PBS, pH 7.4	37	16 h	Yamasaki *et al.* (2010)
Plasma AT	1.0	2 m*M* P14–P9 from AT, 50 m*M* Tris–HCl, 50 m*M* NaCl, pH 8.0	37	48 h	Chang *et al.* (1997)
	1.0	~2 m*M* P14–P8 from AT, 50 m*M* Tris–HCl, 50 m*M* NaCl, 1 m*M* EDTA, pH 7.4	37	24 h	Zhou *et al.* (2004)

additives such as citrate may favor one intermediate over others. This makes it difficult to compare polymerization experiments where different conditions and read outs are used.

Many studies focus on M-α1AT and the pathogenic Z-α1AT variant because of the relevance to disease. Polymerization of Z-α1AT can be induced at slightly lower temperatures than M-α1AT (41 °C), consistent with the small reduction in the stability of Z-αAT. Neuroserpin pathogenic variant S49P readily polymerizes at 37 °C due to its extremely low T_m (Onda *et al.*, 2005). Even wild-type neuroserpin is capable of forming ladders at 37 °C (Onda *et al.*, 2005), indicating an unusual instability of this particular serpin. Recombinantly produced α1ATs in *Escherichia coli* or yeast (within the cytosol) have also been used extensively to study polymerization, and wild type and mutants have reduced stability and tend to polymerize more readily, presumably due to the lack of glycosylation. The general rule seems to be that mutations that reduce the thermal stability of a serpin will allow polymerization at lower temperatures.

2.3. Denaturants

Lomas *et al.* were the first to report *in vitro* polymerization by incubation with 1 *M* GndHCl at 37 °C and showed a decrease in monomer peak and a corresponding increase in high molecular weight species analyzed by analytical gel filtration HPLC (Lomas *et al.*, 1992). Later, the unfolding pathway of M-α1AT in GndHCl was investigated, revealing that it undergoes a three-state transition between native (state one) and denatured (state three), via an intermediate at ~1 *M* GndHCl (James *et al.*, 1999; Fig. 17.4A). Recombinant α1AT similarly polymerizes upon incubation in 0.75 *M* GndHCl at 37 °C, showing a typical laddering in nondenaturing PAGE (Yamasaki *et al.*, 2008). Notably, a polymer specific antibody 2C1 generated with heat-induced polymers reacted with a M-α1AT polymer made in 1 *M* GndHCl at 37 °C (David Lomas, Personal communication). On the other hand, incubation of α1AT in 3 *M* GndHCl or 4 *M* urea at room temperature neither formed a typical laddering on native PAGE nor showed reactivity with the antibody 2C1 (Ekeowa *et al.*, 2010). Therefore, to use heat treatment with a denaturant seems fundamental to obtain long polymers with the kind of linkage that may be related to physiological polymers *in vivo*. Polymerization in other serpins has been induced by detergent, with 100 μ*M* of 1-dodecyl sulfuric acid (SDS) at 37 °C causing PAI-1 and of HCII to polymerize in a concentration-dependent manner. Other amphipathic organic compounds have also been shown to induce polymer formation. Organics such as 10 μ*M* of bis-ANS (4,4′-dianilino-1,1′-binaphthyl-5,5′-disulfonate) or 500 μ*M* of ANS are likely to work in a manner similar to detergents (Pedersen *et al.*, 2003), although it has been postulated that there are distinct binding sites for these compounds that induce polymer formation.

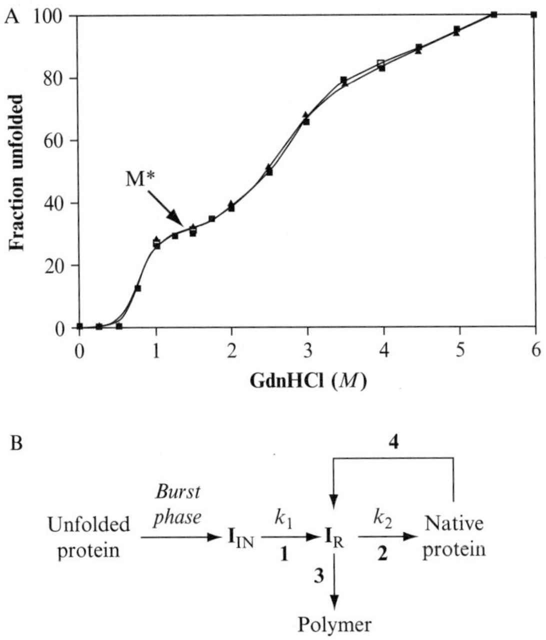

Figure 17.4 Unfolding and refolding of serpins involve intermediates. (A) Unfolding of α1AT upon addition of GndHCl was monitored by change in far UV CD signal, and fit a three-state unfolding profile with an intermediate (M*) populated between 1 and 2 *M* GndHCl (James *et al.*, 1999). (B) A schematic summarizing the results of studies on the refolding of neuroserpin (Takehara *et al.*, 2010), where an initial intermediate is formed in an extremely rapid burst phase (I_{IN}), followed by a slower conversion to an late intermediate (I_R) capable of either conversion to native (step 2) or polymer formation (step 3). The effect of polymerigenic mutations such as S49P or H338R is the slowing of step 2, the conversion of I_R to native. In order to obtain polymers from native neuroserpin it is necessary to repopulate I_R (step 4). This is likely to be the normal means of producing serpin polymers *in vitro* by heat or other denaturing conditions. Figures reproduced with permission.

2.4. Low pH

Similar to the observed amyloid fibril formation of transthyretin in acidic conditions, Devlin *et al.* found that M-α1AT incubated at pH 4.0 and 25 °C converts reversibly to a partially unfolded species and then irreversibly polymerizes via a proto-dimer to high molecular weight species (Devlin *et al.*, 2002). It forms a discrete ladder in nondenaturing PAGE; however, they concluded that the low-pH α1AT polymer is different from the heat-induced polymer, instead resembling the citrate polymer by thioflavin-T binding and CD spectroscopy. It was not reactive to the ATZ11 antibody (Janciauskiene *et al.*, 2004) or the 2C1 antibody (Ekeowa *et al.*, 2010). The key aspect of low-pH-induced polymerization must be the effect of pH on the stability of the individual serpin. By the CD spectral analysis, α1AT showed rapid loss of secondary structure (about 20%) at pH 4.75 (Devlin *et al.*, 2002). Similarly, AT is destabilized below pH 6.0, and polymerization can be induced at physiological temperature (Yamasaki *et al.*, 2008; Zhou *et al.*, 2003). However, neuroserpin showed a strong resistance to polymerization at low pH (Belorgey *et al.*, 2010), and controversially, both native and latent PAI-1 formed discrete ladders in acid nondenaturing PAGE at pH 4.0, but the inter molecular interactions were so weak that the polymers dissociated to the monomeric form when pH was returned to neutral (Zhou *et al.*, 2001). Obviously, the effect of pH on stability depends on the serpin, thus the physiological relevance of low-pH-induced polymers should be evaluated for each individual serpin.

2.5. Chemical modification

Modifying the surface residues of proteins can affect the stability of the native fold, and it is likely that this is what is occurring when plasma AT is citrullinated, by incubation with peptidylarginine deiminase at a 50 times molar excess at 37 °C for 6 h (Ordonez *et al.*, 2009). Citrullination in 10 of the 22 surface Arg residues seemed to be sufficient to destabilize the native fold so that polymerization could occur at 37 °C. Interestingly, polymers formed by this method were similar to those of AT formed at low pH in terms of proteolytic susceptibility in the linker region (Yamasaki *et al.*, 2008).

2.6. Proteolytic cleavage and short peptides

Polymers formed *in vitro* or *in vivo* are composed of intact serpin; however, the first demonstration of intermolecular RCL incorporation between serpin monomers was achieved by proteolytic nicking in the center of the RCL (Mast *et al.*, 1992). Although this method of polymerization is completely artifactual, it has been used in several studies as a model for intact serpin polymerization. Cleavage of α1AT by papaya proteinase IV (PP4) at P10–P9 position, *Crotalus adamanteus* proteinase II at P9–P8 bond, papain at P7–P6 bond, or *Staphylococcus aureus* V8 proteinase at P5–P4 bond, all result in rapid polymer formation, even at vanishingly low serpin concentrations. Once a cleavage in the middle of the RCL has occurred, α1AT instantly polymerizes to form discrete ladders in a native PAGE. The proposed mechanism was that cleavage at P10–P9, for instance, would allow the N-terminal portion of the RCL (P15–P10) to insert into the top of the β-sheet A and result in the opening of the bottom part. The remaining C-terminal portion of the RCL (P9–P3) would be exposed and able to complete sheet A only in another molecule. This mechanism has been confirmed by crystallographic structures of α1AT polymerized by

RCL nicking at P7–P6 bond (Huntington *et al.*, 1999) and at P6–P5 bond (Dunstone *et al.*, 2000). If a serpin does not have an ideal cleavage site in the middle of the RCL, one can be engineered (Yamasaki *et al.*, 2010). Short peptides corresponding to the N-terminus of the RCL can similarly induce polymerization. Although no crystal structure has been solved of these intact polymers, it is reasonable to conclude that the linkage will resemble that of the cleaved polymer. Chang *et al.* showed that a P14–P9 peptide initiated AT polymerization by incubation with a 100-fold molar excess of the peptide at 37 °C (Chang *et al.*, 1997). The peptide does not necessarily have to be derived from the parent serpin; indeed, the P14–P9 peptide derived from AT induces α1AT polymerization more effectively than its own peptide. These methods have been used extensively as a model of loop-sheet polymerization, since the mechanism of polymerization is clear, although both mechanisms are artifactual and do not involve a folding intermediate.

2.7. Refolding

Polymerization upon refolding serpins may be the best method for mimicking how serpins form polymers within the ER of secretory cells. Kim *et al.* denatured α1AT molecules completely in 7 *M* urea for 3 min at 25 °C, refolded in different concentrations of urea after manual mixing with a dead time of 5 s, and characterized the products by the fluorescence (Kim and Yu, 1996). Similarly, Jung *et al.* denatured *E. coli* inclusion bodies in 8 *M* urea, rapidly refolded in 0.8 *M* urea at pH 6.5, purified material in less than 1 h at 4 °C, and characterized their stability by transverse urea PAGE (Jung *et al.*, 2004). Both studies emphasized the persistence of a folding intermediate in mutated α1AT. Recent work on neuroserpin by Takehara *et al.* is in good agreement with what was found for α1AT (Takehara *et al.*, 2010). They found two structurally different folding intermediates, I_{IN} and I_R, during refolding from 5 *M* GndHCl, and were able to trap them by low temperature at 0.5 °C at low concentration of 0.01 mg ml^{-1}. They also showed that I_{IN} is the first intermediate after a burst phase from the unfolded state, and that I_R is a second, long-lived intermediate, just prior to the native conformation. It is of particular interest that the refolding from I_R to a native conformation was 10 times slower in pathogenic S49P mutant compared to wild-type neuroserpin, while the mutation did not affect the speed of the initial burst from I_{IN} to I_R. Incubation of S49P I_R at 20 °C induces polymerization, and the polymers have similar properties to heat-induced polymers from a native neuroserpin. Likewise, native neuroserpin is also able to polymerize by refolding from the unfolded state prepared by 9 *M* urea at pH 2.2 (Onda *et al.*, 2005). Together with α1AT studies, these overwhelmingly indicate that serpin mutations slow down a specific step on the folding pathway and that the persistence of a polymerigenic intermediate is the cause of off-pathway polymerization *in vitro* and *in vivo* (according to the scheme in Fig. 17.4B).

2.8. *In vitro* translation

In a landmark experiment, Yu *et al.* (1995) used an *in vitro* translation system of rabbit reticulocyte lysate to produce Z-α1AT and showed good evidence that the Z mutation retards α1AT folding. A transverse urea gel with a gradient of 0–8 *M* showed that the newly synthesized Z-α1AT polypeptide did not obtain a native conformation in contrast to wild-type α1AT and, instead, was retained as a folding intermediate that readily associated to form polymers when concentration was increased. Further, they found that a prolonged incubation of the polypeptide at 30 °C allowed some of the Z-α1AT to fold into the native conformation, indicating that the Z mutation specifically slowed the folding step from the polymerogenic intermediate to the native state. Consistent with this conclusion is the recent observation that the stability of native Z-α1AT is essentially indistinguishable from that of wild-type α1AT (Levina *et al.*, 2009).

3. Kinetics of Polymerization

3.1. General features and kinetic scheme

Rates of serpin polymerization have been studied using a wide range of techniques over the past 20 years. Work has also focused on α1AT, due to the prevalence of the Z mutation and its relative abundance, although a handful of kinetics studies have been conducted on α1AC and on neuroserpin. In all cases, serpins are isolated in their native state and then partially denatured using high temperature, chaotropic agents, low pH, or a combination of conditions. Any of the conditions detailed in previous section could be applied, but care must be taken to ensure that the rates are neither too fast nor too slow to be detected in the time course of the experiment. Since the seminal paper by Tim Dafforn published in 1999 (Dafforn *et al.*, 1999), the generally accepted mechanism of *in vitro* polymerization is

$$\mathrm{M} \rightarrow \mathrm{M}^{*} \rightarrow \mathrm{P}$$

where M is native, M* is a partially unfolded state, and P is a stable polymer. The first step is reversible, but only the forward rates are described by the kinetic experiments, and the second step is irreversible. The first step was found to be independent of the starting serpin concentration, and the second step was dependent on concentration. This makes sense, but not all reports agree; some find both steps concentration dependent (Devlin *et al.*, 2002) and others find neither to be concentration dependent (James and Bottomley, 1998). This is likely due to differences in conditions used to induce polymerization, the serpin involved and the techniques employed to follow the process. Although useful, the kinetic model is a gross oversimplification. It does not account for what must be an ensemble of partially folded/unfolded species, nor their different possible fates, such as conversion to the stable latent form or nonspecific aggregation. It also considers polymerization to a dimer as an end point, but it is known that polymers can continue to polymerize, self-terminate, or precipitate. In spite of these shortcomings, there is remarkable agreement in reported rates for the two steps of polymerization, with the first step in the order of 10^{-4} s^{-1} and the second around 10^{-5} s^{-1}. These rates are exceedingly slow and are only marginally increased when polymerigenic mutants such as Z, S, and I α1AT are used (Mahadeva *et al.*, 1999). The first report on the rate of Z-α1AT polymerization showed that at 37 or 41 °C using 2 mg ml^{-1} starting concentration, the half-time of conversion to polymer was approximately 75 and 20 h, respectively (Lomas *et al.*, 1992). This was assessed using loss of monomer by size-exclusion chromatography. Assuming second-order kinetics ($k = 1/(t_{1/2} \times [A]_o)$), estimated rate constants of 0.1 and 0.36 M^{-1} s^{-1} can be obtained for the conversion of native Z antitrypsin to polymers at 37 and 41 °C, respectively. Such low rates suggest that polymerization in cells likely proceeds via a folding intermediate, and that, as suggested originally by Yu *et al.* (1995) and later by Takehara *et al.* (2010), mutations cause polymerization by slowing the conversion from a polymerigenic intermediate to the native state, and not by destabilizing the native state as commonly assumed. However, studying rates of serpin polymerization from the native state has provided some useful and interesting information, and details of common techniques are given in the following sections.

3.2. Loss of monomer

3.2.1. Activity

Since serpins are serine protease inhibitors, it is possible to follow polymerization by monitoring loss of inhibitory activity. However, there are other ways that serpins can lose activity, such as latent conversion, nonspecific aggregation, denaturation, and sticking, and therefore, activity can never be a stand-alone method and will require other methods to ensure that all loss of activity can be accounted for by the appearance of polymers (normally by nondenaturing PAGE). However, so long as one is satisfied that polymerization is the *major* cause of loss of inhibitory activity for a particular serpin under the conditions used, activity can provide a reliable and easy method for monitoring the second, slow rate of polymerization. For Z antitrypsin, there was a very good correlation between loss of activity against bovine α-chymotrypsin and the loss of monomer by size-exclusion chromatography, when incubated at 2 mg ml^{-1}, 37 °C for up to 400 h (Lomas *et al.*, 1993a; Fig. 17.5A). And for wild-type plasma α1AT at pH 4, the lag phase and observed rate of loss of inhibitory activity corresponded to fluorescence and turbidity measurements (Devlin *et al.*, 2002; Fig. 17.5B). To determine the fraction residual inhibitory activity, an aliquot is removed from the incubation solution and placed for a period of time in conditions that will allow unpolymerized material to fold back to the native state.

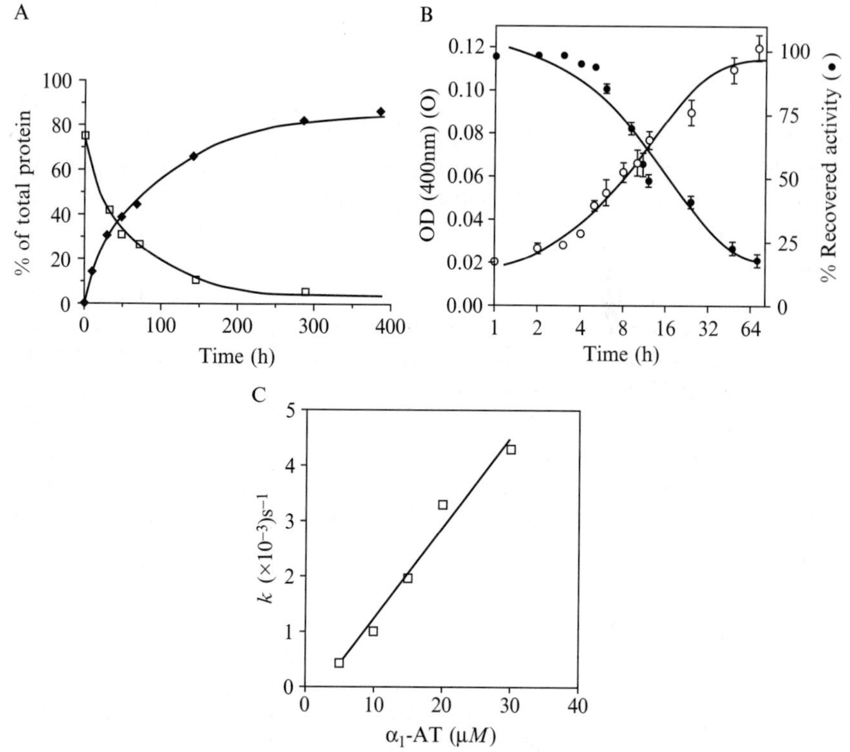

Figure 17.5 Kinetics of polymerization assessed by various means. (A) A good correlation was seen when rates of polymerization (Z-α1AT at 37 °C) were assessed by loss of inhibitory activity (relative to chymotrypsin, open squares) or appearance of high molecular mass species by size-exclusion HPLC chromatography (closed diamonds) (Lomas *et al.*, 1993a). (B) A similar correlation was observed when monitoring activity (against chymotrypsin, closed circles) and turbidity (light scattering, open circles) after incubation of M-α1AT at pH 4 (Devlin *et al.*, 2002). (C) A linear concentration dependence on rate was observed when monitoring thioflavin-T fluorescence after incubating M-α1AT at pH 4 (Devlin *et al.*, 2002). Figures reproduced with permission.

Then a fraction (usually corresponding to the volume of serpin solution initially sufficient for full inhibition) is added to a known amount of protease, incubation until the inhibition reaction has gone to completion, then a large excess of chromogenic substrate (typically 100 μ*M*) is added and residual protease activity is measured as the initial slope relative to that of uninhibited protease (absorbance at 405 nm vs. time). Fractional inhibitory activity is plotted against time and fit to a single exponential decay (Pearce *et al.*, 2007). If data are collected at several initial serpin concentrations, the second-order rate constant can be calculated as the slope of the observed rate versus initial serpin concentration (Devlin *et al.*, 2002), that is, provided a concentration dependency is observed (Fig. 17.5C).

3.2.2. Native PAGE

Native PAGE is an effective technique for visualizing serpin polymerization (Mast *et al.*, 1992), and when stained with Coomassie blue, native gels can be used to quantify the rate of polymerization (James and Bottomley, 1998; Mahadeva *et al.*, 1999). Most native serpins are negatively charged and migrate as a single band on native PAGE, whereas serpin polymers appear as slower migrating species and form a "ladder" of dimers, trimers, tetramers, etc. (James and Bottomley, 1998). Indeed, it is mostly due to this property, ordered laddering on native PAGE, that serpin aggregates have become known as polymers (Schulze *et al.*, 1990). In order to determine rates of polymerization, samples of an incubation mixture are withdrawn at timed intervals and placed under conditions where polymerization does not readily continue. This "quenching" may be achieved by simply allowing the sample to return to room temperature but could require a quenching step like altering pH or placing on ice. However, it is also possible that polymers will continue to grow in spite of "quenching" due to the persistence of donor and acceptor ends (Zhou and Carrell, 2008). An alternative method is to stagger the incubation of several samples so that all reactions (time points) are stopped at the same time; however, this method generally requires more material and is not commonly used. It is good practice, although not always followed, to allow the samples some time under nondenaturing conditions before adding sample buffer and loading the lanes of the gel (Devlin *et al.*, 2002). The reason for this is that the sample will concentrate when in the stacking gel and may actually promote polymerization of any remaining material in the M* conformation. This effect may also lead to polymers that appear larger on a gel than they would in solution, but is not relevant to quantification of polymerization, since this is based solely on the loss of the monomeric band. Coomassie-stained gels can be scanned and densitometry conducted to plot the loss of monomeric serpin with time. The data can then be fit to a single exponential, although may be preceded by a lag phase.

3.2.3. Size-exclusion chromatography

Size-exclusion chromatography was the first method used to obtain data on the rate of serpin polymerization (Lomas *et al.*, 1992), although rate constants were not reported or calculated. The method is simple and can give information on the loss of monomer and the proportion of the native peak to the polymer, allowing an easy visualization of the time to half conversion ($t_{1/2}$). However, this method requires a large quantity of protein and a lot of time. Original studies were conducted using plasma α1AT at 2 mg ml^{-1} in 1 *M* GndHCl at 37 °C, and under these conditions, the $t_{1/2}$ for polymerization was approximately 1.5 h. However, polymerization of Z antitrypsin at 2 mg ml^{-1} at 37 and 41 °C took 75 and 20 h, respectively. For each time point, 40 μg was loaded onto the column, and so 10 samples used nearly half a milligram of protein. In contrast, nondenaturing PAGE or activity use only around 2 μg of protein per time point. In a subsequent paper, data were obtained form 10 μg samples, coming close to matching other methods (Lomas *et al.*, 1993a). The peaks can be quantified by either height or area, plotted versus time, and fit by a single exponential. Care should be taken when using this method to ensure that monomer is lost predominantly to polymer formation, and not the other factors mentioned previously, especially since the fraction maximum polymer formation seems to depend on the starting serpin concentration (Lomas *et al.*, 1993a). This method requires solubility at high concentrations and is best suited for glycosylated plasma-derived material. It is interesting to note that, although there is a difference in apparent rate of M and Z polymerization when induced by temperature, the rates are equivalent when induced by 1 *M* GndHCl at 37 °C (Lomas *et al.*, 1993a).

3.3. Spectroscopic methods

3.3.1. Intrinsic fluorescence

Most serpins contain tryptophan residues and are therefore amenable in principle to the use of intrinsic fluorescence signal to follow conformational change and/or polymerization. Indeed, the kinetic model (above) is predominantly based on the two phases apparent when monitoring the Trp fluorescence of plasma α1AT (0.1 mg ml^{-1}) incubated at 45 °C (exiting at 295 nm and monitoring emission at 340 nm) (Mahadeva *et al.*, 1999). Although fluorescence spectra were not shown, a 2.5-nm blue-shift and a 30% fluorescence enhancement were reported. Approximately one-third of the signal is observed in the first phase, and the rest occurs at a slower time scale (Fig. 17.6A). Rates were derived from an equation with two exponential terms

$$F = A_1\left(1 - e^{-k_1 t}\right) + A_2\left(1 - e^{-k_2 t}\right),$$

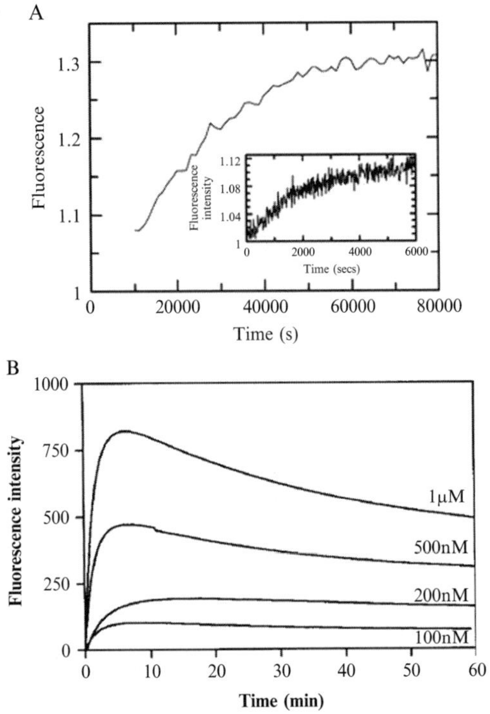

Figure 17.6 Partial unfolding and polymerization kinetics monitored by fluorescence. (A) Following change in intrinsic fluorescence with time (exciting at 295 nm for Trp residues) during an incubation of 0.1 mg ml^{-1} α1AT at 45 °C results in two distinct transitions. The rate of the initial (fast) transition (inset) reflects the unimolecular unfolding event to populate the M* state, and the slower rate coincides with appearance of polymeric forms on native PAGE (Dafforn *et al.*, 1999). (B) Similarly, incubation of α1AT at 65 °C produces two separate transitions when following bis-ANS fluorescence. The fast phase is a fluorescence enhancement, reflecting an increased exposure of hydrophobic areas due to partial unfolding, and the second is a fluorescence quench back to near starting fluorescence of native α1AT (James and Bottomley, 1998). Figures reproduced with permission.

where F is fluorescence intensity, A_1 and A_2 are amplitudes of each phase, t is time, and k_1 and k_2 are rate constants for each phase. The analysis yielded values of $k_1 = 6 \times 10^{-4}$ and $k_2 = 3.7 \times 10^{-5}\ \mathrm{s}^{-1}$, for the first and second phases, respectively. The fluorescence signal appears to be due to the Trp194 in α1AT, which underlies the gap at the top of strands 3 and 5A (Tew and Bottomley, 2001). This Trp residue is highly conserved, and so most serpins are likely to produce an appreciable signal for polymerization. However, α1AC does not undergo any intrinsic fluorescence change, in spite of the conserved Trp. The first phase was independent of starting α1AT concentration, but the second phase showed a near-linear dependence that would be expected of a second-order reaction. Experiments are conducted at 0.1 mg ml^{-1} serpin in 20 m*M* phosphate buffer, 100 m*M* NaCl and 0.1% PEG8000 (to limit sticking to the cuvette), pH 7.4. Phosphate is generally used as a buffer because its pK_as are not particularly sensitive to change in temperature.

3.3.2. Extrinsic fluorescence

Two studies have specifically labeled a Cys residue on a serpin with a fluorescent probe to see if it could report a change in environment with polymerization. Tetramethylrhodamine-5-iodoacetamide (5-TMRIA) was reacted with Cys232 of α1AT (Mahadeva *et al.*, 1999), and *N,N′*-dimethyl-*N*-(iodoacetyl)-*N′*(7-nitrobenz-2-oxa-1,3-diazol-4-yl)ethylenediamine (IANBD) was reacted with Cys238 on α1AC (Crowther *et al.*, 2003). These residues are located on similar positions between strands 1 and 2B, yet they appear to report different events when bound to a fluorophore. On α1AT, TMRIA underwent a fluorescence enhancement and a 2-nm blue-shift but only reported the slow rate, suggesting a more buried environment in the polymer. On the other hand, IANBD on α1AC underwent a sixfold enhancement of fluorescence and reported the fast and slow phases. Intriguingly, neither phase was dependent on starting concentration when following IANBD fluorescence, while both phases were concentration dependent when monitoring light scattering. Another, method of using extrinsic fluorescence to monitor serpin conformational change and polymerization is to add a probe that binds weakly to hydrophobic surfaces. Two successfully used probes are ANSA (8-anilino-1-naphthalenesulfonic acid) and bis-ANS. During serpin polymerization, these probes normally report a rapid fluorescence enhancement followed by a slow reduction in fluorescence. For ANSA, the second phase returns the fluorescence to just below original levels (Mahadeva *et al.*, 1999), while for bis-ANS, the fluorescence levels remain significantly higher for the polymer than for native serpin (James and Bottomley, 1998; Fig. 17.6B). At 45 °C, ANSA fluorescence yielded rates of 6.9×10^{-4} and 4.1×10^{-5} s^{-1}, very similar to what was obtained by intrinsic fluorescence. At 50 °C, bis-ANS reports rates of 3.8×10^{-4} and 5.5×10^{-5} s^{-1}, under similar conditions.

3.3.3. Circular dichroism

CD provides information about protein conformation and can also report quaternary structural changes, and it has been widely used to measure rates of serpin polymerization. Although changes in near UV CD signal have been reported for serpin unfolding and polymerization (Lomas *et al.*, 1993a), all kinetics studies follow signal in the far UV region, normally 216–220 nm. The far UV CD signal change upon polymerization of α1AT is extremely small (5%) and only reports the fast, unimolecular phase (Mahadeva *et al.*, 1999). A similar small signal for α1AC (6%) polymerization has been shown; however, it appears to report only the slow phase (Crowther *et al.*, 2003).

The α1AC used in this study was expressed in *E. coli*, and the loss of CD signal might reflect the loss of soluble serpin due to precipitation. In contrast to the loss of CD signal observed upon incubation of α1AT and α1AC at elevated temperatures, the signal for neuroserpin increases in intensity upon polymerization by 25% in a temperature-dependent manner, but independent of neuroserpin concentration (Belorgey *et al.*, 2002). This implies that CD is reporting only the conformational change that initiates polymerization of neuroserpin, and not polymerization itself.

3.3.4. Light scattering

With light scattering, there is no ambiguity about what is being measured-the formation of a fine precipitate. However, why polymer formation leads to precipitation, or even if polymerization is the main cause of precipitation, is not known. The best way to get nice soluble polymers that ladder on native PAGE is to use glycosylated proteins purified from plasma. Recombinant α1AT and α1AC precipitate at very low concentrations when induced to polymerize (especially by heat), presumably due to lateral association of the polymers. This suggests that polymers expose hydrophobic regions, consistent with the bis-ANS data, and begs the question, 'how', since the hyperstable conformations are generally less prone to aggregation than native serpins. The answer is likely to be found in the structure of the polymers themselves (see below). In any case, light scattering has been used in a handful of papers to determine rates of polymerization. The technique involves the use of a fluorescence instrument with the "excitation" wavelength in the visible range (~400 nm). The light is scattered orthogonally and intensity is detected at the same (or similar) wavelength (Crowther *et al.*, 2003; Pearce *et al.*, 2007). Another method is to align the incident radiation with the detector and measure transmittance (Devlin *et al.*, 2002), since increasing turbidity means less light is transmitted. Long duration experiments require stirring to prevent the settling of precipitate, and commonly have the cuvette sealed or the solution overlayed with mineral oil to prevent evaporation. A lag phase is generally observed, followed by a sigmoidal increase in scattering. Sometimes, the lag phase can be shortened by adding preformed polymers/aggregates, but only if the parent serpins are not glycosylated (Crowther *et al.*, 2003).

4. Effect of Mutations and "Drugs" on Polymerization

4.1. Early studies on the effect of mutations

One of the first mutagenesis studies aimed at repairing the secretion defeat in Z-α1AT focused on the salt bridge between residues 290 and 342. According to the crystal structure of cleaved plasma α1AT (Loebermann *et al.*, 1984), the residue Glu342 has a charged interaction with Lys290. The mutation of Glu342 to Lys (the Z mutation) predictably disrupts this salt bridge and was hypothesized to be the cause of Z-deficiency (Loebermann *et al.*, 1984). Indeed, the additional mutation of Lys290 to Glu successfully rescued the secretion defect of Z-α1AT in the COS-I cell system (Brantly *et al.*, 1988). This result was interpreted as indicating that a pathogenic mutation disrupts interactions around β-sheet A to cause polymerization and suggested that mutations that stabilized sheet A could potentially rescue the secretion defect. Stabilization could be achieved either by charged or by hydrophobic interactions. A good example is the effect of an additional F51L mutation on Z and Siiyama (S53F) α1AT stability and secretion. This mutation was found by a random mutagenesis study, and showed that altering the core packing of hydrophobic residues underlying sheet A can increase the stability of α1AT due to the better tertiary packing (Kwon *et al.*, 1994). Later, Sidhar *et al.* showed the F51L mutation increased Z secretion by two-fold and Siiyama secretion by five-fold (Sidhar *et al.*, 1995). Thus, the basis for serpin polymerization had been explained by the local instability of β-sheet A and the so-called shutter region, resulting in opening β-sheet A and polymerization via a loop-sheet mechanism. However, it seems unlikely to be a general feature of mutations that lead to serpin polymerization. First of all, disruption of the interaction between Glu342 and Lys290 does not necessarily results in polymerization. Using Xenopus oocytes, Foreman's group showed that the mutations of Lys290Glu or Glu342Ala themselves do not affect α1AT secretion, but α1AT only had a problem folding when Glu342 was replaced with charged residues such as Lys (Z mutation) or Arg (Foreman, 1987; Wu and Foreman, 1990). Thus, the cause of the Z deficiency is not the loss of a stabilizing salt bridge in the native form and suggests a specific effect of the charge composition of residue 342 on α1AT folding. Moreover, some mutations, Z included, do not significantly affect the stability of the folded protein.

4.2. Destabilizing and stabilizing mutations

Kim *et al.* expressed and purified 10 dysfunctional α1AT variants from *E. coli*, where mutations were spread throughout the molecule, such as R39C on helix A, S53F (Siiyama variant) on helix B, V55P on helix B, I92N on helix D, G115S on strand 2A, N158K on helix F, E264V (S variant, second most abundant) on helix G, A336T on strand 5A, P369S, or L on turn following strand 4B (Kim *et al.*, 2006; Fig. 17.1A). Among the variants, R39C, G115S, and P369S showed decreased stability of the native conformation, and others did not achieve the native conformation and ran slowly in transverse urea PAGE as a partially folded intermediate, with no inhibitory activity against elastase, similar to *in vitro* translated Z-α1AT (Yu *et al.*, 1995). This suggests that mutations almost anywhere in a serpin can lead to incomplete folding and subsequent polymerization. Further, they found that the defect caused by the mutations can be rescued in combination with stabilizing mutations found by random mutagenesis targeting a hydrophobic core of the molecule (including T68A, A70G, M374I, S381A, and K387R) (Lee *et al.*, 1996). All variants became more stable than wild-type α1AT and recovered activity, with the exception of S53F, located at the very center of the shutter region (Kim *et al.*, 2006). It was difficult to explain the effect of these mutations on the basis of the folded structure. Rather, the change in the side-chain properties seems to affect the folding pathway itself, leading to accumulation of a folding intermediate and polymerization.

4.3. Cavity-filling mutations and drugs

Another approach to stabilize a serpin and inhibit polymerization is to "fill" one of the cavities in the native structure. One such cavity was found in α1AT at the side of the β-sheet A. Lee *et al.* showed that the stability of native α1AT increases with the size of side chain at the residue Gly117 (Lee *et al.*, 2000). Gly117Phe had the largest effect, and it was slightly reduced by the additional mutation of Y160A that increases the size of the cavity, suggesting the state/size of the cavity is related protein stability. On this basis, *in silico* screening of compounds to bind to the cavity to stop *in vitro* polymerization was undertaken (Mallya *et al.*, 2007). Unfortunately, these strategies are unlikely to work to prevent polymerization *in vivo*: the Gly117Phe mutation seems to increase the secretion of both Z-monomer and polymer (Gooptu *et al.*, 2009) and one of the drugs, CG, inhibited the folding of Z-α1AT and led to cellular clearance (Mallya *et al.*, 2007). From a purely theoretical standpoint, such a strategy of targeting a pocket found in the native state presupposes a native to polymer pathway, inconsistent with the preponderance of data showing that polymers are formed from a folding intermediate *in vivo* or a partially unfolded state *in vitro*. Compounds that increase the stability of a serpin *in vitro* will naturally increase the stringency required to form polymers, but this is unlikely to be relevant to blocking polymerization in the ER of cells.

4.4. Inhibition by small molecules

When a native α1AT was incubated with 0.7 *M* sodium citrate at 67 °C for 12 h, it suppressed polymer formation and yielded a high proportion of latent α1AT (Bottomley and Tew, 2000; Lomas *et al.*, 1995). This was based on the observation that near quantitative production of latent AT could be made by adding citrate to the incubation (Wardell *et al.*, 1997). A similar effect was observed for native and L49P neuroserpin when incubated with 0.5 *M* sodium citrate (Onda *et al.*, 2005). Pearce *et al.* illustrated crystallographically that citrate may primarily bind to a pocket in the B/C barrel (the β-barrel formed by sheets B and C) at the top of the molecule and suggested the importance of the intact B/C barrel structure for α1AT unfolding and polymerization (Pearce *et al.*, 2008). Moreover, trimethylamine *N*-oxide (TMAD) was shown to have similar properties, with α1AT stability increasing with TMAD concentration up to 3 *M*, resulting in a delay of polymerization at 60 °C. However, when polymerization was induced by refolding, the addition of 1.5 *M* TMAD into the refolding buffer dramatically *accelerated* polymerization at RT at 0.05 mg ml^{-1}, suggesting TMAD stabilizes a native conformation of α1AT but does not therefore improve the rate of folding to the native state (Devlin *et al.*, 2001). Among the small molecules reported to inhibit polymerization, glycerol seems most general and useful. It succeeded in preventing polymerization both *in vitro* and *in vivo*. At 10% (v/v), glycerol reduced the rate of heat-induced polymerization in Z-α1AT at 41 °C by 2.9-fold (Sharp *et al.*, 2006), and *in vivo* the same concentration of glycerol increased the amount of secreted Z-α1AT by 25% after 12 h at 37 °C from human fibroblast cell engineered to stably expresses Z-α1AT (Burrows *et al.*, 2000). Further, glycerol seems to prevent polymerization of wild type and mutant S49P neuroserpin induced by both heat and refolding, along with other sugar and alcohol molecules, such as glucose, erythritol, or trehalose (Sharp *et al.*, 2006).

4.5. Inhibition by peptides

Schulze *et al.* were the first to demonstrate that binary complex formation of serpins with RCL-derived peptides prevented heat-induced polymerization (Schulze *et al.*, 1990). Subsequently, Mast *et al.* showed that incubation for 15 h at 48 °C with 100–200-fold molar excess of 16-mer peptide derived from α1AT RCL prevented polymerization without preforming the binary complex (Mast *et al.*, 1992). The RCL peptides also inhibit the pathogenic Z-α1AT by the incubation at 41 °C (Lomas *et al.*, 1992). This "inhibition" is certainly due to the incorporation of the full RCL peptide into the strand 4A position in β-sheet A, leading to the hyperstable binary complex. The α1AT RCL peptide can also be effectively incorporated into α1AC and AT (Mast *et al.*, 1992). However, the observed 'inhibition' of polymerization is caused by an off-pathway hyperstable end-state (binary complex formation), and peptide incorporation thus *competes* with polymerization and does not strictly inhibit it.

After the finding that RCL peptides can block polymerization, research focused on which part of the RCL was involved in the presumed loop-sheet polymerization mechanism. In 2002, Mahadeva *et al.* incubated M-α1AT or Z-α1AT at 37 °C with a 100-fold molar excess of 6-mer peptide (P7–P2, FLEAIG) and showed that the peptide was specifically incorporated into Z-α1AT and inhibited polymerization (Mahadeva *et al.*, 2002). This was consistent with the loop-sheet hypothesis that serpin polymerization is induced because the bottom part of β-sheet A is open ('patent') and accepts a part of the RCL of a neighboring molecule (see Section 5.2.2). In such a mechanism, the P7–P2 peptide would block Z-α1AT polymerization by forming a stable binary complex at the strand 4A position. However, they later purified this binary complex and found it was not stable. Indeed, by incubation at 37 °C, the binary complex surprisingly reverted to native Z-α1AT by the release of peptide with a k_{off} value of $2.0 \times 10^6\ s^{-1}$ for N-terminally acetylated hexapeptide and $5.5 \times 10^6\ s^{-1}$ for nonacetylated peptide (Parfrey *et al.*, 2004). This is contrasted to the binary complex of M-α1AT with a 12-mer peptide which did not dissociate after 10 weeks at 37 °C, which seems a general feature of sheet A-expanded serpins. Chang *et al.* performed Ala scanning on the FLEAIG peptide and found that F**A**EAIG, FL**A**AIG, and FLEA**A**G lost preference and formed binary complex almost identically to M-α1AT and Z-α1AT (Chang *et al.*, 2006). In other cases, it was reported that a shorter peptide (TTAI) inhibited polymerization of Z-α1AT (Chang *et al.*, 2009) or that Aβ1-40 peptide inhibited polymerization of neuroserpin (Kinghorn *et al.*, 2006). Whether inhibition for these unrelated peptides is caused by incorporation into sheet A or some other mechanism has not been established. However, crystal structures of AT bound to peptides unrelated to the RCL at the bottom of sheet A (Zhou *et al.*, 2003) suggest that one or multiple copies can occupy the strand 4A position and hence block/prevent polymerization.

4.6. Reversal by peptides

As mentioned previously, polymers are highly stable like the RCL-cleaved conformation and do not denature at temperatures lower than 100 °C in melting temperature studies or by denaturant up to 8 *M* urea at neutral pH by transverse urea PAGE. Thus, conceptually, reversal of polymerization is problematic without resorting to extreme denaturing conditions, such as 10% SDS with boiling. However, in spite of this, a few reports suggest that peptides can revert polymers back to a monomeric state. Zhou *et al.* incubated a 200-fold molar excess of FLEAIG peptide for 21 days at 37 °C with heat-induced Z-α1AT polymer, and a small amount of monomer appeared on the native gel (Zhou *et al.*, 2004; Fig 17.7A). An even more efficient reversal was shown by Chang *et al.* (2009) using a TTAI peptide (Fig. 17.7B). However, there were no controls to show if the material which appeared as the monomer band was actually derived *from* polymers. What is more likely is that the diffuse band (Fig. 17.7A) or discrete band (Fig. 17.7B) that runs cathodal to the native monomer band is unfolded or incorrectly folded monomer, and that prolonged incubation with an RCL peptide results in a small amount of refolding and formation of the binary complex. The "appearance" of a monomer band could thus easily be explained by the compaction of a portion of a diffuse band (Fig. 17.7A) or conversion from a discrete M*-like state (Fig. 17.7B) to binary or ternary

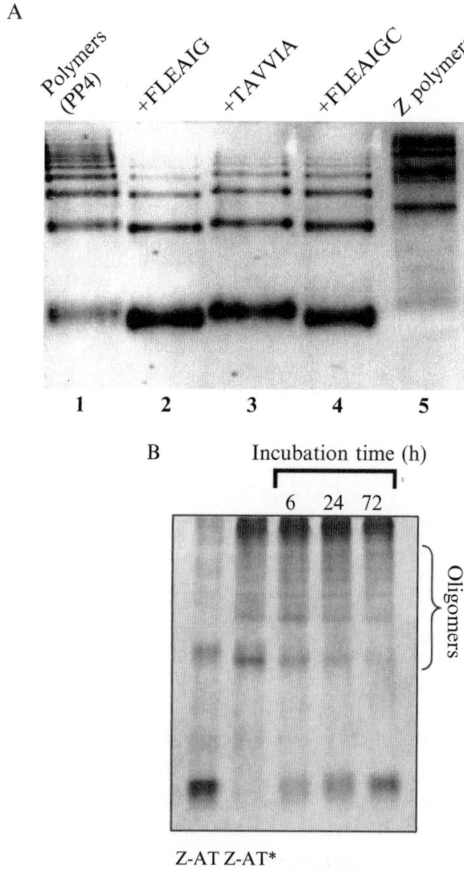

Figure 17.7 Evidence for reversibility of serpin polymers. (A) Polymers of α1AT formed by cleavage at P10 (using protease PP4, lane 1) can be reversed by subsequent incubation with short peptides (lanes 2–4); however, for polymers of Z-α1AT formed by heat treatment (lane 5), reversibility by incubation with peptides is less clear (lane 6) (Zhou *et al.*, 2004). Peptides were incubated with the polymers for 21 days at 37 °C at a ratio of 200:1. In lanes 2–5, there is a clear reduction in the high-order polymers corresponding with an increase in short polymers and monomer. This is what would be expected if the peptide competes with the RCL for insertion into sheet A, and if the intermolecular linkage corresponds to the "lower" (P9–P3) portion of the RCL. In contrast, heat polymers of Z-α1AT do not become smaller, rather they seem to disappear altogether. The only evidence for "reversal" is thus the appearance of a monomer band that is slightly more intense that before the 3-week-long incubation. (B) Similar results were found in a subsequent study using similar polymers, but shorter incubations and a different peptide (Chang et al., 2009). Again, the large polymers are not observed to break down into shorter polymers, indeed the opposite occurs. The appearance of a monomer band corresponds perfectly with the disappearance of the "dimer" band, which may in fact be a partially unfolded intermediate. A control incubation without peptide might have produced a similar result due to the refolding of the intermediate to the native state. Figures reproduced with permission.

peptide complexed monomer. Similarly, the apparent disappearance of the ladders is likely due to the continued polymerization and aggregation of the polymers during the prolonged incubation at 37 °C. If peptides truly reverse serpin polymerization by breaking the intermolecular linkage, then one would expect each linkage to be equally susceptible to peptide annealing. Therefore, long polymers should be broken down into smaller polymers before the observation of monomers on the native gel. This is indeed what is observed when peptides reverse PP4-cleaved polymers (Fig. 17.7A). However, no such accumulation of smaller species is observed when using heat-induced polymers (Fig. 17.7). Other reports suggest polymer reversibility (Mikus and Ny, 1996), but care should be taken in design and interpretation to ensure that large polymers are actually broken down into smaller polymers, and that the appearance of monomer is due to the breakdown of small polymers.

5. Mechanisms of Polymerization

5.1. Introduction

It is not known how serpins polymerize inside the endoplasmic reticulum during the folding process, but it has been assumed that what occurs through the partial unfolding of a native serpin is a good approximation of the *in vivo* process. Of central importance is how the protomers are connected in a noncovalent, yet highly stable fashion. The demonstrated ability of serpins to incorporate RCL-like peptides into the strand 4A position led to the first proposal that the intermolecular insertion of a portion of the RCL into sheet A of another monomer was the basis of the stable linkage (Schulze *et al.*, 1990). This "loop-sheet" mechanism, as with most other proposed mechanisms, amounts to a small "domain swap." Traditionally, domain swapping has been demonstrated for proteins by either crystallographic structures or disulfide trapping of the higher-order species. In this section, we introduce the structures and models of serpin polymers and the techniques used to distinguish between them. Here, we only deal with stable serpin polymers, and not "edge-on" crystal-contact polymers which are not stable on nondenaturing PAGE (McGowan *et al.*, 2006; Nar *et al.*, 2000; Zhang *et al.*, 2008), and are thus clearly unrelated to physiological polymerization.

5.2. Structures and models

5.2.1. Cleaved polymers

There are two crystal structures of "cleaved" polymers, formed by serendipitous cleavage in the N-terminal portion of the RCL during the crystallization process (Fig. 17.8A). In both cases, the P4–P3′ region of Pittsburgh

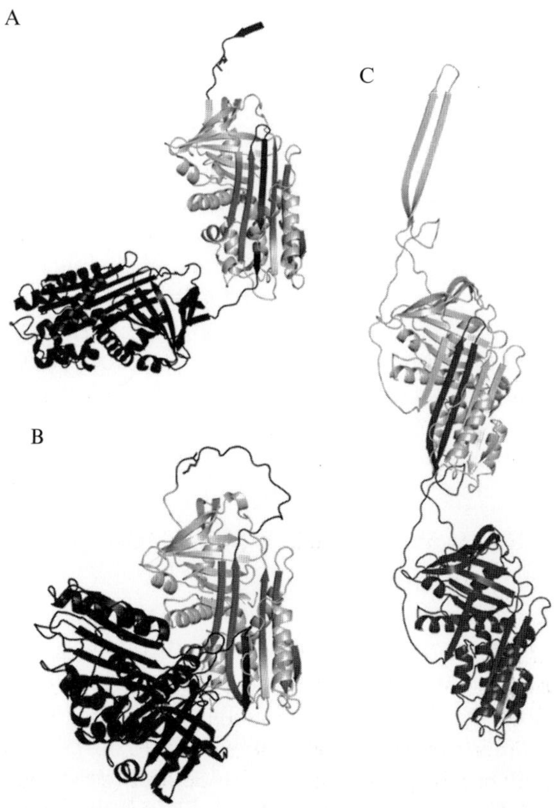

Figure 17.8 Models of cleaved, loop-sheet, and domain-swapped dimers. (A) Cleavage at the P7 position by an unknown protease during crystallization produced an infinite loop-sheet-type polymer within the crystal, a dimer of which is shown here (1QMB; Huntington *et al.*, 1999). The top monomer is shaded and represented as in Fig. 17.1, and the bottom monomer is black. (B) A model of an intact loop-sheet dimer is shown, colored as above (Huntington and Whisstock, 2010). The only intermolecular contact is the in-register insertion of the P8–P3 portion of the RCL of the bottom monomer into β-sheet A of the top monomer. (C) A model of an open domain-swapped (β-hairpin) dimer was generated based on a crystal structure of a self-terminating (closed) dimer (Yamasaki *et al.*, 2008). The top monomer is colored light gray and the bottom monomer is black, to illustrate the extent of the swapped region, including both strands 5 and 4A (the RCL is strand 4A when incorporated into β-sheet A).

α1AT was protected from cleavage by docking with an inactive protease (S195A thrombin). Somehow during the incubation process, a protease in England cut at the P7–P6 bond (Huntington *et al.*, 1999) and protease in Australia cut at the P6–P5 bond (Dunstone *et al.*, 2000). The resulting crystal structures validated the prediction of Mast *et al.* that cleavage on the N-terminal portion of the RCL will induce insertion in *cis* up to the N-terminal side of the scissile bond, leaving a gap in sheet A that could only be filled by the C-terminal portion of the RCL from another protomer (Mast *et al.*, 1992). Cleaved polymers can be made using native serpins and proteases that cleave at only a single position in the RCL, or by mutating the RCL to prevent cleavage at P1 and to induce cleavage N-terminal to P1 (e.g., P10; Yamasaki *et al.*, 2010). Long polymers can be formed by this method at high concentration, while at low concentration the polymers have a tendency to circularize. The conformational equivalent to the polymerigenic M* state would be the cleaved form of the monomer with the N-terminal portion of the RCL self-inserted. This polymerization mechanism is now well established, but its relevance to intact serpin polymerization remains unclear (discussed later). Nevertheless, cleaved polymers have been used as a model system for investigating the properties of polymers, especially with respect to the ability of peptides to block and reverse polymerization (Zhou *et al.*, 2004; Fig. 17.7A).

5.2.2. Loop-sheet mechanism

Exogenous peptides corresponding to the hinge region (P14–P9) of AT can anneal to serpins and create a similar conformer to that achieved by cleavage in the N-terminal portion of the RCL. Incubation of AT and α1AT with large excess of the P14–P9 peptide thus induces polymerization. This observation further supported the idea that creating a patent A sheet was a critical step in the *in vitro* and perhaps *in vivo* polymerization process (Lomas *et al.*, 1992) and was the first demonstration that intact loop-sheet polymers could actually form through intermolecular insertion of the P8–P3 region of the RCL into β-sheet A. Although there is no crystal structure of a peptide-induced loop-sheet polymer, there is little doubt that it will resemble the structure of the cleaved polymer (Fig. 17.8B), however, with less flexibility between protomers due to the return of the P′ side to the monomer donating its loop (Chang *et al.*, 1997; Huntington and Whisstock, 2010). Although in principle it should be possible to induce polymerization of any inhibitory serpin using hinge-region peptides, it has only been demonstrated for α1AT and AT. In contrast to heat induction or the cleavage method, polymers formed by this peptide method are generally short, composed predominantly of dimers and trimers, and always result in a large amount of unpolymerized monomeric material (Yamasaki *et al.*, 2010). The reason for this is likely to be competition of the peptide for the lower portion of sheet A, as seen in the crystal structure of PAI-1 bound to a P14–P10 peptide in register, and additionally in the P6–P2 position (Xue *et al.*, 1998). In spite of these limitations, peptide induction has been widely used as a general model of serpin polymerization.

5.2.3. Domain swap or β-hairpin

The loop-sheet mechanisms described above can be considered domain swaps, since a portion of one monomer is incorporated in *trans* as it would in *cis* upon RCL cleavage (Fig. 17.1B) or conversion to the latent state (Fig. 17.1C). However, it is a stretch to use the term *domain* to describe the P8–P3 portion of the RCL. We recently solved a structure of an intact dimer of AT that revealed a 51 residue domain swap, including the coiled region at the bottom of the serpin and both strands 4 and 5A (Yamasaki *et al.*, 2008). Although the region that swapped is not technically a protein domain, it does fit with the accepted usage of the term 'domain swap' and constitutes a large fraction of the protein structure interacting in exactly the same fashion in *trans* as it would in *cis*. This dimer was formed from native AT at pH 5.7 and 37 °C and purified by size-exclusion chromatography. Presumably, the low pH and slightly elevated temperature resulted in the extraction of strand 5 from β-sheet A to form a partially unfolded polymerigenic monomeric (M*) state. Polymers would then form by the intermolecular, and perhaps simultaneous incorporation of strands 4 (RCL) and 5 into sheet A, resulting in a hyperstable linkage. The top and bottom protomers in the linear dimer (Fig. 17.8C) would be locked into donor and acceptor (patent) conformations, respectively, capable of only two fates: elongation or self-termination (circularization). Low-pH conditions were chosen because the fraction of self-terminating dimer was higher than obtained using heat at neutral pH (Zhou *et al.*, 2003), but there is no reason to suspect that the observed domain swap is particular to the method of partial unfolding. Similar methods could in theory be used to obtain self-terminating domain-swapped dimers, trimers, etc. of any serpin, but conditions will likely be serpin specific. For instance, low pH is effective at inducing polymerization of α1AT and AT (Zhou *et al.*, 2003), but not of neuroserpin (Belorgey *et al.*, 2010). The reason for this may be the positioning of titratable groups, such as the conserved His residue on strand 5A (Zhou *et al.*, 2003).

5.3. Biochemical methods for distinguishing between models

Until the recent structure of the intact domain-swapped serpin dimer (Yamasaki *et al.*, 2008), there was no alternative to the loop-sheet hypothesis to account for stable serpin polymers. Based on what must now be considered circumstantial evidence, the field had come to accept that the only way to get a stable serpin polymer was to insert a portion of the RCL of one molecule into the A sheet of another. The two current models of serpin polymerization, the loop-sheet mechanism and the "β-hairpin" mechanism, should have different structural features that can be distinguished by biochemical methods.

5.3.1. Disulfide trapping

The traditional method for demonstrating domain swapping is to engineer a disulfide bond that would be intramolecular in the monomeric state, but intermolecular in the domain-swapped/polymeric state. Formation of the disulfide is easily verified by SDS-PAGE, with oxidized monomer running slightly faster than reduced monomer, and covalently linked ladders indicating polymers. The first instance where this was used for serpins placed a disulfide between strands 5 and 6A on α1AT to determine if strand 5A is part of the swap in α1AT polymers (Yamasaki *et al.*, 2008). Polymerization was induced by 0.75 *M* GndHCl and 37 °C. Under these conditions, the disulfide bond between strands 5 and 6A prevented polymerization, supporting the idea that strand 5 release is critical for polymerization. However, when DTT was added, the mutant polymerized in a manner identical to the control. The samples were reoxidized by adding an excess of TMAD before denaturation in SDS sample buffer and running on SDS-PAGE. Disulfide-linked ladders were observed for about half of the α1AT, indicating that in some of the polymers strands 5 and 6A come from different protomers (Fig. 17.9A). No studies have been reported to test the extent of involvement of the RCL, but the β-hairpin model has the full intermolecular incorporation of the RCL and the loop-sheet model has only the lower portion (P8–P3) of the RCL inserting in *trans*. The advantage of the disulfide trapping technique over others is that it can be used equally well *in vivo* as *in vitro*, since disulfide bonds are not formed until folding to the native or polymer state is achieved (the cytosol is a reducing environment and the ER allows rapid disulfide isomerization). Of course, it is critical that the appropriate controls are run to ensure that the higher-order species on SDS-PAGE is actually composed exclusively of polymers. A negative control can be made by shifting the cysteines by one or two positions so that a native disulfide should not form, or by placing a disulfide bond between two structural elements that do not swap (e.g., strands 1 and 2A). However, care must be taken in choosing the residues to mutate, since it is possible to have unwanted effects on the folding and/or structure of the serpin.

5.3.2. Limited proteolysis

One of the implications of the β-hairpin model is that a portion of the serpin must remain unfolded to allow the head-to-tail configuration thought to be important for polymer propagation and flexibility, whereas in the loop-sheet model only strand 1C is unfolded relative to the native state (Chang *et al.*, 1997; Huntington and Whisstock, 2010). Thus, one would expect the two mechanisms to result in polymers of radically different proteolytic susceptibilities, with loop-sheet polymers highly resistant to proteolysis and domain-swapped polymers susceptible to proteolysis in regions unavailable in the native serpin. We showed that heat-generated polymers of AT

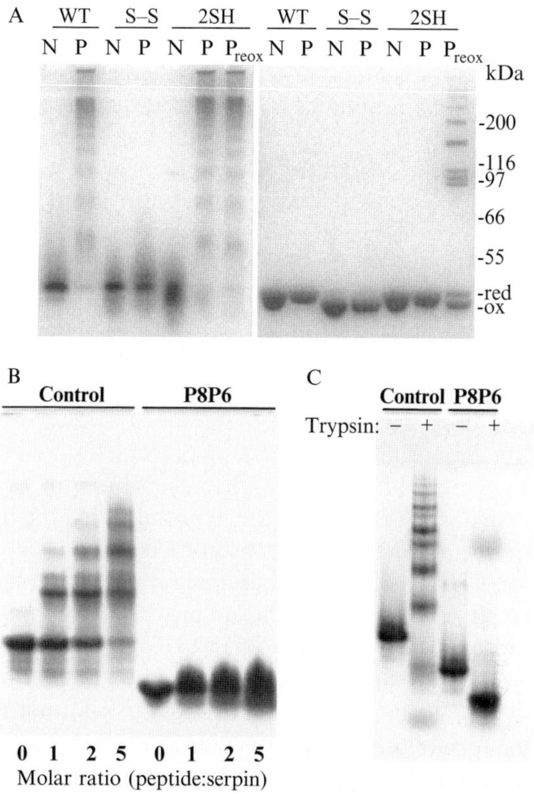

Figure 17.9 Methods to distinguish between possible polymerization mechanisms. (A) Since all proposed mechanisms of serpin polymerization are essentially domain swaps, it is possible to engineer disulfide bonds to covalently trap polymers into SDS-stable ladders. Such an approach is illustrated here, where WT indicates control with no disulfide, S–S indicates an oxidized disulfide bridge in the monomer between strands 5 and 6A, and 2SH indicates the reduced form. Polymers of α1AT (N for before and P for after incubation) were generated by 0.75 *M* GndHCl at 37 °C, and formed only if strand 5 was not covalently linked to strand 6 (native gel on left). Polymers were returned to the monomeric state upon denaturation by SDS-PAGE, unless strands 5 and 6A were donated by different monomers to yield SDS-stable polymers. P_{reox} indicates an oxidation step after polymerizing the reduced strands 5A and 6A disulfide-bonded monomer (Yamasaki *et al.*, 2008). (B) As predicted from the dimer models in Fig. 17.8, mutating the hydrophobic stretch from P8–P6 into Asp residues blocks the ability of α1AT to form loop-sheet polymers upon incubation with small peptides corresponding to the P14–P9 region, while the control forms polymers at the molecular ratios used. (C) A similar result was found when α1AT was mutated to ensure exclusive trypsin cleavage at the P10 position. Cleavage of the control leads to polymers, as in Fig. 17.8A, but the P8–P6 Asp mutations completely inhibit polymerization by the loop-sheet mechanism. In contrast, the P8–P6 mutations had no effect on the ability to polymerize *in vitro* using heat or GndHCl, or *in vivo* when coupled to polymerigenic mutations (Yamasaki *et al.*, 2010). Figures reproduced with permission.

and α1AT were digested by AspN, while the native serpins were resistant to cleavage, and found that some cleavage sites mapped to the predicted linker regions (on or adjacent to helix I: Yamasaki *et al.*, 2008). Similar results were found for polymers of AT induced by chemical modification, polymers of PAI-1 formed by incubation with small organic compounds (Pedersen *et al.*, 2003), and polymers of neuroserpin formed either by heat or upon refolding (Takehara *et al.*, 2010). Interestingly, cleavage sites can also be found in strand 5A, perhaps reflecting the conformation of the "top" monomer in linear polymers that would be highly represented in short polymers (Pedersen *et al.*, 2003). A recent report showed that polymers formed under near physiological conditions (41 °C) from Z-α1AT were specifically cleaved by LysC at position 310 (adjacent to helix I) but found no evidence of cleavage within helix I itself, at residue 300 (Ekeowa *et al.*, 2010). However, the authors did not address the possibility of cleavage at both sites, creating a fragment (301–310) too small to pick up on an SDS gel for N-terminal sequencing. Nevertheless, the accessibility of Lys310 to cleavage in polymers of Z-α1AT supports the hypothesis that the helix I region is unfolded to provide a linker between protomers.

Care must be taken when conducting and interpreting limited proteolysis experiments. It is important to choose proteases and conditions that do not result in cleavage of the native serpin so that any sequenced band reflects the difference in conformation between native and polymeric states and that no conformational changes are induced (i.e., RCL insertion or global destabilization). Also, the size of the polymer and contamination with unpolymerized monomer can affect the results. Ideally, only large purified polymers will be used to test the proteolytic susceptibility of linker regions. Conditions should be chosen that create sufficient product to observe by Coomassie staining on an SDS-PAGE and to prevent over-digestion of the protein. Generally, an overnight incubation at 4 °C at neutral pH with a low concentration of protease should be used (Takehara *et al.*, 2010; Yamasaki *et al.*, 2008). Either the cleavage sites can be sequenced after blotting bands from an SDS gel onto PVDF membrane or the reaction mixture can be used directly on certain types of mass spectrometry to obtain molecular masses of peptides of various sizes. Generally, N-terminal sequencing will only work to identify major cleavage sites resulting in large fragments, while mass spectrometry is best suited to small peptides, so the techniques are complementary. Note, however, that even usage of both techniques will not guarantee that all cleavage events will be observed.

5.3.3. Mutations in RCL

The RCL is thought to play an integral part in both of the current hypotheses of serpin polymerization. In the loop-sheet mechanism, the hinge region (N-terminal part) inserts into the top portion of sheet A to generate a patent, open configuration at the bottom of sheet A, into which the P8–P3 region (C-terminal part) of another monomer inserts to link the two protomers together (Lomas *et al.*, 1992; Mahadeva *et al.*, 2002). In the domain swap (β-hairpin) model, the entire RCL and adjacent strand 5A insert to form a significantly larger intermolecular contact (6 residues for loop-sheet and over 30 residues for β-hairpin). Since we know that it is possible to simulate the loop-sheet mechanism through cleavage in the center of the RCL or by incubation with short peptides, it is possible to alter the features of the P8–P3 region and test the effect of the mutations on the ability to polymerize via the loop-sheet mechanism, by heat, chaotropic agents, or even in cells. We recently mutated the hydrophobic P8–P6 region of α1AT from MFL to DDD in the expectation that charged side chains would slow or prevent loop-sheet polymerization (Yamasaki *et al.*, 2010). The mutations prevented polymerization induced by the AT P14–P9 peptide (Fig. 17.9B); however, this method results in only a small amount of polymerization for the control and is plagued by the competition of the peptide for the gap at the bottom of sheet A. To demonstrate that the mutations prevent loop-sheet polymerization, it is necessary to use a method where polymers form even at very low concentrations, with no possibility of competition for occupying sheet A. We therefore mutated the P1 residue to Asp and the P10 residue to Arg on the wild type and P8–P6 Asp backgrounds to ensure that cleavage by trypsin would exclusively occur at P10 (this was verified by N-terminal sequencing and mass spectrometry). No polymers were formed for the P10-cleaved P8–P6 Asp variant, even for prolonged incubations at high concentration (Fig. 17.9C). However, this mutant polymerized like control when induced by heat or GndHCl, or in cells when coupled with polymerigenic mutations, suggesting either that the P8–P6 region is not involved in the domain swap, or that it is only a small part of an extensive domain swap, as predicted by the β-hairpin model. Had polymerization *in vitro* and in cells been prevented or even significantly slowed by the mutations, it would have been necessary to conclude that the P8–P6 is a significant part of the intermolecular linkage and that the loop-sheet mechanism was likely to be the predominant mode of serpin polymerization. However, based on the current data, we are forced to conclude that polymers utilize a larger domain swap that does not rely on the P8–P6 region for either the rate of formation or the stability of the polymeric linkage.

6. Conclusions

In this chapter, we attempted to summarize more than 20 years of published work on how to create and study serpin polymers *in vitro*. Needless to say, we have not mentioned every paper ever published on this topic nor every experiment ever conducted. In some cases, the omission is accidental and in others it was intentional, either because we felt that the topic was already adequately covered or because we considered the study to be peripheral to our remit or did not contribute in a significant way to our understanding of serpin polymerization. Undoubtedly, some of these judgments were flawed. However, it is our hope that this chapter provides a sufficiently concise and detailed account of the background and techniques of studying serpin polymerization for both the novice and seasoned serpinologist. Indeed, this is an exciting time to be studying serpin folding, misfolding, and polymerization because, in the past few years, the field has moved from a state of stasis, where the loop-sheet hypothesis went unchallenged, to the dynamic state we now find ourselves in, where all conceivable domain swaps are possible and the tools for testing them *in vitro* and in cells are readily available.

From the studies summarized in this chapter, we can conclude that *in vitro* polymerization is not a single phenomenon, but a complex set of ensemble conformations with multiple possible end products, including nonspecific aggregates, linear polymers by several mutually exclusive domain-swapping mechanisms, laterally associated aggregates of linear polymers, circularization of linear polymers, latent monomers, and stable or transiently stable partially folded monomers. However, the loop-sheet polymer is not one of the end points, except under specific artifactual conditions (cleavage in the center of RCL or small peptide annealing) where partial unfolding of the native serpin is not induced.

Several important questions remain to be addressed concerning the basis of polymerization and the relationship between *in vitro* produced polymers and those that accumulate within cells. It is likely that there are multiple domain-swapping mechanisms possible, dependent on the serpin and conditions used, and therefore that multiple polymeric forms are produced *in vitro*, but it is not clear if this is also true *in vivo*. If there are multiple serpin polymer forms, are some toxic and others benign? Does the serpin or its polymerigenic mutation determine the predominant polymer form, and thus the toxicity? Is serpin polymerization drugable? The answer to these and other questions will be found in part by the continued study of serpin polymerization *in vitro*.

ACKNOWLEDGMENTS

J. A. H. is a senior MRC nonclinical fellow, and funding for this research was provided by a project grant from the Medical Research Council (UK).

REFERENCES

Belorgey, D., Crowther, D. C., Mahadeva, R., and Lomas, D. A. (2002). Mutant Neuroserpin (S49P) that causes familial encephalopathy with neuroserpin inclusion bodies is a poor proteinase inhibitor and readily forms polymers in vitro. *J. Biol. Chem.* **277,** 17367–17373.

Belorgey, D., Hagglof, P., Onda, M., and Lomas, D. A. (2010). pH-dependent stability of neuroserpin is mediated by histidines 119 and 138; implications for the control of beta-sheet A and polymerization. *Protein Sci.* **19,** 220–228.

Bottomley, S. P., and Tew, D. J. (2000). The citrate ion increases the conformational stability of alpha(1)-antitrypsin. *Biochim. Biophys. Acta* **1481,** 11–17.

Brantly, M., Courtney, M., and Crystal, R. G. (1988). Repair of the secretion defect in the Z form of alpha 1-antitrypsin by addition of a second mutation. *Science* **242,** 1700–1702.

Burrows, J. A., Willis, L. K., and Perlmutter, D. H. (2000). Chemical chaperones mediate increased secretion of mutant alpha 1-antitrypsin (alpha 1-AT) Z: A potential pharmacological strategy for prevention of liver injury and emphysema in alpha 1-AT deficiency. *Proc. Natl. Acad. Sci. USA* **97,** 1796–1801.

Chang W. S., and Lomas D. A. (1998). Latent alpha1-antichymotrypsin. A molecular explanation for the inactivation of alpha1-antichymotrypsin in chronic bronchitis and emphysema. *J. Biol. Chem.* **273**(6), 3695–3701.

Chang, W. S., Whisstock, J., Hopkins, P. C., Lesk, A. M., Carrell, R. W., and Wardell, M. R. (1997). Importance of the release of strand 1C to the polymerization mechanism of inhibitory serpins. *Protein Sci.* **6,** 89–98.

Chang, Y. P., Mahadeva, R., Chang, W. S., Shukla, A., Dafforn, T. R., and Chu, Y. H. (2006). Identification of a 4-mer peptide inhibitor that effectively blocks the polymerization of pathogenic Z alpha1-antitrypsin. *Am. J. Respir. Cell Mol. Biol.* **35,** 540–548.

Chang, Y. P., Mahadeva, R., Chang, W. S., Lin, S. C., and Chu, Y. H. (2009). Small-molecule peptides inhibit Z alpha1-antitrypsin polymerization. *J. Cell. Mol. Med.* **13,** 2304–2316.

Crowther, D. C., Serpell, L. C., Dafforn, T. R., Gooptu, B., and Lomas, D. A. (2003). Nucleation of alpha 1-antichymotrypsin polymerization. *Biochemistry* **42,** 2355–2363.

Dafforn, T. R., Mahadeva, R., Elliott, P. R., Sivasothy, P., and Lomas, D. A. (1999). A kinetic mechanism for the polymerization of alpha1-antitrypsin. *J. Biol. Chem.* **274,** 9548–9555.

Devlin, G. L., Parfrey, H., Tew, D. J., Lomas, D. A., and Bottomley, S. P. (2001). Prevention of polymerization of M and Z alpha1-antitrypsin (alpha1-AT) with trimethylamine N-oxide. Implications for the treatment of alpha1-at deficiency. *Am. J. Respir. Cell Mol. Biol.* **24,** 727–732.

Devlin, G. L., Chow, M. K., Howlett, G. J., and Bottomley, S. P. (2002). Acid denaturation of alpha1-antitrypsin: Characterization of a novel mechanism of serpin polymerization. *J. Mol. Biol.* **324,** 859–870.

Dunstone, M. A., Dai, W., Whisstock, J. C., Rossjohn, J., Pike, R. N., Feil, S. C., Le Bonniec, B. F., Parker, M. W., and Bottomley, S. P. (2000). Cleaved antitrypsin polymers at atomic resolution. *Protein Sci.* **9,** 417–420.

Ekeowa, U. I., Freeke, J., Miranda, E., Gooptu, B., Bush, M. F., Perez, J., Teckman, J., Robinson, C. V., and Lomas, D. A. (2010). Defining the mechanism of polymerization in the serpinopathies. *Proc. Natl. Acad. Sci. USA* **107,** 17146–17151.

Elliott, P. R., Pei, X. Y., Dafforn, T. R., and Lomas, D. A. (2000). Topography of a 2.0 Å structure of alpha1-antitrypsin reveals targets for rational drug design to prevent conformational disease. *Protein Sci.* **9,** 1274–1281.

Foreman, R. C. (1987). Disruption of the Lys-290–Glu-342 salt bridge in human alpha 1-antitrypsin does not prevent its synthesis and secretion. *FEBS Lett.* **216,** 79–82.

Gerbod, M. C., Janciauskiene, S., Jeppsson, J. O., and Eriksson, S. (1998). The in vitro effect of lithocholic acid on the polymerization properties of PiZ alpha-1-antitrypsin. *Arch. Biochem. Biophys.* **351**(2), 167–174.

Gooptu, B., Miranda, E., Nobeli, I., Mallya, M., Purkiss, A., Brown, S. C., Summers, C., Phillips, R. L., Lomas, D. A., and Barrett, T. E. (2009). Crystallographic and cellular characterisation of two mechanisms stabilising the native fold of alpha1-antitrypsin: Implications for disease and drug design. *J. Mol. Biol.* **387,** 857–868.

Huntington, J. A., and Whisstock, J. (2010). Molecular contortionism—On the physical limits of serpin loop-sheet polymers. *Biol. Chem.* **391**(8), 973–982.

Huntington, J. A., Pannu, N. S., Hazes, B., Read, R. J., Lomas, D. A., and Carrell, R. W. (1999). A 2.6 A structure of a serpin polymer and implications for conformational disease. *J. Mol. Biol.* **293,** 449–455.

James, E. L., and Bottomley, S. P. (1998). The mechanism of alpha 1-antitrypsin polymerization probed by fluorescence spectroscopy. *Arch. Biochem. Biophys.* **356,** 296–300.

James, E. L., Whisstock, J. C., Gore, M. G., and Bottomley, S. P. (1999). Probing the unfolding pathway of alpha1-antitrypsin. *J. Biol. Chem.* **274,** 9482–9488.

Janciauskiene, S., Eriksson, S., Callea, F., Mallya, M., Zhou, A., Seyama, K., Hata, S., and Lomas, D. A. (2004). Differential detection of PAS-positive inclusions formed by the Z, Siiyama, and Mmalton variants of alpha1-antitrypsin. *Hepatology* **40,** 1203–1210.

Jung, C. H., Na, Y. R., and Im, H. (2004). Retarded protein folding of deficient human alpha 1-antitrypsin D256V and L41P variants. *Protein Sci.* **13,** 694–702.

Kim, D., and Yu, M. H. (1996). Folding pathway of human alpha 1-antitrypsin: Characterization of an intermediate that is active but prone to aggregation. *Biochem. Biophys. Res. Commun.* **226,** 378–384.

Kim, M. J., Jung, C. H., and Im, H. (2006). Characterization and suppression of dysfunctional human alpha1-antitrypsin variants. *Biochem. Biophys. Res. Commun.* **343,** 295–302.

Kinghorn, K. J., Crowther, D. C., Sharp, L. K., Nerelius, C., Davis, R. L., Chang, H. T., Green, C., Gubb, D. C., Johansson, J., and Lomas, D. A. (2006). Neuroserpin binds Abeta and is a neuroprotective component of amyloid plaques in Alzheimer disease. *J. Biol. Chem.* **281,** 29268–29277.

Kwon, K. S., Kim, J., Shin, H. S., and Yu, M. H. (1994). Single amino acid substitutions of alpha 1-antitrypsin that confer enhancement in thermal stability. *J. Biol. Chem.* **269,** 9627–9631.

Lee, K. N., Park, S. D., and Yu, M. H. (1996). Probing the native strain iin alpha1-antitrypsin. *Nat. Struct. Biol.* **3,** 497–500.

Lee, C., Park, S. H., Lee, M. Y., and Yu, M. H. (2000). Regulation of protein function by native metastability. *Proc. Natl. Acad. Sci. USA* **97,** 7727–7731.

Levina, V., Dai, W., Knaupp, A. S., Kaiserman, D., Pearce, M. C., Cabrita, L. D., Bird, P. I., and Bottomley, S. P. (2009). Expression, purification and characterization of recombinant Z alpha(1)-antitrypsin—The most common cause of alpha(1)-antitrypsin deficiency. *Protein Expr. Purif.* **68,** 226–232.

Loebermann, H., Tokuoka, R., Deisenhofer, J., and Huber, R. (1984). Human alpha 1-proteinase inhibitor. Crystal structure analysis of two crystal modifications, molecular model and preliminary analysis of the implications for function. *J. Mol. Biol.* **177,** 531–557.

Lomas, D. A., Evans, D. L., Finch, J. T., and Carrell, R. W. (1992). The mechanism of Z alpha 1-antitrypsin accumulation in the liver. *Nature* **357,** 605–607.

Lomas, D. A., Evans, D. L., Stone, S. R., Chang, W. S., and Carrell, R. W. (1993a). Effect of the Z mutation on the physical and inhibitory properties of alpha 1-antitrypsin. *Biochemistry* **32,** 500–508.

Lomas, D. A., Finch, J. T., Seyama, K., Nukiwa, T., and Carrell, R. W. (1993b). Alpha 1-antitrypsin Siiyama (Ser53- > Phe). Further evidence for intracellular loop-sheet polymerization. *J. Biol. Chem.* **268,** 15333–15335.

Lomas, D. A., Elliott, P. R., Chang, W. S., Wardell, M. R., and Carrell, R. W. (1995). Preparation and characterization of latent alpha 1-antitrypsin. *J. Biol. Chem.* **270,** 5282–5288.

Mahadeva, R., Chang, W. S., Dafforn, T. R., Oakley, D. J., Foreman, R. C., Calvin, J., Wight, D. G., and Lomas, D. A. (1999). Heteropolymerization of S, I, and Z alpha1-antitrypsin and liver cirrhosis. *J. Clin. Invest.* **103,** 999–1006.

Mahadeva, R., Dafforn, T. R., Carrell, R. W., and Lomas, D. A. (2002). 6-mer peptide selectively anneals to a pathogenic serpin conformation and blocks polymerization. Implications for the prevention of Z alpha(1)-antitrypsin-related cirrhosis. *J. Biol. Chem.* **277,** 6771–6774.

Mallya, M., Phillips, R. L., Saldanha, S. A., Gooptu, B., Brown, S. C., Termine, D. J., Shirvani, A. M., Wu, Y., Sifers, R. N., Abagyan, R., and Lomas, D. A. (2007). Small molecules block the polymerization of Z alpha1-antitrypsin and increase the clearance of intracellular aggregates. *J. Med. Chem.* **50,** 5357–5363.

Mast, A. E., Enghild, J. J., and Salvesen, G. (1992). Conformation of the reactive site loop of alpha 1-proteinase inhibitor probed by limited proteolysis. *Biochemistry* **31,** 2720–2728.

McGowan, S., Buckle, A. M., Irving, J. A., Ong, P. C., Bashtannyk-Puhalovich, T. A., Kan, W. T., Henderson, K. N., Bulynko, Y. A., Popova, E. Y., Smith, A. I., Bottomley, S. P., Rossjohn, J., *et al.* (2006). X-ray crystal structure of MENT: Evidence for functional loop-sheet polymers in chromatin condensation. *EMBO J.* **25,** 3144–3155.

Mikus, P., and Ny, T. (1996). Intracellular polymerization of the serpin plasminogen activator inhibitor type 2. *J. Biol. Chem.* **271,** 10048–10053.

Miranda, E., PÈrez, J., Ekeowa, U. I., Hadzic, N., Kalsheker, N., Gooptu, B., Portmann, B., Belorgey, D., Hill, M., Chambers, S., Teckman, J., Alexander, G. J., *et al.* (2010). A novel monoclonal antibody to characterize pathogenic polymers in liver disease associated with alpha-1-antitrypsin deficiency. *Hepatology* **52,** 1078–1088.

Mottonen, J., Strand, A., Symersky, J., Sweet, R. M., Danley, D. E., Geoghegan, K. F., Gerard, R. D., and Goldsmith, E. J. (1992). Structural basis of latency in plasminogen activator inhibitor-1. *Nature* **355,** 270–273.

Nar, H., Bauer, M., Stassen, J. M., Lang, D., Gils, A., and Declerck, P. J. (2000). Plasminogen activator inhibitor. 1. Structure of the native serpin, comparison to its other conformers and implications for serpin inactivation. *J. Mol. Biol.* **297,** 683–695.

Onda, M., Belorgey, D., Sharp, L. K., and Lomas, D. A. (2005). Latent S49P neuroserpin forms polymers in the dementia familial encephalopathy with neuroserpin inclusion bodies. *J. Biol. Chem.* **280,** 13735–13741.

Ordonez, A., Martinez-Martinez, I., Corrales, F. J., Miqueo, C., Minano, A., Vicente, V., and Corral, J. (2009). Effect of citrullination on the function and conformation of antithrombin. *FEBS J.* **276,** 6763–6772.

Parfrey, H., Dafforn, T. R., Belorgey, D., Lomas, D. A., and Mahadeva, R. (2004). Inhibiting polymerization: New therapeutic strategies for Z alpha1-antitrypsin-related emphysema. *Am. J. Respir. Cell Mol. Biol.* **31,** 133–139.

Pearce, M. C., Cabrita, L. D., Ellisdon, A. M., and Bottomley, S. P. (2007). The loss of tryptophan 194 in antichymotrypsin lowers the kinetic barrier to misfolding. *FEBS J.* **274,** 3622–3632.

Pearce, M. C., Morton, C. J., Feil, S. C., Hansen, G., Adams, J. J., Parker, M. W., and Bottomley, S. P. (2008). Preventing serpin aggregation: The molecular mechanism of citrate action upon antitrypsin unfolding. *Protein Sci.* **17,** 2127–2133.

Pedersen, K. E., Einholm, A. P., Christensen, A., Schack, L., Wind, T., Kenney, J. M., and Andreasen, P. A. (2003). Plasminogen activator inhibitor-1 polymers, induced by inactivating amphipathic organochemical ligands. *Biochem. J.* **372,** 747–755.

Purkayastha, P., Klemke, J. W., Lavender, S., Oyola, R, Cooperman, B. S., and Gai, F. (2005). Alpha 1-antitrypsin polymerization: a fluorescence correlation spectroscopic study. *Biochemistry.* **44**(7), 2642–2649.

Schulze, A. J., Baumann, U., Knof, S., Jaeger, E., Huber, R., and Laurell, C. B. (1990). Structural transition of alpha 1-antitrypsin by a peptide sequentially similar to beta-strand s4A. *Eur. J. Biochem.* **194,** 51–56.

Sharp, L. K., Mallya, M., Kinghorn, K. J., Wang, Z., Crowther, D. C., Huntington, J. A., Belorgey, D., and Lomas, D. A. (2006). Sugar and alcohol molecules provide a therapeutic strategy for the serpinopathies that cause dementia and cirrhosis. *FEBS J.* **273,** 2540–2552.

Sidhar, S. K., Lomas, D. A., Carrell, R. W., and Foreman, R. C. (1995). Mutations which impede loop/sheet polymerization enhance the secretion of human alpha 1-antitrypsin deficiency variants. *J. Biol. Chem.* **270,** 8393–8396.

Takehara, S., Zhang, J., Yang, X., Takahashi, N., Mikami, B., and Onda, M. (2010). Refolding and polymerization pathways of neuroserpin. *J. Mol. Biol.* **403,** 751–762.

Tew, D. J., and Bottomley, S. P. (2001). Probing the equilibrium denaturation of the serpin alpha(1)-antitrypsin with single tryptophan mutants; evidence for structure in the urea unfolded state. *J. Mol. Biol.* **313,** 1161–1169.

Tsutsui Y., Kuri B., Sengupta T., and Wintrode P. L. (2008). The structural basis of serpin polymerization studied by hydrogen/deuterium exchange and mass spectrometry. *J. Biol. Chem.* **283**(45), 30804–30811.

Wardell, M. R., Chang, W. S., Bruce, D., Skinner, R., Lesk, A. M., and Carrell, R. W. (1997). Preparative induction and characterization of L-antithrombin: A structural homologue of latent plasminogen activator inhibitor-1. *Biochemistry* **36,** 13133–13142.

Wu, Y., and Foreman, R. C. (1990). The effect of amino acid substitutions at position 342 on the secretion of human alpha 1-antitrypsin from Xenopus oocytes. *FEBS Lett.* **268,** 21–23.

Xue, Y., Bjorquist, P., Inghardt, T., Linschoten, M., Musil, D., Sjolin, L., and Deinum, J. (1998). Interfering with the inhibitory mechanism of serpins: Crystal structure of a complex formed between cleaved plasminogen activator inhibitor type 1 and a reactive-centre loop peptide. *Structure* **6,** 627–636.

Yamasaki, M., Li, W., Johnson, D. J., and Huntington, J. A. (2008). Crystal structure of a stable dimer reveals the molecular basis of serpin polymerization. *Nature* **455,** 1255–1258.

Yamasaki, M., Sendall, T. J., Harris, L. E., Lewis, G. M., and Huntington, J. A. (2010). Loop-sheet mechanism of serpin polymerization tested by reactive center loop mutations. *J. Biol. Chem.* **285,** 30752–30758.

Yu, M. H., Lee, K. N., and Kim, J. (1995). The Z type variation of human alpha 1-antitrypsin causes a protein folding defect. *Nat. Struct. Biol.* **2,** 363–367.

Zhang, Q., Law, R. H., Bottomley, S. P., Whisstock, J. C., and Buckle, A. M. (2008). A structural basis for loop C-sheet polymerization in serpins. *J. Mol. Biol.* **376,** 1348–1359.

Zhou, A., and Carrell, R. W. (2008). Dimers initiate and propagate serine protease inhibitor polymerisation. *J. Mol. Biol.* **375,** 36–42.

Zhou, A., Faint, R., Charlton, P., Dafforn, T. R., Carrell, R. W., and Lomas, D. A. (2001). Polymerization of plasminogen activator inhibitor-1. *J. Biol. Chem.* **276,** 9115–9122.

Zhou, A., Stein, P. E., Huntington, J. A., and Carrell, R. W. (2003). Serpin polymerization is prevented by a hydrogen bond network that is centered on his-334 and stabilized by glycerol. *J. Biol. Chem.* **278,** 15116–15122.

Zhou, A., Stein, P. E., Huntington, J. A., Sivasothy, P., Lomas, D. A., and Carrell, R. W. (2004). How small peptides block and reverse serpin polymerisation. *J. Mol. Biol.* **342,** 931–941.

CHAPTER EIGHTEEN

The Serpinopathies: Studying Serpin Polymerization *In Vivo*

James A. Irving,* Ugo I. Ekeowa,* Didier Belorgey,* Imran Haq,* Bibek Gooptu,† Elena Miranda,‡ Juan Pérez,§ Benoit D. Roussel,* Adriana Ordóñez,* Lucy E. Dalton,* Sally E. Thomas,* Stefan J. Marciniak,*,¶ Helen Parfrey,¶ Edwin R. Chilvers,¶ Jeffrey H. Teckman,|| Sam Alam,¶ Ravi Mahadeva,¶ S. Tamir Rashid,*,# Ludovic Vallier,# *and* David A. Lomas*,¶

Contents

1. Introduction to Serpin Polymers and the Serpinopathies—David Lomas 423
2. Biophysical Techniques to Assess Serpin Polymers Formed *In Vivo*—James Irving, Ugo Ekeowa, Didier Belorgey, and Imran Haq 424
   2.1. Quantitation, glycosylation, and proteolytic digestion 425
   2.2. Monomers and polymers 425
   2.3. Stable, native, and polymeric states 427
3. Assessment of Serpin Polymers by Electron Microscopy—Bibek Gooptu 428
   3.1. Electron microscopy 429
   3.2. Image collection, particle picking, and data processing 434
   3.3. Image processing in serpin polymer EM—Challenges and progress 435
4. Development of mAbs to Aberrant Conformers of α_1-Antitrypsin and Neuroserpin—Elena Miranda and Juan Pérez 436

* Department of Medicine, Cambridge Institute for Medical Research, University of Cambridge, Cambridge, United Kingdom
† ISMB/Birkbeck, Crystallography, Department of Biological Sciences, Birkbeck College, London, United Kingdom
‡ Dipartimento di Biologia e Biotecnologie 'Charles Darwin', Universitá di Roma La Sapienza, Piazzale Aldo Moro 5, Rome, Italy
§ Departamento de Biología Celular, Genética y Fisiología, Universidad de Málaga, Facultad de Ciencias, Campus de Teatinos, Malaga, Spain
¶ Respiratory Medicine Division, Department of Medicine, University of Cambridge School of Clinical Medicine, Addenbrooke's and Papworth Hospitals, Cambridge, United Kingdom
|| St. Louis University School of Medicine, Cardinal Glennon Children's Hospital, St. Louis, Missouri, USA
Laboratory for Regenerative Medicine, University of Cambridge, Cambridge, United Kingdom

Methods in Enzymology, Volume 501
ISSN 0076-6879, DOI: 10.1016/B978-0-12-385950-1.00018-3

5. Development of Cell Models to Assess the Polymerization of Antitrypsin—Adriana Ordóñez 441
5.1. Construction of α_1-antitrypsin expression plasmids 441
5.2. Cell cultures and DNA tranfections 441
5.3. Generation and characterization of stable CHO-K1 Tet-On cell lines expressing α_1-antitrypsin 442
5.4. Screening of surviving clones 442
5.5. Sandwich ELISA 443
6. Development of Cell Models to Assess the Polymerization of Neuroserpin—Elena Miranda, Juan Perez, and Benoit Roussel 443
6.1. Transient transfection of COS-7 cells 443
6.2. Stable cell lines overexpressing polymerogenic serpin mutants 444
7. Detection of the UPR and the OPR in the Serpinopathies—Lucy Dalton, Sally Thomas, Benoit Roussel, and Stefan Marciniak 445
7.1. Detection of the UPR 445
7.2. Detection of the OPR 447
8. Characterization of the Interaction Between Serpin Polymers and Neutrophils—Helen Parfrey and Edwin Chilvers 448
8.1. Neutrophil preparation 448
8.2. Neutrophil functional assays 449
9. The Use of Transgenic Mice to Assess the Hepatic Consequences of Serpin Polymerization. Jeff Teckman 451
10. The Use of Transgenic Mice to Assess the Pulmonary Consequences of Serpin Polymerization—Sam Alam and Ravi Mahadeva 453
11. Characterization of Serpin Polymerization Using iPS to Generate Hepatocyte-Like Cell Lines—Tamir Rashid and Ludovic Vallier 457
11.1. Introduction 457
11.2. Obtaining dermal fibroblasts from patient 457
11.3. Derivation and maintenance of human IPS cells 459
11.4. Characterization of hIPSCs 460
11.5. Generation of hepatocyte-like cells from hIPSCs 460
11.6. Validation of hepatic-like cells 461
Acknowledgments 461
References 461

Abstract

The serpinopathies result from point mutations in members of the serine protease inhibitor or serpin superfamily. They are characterized by the formation of ordered polymers that are retained within the cell of synthesis. This causes disease by a "toxic gain of function" from the accumulated protein and a "loss of function" as a result of the deficiency of inhibitors that control important proteolytic cascades. The serpinopathies are exemplified by the Z (Glu342Lys) mutant of α_1-antitrypsin that results in the retention of ordered polymers within the endoplasmic reticulum of hepatocytes. These polymers form the intracellular inclusions that are associated with neonatal hepatitis, cirrhosis, and hepatocellular carcinoma. A second example results from mutations in the neurone-specific serpin–neuroserpin to form ordered polymers that are retained as inclusions within subcortical neurones as Collins' bodies. These inclusions underlie the autosomal dominant dementia familial encephalopathy with neuroserpin inclusion bodies or FENIB. There are different pathways to polymer formation *in vitro* but not all form polymers that are relevant *in vivo*. It is therefore essential that protein-based structural studies are interpreted in the context of human samples and cell and animal models of disease. We describe here the biochemical techniques, monoclonal antibodies, cell biology, animal models, and stem cell technology that are useful to characterize the serpin polymers that form *in vivo*.

1. Introduction to Serpin Polymers and the Serpinopathies—David Lomas

The conformational diseases are characterized by the aggregation and tissue deposition of aberrant conformations of protein (Carrell and Lomas, 1997; Kopito and Ron, 2000). This is the predominant pathology in a group of diseases that we have termed the serpinopathies (Lomas and Mahadeva, 2002). The serpinopathies result from point mutations in members of the serine protease inhibitor or serpin superfamily. They are characterized by the formation of ordered polymers that are retained within the cell of synthesis. This causes disease by a "toxic gain of function" from the accumulated protein and a "loss of function" as a result of the deficiency of inhibitors that control important proteolytic cascades. The serpinopathies are exemplified by the Z (Glu342Lys), King's (His334Asp; Miranda *et al.*, 2010), Siiyama (Ser52Phe; Lomas *et al.*, 1993b), and Mmalton (53Phe del; Lomas *et al.*, 1992) mutations of α_1-antitrypsin that result in the retention of ordered polymers within the endoplasmic reticulum (ER) of hepatocytes. These polymers form the periodic acid Schiff (PAS)-positive, diastase-resistant inclusions that are associated with neonatal hepatitis, cirrhosis, and hepatocellular carcinoma (Eriksson *et al.*, 1986; Le *et al.*, 1992; Sveger, 1976). A second example results from mutations in the neurone-specific serpin–neuroserpin. This leads to the formation of ordered polymers that are retained within subcortical neurones to form PAS-positive, diastase-resistant inclusions that are termed Collins' bodies. These inclusions underlie the autosomal dominant dementia familial encephalopathy with neuroserpin inclusion bodies or familial encephalopathy with neuroserpin inclusion bodies (FENIB; Davis *et al.*, 1999, 2002).

There is a direct relationship between the rate of polymerization, the magnitude of the intracellular accumulation of a serpin, and the severity of disease (Dafforn *et al.*, 1999; Davis *et al.*, 2002; Miranda *et al.*, 2008). This is best illustrated for individuals with the dementia FENIB. In the original family with Ser49Pro neuroserpin (neuroserpin Syracuse), the affected family members had diffuse small intraneuronal inclusions of neuroserpin with an onset of dementia between the ages of 45 and 60 years. However, in a second family, with a conformationally more severe mutation (neuroserpin Portland; Ser52-Arg) and larger inclusions, the onset of dementia was in early adulthood, and in a third family, with yet another mutation (His338Arg), there were more inclusions and the onset of dementia in adolescence. The most striking example was the family with the most "polymerogenic" mutation of neuroserpin, Gly392Glu. This replacement of a conserved residue in the shutter region resulted in large multiple inclusions in every neurone, with affected family members dying by age 20 years. More recently, a fifth mutation has been described (Gly392Arg) that caused a profound intellectual decline in an 8-year-old girl, seizures and electrical brain activity in keeping with "epilepsy of slow-wave sleep (ESES)" (Coutelier *et al.*, 2008).

The characterization of conformers of neuroserpin or α_1-antitrypsin associated with disease requires an understanding of the folding pathway, transitions to the polymeric and latent conformers, and the mechanisms by which these intermediates and folded conformers are handled *in vivo*. There is strong evidence that mutants of both neuroserpin and α_1-antitrypsin are associated with a significant delay in folding (Baek *et al.*, 2007; Takehara *et al.*, 2010). This folding defect can be studied by treating the folded protein with urea or guanidine. However, in the case of α_1-antitrypsin, the polymers that form as a consequence of treatment with these agents are not recognized by the 2C1 monoclonal antibody (mAb) that recognizes the pathological polymers that form *in vivo* (Ekeowa *et al.*, 2010). This implies that the long-lived folding intermediate is efficiently degraded by the proteasome via the pathway of endoplasmic reticulum-associated degradation (ERAD; Liu *et al.*, 1997). Moreover, experiments performed *in vitro* with guanidine or urea produce conformers that, while interesting (Yamasaki *et al.*, 2008), are unlikely to be relevant to disease (Ekeowa *et al.*, 2010). Studies of human samples and cell and animal models of disease are essential to relate findings from the structural studies of proteins to the pathological species that cause disease. We have recently reviewed the structural studies required to characterize serpin polymers (Belorgey *et al.*, 2011). We now describe the biochemical techniques, monoclonal antibodies, cell biology, animal models, and stem cell technology that are useful to characterize the serpin polymers that form *in vivo*.

2. Biophysical Techniques to Assess Serpin Polymers Formed *In Vivo*—James Irving, Ugo Ekeowa, Didier Belorgey, and Imran Haq

Heterogeneous mixtures, such as biological fluids and tissue extracts, preclude the use of several biophysical techniques that are otherwise eminently suited to analysis of the conformational state of serpins in highly purified material (Dafforn *et al.*, 2004). The underlying problem with spectroscopic techniques, for example, is the inability to distinguish the contribution made by the protein of interest from other components of the mixture. In this regard, polyacrylamide gel electrophoresis (PAGE)-based techniques have particular utility when detection of the serpin of interest is possible by Western blot. It is possible to effectively evaluate the spectrum of serpin conformational states by subjecting biological material to a panel of electrophoresis systems: sodium dodecyl sulfate (SDS)–PAGE, nondenaturing (native) PAGE, and transverse urea gel PAGE. Crude source material that has been evaluated in this way includes bronchoalveolar lavage fluid (BALF) for α_1-antitrypsin (Elliott *et al.*, 1998; Mulgrew *et al.*, 2004) and Collins' bodies for neuroserpin (Onda *et al.*, 2005). Where antibodies for Western blot are not available, enrichment of the serpin component of the sample may still permit a gel-based evaluation.

2.1. Quantitation, glycosylation, and proteolytic digestion

The SDS–PAGE system, whether based on the Laemmli Tris/glycine recipe (Laemmli, 1970) or more modern Bis–Tris/MOPS/MES buffers, should be used as a starting point. As most serpins are in the 45–55 kDa molecular weight range, a 10% (w/v) acrylamide or 7.5–15% (w/v) acrylamide gradient gel, with a reducing agent such as dithiothreitol (DTT), gives good results. By including standards of known concentration on the gel, the amount of serpin present in the sample can be estimated. The glycosylation state of the protein can also be evaluated by reference to a purified standard that has been subjected to PNGase digestion. The SDS detergent used in the sample buffer dissociates serpin polymers and unfolds monomers, and the DTT reduces disulfide bridges. This enables the approximate size of the monomeric protein to be determined; generally this will be around 45 kDa for the nonglycosylated form of α_1-antitrypsin and around 7.5 kDa larger for the glycosylated form. An increase in size of 20 kDa or more under these conditions is indicative of a serine protease–serpin complex, and a decrease of 4 kDa (with the concomitant appearance of a 4 kDa fragment) indicates the presence of the reactive center-loop cleaved form (Fig. 18.1B). This gel system will not distinguish between conformations such as active monomer, latent monomer, or polymer.

2.2. Monomers and polymers

Native PAGE buffers lack the denaturing anionic detergent component SDS and separate proteins according to both their endogenous charge and volume. It should be noted that due to the reliance on intrinsic charge, a difference in migration pattern is only meaningful when the comparison is made between samples containing the same protein. Nondenaturing gels can be used to distinguish between monomers and oligomers of increasing

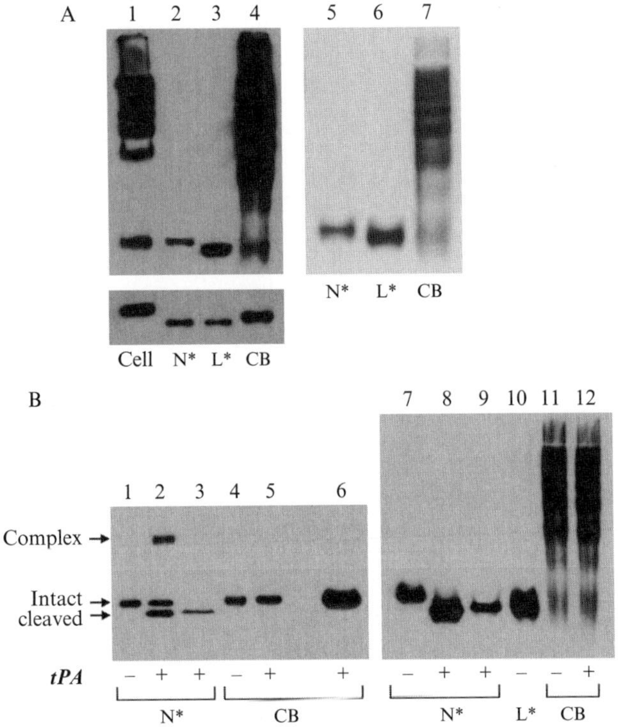

Figure 18.1 Latent neuroserpin in Collins' bodies. (A) Collins' bodies were solubilized by sonication in a Tris–HCl buffer containing 1% (v/v) Triton X-100 and analyzed by nondenaturing PAGE (upper left and right) and SDS–PAGE (lower left). Proteins were visualized by Western blot analysis with a polyclonal antineuroserpin antibody. The upper and lower left results were obtained by the use of a more sensitive developing substrate than the upper right result. Lane 1, S49P secreted from COS-7 cells (*cell*); lanes 2 and 5, native S49P prepared *in vitro* (*N**); lanes 3 and 6, latent S49P prepared *in vitro* (*L**); lanes 4 and 7, S49P extracted from Collins' bodies (CB). (B) Inhibitory activity of neuroserpin in Collins' bodies extraction was assessed by SDS (left) and nondenaturing (right) PAGE. Proteins were visualized by Western blot analysis. 0.2 mg/ml native S49P (lanes 2 and 8), 0.002 mg/ml native S49P (lanes 3 and 9), and 2 mg/ml extracted Collins' bodies (lanes 5, 6, and 12) were incubated with 1.7 μM tPA at 25 °C for 3 min (lanes 2 and 8) or 30 min (lanes 3, 5, 6, 9, and 12). For comparison, native S49P (lanes 1 and 7), latent S49P (lane 10), and extracted Collins' bodies (lanes 4 and 11) prior to treatment with tPA were loaded onto the gels. Neuroserpin and extracted Collins' bodies were loaded onto the gels at 0.04 μg (lanes 3 and 9), 0.1 μg (lanes 1, 7, and 10), 0.15 μg (lanes 4, 5, and 8), 0.2 μg (lanes 2, 11, and 12), and 1 μg (lane 6). Reproduced from Onda *et al.* (2005).

size due to a relatively consistent charge-to-size ratio. A subtle change of conformation can also lead to a shift in position on the gel. Native PAGE analysis has been used on extracts of Collins' bodies from patients with

FENIB, permitting the separation of polymers from the native and latent conformations (Fig. 18.1A; Onda *et al.*, 2005).

With nondenaturing PAGE, care should be taken to ensure that the running buffer is at the appropriate pH. For neuroserpin and α_1-antitrypsin, a basic 7.5% (w/v) acrylamide gel can be cast using the Laemmli recipe but lacking SDS. A 53 m*M* Tris, 68 m*M* glycine, pH 8.9, buffer is used at the cathode and 0.1 *M* Tris, pH 7.8, buffer at the anode. Different conditions should be used for neutral or basic serpins (Dafforn *et al.*, 2004). A method that can assist in the electrophoresis of basic proteins (Wittig *et al.*, 2006) utilizes Serva Blue G, which permits the separation of proteins according to their oligomeric state.

2.3. Stable, native, and polymeric states

A transverse urea gradient (TUG) gel is a native PAGE gel with a gradient of urea, ranging from 0 to 8 *M*, perpendicular to the direction of electrophoresis (Dafforn *et al.*, 2004). It is an approach that distinguishes components based on their relative stability to increasing concentrations of denaturant. Forms with marked stability, including the latent and cleaved conformations, remain compact and migrate a similar distance into the gel across a broad range of urea concentration. The native conformation of mammalian serpins is compact and migrates to a similar extent at very low concentrations of urea; however, this form undergoes a transition to a poorly migrating denatured species at higher concentrations of urea. Polymers remain both multimeric and stable in urea, retain their characteristic ladder appearance, and exhibit intermediate electrophoretic mobility (Figs. 18.2 and 18.3).

TUG gels are particularly useful for identification of serpin variants with altered stability by their corresponding change in migration profile. This is exemplified by a comparison of wild-type neuroserpin and the S49P Syracuse mutant, which showed a difference in behavior for both native and latent forms (Fig. 18.2). TUG gels have also been used for the characterization of intrapulmonary polymers of α_1-antitrypsin. In this case, BALF from a patient with Z α_1-antitrypsin clearly showed the appearance of polymers that were not present in a control M patient sample (Elliott *et al.*, 1998; Fig. 18.3).

If multiple samples are to be compared simultaneously, an 8 *M* urea native PAGE gel can be utilized instead. It has the additional advantage that it is simpler to make than a TUG gel, yet still permits the separation of stable serpin conformations from native and polymeric states. In this case, urea is simply dissolved in the acrylamide solution at the time of casting. The latent and cleaved forms of the serpin migrate the most rapidly, the polymer ladder to an intermediate extent, and the denaturant-sensitive native form the most slowly.

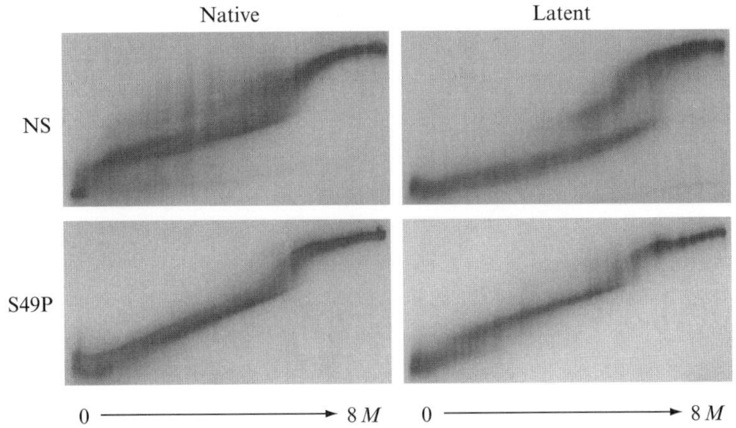

Figure 18.2 Transverse urea gradient PAGE of native and latent neuroserpin. Native neuroserpin (upper left), latent neuroserpin (upper right), native S49P (lower left), and latent S49P (lower right). The left and right of each gel represent 0 and 8 m urea, respectively. Reproduced from Onda *et al.* (2005).

3. Assessment of Serpin Polymers by Electron Microscopy—Bibek Gooptu

Electron microscopy (EM) has been used to characterize serpin polymers *ex vivo* and *in vitro* over two decades. Scanning EM has directly visualized the pathognomonic intracellular inclusion bodies of α_1-antitrypsin deficiency and FENIB. It shows how these bodies, composed entirely of mutant α_1-antitrypsin (Lomas *et al.*, 1992) and neuroserpin (Davis *et al.*, 1999), respectively, grossly distort the ER to form cisternae (Fig. 18.4A). Following disruption of such material by gentle sonication and higher salt conditions, transmission EM studies at much greater magnifications demonstrate that the protein molecules are self-associated into polymer chains (Fig. 18.4B). Subsequently, studies of plasma samples from patients with α_1-antitrypsin deficiency (Lomas *et al.*, 1993b) confirmed the presence of systemically circulating polymers sharing the characteristic "beads on a string" appearance (Fig. 18.4C).

Transmission EM can also provide 3D structural data, avoiding some limitations of other structural or biophysical techniques in studying serpin polymers. These include difficulties in separating oligomers into stable, homogeneously sized samples. Single particle-based EM techniques, in contrast, are able to analyze polydisperse samples in terms of different populations through a number of approaches. EM therefore offers a powerful approach that will likely be most useful when integrated with complementary data from other biochemical, biophysical, and structural studies.

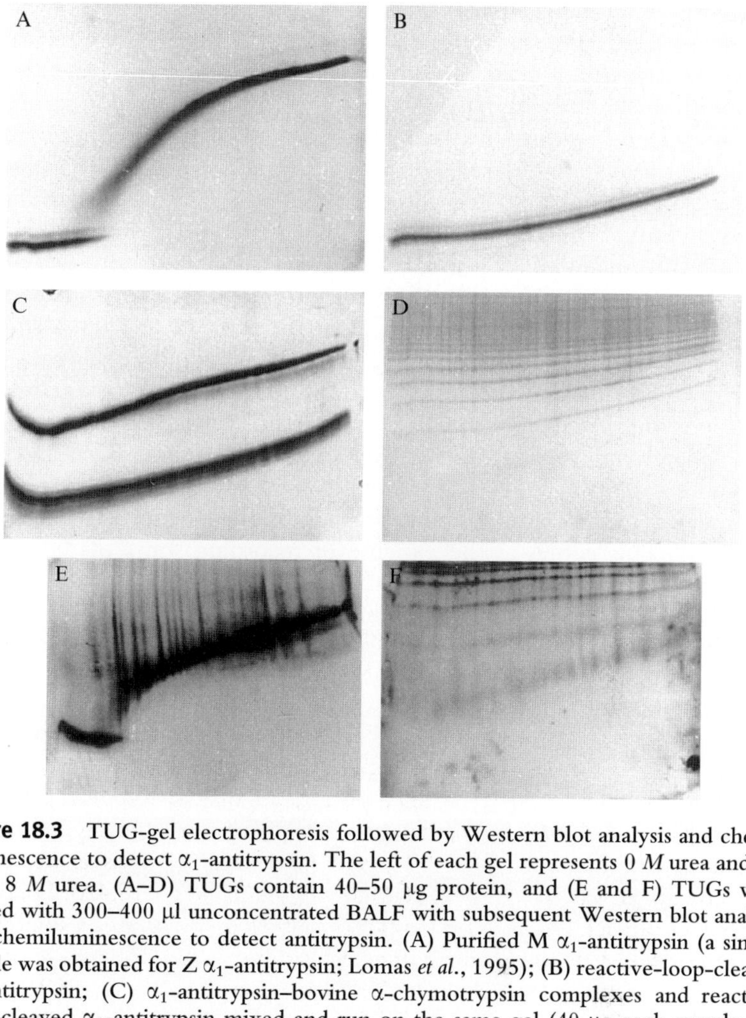

Figure 18.3 TUG-gel electrophoresis followed by Western blot analysis and chemiluminescence to detect α_1-antitrypsin. The left of each gel represents 0 *M* urea and the right 8 *M* urea. (A–D) TUGs contain 40–50 μg protein, and (E and F) TUGs were loaded with 300–400 μl unconcentrated BALF with subsequent Western blot analysis and chemiluminescence to detect antitrypsin. (A) Purified M α_1-antitrypsin (a similar profile was obtained for Z α_1-antitrypsin; Lomas *et al.*, 1995); (B) reactive-loop-cleaved α_1-antitrypsin; (C) α_1-antitrypsin–bovine α-chymotrypsin complexes and reactive-loop-cleaved α_1-antitrypsin mixed and run on the same gel (40 μg each sample; the upper band represents complexed α_1-antitrypsin, and the lower band represents reactive-loop-cleaved α_1-antitrypsin); (D) M α_1-antitrypsin polymer control generated by heating M α_1-antitrypsin at 60 °C for 3 h; (E) a normal α_1-antitrypsin unfolding transition in a bronchoalveolar lavage fluid specimen from a control M α_1-antitrypsin homozygote investigated for chronic cough; (F) characteristic profile of loop-sheet polymers in a BALF specimen from a Z α_1-antitrypsin homozygote with emphysema. Reproduced from Elliott *et al.* (1998).

3.1. Electron microscopy

EM is analogous to light microscopy, with the visible light beam substituted for an electron beam that travels through an evacuated column (Ruska, 1987). The series of glass lenses (condenser, objective, projector) are replaced by equivalent annular electromagnetic lenses. Biological samples are laid on grids inserted into the column and images focused onto a fluorescent screen using the objective lens. The interaction of the electron beam with the atoms within proteins is only slightly different to its interaction with atoms within the background (carbon or vitreous solution). Resultant images are therefore low contrast and have a low signal-to-noise ratio. For electron crystallography (Amos *et al.*, 1982) or helical reconstruction (DeRosier and Klug, 1968) techniques, the combination of phase information in the image with electron diffraction data allows high-resolution structure solution. Most samples readily form neither 2D crystals nor helices, and single particle techniques (for a detailed understanding, see Frank, 2006) may be used to generate a 3D volume analogous to an electron density map in the absence of diffraction data. To overcome the limitations of noisy image data, large numbers of particles must be selected (particle picking) representing projections of as many different object orientations as possible. To aid particle picking, contrast is usually enhanced by recording images within a range of slight defocus (e.g., 1.0–3.5 μm for cryo-EM). This comes at the expense of recording some of the highest resolution data available at true focus.

Initial EM visualization of protein samples is usually performed using negative staining, that further delineates the sample by staining of background to a greater extent than protein. This is a simpler technique than cryo-EM and typically provides a rapid qualitative assessment of sample

A

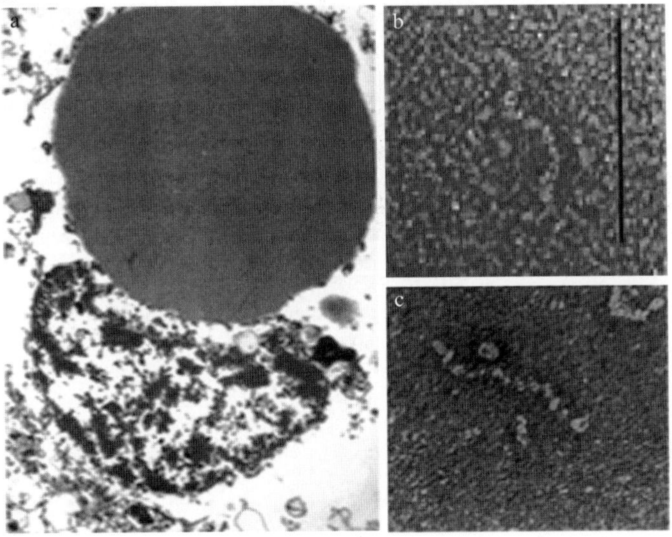

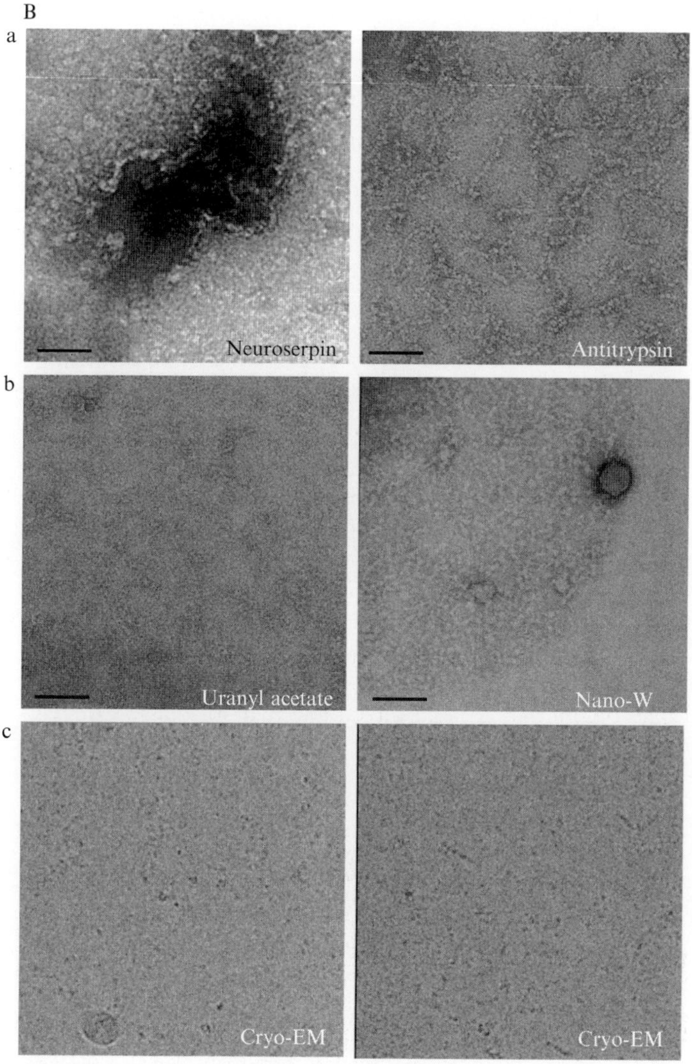

Figure 18.4 (Continued)

quality and particle characteristics. Moreover, small samples will tend to project with very low contrast and are therefore likely to be far better visualized in negative stain. Typically, ~4 μl of the sample solution is pipetted onto glow-discharged continuous carbon film grids and allowed to stand for between several seconds and 1 min before blotting with filter paper touched to the edge of the droplet. Four microliters of filtered stain

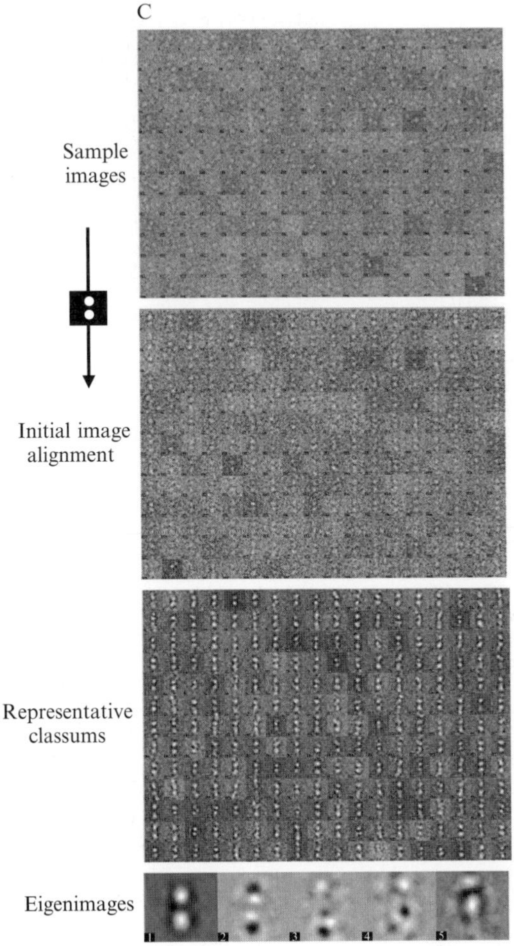

Figure 18.4 (A) EM visualization of *ex vivo* serpin polymer samples. (a) SEM of a Collins body, the hallmark lesion of familial encephalopathy with neuroserpin inclusion bodies (FENIB), causing cisterna formation within the rough endoplasmic reticulum; (b) material from purified inclusion bodies of Ser49Pro neuroserpin viewed by TEM after gentle sonication and phosphotungstate negative staining, using a Philips 208 electron microscope operating at 80 kV (scale bar represents 100 nm); (c) polymer of α_1-antitrypsin Siiyama, purified from plasma, viewed on the same apparatus using rotary shadowing. (B) Negative stain and cryo-EM of serpin polymers formed *in vitro*. Images recorded on CCD, using a Tecnai T12 electron microscope operating at 120 kV (a and b) and a Tecnai F20 operating at 200 kV (c). (a) Polymers of recombinant wild-type neuroserpin (left) and M α_1-antitrypsin (right), negatively stained using 2% (w/v) uranyl acetate. Images taken at relatively high defocus (~1 μm) to increase contrast. (b) Polymers of M α_1-antitrypsin visualized at ~850 nm defocus using 2% (w/v) uranyl acetate (left) and Nano-W® (methylamine tungstate, Nanoprobes Inc., NY, USA;

solution is then applied for a similar timespan before blotting dry as before. Precise timings for these steps are sample dependent and best optimized by trying a number of alternatives. Staining may sometimes be optimized by repeating the application. Buffer constituents that can complicate EM studies include sugars, glycerol, phosphate ions, high salt, denaturant, and detergents. These can be addressed by predialysis, or by blotting and washing on the grid with a more suitable buffer. It is advisable to screen a range of sample concentrations (e.g., 5- or 10-fold serial dilutions from 0.1 mg/ml) and a range of different stains. Varying glow discharge conditions for carbon grids can also aid sample adhesion to the grid.

Polymers of glycosylated serpins appear slightly less susceptible to nonspecific lateral aggregation than polymers formed from nonglycosylated species. Figure 18.4B(a) compares 2% (w/v) uranyl acetate staining of recombinant, nonglycosylated wild-type neuroserpin and glycosylated α_1-antitrypsin polymers at high defocus. α_1-Antitrypsin polymers are not easily visible in negative stain, although Nano-W appears to delineate them more readily than uranyl acetate (Fig. 18.4B(b)). Visualization is more effective in areas of stain pooling: near corners of grid squares or where polymer chains intersect in meshworks on the grid (Fig. 18.4B(b), right panel). Despite the different pH of the uranyl acetate (pH 4.0) and Nano-W (pH 6.8), the overall polymer morphology appears similar. However, only sections of single chains are useful for particle picking, so much of the meshed material cannot be utilized. In negative stain, the views indicate a preferential orientation on the grid surface for bead subunits along their long axis, projecting bead side views.

Negative staining is limited as a technique by dependence of the maximum resolution upon the stain molecule grain size (inverse relationship), effects on the sample arising from the pH of the stain solution, and sample

right). (c) Cryo-EM of M α_1-antitrypsin polymers on C-flat^{TM} holey carbon grids (Protochips Inc., PA, USA), images taken at $\sim$3 μm defocus and median filtered in BOXER. (C) Cryo-EM M α_1-antitrypsin polymer bead pair dataset and initial processing. Sample images: representative gallery of polymer segments including at least two polymer beads picked using BOXER, phase corrected according to contrast transfer function and grayscale inverted so beads appear light gray, and band-pass filtered using SPIDER; Initial alignment: Images following alignment to a simple model (two softened white circles) using SPIDER; representative classums: following alignment, images are binned into a prespecified number of subclasses according to multivariate statistical analysis (IMAGIC MSA). These are represented by the sum of all images within the class (classum). Variance across the dataset is illustrated by the eigenimages. The first image shown is the sum of all images in dataset (totalsum), subsequent images demonstrate principal component of variance followed by decreasingly major components. Variance around particles may represent different views, suboptimal alignment, or sample heterogeneity. (For color version of this figure, the reader is referred to the web version of this chapter.)

drying and flattening effects. Cryo-EM allows a hydrated sample to be visualized and avoids these limitations, usually at the expense of contrast between sample and background. The benefits may therefore be partly offset by the need to record data at higher defocus ranges (e.g., 1–3.5 μm) than required for negative stain EM. In this technique, the sample is pipetted onto grids without a continuous overlying surface. Excess solution is blotted from the grid while leaving it covered with the solution and the grid is then rapidly cooled in liquid helium that is in turn temperature-controlled by liquid nitrogen and a heating element. The solution vitrifies (cools as a glass without forming ice crystals) on the grid, preventing damage to protein structures and allowing clear visualization of particles where they lie in the grid spaces (usually circular holes). If blotting of excess solution is insufficient or excessive, the vitrified layer may be too thick or thin, respectively. The blotting step may therefore be carried out in an automated and more precisely reproducible manner using a system such as the Vitrobot (FEI, Oregon, USA) that regulates ambient humidity and blotting duration. Although serpin polymers might be predicted to be near invisible in cryo-EM conditions, in fact polymers of plasma-derived M α_1-antitrypsin can be seen at 0.1 mg/ml (Fig. 18.4B(c)). In cryo-EM, bead side views are less evident, with the majority of views closer to the short axes (tip or end views).

3.2. Image collection, particle picking, and data processing

Images are recorded on photographic film or charge-coupled device (CCD) plates. Film currently provides the highest resolution record and images a far larger field at the same magnification than CCD use, where plate size and detector spacing are limiting factors. However, film images must be digitized by high-resolution scanning before further processing, potentially introducing artifacts and other errors, while data is recorded from CCD plates directly into digital format.

Particle picking can be carried out in a manual (interactive) or more automated manner using programs including BOXER (EMAN suite; Ludtke *et al.*, 1999) and LABEL (MRC program suite; Crowther *et al.*, 1996) in conjunction with XIMDISP (Smith, 1999) viewing or FINDEM (Roseman, 2003). Automation of particle picking typically results in the selection of a significant proportion of random noise or unsuitable (e.g., overcrowded) particles. Once acknowledged, this may be addressed manually or, in a few cases, later during statistical sorting. Because individual transmission EM images have low signal:noise ratios, 3D reconstruction typically requires the picking of 10^3–10^5 particles for the dataset. Picked particles can be compiled as a series of numbered images into a "stack file" in .spd, .img, or .mrc formats for further image processing in programs such as SPIDER (Frank *et al.*, 1996), IMAGIC (van Heel *et al.*, 1996), or EMAN (Tang *et al.*, 2007). Since the different programs have different strengths, these can be used in combination so long as image formats are converted appropriately, and different conventions (e.g., differing coordinate axis labels) are accounted for where necessary.

Images collected via any optical detector are influenced by a function of the lens system known as the contrast transfer function (CTF) that modifies phase and amplitude in the observed image relative to the object from which it is formed. It can be derived computationally using the program CTFFIND3 (Mindell and Grigorieff, 2003) that requires a number of known parameters as inputs to analyze the Fourier transform of the image. The output can then be used for CTF correction. In all but particularly high-resolution EM datasets, there is no benefit to correction beyond that required to convert the sign of negative phase data (phase flipping).

3.3. Image processing in serpin polymer EM—Challenges and progress

An ideal subject for EM single particle reconstruction would be rigid (allowing alignment into highly homogeneous classes), large (several hundred kDa), project with clearly identifiable features (aiding accurate alignment), and possessing a high order of symmetry (each particle image represents *n* equivalent views for an *n*-fold symmetric object). By these standards, serpin polymers are highly flexible, with very small (~50 kDa) and thin (~40 Å width) subunits. The structural and asymmetric features of a serpin monomer at the practical resolution range (15–30 Å) are very subtle (Fig. 18.4C(a)) and will be harder to discern in real (noisy) images than is apparent in model projections.

Notwithstanding these very real challenges, we have found progress can be made using single particle techniques. First, equivalent segments of the polymer chain must be selected by boxing them out from the initial image. To limit the problems caused by polymer flexibility, segments should be chosen such that beads in a boxed segment align linearly with each other. Since such linear alignment is rarely observed for segments longer than four bead subunits, we have tended to work with short segments of four or even two beads selected by interactive particle picking in BOXER. This is currently more efficient than automated particle picking and manual dataset cleaning for these segments.

Initial alignment of selected particles works surprisingly well for datasets of α_1-antitrypsin polymer segments using simple schematic references such as a rectangle or a pair of ellipses or circles in SPIDER (Fig. 18.4C). Subsequent classification in IMAGIC demonstrates a wide range of angular orientations of bead projections, particularly in the case of cryo-EM data. Interbead spacing in the projected images is also a major source of variation across the dataset as demonstrated in the initial eigenimage (Fig. 18.4C, lowest panel, second image).

In further image processing steps, the heterogeneity of angular relationships between adjacent beads is likely to be a major factor limiting the resolution of 3D volume reconstructions. However, a number of low resolution models may be achievable that represent more strongly populated relative orientations and/or extremes. Moreover, quantifying the distributions of the angular relationships (and maximal observed interbead distances) observed in class averages will likely provide useful information in evaluating different structural models. On this basis, despite the technical challenges in applying single particle techniques to serpin polymers, our experience indicates the potential of such methods to provide important insights into the structural assembly of polymers. An example of this would be in quantifying the dynamic range of subunit orientation observed in cryo-EM data.

4. Development of mAbs to Aberrant Conformers of α_1-Antitrypsin and Neuroserpin—Elena Miranda and Juan Pérez

As part of our studies on serpin polymerization and disease, we have produced mAbs specific to polymers that constitute a valuable tool to detect and quantify neuroserpin and α_1-antitrypsin polymers in a variety of biological samples and techniques (Miranda *et al.*, 2008, 2010; Fig. 18.5).

When producing mAbs, we pay special attention to the type and quality of the antigen for the immunization and screening steps, to increase the chances of obtaining antibodies with the desired specificity (Kummer *et al.*, 2004). In order to produce our polymer-specific antibodies, we have used two types of antigen: polymers of Ser49Pro neuroserpin, made by heating purified protein obtained from *Escherichia coli* (Belorgey *et al.*, 2002; Onda *et al.*, 2005), and polymers of Z α_1-antitrypsin made by heating plasma purified protein at 0.2 mg/ml and 60 °C for 1 h in PBS buffer (Lomas *et al.*, 1993a).

1. *Immunization.* We use 7–10-week-old female Balb/c mice. Before starting the immunization, obtain a sample of preimmune serum from each animal. Mice are injected intraperitoneally with 10 μg of antigen emulsified with Freund's Complete Adjuvant (Sigma F5881), followed by four boost doses of 10 μg each emulsified in Freund's Incomplete Adjuvant (Sigma F5506) at 3 weeks intervals through the same route. Serum samples are collected from the tail veins 5 days after each injection, so the antibody title can be followed by antigen-mediated ELISA (Belorgey *et al.*, 2011). The prefusion boosts are given 3 days before spleen collection, by intravenous (morning) and intraperitoneal (evening) injection of 10 μg of antigen in PBS each, and 2 days before culling the mice by intraperitoneal injection only. At the end of the immunization procedure, mice are sacrificed by cervical dislocation, bled for the terminal serum sample, and spleen cells are

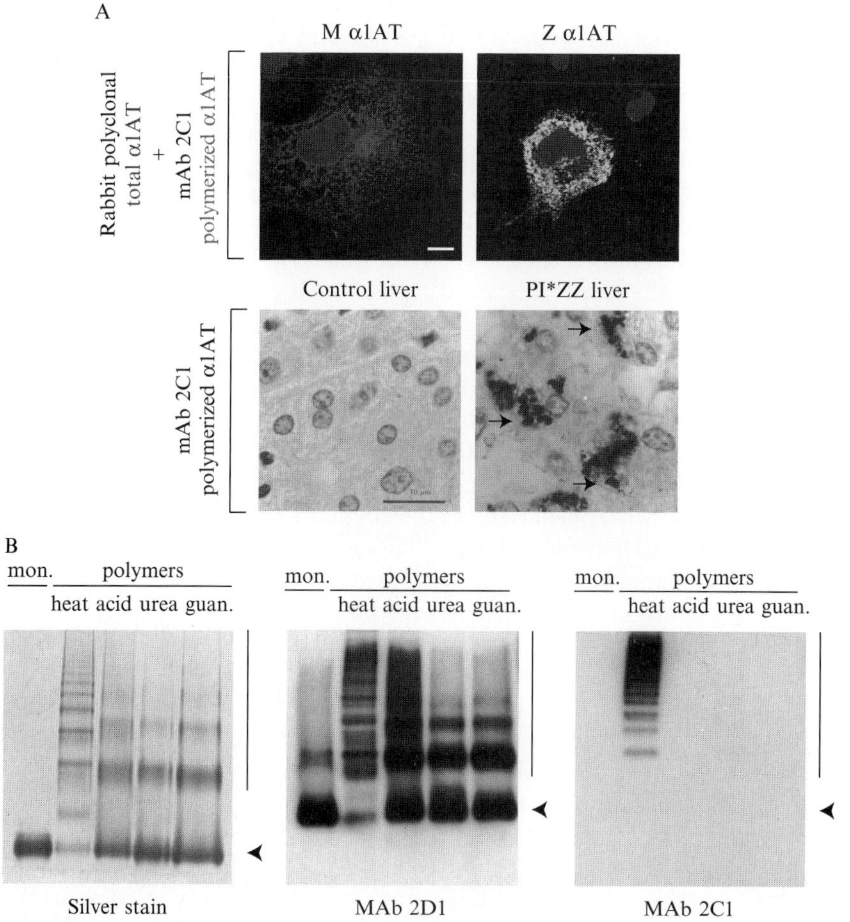

Figure 18.5 Detection of polymers with the 2C1 monoclonal antibody. The mAb 2C1 detects polymers of α_1-antitrypsin in immunocytochemistry. In (A) and (B), COS-7 cells transiently transfected with M or Z α_1-antitrypsin (a1AT) were fixed and immunostained with mAb 2C1 (green) and with a polyclonal antibody that recognizes all forms of α_1-antitrypsin (red). Only the merged panels are shown, with overlapping signals in yellow. The DNA is stained blue (DAPI). The polyclonal antibody detected α_1-antitrypsin in cells expressing the M and Z protein (both panels), while mAb 2C1 only reacted with Z α_1-antitrypsin contained in perinuclear inclusions (green and yellow staining in the right panel). In the lower panels, immunostaining of paraffin embedded liver sections with mAb 2C1. The mAb-detected polymers of α_1-antitrypsin within hepatocytes of a PI*ZZ individual (right panel, black arrows) but gave no signal in a control liver (left panel). Scale bars: 10 μm. (B) MAb 2C1 recognizes polymers of M α_1-antitrypsin prepared by heating but not by other denaturing conditions. Equal amounts of total protein were loaded in nondenaturing PAGE and analyzed either by silver stain (left panel) or by Western blot, either with mAb 2D1 (middle panel) or with mAb 2C1 (right panel) in the same membrane. The 2C1 mAb only recognized polymers prepared by heating of M α_1-antitrypsin. It gave no signal in polymers prepared by

aseptically collected in serum-free DMEM (Dulbeco's modified Eagle's medium, Sigma D5796). They are then frozen in freezing medium (8%, v/v DMSO in DMEM plus 20%, v/v fetal bovine serum, Sigma F2442) and kept in liquid nitrogen until needed.

2. On the fusion day, the spleen cells are thawed in DMEM-20 [DMEM plus sodium pyruvate (Sigma P3662), HEPES (Sigma H0887), and 20%, v/v fetal bovine serum (Sigma F2442)], and washed with DMEM before mixing with the myeloma cells (mouse myeloma cells, line P3-X63-Ag8-653 also washed with DMEM) at a ratio of four spleen cells per one myeloma cell (about 100×10^6 spleen cells and 25×10^6 myeloma cells). Cells are fused by dropwise addition of 1 ml 50% (w/v) polyethylene glycol (Sigma P7181) to the cell pellet at 37 °C with gentle stirring. The fusion products are then diluted in HAT-20 (DMEM-20 plus HAT Media Supplement, Sigma H0262) plus 5% (v/v) cloning supplement (PAA Ltd., F05-009) and distributed in 96-well plates at different densities. Nonfused myeloma and spleen cells are plated in control wells to assess the selectivity of the medium and the background IgG production, respectively, during the screening step. The use of cloning supplement avoids the need of using feeder cells to support the growth of the hybridoma cells at critical steps, such as fusion and cloning. Gradual removing of this (or the feeder cells) in posterior expansion steps should be fine for most hybridoma clones.
3. *Selection.* After 4–5 days in HAT-20, most myeloma cells should die and the first hybridoma colonies should be visible. On day 5 after fusion, add 50 μl/well of fresh medium, then feed the cells every other day by replacing half of the volume with fresh medium.
4. *Screening.* The primary screening is aimed to identify wells containing hybridoma colonies that produce specific antibodies. It can be done at about day 8–10 after fusion, if most of the colonies are large enough. We do this by antigen-mediated ELISA as described in Belorgey *et al.* (2011), using the same antigen used as immunogen. The positive wells are expanded into 24-well plates, grown for 1–2 days until 25–50% confluent, and subjected to secondary screening.
5. *Secondary screening.* This is the step where a correct strategy is needed in order to find mAb with the desired specificity. We use antigen-mediated ELISA to confirm the positivity of the expanded wells followed by sandwich ELISA using different serpin conformers as the antigen (native

treating M α_1-antitrypsin at low pH (4.5) or 3 *M* guanidine (guan.), and only a low signal was detected with longer exposures of polymers of M α_1-antitrypsin prepared by incubation with 4 *M* urea (results not shown). Vertical line: polymers; black arrowhead: monomers. Reproduced from Miranda *et al.* (2010) and Ekeowa *et al.* (2010). (See Color Insert.)

monomer, polymer, serpin–protease complex, cleaved serpin, latent monomer), as described below and in Belorgey *et al.* (2011). In this phase, the culture medium supernatant can also be tested in other techniques of interest, such as immunocytochemistry or Western blot. Cells at this stage are frozen as backup.

6. The cells from selected wells are next cloned by limiting dilution in HT-20 medium containing 10% (v/v) cloning supplement. We use the "diagonal cloning" method, where 25–250 μl (depending on cell density) of cells from a well of the 24-well plate are added to well 1A of a 96-well plate, then serially diluted 1:2 down the first column, and all wells from the first column serially diluted 1:2 along the rows. The cloning plates are screened after 10 days by antigen-mediated ELISA, and positive wells from the lowest density area are selected and expanded into 24-well plates in HT-20 medium (DMEM-20 plus HT Media Supplement, Sigma H0137) with cloning supplement, and subsequently into six-well plates, where cells are passed from HT-20 to HT/DMEM-20 and then to DMEM-20. The positivity and specificity of the hybridoma clones is monitored by ELISA during this expansion process.
7. Expansion of the selected clones into cell lines is achieved by culturing into bigger vessels and passing gradually to DMEM-10, low serum medium (down to 1%, v/v) and serum-free medium (serum-free hybridoma culture medium, Sigma 14610C). Frozen aliquots from these cultures are prepared and stored in liquid nitrogen as backups.

Once produced, the mAb are tested for use in ELISA, immunocytochemistry, immunoprecipitation, and Western blot. The protocols for these techniques are those commonly used and, except for the immunoprecipitation, have been described in detail elsewhere (Belorgey *et al.*, 2011). However, a few characteristics particular to serpin polymers are worth considering here.

Two formats are used for the ELISA assays: antigen-mediated and sandwich ELISA. The first one is used as the primary screening technique during mAb production, while the second is used to characterize the mAb against different conformers of the serpins (native, complex, cleaved, latent, polymer) and, once this is determined, to detect and quantify the antigen in a range of biological samples, such as cell lysates, cell culture medium supernatant, patient serum, and cerebrospinal fluid (Kennedy *et al.*, 2007; Miranda *et al.*, 2008, 2010; Nielsen *et al.*, 2007). In setting up this type of ELISA, we use our own antigen-purified polyclonal antibodies to coat the plates, and we produce our own purified antigen (neuroserpin or α_1-antitrypsin) for the standard curve. This means that all determinations of serpin concentration will depend on the value obtained for the purified standard by other means (Bradford assay absorbance at 280 nm) and hence is only a relative value. In our serpin polymer assays, the binding antibody is an antipolymer mAb against either neuroserpin (7C6; Miranda *et al.*, 2008) or

α_1-antitrypsin (2C1; Miranda *et al.*, 2010). We have also observed that the same mAb, when able to bind both monomeric and polymeric serpin, usually gives a higher signal with polymers, making it difficult to determine the exact amount of total serpin in samples where a mix of monomeric and polymeric serpin is present. This is also true in other techniques such as Western blotting, where the immunoreactivity of polymers is higher than that of the monomer after running a nondenaturing PAGE with equal amounts of protein in each lane.

Although polymers of the serpins produced under different conditions look similar on nondenaturing PAGE, the use of the 2C1 mAb against α_1-antitrypsin polymers has allowed us to demonstrate that polymers made from purified protein by heating are similar to those found in patients of α_1-antitrypsin deficiency and different from polymers produced *in vitro* by denaturants such as urea or guanidine (Ekeowa *et al.*, 2010; Fig. 18.5). These results have made it clear that several types of polymers can be formed from the same serpin and care should be taken when generalizing the structural features of a particular type of polymer.

Our antipolymer mAb have also been useful in immunocytochemistry, allowing us to determine the subcellular localization of polymers in colocalization studies using a panel of antibodies against resident proteins of different compartments of the secretory pathway (Miranda *et al.*, 2008, 2010; Fig. 18.5). When comparing the distribution of polymerized versus total serpin, we use our polymer-specific mAb first, followed by the corresponding secondary antibody and a fixation step (4%, v/v paraformaldehyde for 15 min at room temperature), in order to avoid its displacement by the polyclonal antibody used next to detect all conformations of the same serpin. It is important to know the specificity of each antibody and its limitations, as well as including appropriate controls for specificity, in order to extract the correct information. For example, our antipolymer mAb for neuroserpin (7C6) has a much greater affinity for polymers than for monomers of neuroserpin, as seen by sandwich ELISA, and it detects polymers only in immunocytochemistry if used at the right concentration (Miranda *et al.*, 2008). When imaging serpin polymers by immunofluorescence, the intensity of polymers is again higher and this should be taken into account when adjusting the gain.

Polymers of neuroserpin and α_1-antitrypsin can be specifically immunoprecipitated from lysates of cells overexpressing a polymerogenic variant. This is achieved by mixing the cell lysate [prepared in 10 m*M* Tris, 150 m*M* NaCl, pH 7.4, 1%, v/v Nonidet P40, 1 m*M* PMSF, EDTA-free protease inhibitor cocktail (Roche)] with the mAb bound to protein G Sheparose beads (Sigma P3296), from a few hours to overnight at 4 °C in a rotator mixer. We usually prepare the beads by mixing 1 μg of mAb per 30 μl of 50% slurry in 500 μl of PBS, then incubating at 4 °C for 2 h in a rotator mixer and washing them twice with PBS and once with lysis buffer. This amount of beads is enough to immunoprecipitate radiolabelled polymers from a 35 mm dish after a 15 min pulse with 1.3 MBq ^{35}S methionine and cysteine. When doing immunoprecipitation from steady state cultures, we scale this up and use ~100 μl of mAb beads for one 35 mm dish.

5. Development of Cell Models to Assess the Polymerization of Antitrypsin—Adriana Ordóñez

Point mutations in α_1-antitrypsin induce the protein to form polymers that are retained within the ER and associated with liver and pulmonary disease by a toxic gain/loss of function (Gooptu and Lomas, 2009). The most representative and best characterized is the common severe Z mutation (E342K) of α_1-antitrypsin (Lomas *et al.*, 1992). We have recently described a novel variant, α_1-antitrypsin King's (H334D) (Miranda *et al.*, 2010), that forms polymers even more rapidly than the Z variant. In this section, we detail the specific steps to develop stable transfected cells lines expressing these two polymerogenic mutations of α_1-antitrypsin.

5.1. Construction of α_1-antitrypsin expression plasmids

Plasmids expressing the cDNA for human E342K and H334D α_1-antitrypsin are generated from wild-type α_1-antitrypsin in the pcDNA3.1 vector by site-directed mutagenesis with the QuikChange Site-Directed Mutagenesis Kit (Stratagen, La Jolla, CA, USA). For generation of stable CHO-K1 Tet-On α_1-antitrypsin cell lines, wild-type, Z (E342K) and H334D α_1-antitrypsin are subsequently cloned into the pL1180 plasmid using the *Kpn*I and *Not*I restriction sites, and then subcloned into the pTRE2-hyg vector (Clontech, BD Biosciences, Oxford, UK) using the *Not*I and *Bam*HI sites.

5.2. Cell cultures and DNA tranfections

Purchase CHO-K1 Tet-On cell lines from Clontech, BD Biosciences. Grow the cells in the recommended cell culture medium: 90% DMEM high glucose (4.5 g/l), supplemented with 10% (v/v) Tet System Approved FBS (fetal bovine serum), 4 m*M* L-glutamine, 1% (v/v) nonessential amino acids, 1% (w/v) penicillin/streptomycin, and 200 μg/ml G418, at 37 °C and 5% (v/v) CO_2 in a humidified incubator.

To transfect cells, we use the following protocol:

1. The day before transfection, plate 2×10^5 cells/well in a six-well plate containing 2 ml of the appropriate complete growth medium. Cell density should be 70–80% confluent on the day of transfection.

2. For each well of cells to be transfected, dilute 1 μg of DNA (pTRE-2hyg plasmid with wild-type, Z or H334D α_1-antitrypsin) into 500 μl of Opti-MEM® I Reduced Serum Medium without serum, and then add 4 μl of Lipofectamine™ LTX. Mix gently and incubate for 25 min at room temperature to allow the formation of DNA–Lipofectamine™ LTX complexes.
3. Remove growth medium from cells and replace with 2 ml of growth medium (without G418). Add 500 μl of the DNA–Lipofectamine™ LTX complexes directly to each well containing cells and mix gently.
4. Incubate the cells at 37 °C in a CO_2 incubator for 18–24 h posttransfection.

5.3. Generation and characterization of stable CHO-K1 Tet-On cell lines expressing α_1-antitrypsin

The following protocol describes the development of Stable Tet-On cell lines.

1. Twenty-four hours after transfection, remove the transfection medium from the cells and wash twice in 1 ml of PBS.
2. Trypsinize cells and perform limiting dilutions (from 1:10 to 1:10^5) of transfected cells in 10-cm culture dishes, each containing 10 ml of complete growth medium (with G418).
3. Allow cells to grow during 48–72 h, and then add the appropriate selection antibiotic, in this case hygromycin (500 μg/ml).
4. Replace medium with fresh complete medium containing the selection antibiotic (hygromycin) every 4 days. After about 5 days, cells that have not taken up the plasmid should start to die.
5. Pick isolated and healthy clones and transfer them to individual 24-well plates, containing 0.5 ml of complete growth medium.

5.4. Screening of surviving clones

Once the cell clones are ready, the last step is to analyze the expression of α_1-antitrypsin within the cells.

1. Screen clones once they reach 50–80% confluence in a six-well plate.
2. α_1-Antitrypsin expression is induced typically for 2 days with doxycycline (1 μg/ml).
3. Expression levels are measured by sandwich ELISA (see below).
4. Select clones with the highest fold induction (highest expression with lowest background).
5. Freeze stocks of each clone as soon as possible after expanding the culture.

5.5. Sandwich ELISA

The expression levels and conformer of α_1-antitrypsin expressed by the stable clones can be assessed by sandwich ELISA, using the specific antibodies used for the detection step, as previously described (Belorgey *et al.*, 2011; Miranda *et al.*, 2010).

6. Development of Cell Models to Assess the Polymerization of Neuroserpin—Elena Miranda, Juan Perez, and Benoit Roussel

We have used cells overexpressing serpins to analyze the effects of mutations of neuroserpin that cause human disease (Davies *et al.*, 2009; Kröger *et al.*, 2009; Miranda *et al.*, 2004, 2008) and other artificially designed variants aimed to test hypotheses on serpin polymerization. We use two main types of cell systems in our studies: transient overexpression of serpin mutants in COS-7 cells and stable cell lines in which neuroserpin is overexpressed under the control of an inducible promoter.

6.1. Transient transfection of COS-7 cells

We use the pcDNA 3.1 plasmid (Invitrogen) encoding wild-type or mutant variants of neuroserpin.

1. Prepare six well plates by coating them with poly-L-lysine. Incubate the wells with 0.5 ml of poly-L-lysine (Sigma P1524) solution at 0.1 mg/ml in water for 5 min, rinse with sterile water, and leave to dry for 2 h in a sterile air flow cabinet. Once dry, the plates can be stored for several days at room temperature in sterile conditions.
2. Plate the cells so they will be around 90% confluent at the moment of transfection. We plate 200,000 cells in 2 ml of DMEM (Sigma 6429) plus 20% (v/v) FBS (Sigma) per well, and transfect them the day after plating.
3. *Cell transfection.* We use Lipofectamine 2000 (Invitrogen 11668027), following the manufacturer's protocol. Briefly, for each well mix 4 μg of DNA and 10 μl of Lipofectamine into two separate tubes with 250 μl of Opti-MEM I each (Invitrogen 31985047), then mix together and let react for 20 min at room temperature. Add dropwise to the cells, over the normal culture medium, and incubate for 4–6 h at 37 °C before changing the medium to 2 ml of Opti-MEM I per well. Transfected cells and conditioned culture media are collected after 1, 2, or 3 days for analysis. The use of serum-free medium allows the analysis of the

supernatant without the interference of serum proteins. A fraction of the total 4 μg DNA per well can be substituted with a plasmid expressing a control protein (e.g., luciferase) that can be detected by Western blot or any other appropriated technique to assess the efficiency of transfection.

6.2. Stable cell lines overexpressing polymerogenic serpin mutants

Since neuroserpin polymerization causes human dementia (Davis *et al.*, 1999) through a gain of toxic function, our strategy has been to create cell lines overexpressing polymerogenic variants of the protein under the control of an inducible promoter. We use the Tet-On system with the pTRE-Tight vector (both from Clontech), which allows good expression levels upon treatment with doxycycline with minimum or no leakiness. Our cell line of choice was the premade Tet-On PC12 (rat pheochromocytoma) from Clontech, since these cells can be differentiated into a neuronal phenotype by culturing them over collagen and treating them with nerve growth factor (NGF).

1. PC12 Tet-On cells are plated on poly-L-lysine and transfected using Lipofectamine 2000 in a similar manner to that described above for transient transfection of COS-7 cells. The day after transfection, cells are replated at low density in 10 cm dishes.
2. Cell lines are prepared following the instructions of the Tet-On/Tet-Off system manual. Briefly, after waiting for two cell cycles, cells are cultured in selective medium, using the antibiotic of choice that was cotransfected with the pTRE-Tight vector containing the protein of interest. In our case, this is hygromycin. It is recommended to isolate around 50 clones for each cell line that can be screened for neuroserpin expression by sandwich ELISA (Belorgey *et al.*, 2011). The positive clones are then evaluated by Western blot and two to four clones of each line are expanded and cell aliquots frozen for future use.

These cell lines have been used to show the formation of neuroserpin polymers by mutants that cause dementia, to a degree that correlates with the disease phenotype associated to each mutant (Miranda *et al.*, 2008). They have also been used to show that polymer accumulation activates the ER overload response (EOR) or ordered protein response (OPR) without eliciting the canonical unfolded protein response (UPR; Davies *et al.*, 2009), and to assess the roles of the proteasome and autophagy pathways in the degradation of wild-type and mutant neuroserpin (Kröger *et al.*, 2009).

7. Detection of the UPR and the OPR in the Serpinopathies—Lucy Dalton, Sally Thomas, Benoit Roussel, and Stefan Marciniak

The signals elicited by ER dysfunction are determined by the nature, intensity, and duration of aberrant protein accumulation (Ekeowa *et al.*, 2009; Marciniak and Ron, 2006). When improperly folded proteins sequester the chaperone BiP away from the ER stress signaling molecules PERK, IRE1, and ATF6, a homeostatic mechanism called the UPR is activated (Marciniak and Ron, 2006). This occurs, for example, when a truncated version of neuroserpin is expressed to high levels in the cell (Davies *et al.*, 2009). Treatment with the glycosylation inhibitor tunicamycin at 2 μg/ml for 8 h serves as a specific positive control for ER stress and the UPR (Marciniak *et al.*, 2004). In contrast, when proteins accumulate in the ER but fail to sequester chaperones, for example, polymers of Z α_1-antitrypsin or polymers of G392E neuroserpin, we have not detected UPR activation (Davies *et al.*, 2009). However, we do detect an NFκB signal reminiscent of the EOR (Pahl and Baeuerle, 1995), for which we have proposed to name the term OPR (Davies *et al.*, 2009).

The proximal signals of ER stress include the splicing by IRE1 of the mRNA of *XBP1*, a UPR transcription factor, and the phosphorylation of eIF2a by PERK (Marciniak and Ron, 2006). Splicing of *XBP1* mRNA can be detected by RT-PCR across the spliced intron, while eIF2α phosphorylation can be detected by Western blot. The UPR can be followed by detecting changes in the expression levels of ER stress signaling mediators (e.g., GADD34 and CHOP) and of ER chaperones (e.g., BiP and Grp94); however, these methods require high levels of ER stress, which are not readily achieved by the overexpression of misfolded protein. For this reason, it is often necessary to use more sensitive reporter assays (Davies *et al.*, 2009).

7.1. Detection of the UPR

7.1.1. RT-PCR and XBP1 splicing assay

1. Total RNA can be isolated from mammalian cells using commercial kits (e.g., Qiagen RNeasy Isolation kit, Qiagen). One microgram of total RNA is used as the template for first strand total cDNA synthesis (Promega).
2. This cDNA is used as the template from which to amplify a portion of the XBP1 cDNA using the primers XBP.1S (AAACAGAGTAGCAG-CACAGACTGC) and XBP1.AS (TCCTTCTGGGTAGACCTC-TGGGAG) (Davies *et al.*, 2009; Marciniak *et al.*, 2004). The PCR protocol used is anneal at 68 °C × 30 s, extend 72 °C × 30 s, denature 95 °C × 10 s 35 cycles.

3. The unspliced *XBP1* mRNA generates a PCR product of 480 bp and the spliced cDNA is 454 bp. These can be resolved on a 2% w/v TAE agarose gel.

7.1.2. Western blot for UPR mediators and targets

1. Cells are collected in PBS 1 m*M* EDTA and washed with PBS.
2. The cell pellet is resuspended in 5 volumes of buffer H (10 m*M* HEPES, pH 7.9, 50 m*M* NaCl, 500 m*M* sucrose, 0.1 m*M* EDTA and 0.5%, v/v Triton X-100, 1 m*M* DTT, 1 m*M* PMSF, 1× Complete^{TM} protease inhibitor cocktail (Roche)) and incubated on ice for 5 min. When phosphorylation of eIF2α is to be detected, one must include phosphatase inhibitors in buffer H (10 m*M* tetrasodium pyrophosphate, 17.5 m*M* *b*-glycerophosphate, 100 m*M* NaF; Harding *et al.*, 2000; Novoa *et al.*, 2001).
3. Nuclei are pelleted by centrifugation at 800×*g* for 10 min. The supernatant is taken as the cytosolic fraction.
4. Nuclei are washed in buffer A (10 m*M* HEPES, pH 7.9, 10 m*M* KCl, 0.1 m*M* EDTA, 0.1 EGTA, 1 m*M* DTT with protease inhibitors) and then soluble nuclear proteins are extracted in four pellet volumes of buffer C (10 m*M* HEPES pH 7.9, 500 m*M* NaCl, 0.1 m*M* EDTA, 0.1 m*M* EGTA, 0.1% NP40, 1 m*M* DTT, and protease inhibitors). The nuclear proteins are best extracted by vortexing for 15 min at 4 °C in the cold room. The nonextractable proteins and DNA are pelleted by centrifugation at 16,000×*g* for 10 min and the supernatant is taken as extractable nuclear proteins.
5. Fifty milligrams of cytosol and nuclear extracts are subjected to SDS–PAGE and transferred onto nitrocellulose membrane.
6. The membrane of the cytoplasmic proteins is then probed with antibodies to KDEL (cat# SPA-827, Stressgen), which detects the chaperones BiP/Grp78 and Grp94; to GADD34 (PP1R15a, cat# PAB6009, Abnova); and to phosphorylated eIF2α (cat# 9721, Cell Signaling). The nuclear extracts are probed for CHOP/GADD153 (cat# 2895, Cell Signaling).

7.1.3. Luciferase assay for ATF6 signaling

1. Cells are plated at 2.5 × 10^5 cells/well in six-well plate precoated with 0.1 mg/ml poly-L-lysine.
2. The following day, media is changed for 2 ml of Opti-Mem and the cells are transfected with 50 ng of pRL-TK (renilla-luciferase) plasmid and 2 μg of p(5×)ATF6-luc (firefly-luciferase) plasmid per well using Lipofectamine 2000. These plasmids were gifts from Dr. Timothy Weaver (Cincinnati Children's Hospital Medical Center, Ohio, USA). The

plasmids are diluted in 250 μl of Opti-Mem and 6.5 μl of Lipofectamine 2000 is diluted in 250 μl of Opti-Mem. After 10 min, plasmids and Lipofectamine solutions are mixed dropwise and left at room temperature for 30 min before being added to the cells for between 6 and 9 h. The cells are then transferred back to serum-containing growth media.
3. Twenty-four hours after transfection, cells are lysed using the Dual-Luciferase Reporter (Promega) assay protocol.
4. Both firefly- and renilla-luciferase activities are measured using a Glomax Luminometer (Promega). Firefly activity is calculated relative to renilla transfection efficiency and taken as a reporter of ATF6 activity.

7.2. Detection of the OPR

7.2.1. Ca^{2+} flux measurements

1. PC12 cells are plated in 10 cm dishes at 10^6 cells/dish. The media is removed and the cells are washed with calcium-free PBS. Four point five milliliters of HBBSS buffer (HEPES and bicarbonate-buffered saline solution: NaCl 116 m*M*, KCl 5.4 m*M*, $CaCl_2$ 1.8 m*M*, $MgSO_4$ 0.8 m*M*, NaH_2PO_4 1.3 m*M*, HEPES 12 m*M*, D-Glucose 5.5 m*M*, $NaHCO_3$ 25 m*M*, MS-glycine 0.1%, pH 7.45) is then added and the cells are loaded with 10 μ*M* of Fura-2AM (Molecular Probes) diluted in 500 μl total of HBBSS. This mixture is added dropwise onto cells and incubated for 30–40 min at 37 °C. The cells are washed with calcium-free PBS, harvested, and pelleted. We resuspend cells in 300 μl of HBBSS and use 100 μl per measurement.
2. The cells are analyzed fluorimetrically (excitation: 340, 380 nm; emission: 510 nm).
3. Calcium is mobilized by the addition of 1 μ*M* thapsigargin and the measurements continued to a relative increase in fluorescence that was stable over time. When cells expressing wild-type or polymerogenic neuroserpin are analyzed in this way, the cells expressing polymer-forming mutants show lower levels of thapsigargin-induced calcium release, which we interpret as depletion of ER calcium stores. All experiments must be performed in triplicate.

7.2.2. Luciferase assay for NFκB ATF6 signaling

1. This assay is performed similar to the ATF6 reporter assay (Section 7.1.3), but in place of the p(5×)ATF6-luc reporter plasmid, we use 2 μg of pELAM1-Luc (firefly-luciferase) plasmid.
2. In this way, it is possible to demonstrate that NFkB signaling from cells expressing polymer-forming mutants can be abolished by the

pretreatment with cell permeable calcium chelators, for example, BAPTA-AM (Sigma).

8. Characterization of the Interaction Between Serpin Polymers and Neutrophils—Helen Parfrey and Edwin Chilvers

Neutrophils are critical effector cells of the innate immune response, with specialized characteristics that allow them to migrate rapidly from the circulation to sites of tissue injury or infection. Here, they are essential for the killing of invading pathogens through release of their toxic granules and subsequent restoration of the normal tissue environment (Cowburn *et al.*, 2008). Neutrophils are also recognized to play a key role in the adaptive immune response, for example, in antigen processing and in the immunobiology of tumors (Müller *et al.*, 2009). As a consequence, the production, release, homing, priming, and removal of these cells are events that are all tightly regulated. Many inflammatory conditions are characterized by dysregulated neutrophil accumulation and activation. Unsurprisingly, serpin–protease complexes have been reported to be chemotactic for human neutrophils (Banda *et al.*, 1988) and are proinflammatory in cell culture (Kurdowska and Travis, 1990). Moreover, polymers of α_1-antitrypsin, which are present in the lung lining fluid of patients with Z α_1-antitrypsin deficiency (Elliott *et al.*, 1998), also display neutrophil activating properties *in vitro* (Mulgrew *et al.*, 2004; Parmar *et al.*, 2002) and *in vivo* (Mahadeva *et al.*, 2005).

The methods below detail a way of preparing basally unprimed pure human neutrophils and initial options for testing the capacity of serpin polymers to interact with these cells. More extensive characterization would involve assessing activation of the NADPH oxidase, alterations in the receptor and adhesion molecule repertoire on the neutrophil surface, pathogen killing, and changes in the rate of constitutive apoptosis. One important consideration in all such assays is the potential for contamination by lipopolysaccharide (LPS), which has a profound effect on most aspects of neutrophil biology. Stringent testing and elimination of LPS is essential (Parmar *et al.*, 2002).

8.1. Neutrophil preparation

The following preparation is recommended, performed under sterile conditions and, unless otherwise stated, at room temperature (Haslett *et al.*, 1985).

1. Whole blood is added to 3.8% (w/v) sterile sodium citrate (Martindale Pharmaceuticals) to a final concentration of 10% (v/v) and centrifuge at 300×*g* for 20 min to separate the cells from plasma. Avoid contact with heparin, glass, or polystyrene.
2. Transfer the platelet-rich plasma (PRP) to clean 50 ml Falcon tubes for the preparation of platelet-poor plasma (PPP) and autologous serum. For autologous serum, transfer 10 ml of PRP to a sterile glass vial, add 220 μl of 10 m*M* $CaCl_2$, and incubate at 37 °C for up to 1 h. Centrifuge the remaining PRP at 1400×*g* for 20 min and decant the resulting PPP.
3. For each 10 ml of the cell pellet, add 2.5 ml of 6% (w/v) Dextran T500 (Pharmacosmos A/S), gently mix, and allow to sediment for 30–45 min. Without disturbing the red cell pellet, carefully aspirate the leukocyte-rich supernatant and centrifuge at 256×*g* for 5 min.
4. Separate the granulocytes from the total leukocytes using a 51–42% discontinuous plasma/Percoll gradient. Dilute cold Percoll (GE Healthcare) with 10% (v/v) cold sterile saline and use as follows: 42% plasma/Percoll—840 μl 90% Percoll and 1160 μl PPP, 51% plasma/Percoll—1020 μl 90% Percoll and 980 μl PPP. Vortex to mix. Resuspend the leukocyte-rich cell pellet (step 3) in 1 ml of PPP, place in a clean 15 ml Falcon tube, and underlay first the 42%, then the 51% plasma/Percoll solutions with a glass Pasteur pipette. Centrifuge at 150×*g* for 14 min with both the brake and acceleration speeds set to zero so as not to perturb the gradient. Harvest the neutrophils from the 42%/51% plasma/Percoll interface with a plastic Pasteur pipette.
5. Wash the neutrophils with PPP and centrifuge at 256×*g* for 5 min. This is followed by Dulbecco's PBS (without calcium and magnesium, Sigma, UK) and finally Dulbecco's PBS (with calcium and magnesium, Sigma, UK). Resuspend the neutrophils as indicated.
6. Prepare a cytospin to assess neutrophil purity and morphology by light microscopy.

8.2. Neutrophil functional assays

Neutrophils are highly motile and malleable. These properties are essential for their migration through and between vascular endothelial cells and into infected or injured tissues (Ley *et al.*, 2007). Cell polarization, chemokinesis, and chemotaxis are all essential components of this response and can be triggered readily by agents such as IL8, fMLP, and TNFα (Cowburn *et al.*, 2008). Neutrophil shape change is an excellent and easily detected indicator of neutrophil priming and activation, chemotaxis assays indicate specific receptor-mediated neutrophil interactions, and the MPO assay allows assessment of neutrophil degranulation.

8.2.1. Neutrophil shape change

1. Prepare a concentration range of the test substance and a positive control (e.g., 100 n*M* fMLP) in Dulbecco's PBS with calcium and magnesium (Sigma, UK) (PBS+). Add 50 μl of each test solution in triplicate to BD Falcon™ flexible 96-well plates.
2. Add 100 μl of purified neutrophils, resuspended at 7.5 × 10^6 ml^{-1} in PBS+ to each well. Incubate at 37 °C, 5% (v/v) CO_2 for 10–90 min.
3. Terminate the reaction by adding 250 μl of ice cold 1× CellFix™ (BD Biosciences).
4. The proportion of neutrophils that have undergone polarization is determined by light microscopy and/or flow cytometry (BD FACS Calibur™) as detailed (Qu *et al.*, 1995).

8.2.2. Neutrophil chemotaxis

Chemotaxis describes the directional migration of a neutrophil in response to a stimulus gradient. Several methods have been developed to assess chemotaxis including gel invasion assays, shear flow chambers with an endothelial cell monolayer, migration under agarose, and direct observation cell tracking chambers (Frow *et al.*, 2004). While these offer the advantage of a more physiologically relevant environment to assess cell migration, these methods are limited as they are unsuitable for high-throughput screening, are labor intensive, and may require specialist equipment for cell imaging. An alternative and more commonly used approach is the filter migration assay, based upon a modified Boyden Chamber (Falk *et al.*, 1980). This system allows rapid screening and characterization of chemoattractants using small volumes of cells and reagents.

1. Prepare a concentration range of the test chemoattractant in IMDM (Sigma, UK) containing 1% (v/v) autologous serum. In addition, include maximally effective concentrations of one or more known neutrophil chemoattractants IL8 (100 ng/ml), fMLP (10 n*M*), or C5a (100 n*M*) for comparison. Add 29 μl aliquots of the test solutions in triplicate to the 96-well base plate.
2. Place ChemoTx® 206-5 polyvinylpyrrolidone (PVP)-coated filter (Neuro Probe Inc., USA) over the plate ensuring contact with the chemoattractant.
3. Resuspend the purified neutrophils at 5 × 10^6 ml^{-1} in IMDM containing 1% (v/v) autologous serum and add 50 μl to the filter upper well. Cover and incubate at 37 °C, 5% (v/v) CO_2 for typically 90 min.
4. At the end of the incubation, carefully remove the filter and transfer the contents of the lower wells to clean Eppendorf tubes. Wash the lower wells with 29 μl of 1× trypsin/EDTA (Invitrogen) to remove any adherent neutrophils, add to the Eppendorf, and repeat.

5. Manually count the number of migrated cells using a standard hemocytometer.
6. Chemokinesis (nondirectional migration) must be assessed independently within the same assay to exclude a nonspecific effect on cell motility. This is measured by adding an identical concentration of the test solution to both the upper and lower wells simultaneously to obliterate any chemotactic gradient.

8.2.3. Myeloperoxidase assay

This is a measure of neutrophil degranulation, which is the principal constituent of the primary or azurophil granule.

1. Preincubate purified neutrophils (12.5×10^6 ml^{-1} in PBS+) in the presence and absence of a suitable priming agent (e.g., TNFα 200 U/ml) for 30 min at 37 °C, 5% (v/v) CO_2, prior to stimulation with 100 n*M* fMLP (as a positive control) or a concentration range of the test substance. As with extracellular superoxide anion release, neutrophil degranulation requires the cells to be primed prior to addition of the activating agent even in the case of well-recognized high efficacy agonists.
2. Incubate 5% (v/v) CO_2 for up to 90 min at 37 °C and quench the reaction by the addition of 100 μl ice cold Dulbecco's PBS+ to the samples on ice.
3. Pellet the neutrophils, decant supernatant into a UV cuvette, and determine the change in absorbance at $\lambda = 460$ nm induced by the oxidation of O-dianisidine in the presence of H_2O_2 (Gabay *et al.*, 1986).
4. MPO release is expressed as a percentage of total cellular MPO release following addition of 4 μl 10% (v/v) Triton X-100.

Summary: The above methods have been used to demonstrate effects of α_1-antitrypsin polymers on human neutrophils *in vitro* and would allow similar investigation of other serpin polymers.

9. The Use of Transgenic Mice to Assess the Hepatic Consequences of Serpin Polymerization. Jeff Teckman

The PiZ mouse is an animal model transgenic for the human α_1-antitrypsin mutant Z gene, which recapitulates many aspects of the liver lesions seen in PIZZ human liver (Carlson *et al.*, 1988). This model is the most widely used and best documented, although several other transgenic models have been reported (Geller *et al.*, 1994; Hidvegi *et al.*, 2005). PiZ mice exhibit accumulation of α_1-antitrypsin mutant Z protein within hepatocytes, in both monomeric and polymeric conformations, similar to human liver, and globular inclusions in the ER with similar histologic and ultrastructural characteristics as those seen in human liver (An *et al.*, 2005; Carlson *et al.*, 1988; Teckman and Perlmutter, 2000; Teckman *et al.*, 2002). The intracellular processing of α_1-antitrypsin mutant Z protein, and the intracellular injury pathways triggered by accumulation of the mutant Z protein are also similar in the PiZ mice compared to other model systems and to studies of human tissue. This includes ERAD involving the proteasome, activation of autophagy, hepatocellular proliferation involving hepatic progenitor cells, the development of dysplasia, hepatic fibrosis with age, and the development of hepatocellular carcinoma with age (Brunt *et al.*, 2010; Lindblad *et al.*, 2007; Marcus *et al.*, 2010; Rudnick *et al.*, 2004; Teckman and Perlmutter, 2000). Analysis of the mechanisms of cell death and apoptosis, which appears to be a critical mechanism in the liver injury related to α_1-antitrypsin deficiency, is also well described in the PiZ mouse (Hidvegi *et al.*, 2007; Lawless *et al.*, 2004; Lindblad *et al.*, 2007; Miller *et al.*, 2007).

A variety of studies from several laboratories have shown how there is likely a critical role for the unique polymer conformation of α_1-antitrypsin deficiency mutant Z protein in triggering the pathways for cell injury described above (Ekeowa *et al.*, 2010; Kaushal *et al.*, 2010; Lindblad *et al.*, 2007). As part of these studies, it became clear that a quantitative method to isolate α_1-antitrypsin mutant Z protein polymers from human or mouse model liver tissue would be needed in order to fully elucidate the role of the polymer in liver disease. Using an empiric development process and employing analytic techniques, such a process was successfully developed and validated. The primary obstacle to overcome was the fact that α_1-antitrypsin mutant Z polymers in the *in vivo* liver are very large and are segregated into insoluble masses. However, the chains of polymers are not joined by covalent bonds, so that if the polymers are denatured they revert to monomers, which are indistinguishable from any other monomeric α_1-antitrypsin mutant Z molecule. The quantitative isolation of α_1-antitrypsin mutant Z polymers from liver tissue is, therefore, based on the concept that these insoluble protein aggregate polymers in a nondenatured liver homogenate are suspended, but not dissolved.

Quantitative isolation of α_1-antitrypsin mutant Z protein polymers from human or mouse model liver tissue, as published, first involves making a liver homogenate in a nondenatured buffer in the presence of only mild detergent (An *et al.*, 2005). In this buffer environment, enough physical force is then applied to disrupt lipid bilayers, but the α_1-antitrypsin mutant Z protein polymers are not denatured. These polymers remain suspended in the homogenate such that they can be pelleted by moderate centrifugation. Monomeric α_1-antitrypsin mutant Z molecule are in solution, do not pellet under these conditions, and can then be separated from the polymers. The separate monomer and polymer fractions can then be quantitatively analyzed. If the polymer fraction is denatured, then the number of monomeric molecules that result can be quantitatively compared to the number of molecules in the monomer fraction by ELISA or by quantitative immunoblot. Specifically, the steps in the protocol involve isolation of 10 mg of whole liver (human or model mouse) added to 2 ml buffer at 4 °C (50 m*M* Tris–HCl, pH 8.0, 150 m*M* NaCl, 5 m*M* KCl, 5 m*M* $MgCl_2$, 0.5%, v/v Triton X-100, 80 μl of Complete^{TM} protease inhibitor stock). The tissue is homogenized in a prechilled dounce homogenizer by 30 repetitions, then vortexed vigorously. A 1-ml aliquot is passed through a 28-gauge needle 10 times. The total protein concentration of the sample is determined and a 5 μg total liver protein sample is aliquoted, and centrifuged at 10,000×*g* for 30 min at 4 °C. Supernatant (soluble, S, fraction) is immediately removed into fresh tubes being extremely careful not to disturb the pellet (insoluble, I, fraction). The insoluble polymers pellet (I fraction) can then be denatured and solubilized by addition of 10 μl chilled cell lysis buffer (1%, v/v Triton X-100, 0.05%, w/v deoxycholate, 10 m*M* EDTA in PBS), vortexed 30 s, sonicated on ice 10 min, and vortexed. If further analysis by quantitative immunoblot is planned, then 2.5× sample buffer (50%, v/v 5× sample buffer [5%, w/v SDS, 50%, v/v glycerol, 0.5 *M* Tris, pH 6.8], 10%, v/v β-mercapto-ethanol, 40% ddH_2O) is added to each soluble and insoluble sample at a volume of 50% of the sample volume. Samples are boiled and loaded for SDS–PAGE (with equal amounts of total liver protein loaded per soluble/insoluble pair in quantitative experiments). Quantitative immunoblot with anti-α_1-antitrypsin antibody can then compare lanes from the soluble (monomer) and insoluble (polymer) fractions. This protocol has been validated in multiple published studies and has been shown to be highly reproducible at quantitating α_1-antitrypsin mutant Z protein polymers present in human or mouse model liver tissue (An *et al.*, 2005; Kaushal *et al.*, 2010; Lindblad *et al.*, 2007). Since the polymers in the *in vivo* liver are large and insoluble, and since denaturation breaks up the polymers, it has therefore been found that direct analysis of polymers based on antipolymer antibodies, antipolymer ELISA, or direct immunoprecipitation of polymers is not effective or reproducible for quantitative studies of liver tissue.

10. The Use of Transgenic Mice to Assess the Pulmonary Consequences of Serpin Polymerization—Sam Alam and Ravi Mahadeva

The pathogenesis of emphysema in Z α_1-antitrypsin homozygotes is thought to arise mainly from severe deficiency of the proteinase inhibitor. Previous studies demonstrated that polymeric Z α_1-antitrypsin was not only present in hepatocytes but also in BALF from α_1-antitrypsin homozygotes (Elliott *et al.*, 1998; Mulgrew *et al.*, 2004). Interestingly, polymeric α_1-antitrypsin is also chemotactic to neutrophils *in vitro* (Parmar *et al.*, 2002).

A series of studies applying biochemical and cellular techniques to human lung tissue and in murine models were designed to explore these observations further. Tissue bank samples of paraffin lung tissues from explanted lungs from individuals with M and Z α_1-antitrypsin emphysema were assessed using a succession of histochemical techniques (Mahadeva *et al.*, 2005). A variety of primary antibodies to conformations of α_1-antitrypsin were used to characterize the presence and distribution of α_1-antitrypsin conformers. The specificities of antibodies were defined in parallel using Western blot PAGE and ELISA analysis. In all cases, the signal was optimized using antigen retrieval, quenching of endogenous peroxidases and the streptavidin–biotin complex, and appropriate counterstaining, followed by examination under light microscopy (Mahadeva *et al.*, 2005). Immunostaining demonstrated a positive signal for Z α_1-antitrypsin polymers within the alveolar wall localizing around type II cells exclusively in Z α_1-antitrypsin lungs. No signal was detected in M α_1-antitrypsin alveoli. More detailed analysis confirmed the specificity of immunostaining for the polymeric conformation, that is, the absence of a signal in sections incubated with isotype control and omission of the primary antibody. Importantly, preincubation of polymers, but not the monomeric species with the antipolymer (ATZII) antibody abolished staining within the alveolar epithelium in Z α_1-antitrypsin. The ATZII antibody recognizes the α_1-antitrypsin–elastase complex and polymeric α_1-antitrypsin, but not the native or cleaved form of α_1-antitrypsin (Janciauskiene *et al.*, 2004). Therefore to distinguish between the polymeric α_1-antitrypsin and the α_1-antitrypsin–elastase complex, the explanted lungs were also examined with a mouse antihuman neutrophil elastase (NE) antibody. The numbers of neutrophils present in the alveolar wall were quantified by manual counting of six randomly selected high-power fields. In both M α_1-antitrypsin and Z α_1-antitrypsin emphysematous lungs, the anti-NE antibody stained for neutrophils and did not reveal any extracellular staining to suggest α_1-antitrypsin–elastase complexes. Neutrophils were primarily localized within the alveolar wall in Z α_1-antitrypsin homozygotes, and the mean counts per high-power field in the alveolar wall were at least fourfold higher in the Z α_1-antitrypsin group than the M α_1-antitrypsin group ($P < 0.001$).

Colocalization for polymeric α_1-antitrypsin and neutrophils was performed using double-immunocytochemistry analysis and visualized using DakoCytomation Envision system, which is a biotin-free dextran polymer secondary antibody. To allow for differentiation, neutrophils were visualized using 3-amino-9-ethylcarbazole, producing a red reaction product. Double staining for localization of NE and polymeric α_1-antitrypsin show that in Z α_1-antitrypsin lungs, neutrophils within the alveolar wall were colocalized with polymers surrounding them. These data suggest that polymers of α_1-antitrypsin are present in the alveolar wall of Z α_1-antitrypsin emphysematous lungs in association with an excess of potentially deleterious neutrophils.

The colocalization studies confirmed the possibility of a direct relationship between polymeric α_1-antitrypsin and neutrophilic inflammation and a novel hypothesis of lung disease in Z α_1-antitrypsin homozygotes (Lomas and Mahadeva, 2002). A series of murine experiments were designed to directly assess the mechanism. C57BL/6J mice were anesthetized and intubated and received either pure polymeric α_1-antitrypsin or native α_1-antitrypsin free of significant endotoxin. Instillation of polymeric α_1-antitrypsin produced a significant neutrophil influx into the BALF at 4 and 24 h compared with native α_1-antitrypsin. By 72 h, neutrophil counts had returned to baseline (Fig. 18.2). To substantiate this, NE activity was assessed after lysis of cells recovered from BAL using a specific substrate for NE; MeOSuc-Ala-Ala-Pro-Val-AFC (Owen *et al.*, 1997). NE levels were detectable only after polymer instillation.

These data generated by a number of different models and methodologies confirmed that polymeric Z α_1-antitrypsin was present in lung tissue and was chemotactic to neutrophils *in vivo*. This was highly suggestive of an additional mechanism that would exaggerate deficiency of Z α_1-antitrypsin further and contribute to increased neutrophilic inflammation in the lungs in Z α_1-antitrypsin individuals with COPD. What was not known was the *in vivo* molecular driver for polymerization. Epidemiological observations indicate that Z-AT homozygotes who smoked developed rapidly progressive emphysema at a young age. Therefore, transgenic mice for human Z α_1-antitrypsin (Z mice) and M α_1-antitrypsin (M mice) (Ali *et al.*, 1994) were assessed in a model of acute cigarette smoke (CS) exposure (Hautamaki *et al.*, 1997). The cellular component of inflammatory cells within lung homogenates and BALF was characterized and quantified on cytospins, and free NE activity was assessed as described above (Alam *et al.*, 2011; Li *et al.*, 2009). Polymeric α_1-antitrypsin was assessed using ATZII by specific ELISA and Western blot of nondenaturing PAGE. CS-Z mice developed a significant increase in pulmonary polymeric α_1-antitrypsin and neutrophil influx versus controls (Fig. 18.6). Neutrophil numbers were increased in CS-Z lungs and tightly correlated with polymeric AT concentrations, $r^2 = 0.93$ (Alam *et al.*, 2011). It is known that CS contains oxidants that can produce oxidized AT (Ox-AT; Rahman and MacNee, 1999). Plasma-purified Z α_1-antitrypsin was exposed to CS extract (CSE) prepared by bubbling CS through PBS and pH normalized. Samples were assessed for the temporal relationship between the oxidized and polymeric conformers using a specific ELISAs (Janciauskiene *et al.*, 2002; Ueda *et al.*, 2002). CSE rapidly induced oxidation and a 10-fold increase in polymers of Z α_1-antitrypsin when compared to control Z α_1-antitrypsin, with the production of oxidized polymers. Detailed analysis revealed that oxidation of Z α_1-antitrypsin preceded

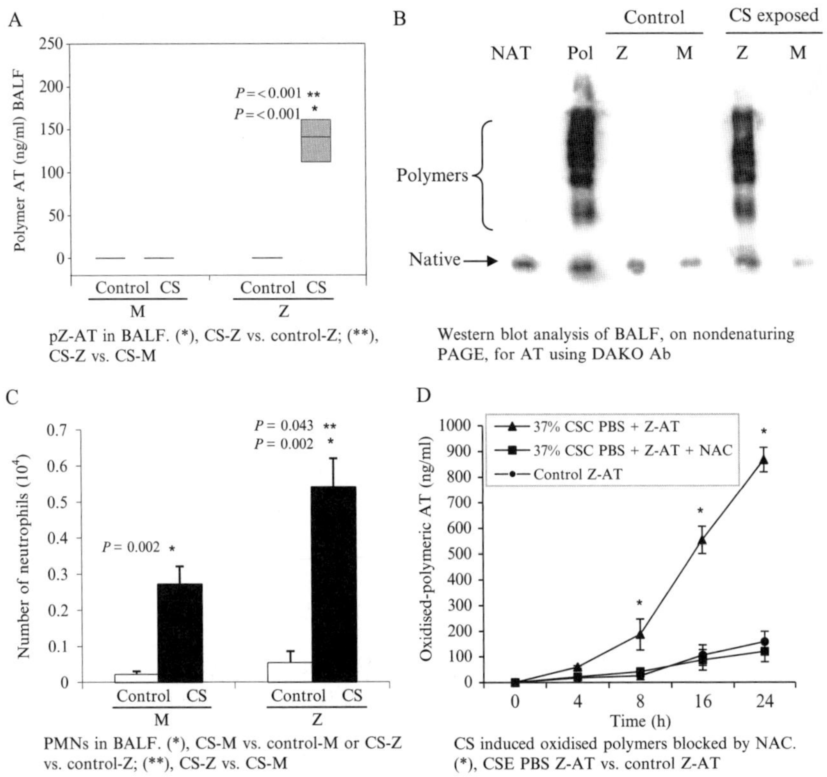

Figure 18.6 Effect of acute cigarette smoke exposure on transgenic M α_1-antitrypsin and Z α_1-antitrypsin mice. Female mice transgenic for M α_1-antitrypsin or Z α_1-antitrypsin were exposed to four 1R3F cigarettes' smoke (CS) per day for 5 days and BALF was collected. (A) ELISA showed there was significant increase in polymeric AT in BALF of CS-Z when compared to control Z mice; $P = 0.002$. (B) Western blot confirmed classical ladders of polymeric α_1-antitrypsin in CS-Z mice. (C) There was a significant influx of neutrophils in CS-Z mice compared to controls and CS-M mice, and (D) polymeric Z-AT formation can be prevented by pretreatment with an antioxidant *N*-acetyl cysteine (NAC).

polymerization of the protein, suggesting that oxidation of Z α_1-antitrypsin by CS promoted accelerated polymerization of Z α_1-antitrypsin. These data were strongly supported by the finding that oxidation by the oxidant *N*-chlorosuccinimide also accelerated polymerization of Z α_1-antitrypsin. Further, CS-induced polymerization of Z α_1-antitrypsin could be prevented by the antioxidant *N*-acetyl cysteine (NAC; Fig. 18.6D).

In conclusion, CS accelerates polymerization of Z α_1-antitrypsin by oxidative modification, which in so doing further reduces pulmonary defense and increases neutrophil influx into the lungs. These novel findings provide a molecular explanation for the striking association of premature emphysema in ZZ α_1-antitrypsin homozygotes who smoke.

11. Characterization of Serpin Polymerization Using iPS to Generate Hepatocyte-Like Cell Lines—Tamir Rashid and Ludovic Vallier

11.1. Introduction

Overexpressing a cocktail of well-defined embryonic stem cell-associated transcription factors in somatic cells has successfully been used to reprogram adult cells back to a pluripotent stem cell-like state. These cells, known as induced pluripotent stem cells or iPSCs, retain the potential to be expanded in virtually unlimited numbers while remaining capable of forming most of the cell types of the adult body. The possibility of coupling iPSCs derived from patients suffering from specific diseases with subsequent *in vitro* differentiation has opened up the possibility of using this technology for advances in developmental biology, cell-based therapy, and *in vitro* disease modeling.

We have recently demonstrated that this new strategy can successfully be used to derive fetal human hepatocyte cells from patients suffering from a range of inherited metabolic disorders of the liver including PiZ α_1-antitrypsin deficiency (Rashid *et al.*, 2010). Key elements of disease pathophysiology such as protein polymerization and ER entrapment were conserved, therefore suggesting this platform will be a useful tool to extend *in vitro* studies into serpin biology. Here, we provide a guide for generating patient-specific hepatocyte-like cells, starting from a skin biopsy (Fig. 18.7).

11.2. Obtaining dermal fibroblasts from patient

The starting primary cell type used for reprogramming can vary according to local expertise and tissue availability. We prefer to use dermal fibroblasts as outlined below.

(a) Obtain appropriate local/national ethical approval and patient consent.
(b) Take 8 mm skin punch biopsy from patient's inner forearm—procedure should be done by appropriately qualified clinician according to local/national medical regulations.
(c) Place the biopsy specimen immediately into human fibroblast medium (containing antibiotics and antifungals) and transport to the laboratory. Perform all subsequent steps in the tissue culture hood.
(d) Use sterile forceps and dissecting scissors to scrape away all fatty tissue and then mince the skin biopsy into small pieces of $\sim$1 mm^3 size.
(e) Place the skin pieces randomly across the base of a 10 cm petri dish or into the center of a well of a six-well plate with enough human fibroblast medium to cover the bottom of the respective dish/well.
(f) Carefully lower a sterile coverslip onto the pieces to hold them in place against the bottom of the plate and then add human fibroblast

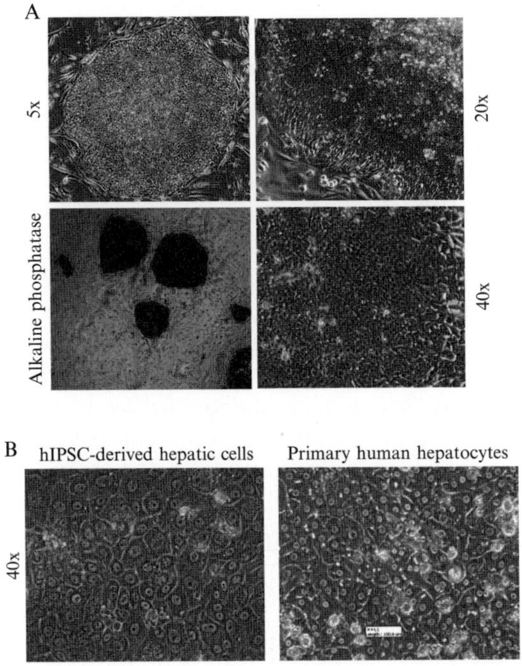

Figure 18.7 (A) Bright field pictures depicting the characteristic hESC like morphology of hIPSC colonies (5×, 20×) and hIPS cells (40×). The colonies also displayed the alkaline phosphatase activity characteristic of pluripotent stem cells. (B) Bright field pictures comparing morphology of hIPSC-derived hepatic cells (left hand panel, 40×) with morphology of hepatocytes derived from adult human liver (right hand panel, 40×). (For color version of this figure, the reader is referred to the web version of this chapter.)

medium to a total of 3 ml for a six-well plate or 10 ml for a 10-cm petri dish. Take care that the biopsy samples to do not float away—floating samples will not yield fibroblasts.

(g) Incubate the plate at 37 °C, 5% (v/v) CO_2 being careful not to disturb for the first 3 days. Carefully change medium on day 4 without disturbing the skin samples.
(h) By 10–14 days, dense outgrowths of fibroblasts appear.
(i) Aspirate the media and wash twice with PBS.
(j) Use sterile forceps to dislodge and lift the coverslip to ensure the media has been washed away with PBS. Fibroblasts will be adhered to the coverslip; take caution not to aspirate these cells.
(k) Add 0.5 ml 0.5% trypsin/EDTA. Use sterile forceps to lift the coverslip to ensure that trypsin reaches the underside of the coverslip where fibroblasts are adhered.
(l) Incubate for 5 min at 37 °C, then using sterile forceps flip the coverslip over and incubate for an additional 5 min at 37 °C.

(m) Using a 5 ml pipette, add 2.5 ml of human fibroblast medium to inactive the trypsin.
(n) By aspirating up and down gently, in addition to scraping the well bottom and coverslip, disperse all the cells. Collect cells into 15 ml falcon tube.
(o) Pass the cells through a 70 μm cell strainer to remove large chunks of tissue.
(p) Centrifuge the cells at 1000 rpm for 4 min and discard the supernatant.
(q) Resuspend the cells with 5 ml of human fibroblast medium (KO-DMEM (Gibco, 10% (v/v) FBS (Hyelone)) and plate into T-25 tissue culture flask (passage 2).
(r) Passage 1:3 every 7 days or until the cells have reached 80% confluence.
(s) For subsequent passages, aspirate medium, wash once with PBS, trypsinize with 0.5% trypsin for 5 min at 37 °C. Continue to passage at a ratio of 1:3.

11.3. Derivation and maintenance of human IPS cells

Several techniques have now been described for reprogramming cells back to IPSCs. In our hands, the most robust and reliable method remains retroviral-based overexpression of the four classic Yamanaka factors as described below (Takahashi *et al.*, 2007; Vallier *et al.*, 2009a).

(a) The day before viral transduction plate dermal fibroblasts out at a density of 100,000 cells per well of a six-well plate in standard human fibroblast medium.
(b) Transduce the cells the following day with Moloney murine leukemia virus-derived vectors containing the coding sequences of each one of the four human genes, *Oct-4, Sox2, c-Myc,* and *Klf4* (obtained from Vectalys Toulouse, France, http://www.vectalys.com). Fibroblasts should be infected at a Multiplicity Of Infection (MOI) of 10 or 50.
(c) The following day, aspirate the medium, wash once with PBS, and replace with human fibroblast medium.
(d) On the fifth day posttransduction, split the cells (as previously described when splitting fibroblasts) and transfer cells from each well of the six-well plate onto a 10-cm dish seeded with irradiated mouse embryonic feeders.
(e) On the seventh day posttransduction, aspirate the medium and replace with standard human embryonic stem cell (hESC) culture medium (knockout [KSR], (Gibco) + FGF2 (4 ng/ml; R&D Systems Inc., Minneapolis, http://www.rndsystems.com)).
(f) From now onwards change medium daily.
(g) Colonies of cells with typical features of hESCs appear between 20 and 40 days posttransduction. These hIPSC colonies should be manually picked and expanded as per standard hESC procedures.

11.4. Characterization of hIPSCs

Routine characterization of hIPSC lines includes (Vallier *et al.*, 2009a)

- QPCR and immunostaining analyses for the expression of pluripotency markers such as Oct-4, Sox2, Nanog, and Tra-1-60
- QPCR analyses for the expression of exogenous transgenes
- QPCR or Southern blot analysis for the number of transgenes inserted into the genome
- Differentiation *in vitro* using the embryoid body formation or directed differentiation in chemically defined medium as described (Vallier and Pedersen, 2008; Vallier *et al.*, 2009b)
- Karyotype analyses

11.5. Generation of hepatocyte-like cells from hIPSCs

For disease modeling, hIPSCs must be differentiated to hepatic cells. Here, we describe a three-step protocol to generate such cells.

(a) hIPSCs are passaged using 5 mg/ml collagenase IV/dispase (0.1%, GIBCO) 1:1 (v/v) mix and then transferred onto plates precoated with FBS (KO-DMEM (Gibco), 10% FBS (Hyclone)) in CDM-PVA or in plates precoated with human fibronectin as previously described (Touboul *et al.*, 2010).

(b) For the first following day, cells are grown in CDM-PVA supplemented with CHIR99021 (3 μ*M*, Stemgent), Ly294002 (10 μ*M*, Calbiochem), Activin (100 ng/ml, R&D Systems), FGF2 (40 ng/ml, R&D Systems), and BMP4 (10 ng/ml, R&D Systems) to drive differentiation of hIPSCs into primitive streak like cells.

(c) The next day, the resulting cells are grown in CDM-PVA supplemented with Ly294002 (10 μ*M*, Calbiochem), Activin (100 ng/ml), FGF2 (40 ng/ml, R&D Systems), and BMP4 (10 ng/ml, R&D Systems) to drive their differentiation toward definitive endoderm.

(d) On the third day, the basal medium is changed to RPMI (Gibco 1640) and supplemented with Activin (100 ng/ml, R&D Systems), FGF2 (40 ng/ml, R&D Systems), and B27 to obtain anterior definitive endoderm cells (ADE).

(e) To induce hepatic endoderm, ADE cells are then cultured for 5 days in the presence of RPMI and B27 (Gibco 1640) supplemented with Activin (50 ng/ml, R&D Systems).

(f) Finally, to mature the resulting hepatic progenitors, cells are grown in a CMRL/Hepatozyme (Invitrogen) basal medium supplemented by HGF (10 ng/ml, Peprotech) and Oncostatin—M (20 ng/ml, R&D) for a further 14–21 days.

11.6. Validation of hepatic-like cells

Routine characterization of hIPSC-derived hepatocyte cells includes (Rashid *et al.*, 2010)

- QPCR and immunostaining analyses for the expression of hepatocyte markers such as albumin, α_1-antitrypsin, and CyP3A4
- ELISA for albumin and α_1-antitrypsin secretion
- PAS staining for glycogen storage
- LDL cholesterol incorporation assay

ACKNOWLEDGMENTS

D. A. L. is supported by grants from the Medical Research Council (UK), the Engineering and Physical Sciences Research Council, the British Lung Foundation, GlaxoSmithKline, and Papworth NHS Trust. U. I. E. is an MRC Clinical Research Training Fellow, B. G. is a Wellcome Trust Intermediate Clinical Fellow, and S. T. R. is a Wellcome Trust Clinical Training Fellow. E. M. is supported by the University of Rome La Sapienza. J. P. is supported by the Ministerio de Ciencia e Innovación (Grants PR 2009-0201 and SAF2010-19087) and the Junta de Andalucía (Grant P07 CVI 03079). A.O. is an eALTA Fellow. S. J. M. is an MRC Clinical Scientist (G0601840) and is supported by Diabetes UK and the British Lung Foundation. H. P. and E. R. C. are funded by the Medical Research Council (UK), Wellcome Trust, Asthma-UK, BBSRC, Papworth Hospital R&D Department, Intensive Care Society, Addenbrooke's Charitable Trust, and the Cambridge NIHR Biomedical Research Centre. S. A. and R. M. are supported by the Wellcome Trust and the Cambridge NIHR Biomedical Research Centre. S. T. R. and L. V. are funded by the Wellcome Trust, Medical Research Council (UK), and the Cambridge NIHR Biomedical Research Centre. We are grateful to Dr. D. K. Clare, Birkbeck College for preparing the cryo-EM grids and recording cryo-EM images.

REFERENCES

Alam, S., Li, Z., Janciauskiene, S., and Mahadeva, R. (2011). Oxidation of Z alpha-1-antitrypsin by cigarette smoke induces polymerization: A novel mechanism of early-onset emphysema. *Am. J. Respir. Cell Mol. Biol.* **45,** 261–269.

Ali, R., Perfumo, S., della Rocca, C., Amicone, L., Pozzi, L., McCullagh, P., Millward-Sadler, H., Edwards, Y., Povey, S., and Tripodi, M. (1994). Evaluation of a transgenic mouse model for alpha-1-antitrypsin (AAT) related liver disease. *Ann. Hum. Genet.* **58,** 305–320.

Amos, L. A., Henderson, R., and Unwin, P. N. (1982). Three-dimensional structure determination by electron microscopy of two-dimensional crystals. *Prog. Biophys. Mol. Biol.* **39,** 183–231.

An, J. K., Blomenkamp, K., Lindblad, D., and Teckman, J. H. (2005). Quantitative isolation of alpha-l-antitrypsin mutant Z protein polymers from human and mouse livers and the effect of heat. *Hepatology* **41,** 160–167.

Baek, J. H., Im, H., Kang, U. B., Seong, K. M., Lee, C., Kim, J., and Yu, M. H. (2007). Probing the local conformational change of α_1-antitrypsin. *Protein Sci.* **16,** 1842–1850.

Banda, M., Rice, A. G., Griffin, G. L., and Senior, R. M. (1988). α_1-proteinase inhibitor is a neutrophil chemoattractant after proteolytic inactivation by macrophage elastase. *J. Biol. Chem.* **263,** 4481–4484.

Belorgey, D., Crowther, D. C., Mahadeva, R., and Lomas, D. A. (2002). Mutant neuroserpin (Ser49Pro) that causes the familial dementia FENIB is a poor proteinase inhibitor and readily forms polymers *in vitro*. *J. Biol. Chem.* **277,** 17367–17373.

Belorgey, D., Irving, J. A., Ekeowa, U. I., Freeke, J., Roussel, B. D., Miranda, E., Pérez, J., Robinson, C. V., Marciniak, S. J., Crowther, D. C., Michel, C. H., and Lomas, D. A. (2011). Characterisation of serpin polymers *in vitro* and *in vivo*. *Methods* **53**(3), 255–266.

Brunt, E. M., Blomenkamp, K., Ahmed, M., Ali, F., Marcus, N., and Teckman, J. (2010). Hepatic progenitor cell proliferation and liver injury in alpha-1-antitrypsin deficiency. *J. Pediatr. Gastroenterol. Nutr.* **51,** 626–630.

Carlson, J. A., Rogers, B. B., Sifers, R. N., Hawkins, H. K., Finegold, M. J., and Woo, S. L. (1988). Multiple tissues express alpha-1-antitrypsin in transgenic mice and man. *J. Clin. Invest.* **82,** 26–36.

Carrell, R. W., and Lomas, D. A. (1997). Conformational disease. *Lancet* **350,** 134–138.

Coutelier, M., Andries, S., Ghariani, S., Dan, B., Duyckaerts, C., van Rijckevorsel, K., Raftopoulos, C., Deconinck, N., Sonderegger, P., Scaravilli, F., Vikkula, M., and Godfraind, C. (2008). Neuroserpin mutation causes electrical status epilepticus of slow-wave sleep. *Neurology* **71,** 64–66.

Cowburn, A. S., Condliffe, A. M., Farahi, N., Summers, C., and Chilvers, E. R. (2008). Advances in neutrophil biology: Clinical implications. *Chest* **134,** 606–612.

Crowther, R. A., Henderson, R., and Smith, J. M. (1996). MRC image processing programs. *J. Struct. Biol.* **116,** 9–16.

Dafforn, T. R., Mahadeva, R., Elliott, P. R., Sivasothy, P., and Lomas, D. A. (1999). A kinetic mechanism for the polymerisation of α_1-antitrypsin. *J. Biol. Chem.* **274,** 9548–9555.

Dafforn, T., Pike, R. N., and Bottomley, S. P. (2004). Physical characterization of serpins conformations. *Methods* **32,** 150–158.

Davies, M. J., Miranda, E., Roussel, B. D., Kaufman, R. J., Marciniak, S. J., and Lomas, D. A. (2009). Neuroserpin polymers activate NF-kappaB by a calcium signalling pathway that is independent of the unfolded protein response. *J. Biol. Chem.* **284,** 18202–18209.

Davis, R. L., Shrimpton, A. E., Holohan, P. D., Bradshaw, C., Feiglin, D., Sonderegger, P., Kinter, J., Becker, L. M., Lacbawan, F., Krasnewich, D., Muenke, M., Lawrence, D. A., *et al.* (1999). Familial dementia caused by polymerisation of mutant neuroserpin. *Nature* **401,** 376–379.

Davis, R. L., Shrimpton, A. E., Carrell, R. W., Lomas, D. A., Gerhard, L., Baumann, B., Lawrence, D. A., Yepes, M., Kim, T. S., Ghetti, B., Piccardo, P., Takao, M., *et al.* (2002). Association between conformational mutations in neuroserpin and onset and severity of dementia. *Lancet* **359,** 2242–2247.

DeRosier, D. J., and Klug, A. (1968). Reconstruction of three-dimensional structures from electron micrographs. *Nature* **217,** 130–134.

Ekeowa, U. I., Gooptu, B., Belorgey, D., Hägglöf, P., Karlsson-Li, S., Miranda, E., Pérez, J., MacLeod, I., Kroger, H., Marciniak, S. J., Crowther, D. C., and Lomas, D. A. (2009). α_1-antitrypsin deficiency, chronic obstructive pulmonary disease and the serpinopathies. *Clin. Sci. (Lond.)* **116,** 837–850.

Ekeowa, U. I., Freekeb, J., Miranda, E., Gooptu, B., Bush, M. F., Pérez, J., Teckman, J., Robinson, C. V., and Lomas, D. A. (2010). Defining the mechanism of polymerization in the serpinopathies. *Proc. Natl. Acad. Sci. USA* **107,** 17146–17151.

Elliott, P. R., Bilton, D., and Lomas, D. A. (1998). Lung polymers in Z α_1-antitrypsin related emphysema. *Am. J. Respir. Cell Mol. Biol.* **18,** 670–674.

Eriksson, S., Carlson, J., and Velez, R. (1986). Risk of cirrhosis and primary liver cancer in alpha$_1$-antitrypsin deficiency. *N. Engl. J. Med.* **314,** 736–739.

Falk, W., Goodwin, R. H., Jr., and Leonard, E. J. (1980). A 48 well microchemotaxis assembly for rapid and accurate measurement of leukocyte migration. *J. Immunol. Methods* **33,** 239–247.

Frank, J. (2006). Three-Dimensional Electron Microscopy of Macromolecular Assemblies. Oxford University Press, Oxford, United Kingdom.

Frank, J., Radermacher, M., Penczek, P., Zhu, J., Li, Y., Ladjadj, M., and Leith, A. (1996). SPIDER and WEB: Processing and visualization of images in 3D electron microscopy and related fields. *J. Struct. Biol.* **116,** 190–199.

Frow, E. K., Reckless, J., and Grainger, D. J. (2004). Tools for anti-inflammatory drug design: In vitro models of leukocyte migration. *Med. Res. Rev.* **24,** 276–298.

Gabay, J. E., Heiple, J. M., Cohn, Z. N., and Nathan, C. F. (1986). Subcellular localisation and properties of bactericidal factors from human neutrophils. *J. Exp. Med.* **164,** 1407–1421.

Geller, S. A., Nichols, S. W., Kim, S., Tolmachoff, T., Lee, S., Dycaico, M. J., Felts, K., and Sorge, J. A. (1994). Hepatocarcinogenesis is the sequel to hepatitis in Z#2 α_1-antitrypsin transgenic mice: Histopathological and DNA ploidy studies. *Hepatology* **19,** 389–397.

Gooptu, B., and Lomas, D. A. (2009). Conformational pathology of the serpins—Themes, variations and therapeutic strategies. *Annu. Rev. Biochem.* **78,** 147–176.

Harding, H. P., Novoa, I., Zhang, Y., Zeng, H., Wek, R., Schapira, M., and Ron, D. (2000). Regulated translation initiation controls stress-induced gene expression in mammalian cells. *Mol. Cell* **6,** 1099–1108.

Haslett, C., Guthrie, L. A., Kopaniak, M. M., Johnston, R. B., Jr., and Henson, P. M. (1985). Modulation of multiple neutrophil functions by preparative methods or trace concentrations of bacterial lipopolysaccharide. *Am. J. Pathol.* **119,** 101–110.

Hautamaki, R. D., Kobayashi, D. K., Senior, R. M., and Shapiro, S. (1997). Requirement for macrophage elastase for cigarette smoke-induced emphysema in mice. *Science* **277,** 2002–2004.

Hidvegi, T., Schmidt, B. Z., Hale, P., and Perlmutter, D. H. (2005). Accumulation of mutant α_1-antitrypsin Z in the endoplasmic reticulum activates caspases-4 and -12, NFkappaB, and BAP31 but not the unfolded protein response. *J. Biol. Chem.* **280,** 39002–39015.

Hidvegi, T., Mirnics, K., Hale, P., Ewing, M., Beckett, C., and Perlmutter, D. H. (2007). Regulator of G signaling 16 is a marker for the distinct ER stress state associated with aggregated mutant alpha 1antitrypsin Z in the classical form of α_1-antitrypsin deficiency. *J. Biol. Chem.* **282,** 27769–27780.

Janciauskiene, S., Dominaitiene, R., Sternby, N. H., Piitulainen, E., and Eriksson, S. (2002). Detection of circulating and endothelial cell polymers of Z and wildtype alpha-1-antitrypsin by a monoclonal antibody. *J. Biol. Chem.* **277,** 26540–26546.

Janciauskiene, S., Eriksson, S., Callea, F., Mallya, M., Zhou, A., Seyama, K., Hata, S., and Lomas, D. A. (2004). Differential detection of PAS-positive inclusions formed by the Z, Siiyama and Mmalton variants of α_1-antitrypsin. *Hepatology* **40,** 1203–1210.

Kaushal, S., Annamali, M., Blomenkamp, K., Rudnick, D., Halloran, D., Brunt, E. M., and Teckman, J. H. (2010). Rapamycin reduces intrahepatic alpha-1-antitrypsin mutant Z protein polymers and liver injury in a mouse model. *Exp. Biol. Med. (Maywood)* **235,** 700–709.

Kennedy, S. A., van Diepen, A. C., van den Hurk, C. M., Coates, L. C., Lee, T. W., Ostrovsky, L. L., Miranda, E., Perez, J., Davies, M. J., Lomas, D. A., Dunbar, P. R., and Birch, N. P. (2007). Expression of the serine protease inhibitor neuroserpin in cells of the human myeloid lineage. *Thromb. Haemost.* **97,** 394–399.

Kopito, R. R., and Ron, D. (2000). Conformational disease. *Nat. Cell Biol.* **2,** 207–209.

Kröger, H., Miranda, E., MacLeod, I., Pérez, J., Crowther, D. C., Marciniak, S. J., and Lomas, D. A. (2009). Endoplasmic reticulum-associated degradation (ERAD) and autophagy cooperate to degrade polymerogenic mutant serpins. *J. Biol. Chem.* **284,** 22793–22802.

Kummer, J. A., Strik, M. C., Bladergroen, B. A., and Hack, C. E. (2004). Production, characterization, and use of serpin antibodies. *Methods* **32,** 141–149.

Kurdowska, A., and Travis, J. (1990). Acute phase protein stimulation by α_1-antichymotrypsin-cathepsin G complexes. Evidence for the involvement of interleukin-6. *J. Biol. Chem.* **265,** 21023–21026.

Laemmli, U. K. (1970). Cleavage of structural proteins during the assembly of the head of bacteriophage T4. *Nature* **227,** 680–685.

Lawless, M. W., Greene, C. M., Mulgrew, A., Taggert, C. C., O'Neill, S. J., and McElvaney, N. G. (2004). Activation of endoplasmic reticulum-specific stress responses associated with the conformational disease Z α_1-antitrypsin deficiency. *J. Immunol.* **172,** 5722–5726.

Le, A., Ferrell, G. A., Dishon, D. S., Quyen-Quyen, A. L., and Sifers, R. N. (1992). Soluble aggregates of the human PiZ α_1-antitrypsin variant are degraded within the endoplasmic reticulum by a mechanism sensitive to inhibitors of protein synthesis. *J. Biol. Chem.* **267,** 1072–1080.

Ley, K., Laudanna, C., Cybulsky, M. I., and Nourshargh, S. (2007). Getting to the site of inflammation: The leukocyte adhesion cascade updated. *Nat. Rev. Immunol.* **7,** 678–689.

Li, Z., Alam, S., Wang, J., Sandstrom, C. S., Janciauskiene, S., and Mahadeva, R. (2009). Oxidized α_1-antitrypsin stimulates the release of monocyte chemotactic protein-1 from lung epithelial cells: Potential role in emphysema. *Am. J. Physiol. Lung Cell. Mol. Physiol.* **297,** L388–L400.

Lindblad, D., Blomenkamp, K., and Teckman, J. (2007). Alpha-1-antitrypsin mutant Z protein content in individual hepatocytes correlates with cell death in a mouse model. *Hepatology* **46,** 1228–1235.

Liu, Y., Choudhury, P., Cabral, C. M., and Sifers, R. N. (1997). Intracellular disposal of incompletely folded human α_1-antitrypsin involves release from calnexin and post-translational trimming of asparagine-linked oligosaccharides. *J. Biol. Chem.* **272,** 7946–7951.

Lomas, D. A., and Mahadeva, R. (2002). Alpha-1-antitrypsin polymerisation and the serpinopathies: Pathobiology and prospects for therapy. *J. Clin. Invest.* **110,** 1585–1590.

Lomas, D. A., Evans, D. L., Finch, J. T., and Carrell, R. W. (1992). The mechanism of Z α_1-antitrypsin accumulation in the liver. *Nature* **357,** 605–607.

Lomas, D. A., Evans, D. L., Stone, S. R., Chang, W.-S. W., and Carrell, R. W. (1993a). Effect of the Z mutation on the physical and inhibitory properties of α_1-antitrypsin. *Biochemistry* **32,** 500–508.

Lomas, D. A., Finch, J. T., Seyama, K., Nukiwa, T., and Carrell, R. W. (1993b). α_1-antitrypsin S_{iiyama} (Ser^{53}ÆPhe); further evidence for intracellular loop-sheet polymerisation. *J. Biol. Chem.* **268,** 15333–15335.

Lomas, D. A., Elliott, P. R., Sidhar, S. K., Foreman, R. C., Finch, J. T., Cox, D. W., Whisstock, J. C., and Carrell, R. W. (1995). Alpha1-antitrypsin Mmalton (52Phe deleted) forms loop-sheet polymers *in vivo*: Evidence for the C sheet mechanism of polymerisation. *J. Biol. Chem.* **270,** 16864–16870.

Ludtke, S. J., Baldwin, P. R., and Chiu, W. (1999). EMAN: Semiautomated software for high-resolution single-particle reconstructions. *J. Struct. Biol.* **128,** 82–97.

Mahadeva, R., Atkinson, C., Li, J., Stewart, S., Janciauskiene, S., Kelley, D. G., Parmar, J., Pitman, R., Shapiro, S. D., and Lomas, D. A. (2005). Polymers of Z α_1-antitrypsin co-localise with neutrophils in emphysematous alveoli and are chemotactic *in vivo*. *Am. J. Pathol.* **166,** 377–386.

Marciniak, S. J., and Ron, D. (2006). Endoplasmic reticulum stress signaling in disease. *Physiol. Rev.* **86,** 1133–1149.

Marciniak, S. J., Yun, C. Y., Oyadomari, S., Novoa, I., Zhang, Y., Jungreis, R., Nagata, K., Harding, H. P., and Ron, D. (2004). CHOP induces death by promoting protein synthesis and oxidation in the stressed endoplasmic reticulum. *Genes Dev.* **18,** 3066–3077.

Marcus, N. Y., Brunt, E. M., Blomenkamp, K., Ali, F., Rudnick, D. A., Ahmad, M., and Teckman, J. H. (2010). Characteristics of hepatocellular carcinoma in a murine model of alpha-1-antitrypsin deficiency. *Hepatol. Res.* **40,** 641–653.

Miller, S. D., Greene, C. M., McLean, C., Lawless, M. W., Taggart, C. C., O'Neill, S. J., and McElvaney, N. G. (2007). Tauroursodeoxycholic acid inhibits apoptosis induced by Z alpha-1 antitrypsin via inhibition of Bad. *Hepatology* **46,** 496–503.

Mindell, J. A., and Grigorieff, N. (2003). Accurate determination of local defocus and specimen tilt in electron microscopy. *J. Struct. Biol.* **142,** 334–347.

Miranda, E., Römisch, K., and Lomas, D. A. (2004). Mutants of neuroserpin that cause dementia accumulate as polymers within the endoplasmic reticulum. *J. Biol. Chem.* **279,** 28283–28291.

Miranda, E., McLeod, I., Davies, M. J., Pérez, J., Römisch, K., Crowther, D. C., and Lomas, D. A. (2008). The intracellular accumulation of polymeric neuroserpin explains the severity of the dementia FENIB. *Hum. Mol. Genet.* **17,** 1527–1539.

Miranda, E., Pérez, J., Ekeowa, U. I., Hadzic, N., Kalsheker, N., Gooptu, B., Portmann, B., Belorgey, D., Hill, M., Chambers, S., Teckman, J., Alexander, G. J., *et al.* (2010). A novel monoclonal antibody to characterise pathogenic polymers in liver disease associated with α_1-antitrypsin deficiency. *Hepatology* **52**(3), 1078–1088.

Mulgrew, A. T., Taggart, C. C., Lawless, M. W., Greene, C. M., Brantly, M. L., O'Neill, S. J., and McElvaney, N. G. (2004). Z α_1-antitrypsin polymerizes in the lung and acts as a neutrophil chemoattractant. *Chest* **125,** 1952–1957.

Müller, I., Munder, M., Kropf, P., and Hänsch, G. M. (2009). Polymorphonuclear neutrophils and T lymphocytes: Strange bedfellows or brothers in arms? *Trends Immunol.* **30,** 522–530.

Nielsen, H. M., Minthon, L., Londos, E., Blennow, K., Miranda, E., Perez, J., Crowther, D. C., Lomas, D. A., and Janciauskiene, S. M. (2007). Plasma and CSF serpins in Alzheimer disease and dementia with Lewy bodies. *Neurology* **69,** 1569–1579.

Novoa, I., Zeng, H., Harding, H. P., and Ron, D. (2001). Feedback inhibition of the unfolded protein response by GADD34-mediated dephosphorylation of eIF2alpha. *J. Cell Biol.* **153,** 1011–1022.

Onda, M., Belorgey, D., Sharp, L. K., and Lomas, D. A. (2005). Latent S49P neuroserpin spontaneously forms polymers: Identification of a novel pathway of polymerization and implications for the dementia FENIB. *J. Biol. Chem.* **280,** 13735–13741.

Owen, C. A., Campbell, M. A., Boukedes, S. S., and Campbell, E. J. (1997). Cytokines regulate membrane-bound leukocyte elastase on neutrophils: A novel mechanism for effector activity. *Am. J. Physiol.* **272**(3 Pt. 1), L385–L393.

Pahl, H. L., and Baeuerle, P. A. (1995). A novel signal transduction pathway from the endoplasmic reticulum to the nucleus is mediated by transcription factor NF-kappa B. *EMBO J.* **14,** 2580–2588.

Parmar, J. S., Mahadeva, R., Reed, B. J., Farahi, N., Cadwallader, K., Bilton, D., Chilvers, E. R., and Lomas, D. A. (2002). Polymers of α_1-antitrypsin are chemotactic for human neutrophils: A new paradigm for the pathogenesis of emphysema. *Am. J. Respir. Cell Mol. Biol.* **26,** 723–730.

Qu, J., Condliffe, A. M., Lawson, M., Plevin, R. J., Riemersma, R. A., Barclay, G. B., McClelland, D. B. L., and Chilvers, E. R. (1995). Lack of effect of recombinant platelet-derived growth factor on human neutrophil function. *J. Immunol.* **154,** 4133–4141.

Rahman, I., and MacNee, W. (1999). Lung glutathione and oxidative stress: Implications in cigarette smoke-induced airway disease. *Am. J. Physiol.* **277**(6 Pt 1), L1067–L1088.

Rashid, S. T., Corbineau, S., Hannan, N., Marciniak, S. J., Miranda, E., Alexander, G., Huang-Doran, I., Griffin, J., Ahrlund-Richter, L., Skepper, J., Semple, R., Weber, A., *et al.* (2010). Modeling inherited metabolic disorders of the liver using human induced pluripotent stem cells. *J. Clin. Invest.* **120,** 3127–3136.

Roseman, A. M. (2003). Particle finding in electron micrographs using a fast local correlation algorithm. *Ultramicroscopy* **94,** 225–236.

Rudnick, D. A., Liao, Y., An, J. K., Muglia, L. J., Perlmutter, D. H., and Teckman, J. H. (2004). Analyses of hepatocellular proliferation in a mouse model of α_1-antitrypsin deficiency. *Hepatology* **39,** 1048–1055.

Ruska, E. (1987). The development of the electron microscope and of electron microscopy. *Rev. Mod. Phys.* **59,** 627–638.

Smith, J. M. (1999). XIMDISP—A visualization tool to aid structure determination from electron microscope images. *J. Struct. Biol.* **125,** 223–228.

Sveger, T. (1976). Liver disease in alpha$_1$-antitrypsin deficiency detected by screening of 200,000 infants. *N. Engl. J. Med.* **294,** 1316–1321.

Takahashi, K., Okita, K., Nakagawa, M., and Yamanaka, S. (2007). Induction of pluripotent stem cells from fibroblast cultures. *Nat. Protoc.* **2,** 3081–3089.

Takehara, S., Zhang, J., Yang, X., Takahashi, N., Mikami, B., and Onda, M. (2010). Refolding and polymerization pathways of neuroserpin. *J. Mol. Biol.* **403**(5), 751–762.

Tang, G., Peng, L., Baldwin, P. R., Mann, D. S., Jiang, W., Rees, I., and Ludtke, S. J. (2007). EMAN2: An extensible image processing suite for electron microscopy. *J. Struct. Biol.* **147,** 38–46.

Teckman, J. H., and Perlmutter, D. H. (2000). Retention of mutant α_1-antitrypsin Z in endoplasmic reticulum is associated with an autophagic response. *Am. J. Physiol. Gastrointest. Liver Physiol.* **279,** G961–G974.

Teckman, J. H., An, J. K., Loethen, S., and Perlmutter, D. H. (2002). Fasting in α_1-antitrypsin deficient liver: Constitutive activation of autophagy. *Am. J. Physiol. Gastrointest. Liver Physiol.* **283,** G1156–G1165.

Touboul, T., Hannan, N. R., Corbineau, S., Martinez, A., Martinet, C., Branchereau, S., Mainot, S., Strick-Marchand, H., Pedersen, R., Di Santo, J., Weber, A., and Vallier, L. (2010). Generation of functional hepatocytes from human embryonic stem cells under chemically defined conditions that recapitulate liver development. *Hepatology* **51,** 1754–1765.

Ueda, M., Mashiba, S., and Uchida, K. (2002). Evaluation of oxidized alpha-1-antitrypsin in blood as an oxidative stress marker using anti-oxidative alpha1-AT monoclonal antibody. *Clin. Chim. Acta* **317**(1–2), 125–131.

Vallier, L., and Pedersen, R. (2008). Differentiation of human embryonic stem cells in adherent and in chemically defined culture conditions. *Curr. Protoc. Stem Cell Biol.* Chapter 1: p. Unit 1D.4.1–1D.4.7.

Vallier, L., Touboul, T., Brown, S., Cho, C., Bilican, B., Alexander, M., Cedervall, J., Chandran, S., Ahrlund-Richter, L., Weber, A., and Pedersen, R. A. (2009a). Signalling pathways controlling pluripotency and early cell fate decisions of human induced pluripotent stem cells. *Stem Cells* **27,** 2655–2666.

Vallier, L., Touboul, T., Chng, Z., Brimpari, M., Hannan, N., Millan, E., Smithers, L. E., Trotter, M., Rugg-Gunn, P., Weber, A., and Pedersen, R. A. (2009b). Early cell fate decisions of human embryonic stem cells and mouse epiblast stem cells are controlled by the same signalling pathways. *PLoS One* **4,** e6082.

van Heel, M., Harauz, G., Orlova, E. V., Schmidt, R., and Schatz, M. (1996). A new generation of the IMAGIC image processing system. *J. Struct. Biol.* **116,** 17–24.

Wittig, I., Braun, H.-P., and Schägger, H. (2006). Blue native PAGE. *Nat. Protoc.* **1**(1), 418–428.

Yamasaki, M., Li, W., Johnson, D. J., and Huntington, J. A. (2008). Crystal structure of a stable dimer reveals the molecular basis of serpin polymerization. *Nature* **455,** 1255–1258.

Author Index

A

Abagyan, R., 148, 151, 152, 154, 155, 160–161, 167–168, 404
Abbott, G. L., 22–23
Abderrahmani, R., 178
Abell, C., 373
Abou-Gharbia, M., 184, 185, 189–190
Aboussouan, L. S., 11–12
Abrahams, J. P., 51, 52, 54, 65, 77, 90, 118–120, 123–124, 141, 142, 182–183, 296–298
Abrahamsson, T., 178
Abts, H. F., 225
Abuknesha, R., 194
Acornley, A., 194
Acton, A., 309
Adams, D. S., 188–189
Adams, J. J., 404–405
Adams, P. D., 53–54, 56, 97
Adams, T. E., 16, 65, 76–77, 110, 120–121, 182–183, 240
Adrain, C., 260–262
Aebersold, R., 287
Aertgeerts, K., 65
Agarwal, S., 15, 16
Ahmad, M., 451–452
Ahmed, M., 451–452
Ahrlund-Richter, L., 167, 168, 457, 459–460, 461
Aiach, M., 121–122
Aihara, K., 121–122, 128–129
Ajzenberg, N., 121
Akaike, M., 121–122, 128–129
Akutsu, T., 241, 256–257
Alam, S., 455–456
Al-Ayyoubi, M., 50–51, 65
Albini, A., 276–277
Albornoz, F., 178
Albrecht, D. W., 257
Aleshkov, S., 182–183
Alessi, M. C., 178
Alex, A., 153
Alexander, G. J., 167, 168, 382, 383–391, 423, 436, 437, 439–440, 441, 443, 457, 461
Alexander, M., 459–460
Ali, F., 451–452
Ali, R., 455–456
Allain, F. H., 37
Allen Annis, D., 156–157
Allen, M. P., 299
Amadei, A., 148–149, 308
Amicone, L., 455–456
Amir, D., 353, 356
Amla, D. V., 15, 16
An, C., 210–211
Andersen, T. B., 121, 189–190
Andersson, L., 108
Andreasen, P. A., 121, 178, 181–182, 183, 184, 185, 186, 189–190, 385, 391, 412–414
Andrew, M., 109
Andries, S., 423–424
Anisowicz, A., 296
An, J. K., 148, 151, 451–453
Annamali, M., 168, 452–453
Annand, R. R., 225
Antonik, M., 369–372
Antrilli, T. M., 178
Aoki, N., 230
Archontis, G., 309
Arcone, R., 110
Arii, Y., 65
Arkin, I. T., 300
Armstrong, R. C., 166
Armstrong, R. N., 113
Arocas, V., 109, 113
Asanuma, K., 130–131
Aschmies, S., 178
Asgedom, M., 145–147
Askew, D. J., 210–211, 238
Askew, Y. S., 210–211
Astola, N., 356, 359–362
Atchison, K., 178
Atchley, W. R., 217
Atha, D. H., 108
Atilgan, A. R., 303, 304
Atkins, J. F., 243–244
Atkinson, C., 448, 454
Augeri, D. J., 166
Austin, R. C., 110
Ausubel, F. M., 42
Auton, T. R., 153
Avagyan, V., 215
Azuma, H., 121–122, 128–129

B

Backovic, M., 164
Badertscher, M., 159
Baek, J. H., 338–340, 343, 424

Baeuerle, P. A., 445
Baglin, T. P., 55, 65, 76–77, 113, 118–120, 125–127
Bahar, I., 298–299, 303, 304
Bahr, U., 156–157
Bailey, T. J., 90–91
Bai, Y., 326–327
Bakajin, O., 356
Bakan, A., 303
Baker, D., 57
Bakowies, D., 299, 309, 310
Balasubramanian, S., 353, 356, 358, 364–366, 367
Baldwin, D., 210–211
Baldwin, P. R., 434–435
Balint, M., 76, 296
Balsera, M., 308
Banaszak, L. J., 149
Banck, M., 309
Banda, M., 448
Baneyx, F., 18
Bangert, K., 226
Banks, J. L., 152, 153
Bannen, R. M., 304
Baranauskiene, L., 156
Barclay, G. B., 450
Bardin, C. W., 90–91
Barfield, R. C., 16
Barkan, D. T., 256–257
Barnes, R. C., 14–15
Barnickel, G., 149
Baron, G. S., 195
Baron, R., 299, 309, 310
Barrett, R. W., 247–248
Barrett, T. E., 149, 167, 334, 341, 404
Bartels, C., 309
Barton, G. J., 258
Bartuski, A. J., 15, 225
Barzegar, S., 128
Bashford, D., 309–311
Bashtannyk-Puhalovich, T. A., 51, 56, 408
Bassel-Duby, R. S., 188–189
Bateman, R. H., 158
Battey, F. D., 181–182
Bauer, M., 65, 408
Baumann, B., 423–424
Baumann, U., 53, 65, 81, 91, 142, 297, 398, 404–405, 408
Baumgartel, P., 356
Baxi, S., 178
Bayly, C. I., 311, 313–314
Bealer, K., 215
Beatty, K., 225
Beauchamp, N. J., 95, 298
Becerra, S. P., 14–15
Becher, A. M., 210–211
Becker, J. W., 260–262
Becker, L. M., 358–359, 423, 428, 444
Beckett, C., 168, 451–452
Beddoe, T. C., 51, 56
Beechem, J. M., 108, 109–110
Beeler, D., 128
Beinrohr, L., 51, 56, 65, 81–82, 298
Bekker, H., 316
Belleri, M., 276–277
Belli, B. A., 166
Bellott, M., 309–311
Belmares, R., 65, 76–77, 240
Belorgey, D., 18, 179–180, 353, 354, 355–356, 358–359, 364–367, 368, 372, 373, 382, 383–391, 392–393, 394–395, 401–402, 404–406, 411, 423, 424–425, 426, 428, 436–440, 441, 443, 444, 445
Belzar, K. J., 50, 52, 55, 65, 118–120
Benarafa, C., 258
Benderitter, M., 178
Ben-Naim, A., 153
Bennett, B., 178
Bentele, C., 210–211
Berendsen, H. J. C., 148–149, 299, 305, 308, 315
Berger, A., 210, 258
Bergeron, J. J., 167
Bergfors, T. M., 51–52
Bergstrom, F., 182–183
Bergstrom, R. C., 65, 76–77, 240
Berkenpas, M. B., 50, 51, 55–56, 65, 81–82, 179–180, 186, 199, 224
Berman, H. M., 50, 311
Bernacki, K., 153
Bernstein, E. F., 178
Berry, C. N., 178
Berry, L., 109
Bertenshaw, R., 93, 95
Bertini, I., 165
Best, C. F., 96–97
Best, R. B., 163–164
Betts, R. L., 186, 187, 188
Betz, A., 109
Betzel, C., 179
Bhiladvala, P., 121–122
Bialas, R. C., 225
Biemond, B. J., 178
Bieth, J. G., 225, 226, 291, 354, 368, 372
Bilican, B., 159–160
Billeter, S. R., 309, 310
Bilton, D., 424–425, 427, 429, 448, 453–454
Binder, B. R., 130
Birch, N. P., 439–440
Bird, C. H., 225, 241–242, 245, 260–262
Bird, P. I., 10, 14, 15, 51, 56, 148, 210–211, 225, 238, 240, 241–242, 245, 257–258, 260–262, 297, 385, 395
Birkett, N. R., 353, 356, 358, 364–366, 367
Bissantz, C., 153
Bissonnette, L., 210
Bitomsky, W., 122–124

Björk, I., 107, 108, 109, 111–113, 142, 182–183, 225, 226, 230, 241
Björquist, P., 50, 65, 81–82, 183, 184, 185, 189–190, 381, 410
Blackshields, G., 253, 258
Blacque, O. E., 14–15
Bladergroen, B. A., 436–439
Blajchman, M., 121–122
Blake, J. T., 242
Bleasby, A., 213
Blennow, K., 439–440
Blombäck, M., 178
Blomenkamp, K., 168, 451–453
Blouse, G. E., 22–23, 178, 181–182, 186, 187, 188, 189–190
Blum, R., 225
Blundell, T. L., 57, 122–123, 148–149
Bock, I., 113
Bock, P. E., 115, 142
Bock, S. C., 109, 113
Bode, W., 53, 65, 81, 187–188, 223–224, 291, 296–298
Bodker, J. S., 178, 181–182
Bohlender, J., 99–100
Bohm, H.-J., 153
Boisclair, M., 121–122
Bojanic, D., 189
Bollen, A., 14–15
Bond, P. J., 299
Bonhomme, E., 178
Bonnet, G., 358, 359–361, 362
Booth, N. A., 14–15, 178
Border, W. A., 178
Bordin, G. M., 178
Boresch, S., 309
Börner, S., 210–211, 217, 240
Bottegoni, G., 154
Bottomley, S. P., 10, 14, 15, 16–17, 18–19, 26–27, 51, 56, 64, 65, 77, 78–79, 80, 118–120, 142, 210–211, 225, 226, 228, 233, 238, 240, 257–258, 296, 298, 299–300, 304, 306, 338–340, 344–345, 353, 368, 383–391, 392–394, 395–398, 399–401, 402, 404–405, 408–410, 424–425, 427
Boucher, L., 113
Boudier, C., 116–117, 226
Boukedes, S. S., 455
Boulynko, I. A., 225
Bouzhir, L., 96–97
Boyce, S. E., 153
Boyd, S. E., 241, 242, 243, 254, 256–257, 260–262
Bradley, P., 57
Bradshaw, C., 358–359, 423, 428, 444
Branchereau, S., 460
Brandal, S., 225
Brandt, K. S., 210–211
Brantly, M. L., 402–403, 424–425, 448, 453–454
Bratton, D., 373
Bräuchle, C., 358
Braun, B. C., 182
Braun, R., 309
Bréfort, J., 309
Brennan, S. O., 90–91, 240
Briand, C., 50, 57, 65
Brimpari, M., 460
Brix, D. M., 121, 189–190
Brix, K., 225
Bromme, D., 225
Brooijmans, N., 148, 152, 153
Brooks, B. R., 309
Brooks, C. L. III., 309, 311, 312
Brooks, P. C., 178
Brooks, S. A., 187
Broos, T., 121, 189–190
Broughton Pipkin, F., 52, 56–57, 58, 96–98, 99–100
Brown, A. J., 117, 178
Brown, N. J., 178
Brown, N. P., 253, 258
Brown, S. C., 334, 341, 404, 459–460
Brozell, S. R., 152
Bruce, D., 404–405
Bruncko, M., 166
Brunger, A. T., 307
Brüning, M., 210–211
Brunt, E. M., 168, 451–453
Bryans, J., 184, 185
Bryant, S. H., 151–152
Bryson, K., 268–269
Buard, V., 178
Buchacher, A., 9–10
Buckle, A. M., 14–15, 18, 51, 56, 65, 77, 78–79, 238, 241, 257–258, 296, 298, 408
Buckley, D. I., 199, 298
Buller, H. R., 168
Bull, H. G., 260–262
Bulynko, Y. A., 408
Bunagan, M. R., 353, 355–356, 358–363, 373
Burckle, C., 96–97
Burgi, R., 309
Burke, D. F., 122–123
Burnelis, V., 156
Burrows, J. A. J., 166, 404–405
Burton, G. J., 98–99
Buschmann, V., 354
Busenlehner, L. S., 113
Bush, M. F., 144, 344–345, 346–347, 385, 391, 392–393, 412–414, 424, 437, 440
Bustamante, C., 367–368, 373
Butler, J. S., 353, 356
Butler, L., 95, 298
Buzza, M., 51, 56

C

Cabral, C. M., 424
Cabrita, L. D., 10, 15, 16–17, 18, 65, 77, 78–79, 296, 298, 299–300, 304, 338–340, 385, 395, 396–398, 402
Caccia, S., 65
Cadane, M., 226
Cadwallader, K., 448, 453–454
Caflisch, A., 309
Cahoon, M., 354, 368, 372
Caldwell, J. W., 311, 313–314
Cale, J. M., 178, 184, 185, 189–190, 192–194
Callea, F., 385, 392–393, 454
Calugaru, S. V., 189, 224–225
Calvin, J., 395–396, 398, 399–402
Camacho, C., 215
Cama, E., 122–123
Cameron, A., 95
Cameron, P. H., 167
Campbell, E. J., 455
Campbell, I. D., 117–118
Campbell, M. A., 455
Camproux, A.-C., 148, 151
Campuzano, I., 158
Cao, Y., 291
Capila, I., 117–118
Carl, S. A., 367–368
Carlson, H. A., 152–153
Carlson, J. A., 423, 451–452
Carmeliet, P., 128, 130
Carrell, R. W., 9–10, 14–15, 18, 20–21, 50–51, 52, 54, 55–57, 58, 64–78, 79, 80, 81–82, 89–106, 118–120, 123–124, 125–127, 141, 142, 144, 145, 148, 179–180, 181–183, 186, 199, 210, 224–225, 238–240, 267–268, 291, 296–298, 299–300, 314, 336, 344–345, 358–359, 368, 370–372, 382, 383–391, 392–394, 395–398, 397, 399, 401–403, 404–411, 412–415, 423–424, 428, 429, 436–439, 441
Carroll, A. R., 184, 185, 189–190
Carr, R. A. E., 161
Carruthers, L. M., 37
Carter, W. J., 122–123
Case, D. A., 152, 153, 310, 311
Cassatella, M. A., 276–277
Casu, B., 116–117
Cataltepe, S., 225, 238
Caughey, B., 195
Caves, L., 309
Cech, A. L., 65, 76–77, 240
Cedervall, J., 459–460
Celerier, J., 98
Chader, G. J., 14–15
Chakraborty, K., 356, 359–362
Chamankhah, M., 210–211
Chambers, S., 382, 383–391, 423, 436, 437, 439–440, 441, 443
Chan, A. K. C., 109
Chance, M. R., 113, 114, 160–161, 300–301, 330, 333
Chandran, S., 459–460
Chandrasekaran, V., 300–301
Chandrasekhar, I., 299
Chang, H. T., 364, 385, 405–406
Chang, W. S. W., 9–10, 65, 79, 142, 144, 145–147, 167, 298, 299–300, 336, 344–345, 383–391, 393–394, 395–398, 399–402, 404–408, 410, 412–414, 436–439
Chang, Y. P., 144, 145–147, 167, 405–408
Chang, Z., 151–152
Chan, H. S., 341, 342
Chan, W. L., 90–93, 95
Chapman, H. A., 225
Charlton, P., 184, 185, 392–393
Chattopadhyay, K., 353, 356
Chaubet, F., 121
Cheatham, T. E. III., 310
Chemla, D. S., 353, 357–358, 369–370
Chen, A. Y., 353, 355–356, 358–359, 364–367, 373
Chen, C. T., 256–257
Cheng, W., 370
Cheng, X. F., 181–182
Chen, H., 353, 356
Chen, J., 195
Chenna, R., 253, 258
Chen, S. H., 283–285
Chen, Y., 194
Chernushevich, I. V., 157–158
Chevet, E., 167
Chicarelli-Robinson, I., 184, 185
Chi Guo, A., 151–152
Chilvers, E. R., 448, 449, 450, 453–454
Chinali, A., 110
Chiou, A., 353, 355–356, 358–359, 364–367, 373
Chipot, C., 309
Chiu, W., 434–435
Chmielewska, J., 187–188
Chng, Z., 460
Choay, J., 53, 65, 81, 108, 113, 241
Cho, C., 459–460
Cho, E., 276–277
Choi, H. J., 65
Chothia, C., 54–55, 181–182
Choudhury, P., 424
Chou, P. Y., 145
Chowdhury, P., 353, 355–356, 358–363, 373
Chowdhury, V., 121–122
Chow, M. K., 344–345, 383–391, 392–393, 395–398, 402
Chow, N. H., 283–285
Chrenek, P., 130
Christen, M., 299, 309
Christensen, A., 121, 178, 181–182, 183, 186, 189–190, 391, 412–414

Christensen, U., 226
Christianson, D. W., 76, 296
Christi, P., 128
Christodoulou, J., 139–175
Church, F. C., 65, 76–77, 114, 120–121, 125–127, 141, 148, 210, 225, 238, 241
Chu, Y. H., 144, 145–147, 167, 405–408
Cieplak, P., 311, 313–314
Cierpicki, T., 179
Cimmperman, P., 156
Clamp, M., 258
Clarke, A., 95
Clarke, D. T., 117
Clarke, R. W., 353, 356, 358, 364–366, 367–368, 373
Clarkson, J., 117–118
Clayton, P. T., 335–336
Cleland, S. J., 96–97
Coates, L. C., 439–440
Coburn, G. A., 14–15
Coetzer, T. H., 225
Coffman, T. M., 96–97
Cohn, Z. N., 451
Colaert, N., 260–262
Coldwell, M., 194
Cole, E. R., 260–262
Cole, J. C., 152
Cole, K., 210–211
Collart, F. R., 354
Collen, D., 128, 178, 187–188, 291
Collins, G. H., 358–359
Collins, M., 184, 185
Colucci, M., 178
Comery, T. A., 178
Condliffe, A. M., 448, 449, 450
Condron, B. G., 243–244
Congreve, M., 161
Connell, J., 109
Constant, A. A., 122–124
Coombs, G. S., 187–188
Cooper, D. N., 121–122
Cooperman, B. S., 14–15, 76, 296, 353, 355–356, 358–363, 364, 366, 373
Cooper, S. T., 120–121
Corbineau, S., 167, 168, 457, 460, 461
Corey, D. R., 65, 76–77, 187–188, 240
Cornell, W. D., 311, 313–314
Coronel, R., 178
Corrales, F. J., 393
Corral, J., 121–122, 385, 393–394, 408–410
Corvol, P., 98, 99–100
Cosentino, C., 117
Couch, G. S., 310
Coughlin, P. B., 14–15, 98, 148, 210, 225, 226, 228, 230, 233, 238, 296
Coughlin, S. R., 260–262
Coulouris, G., 215
Coulter, J., 14–15
Courtney, M., 402–403
Coutelier, M., 423–424
Coutsias, E. A., 303
Covell, D. G., 299
Cowburn, A. S., 448, 449
Cowburn, D., 165
Cox, D. W., 344–345, 429
Cox, J. H., 276–277
Craig, P. A., 113, 241
Craik, C. S., 256–257, 260–262, 297
Crain, P., 99–100
Crandall, D. L., 178, 182–183, 184, 185, 186–187, 189–190
Cravador, A., 14–15
Crawford, A. M., 373
Crowther, D., 366
Crowther, D. C., 18, 167, 364, 367, 385, 401–402, 404–406, 423–424, 436–440, 443, 444, 445
Crowther, R. A., 434–435
Cruz, M., 309
Crystal, R. G., 402–403
Cui, Q., 299, 304
Cullen, S. P., 260–262
Cummings, R. T., 194–195
Curtis, D., 309
Cwirla, S. E., 247–248
Cybulsky, M. I., 449

D

Dafforn, T. R., 65, 79, 95, 142, 144, 145–147, 148, 149, 154, 164, 180, 296, 297, 298, 300–301, 314–315, 326, 332–333, 344–345, 381, 385, 392–393, 395–396, 398, 399–402, 405–406, 414–415, 423–425, 427
Dahan, M., 353, 357–358, 369–370
Dahlen, J. R., 225
Dai, W., 10, 15, 65, 80, 142, 225, 298, 299–300, 344–345, 385, 393–394, 395, 408–410
Dale, I., 194
Dalke, A., 310, 316
Dalton, L. E., 421–467
Dalvit, C., 165, 166
Daly, M., 95, 298
Dan, B., 423–424
Dang, A. T., 187–188
Daniel, J. M., 158
Danielli, A., 210–211
Danielsson, Å ., 107, 108
Danley, D. E., 51, 54, 55–56, 65, 78, 81–82, 90, 179–180, 298, 381
Darden, T., 310
Das, R., 57
Daudi, I., 276–277
Daura, X., 299
Dauter, Z., 179
Davey, C. A., 30

Davidson, M., 353, 373
Davies, M. J., 423–424, 436, 439–440, 443, 444, 445
Davis, B., 179
Davis, R. L., 358–359, 364, 405–406, 423–424, 428, 444
Day, D. E., 22–23, 179–180, 181–182, 186–187, 188–189, 198, 224–225, 368
Dayhoff, M. O., 296
Day, T., 155
de Agostini, A. I., 128
Dean, R. A., 276–277, 291
De Bock, P. J., 260–262
de Boer, B., 52, 65, 81, 90
De Bondt, H. L., 65
Debrock, S., 186–187, 189–190
Deckwerth, T. L., 166
Declerck, P. J., 184, 186–187, 189–190, 408
Declercq, W., 256–257
Deconinck, N., 423–424
De Dreu, P., 225
Deerfield, D. W. II., 213
de Faire, U., 178
De Fatima, A., 121
de Groot, B. L., 148–149, 308
Deinum, J., 50, 65, 81–82, 116, 183, 184, 185, 189–190, 381, 410
Deisenhofer, J., 53, 54, 57–58, 64, 65, 81, 89–90, 381, 402–403
Dejong, T. A., 225
de la Banda, M. G., 257
Delarue, F., 96–97
Delarue, M., 53, 65, 81
de, l. C., 113
della Rocca, C., 455–456
Demello, A. J., 373
deMello, A. J., 373
Dementiev, A., 65, 76–78, 81–82, 118–120, 123–124, 179–180, 297, 368, 370–372
Demirel, M. C., 303, 304
Denault, J. B., 210
Deng, Y. Z., 337
Deniz, A. A., 353, 357–358, 369–370
Dennison, C., 225
Denton, J., 296–298
Deperthes, D., 243, 244
DePristo, M. A., 148–149, 163–164
Derakhshan, B., 100
DeRosier, D. J., 429–430
Desai, U. R., 108, 109, 111–113, 123–124, 125–127
De Taeye, B., 178
Devlin, G. L., 344–345, 353, 356, 358, 364–366, 367, 383–391, 392–393, 395–398, 402, 404–405
de Vries, A. H., 302, 312
Dewerchin, M., 128, 130
Dewilde, M., 65
Dhahri, M., 121
Diamond, S. L., 243
Diaz, L. A., 291
Dibo, G., 145–147
Dickson, M. E., 96–97
Di Giusto, D. A., 65
Dijkema, R., 52, 65, 81, 90
Dilley, R. B., 178
Dill, K. A., 340, 341, 342
Dimitrov, S., 30
Dinges, J., 166
Ding, Y., 291
DiNola, A., 315
Dinola, A., 305
Dippold, C., 168
Di Santo, J., 460
Dishon, D. S., 423
Dixon, J. D., 178
Dobó, J., 51, 56, 65, 76, 81–82, 179–180, 298, 368, 370–372
Dobson, C. M., 163–164, 353, 356, 358, 364–366, 367
Dominaitiene, R., 455–456
Dominy, B. W., 161
Doolittle, R. F., 90–91
Doose, S., 356
Doucet, A., 276, 277, 284
Dower, W. J., 247–248
Drenth, J., 50
Dress, A., 217
Drew, H., 91–93
Duchesne, L., 117
Dudek-Wojciechowska, G., 90–91
Dufour, E. K., 210
Duke, R. E., 300–301
Dülcks, T., 156–157
Dunbar, P. R., 439–440
Dunbrack, R. L., 309–311
Dunne, P. D., 353, 355–356, 358–359, 364–367, 373
Dunstone, M. A., 63–89, 65, 142, 298, 344–345, 393–394, 408–410
Dupont, D. M., 121, 178, 181–182, 189–190
Durand, M. K., 178
Durell, S. R., 303, 304
Duriez, P. J., 260–262
Duyckaerts, C., 423–424
Dycaico, M. J., 451–452

E

Earnest, T., 297
Eastman, P., 309–310
Eaton, W. A., 353, 356, 373
Edel, J. B., 373
Edholm, O., 305
Edwards, Y., 455–456
Egelund, R., 181–182, 183, 184, 185
Ehnebom, J., 183, 184, 185, 189–190

Ehrlich, H. J., 240–241
Eid, J. S., 194
Einarsson, R., 108
Einholm, A. P., 183, 186, 385, 391, 412–414
Eising, A. A., 309, 310
Eitzman, D. T., 128–129
Ekeowa, U. I., 144, 344–345, 346–347, 367, 382, 383–391, 392–393, 412–414, 423, 424, 436, 437, 438, 439–440, 441, 443, 444, 445
El, A. S., 178
Eldaw, A., 194
Eldridge, M. D., 153
Elg, S., 116
Eliseenkova, A. V., 122–123
Elisen, M. G. L. M., 130–131
Elliott, P. R., 9–10, 65, 148, 149, 154, 180, 296, 297, 298, 300–301, 314–315, 332–333, 344–345, 358–359, 381, 385, 395–396, 404–405, 423–425, 427, 429, 448, 453–454
Ellisdon, A. M., 396–398, 402
Elmore, S. W., 166
Elokdah, H., 178, 182–183, 184, 185, 186–187, 189–190
Elson, E. L., 353, 356, 357–358, 359–361, 363
Emal, C. D., 184, 185, 189–190, 192–194
Emmerich, J., 121–122
Emod, I., 276–277
Engelborghs, Y., 367–368
Engen, J. R., 327
Enghild, J. J., 291, 298, 383, 384, 385, 393–394, 398, 404–405, 408–410
Engh, R. A., 65, 117, 121
Eng, J., 287
Englander, S. W., 326–327
Enjyoji, K., 128
Ens, W., 157–158
Eren, M., 178
Ericson, J., 179–180
Eriksson, S., 385, 392–393, 423, 454, 455–456
Eriksson, U., 187–188
Erlandson, M., 210–211
Erlanson, D. A., 160–161
Esmon, C. T., 65, 76–77, 118–121, 123–124, 125–127, 240–241, 296–298
Etchebest, C., 303
Evans, D. L. I., 89–90, 91, 142, 144, 238–240, 298, 299–300, 314, 358–359, 383–391, 395–398, 399, 401–402, 405, 410, 414–415, 423, 428, 436–439, 441
Evanseck, J. D., 309–311
Ewing, M., 168, 451–452
Eyal, E., 304

F

Faham, S., 122–123
Faint, R., 184, 185, 385, 392–393
Falchetto, R., 156–157
Falcone, D. J., 276–277
Falk-Marzillier, J., 276–277
Falk, W., 450–451
Fa, M., 182–183, 224
Fan, B., 109
Fang, F. C., 100
Fan, K., 182–183, 186–187
Farahi, N., 448, 449, 453–454
Fareed, J., 116
Farid, R., 155
Farris, D., 178
Fasman, G. D., 145
Fauman, E., 297
Fay, W. P., 178
Fears, C. Y., 276–277
Feeney, P. J., 161
Feiglin, D., 358–359, 423, 428, 444
Feig, M., 311, 312
Feil, S. C., 18–19, 26–27, 65, 80, 142, 298, 344–345, 353, 393–394, 404–405, 408–410
Felekyan, S., 369–372
Felts, A. K., 148, 309, 310
Felts, K., 451–452
Ferguson, D. M., 311, 313–314
Fermi, G., 52, 65, 77, 81, 90, 118–120, 296–298
Fernandes, S. A., 121
Fernig, D. G., 117
Ferreira-da-Silva, J. M. S., 307
Ferrell, G. A., 423
Ferrin, T. E., 310
Fersht, A. R., 179
Fetterer, R. H., 16
Fieber, C., 187–188
Field, M. J., 309–311
Filippis, V. D., 110
Filipuzzi, I., 156–157
Finch, J. T., 64, 65, 81, 89–90, 142, 298, 299–300, 314, 344–345, 358–359, 382, 385, 391, 395–396, 399, 405, 410, 414–415, 423, 428, 429, 441
Finegold, M. J., 451–452
Fingerhut, A., 93, 182
Fischer, S., 309–311
Fitton, H. L., 95, 142, 144, 298
Flatow, D., 299
Fletterick, R. J., 260–262
Flink, I. L., 90–91
Flynn, A. M., 296–298
Fogel, D. B., 53–54
Fogo, A. B., 178
Foidart, J. M., 276–277
Folkers, G., 153
Foreman, R. C., 154, 164, 344–345, 395–396, 398, 399–403, 429
Forsyth, S., 210
Fortenberry, Y. M., 225
Foster, D. C., 225
Fowler, A., 194
Fox, D. M., 110, 225
Fox, T., 311, 313–314

Fraaije, J. G. E. M., 316
Francis-Chmura, A. M., 109, 181–182, 198, 368
Francois, A., 178
Frangione, B., 276–277
Frank, J., 429–430, 434–435
Freddolino, P. L., 302
Fredenburgh, J. C., 109, 110
Frederix, L., 178
Fredriksson, L., 187–188
Freekeb, J., 424, 437, 440
Freeke, J., 144, 344–345, 346–347, 367, 385, 391, 392–393, 412–414, 424, 436, 438, 439, 443, 444
Freisner, R. A., 155
Frieden, C., 353, 356
Friedrichs, G. S., 178
Friesner, R. A., 148, 152, 153, 312
Friess, S. D., 158
Froelich, C., 368
Frohnert, P. W., 117, 121
Froman, G., 246, 247
Fromm, D. P., 355–356
Fromm, J. R., 122–123
Frow, E. K., 450–451
Fuhrmann, U., 182
Fukuda, T., 128–129
Fulton, K. F., 65
Furie, M. B., 276–277
Furka, A., 145–147
Furnham, N., 148–149
Furtmüller, M., 130
Futamura, A., 109–110

G

Gabay, J. E., 451
Gabazza, E. C., 130–131
Gadea, F. X., 304
Gaiduk, A., 369–372
Gai, F., 353, 355–356, 358–364, 366, 373
Gaines, P. J., 210–211
Gallagher, J., 113
Gallicchio, E., 309, 310
Gál, P., 51, 56, 65, 81–82, 298
Ganem, B., 159–160
Gao, J., 309–311
Garcia, A. E., 306
Garcia-Calvo, M., 260–262
Garcia de la Banda, M., 242, 243, 254, 256–257
Gardell, S. J., 178
Gardner, H. A., 276–277
Garrels, J. I., 296
Gasiunas, N., 113
Gassman, N. R., 354
Gee, P., 299
Geerke, D. P., 299, 309
Gehlhaar, D. K., 53–54
Geiger, M., 130
Geissler, W. M., 260–262
Geller, S. A., 451–452
Genov, N., 179
Geoghegan, K. F., 51, 54, 55–56, 65, 78, 81–82, 90, 179–180, 298, 381
Georgel, P. T., 37, 40
Gerard, R. D., 51, 54, 55–56, 65, 78, 81–82, 90, 179–180, 188–189, 298, 381
Gerber, S., 166
Gerewitz, J., 210
Gerhard, L., 423–424
Gershenson, A., 109, 113, 160–161, 331, 353, 355–356, 358, 368, 370–372, 371
Gesteland, R. F., 243–244
Gething, M. H., 188–189
Gettins, P. G. W., 21–22, 50–51, 52, 65, 76–78, 81–82, 91–93, 108, 109–110, 113, 118–120, 121, 122, 123–124, 142, 148, 164–165, 179–180, 182–183, 210, 224–225, 238, 269, 297, 326, 333, 334–335, 348, 368, 370–372
Gevaert, K., 256–257, 260–262
Ghariani, S., 423–424
Ghelis, C., 296
Ghetti, B., 423–424
Giannelli, G., 276–277
Gibson, T. J., 253, 258
Giegel, D. A., 225
Giles, K., 158
Giller, T., 96–97
Gils, A., 184, 189–190, 408
Gilson, M., 373
Gimenez-Roqueplo, A. P., 98
Ginsburg, D., 50, 51, 55–56, 65, 81–82, 178, 179–180, 181–182, 186, 198, 199, 224–225, 368
Giraldo, R., 121–122
Giralt, E., 165
Girardot, C., 178
Giri, T. K., 129–130
Girma, M., 130–131
Gladson, C. L., 276–277
Glattli, A., 299
Gliemann, J., 181–182
Glykos, N. M., 53–54
Goddard, T. D., 310
Godfraind, C., 123–124
Goethals, M., 260–262
Goga, N., 302
Gohlke, H., 153, 310
Golcher, H. M., 93, 182
Goldbach, C., 168
Goldfinger, L. E., 276–277
Gold, L. I., 276–277
Goldner, L. S., 373
Goldsmith, E. J., 51, 54, 55–56, 65, 76–77, 78, 81–82, 90, 179–180, 188–189, 198, 240, 298, 381
Golubkova, N. V., 291

Golubkov, V. S., 291
Gongora, M. C., 98–99
Gonzales, C., 178
Gonzalez-Conejero, R., 121–122
Goodman, L. J., 253–254
Goodwin, R. H. Jr., 450–451
Gooptu, B., 65, 79, 142, 144, 149, 151, 155, 160–161, 167–168, 334, 341, 344–345, 346–347, 358–359, 382, 383–391, 392–393, 401–402, 404, 412–414, 423, 424, 436, 437, 439–440, 441, 443, 445
Gopich, I. V., 369–372
Goping, I. S., 90–91
Gordon, N. C., 21–22, 372
Gore, M. G., 338–340, 391, 392
Gorlatova, N., 181–182
Gorlatova, N. V., 182–183, 186–187
Gorman, C. M., 253–254
Gornstein, E. R., 225
Gould, I. R., 311, 313–314
Goulet, B., 31
Graf, L., 76, 296
Graham, H., 65, 199, 298
Grainger, D. J., 450–451
Gramling, M. W., 141
Grant, S. K., 194–195
Grasberger, H., 93, 182
Gratton, E., 194
Graves, A. P., 153
Greenblatt, D. M., 310
Green, C., 364, 405–406
Greene, C. M., 424–425, 448, 451–452, 453–454
Greene, L. H., 158
Green, J. J., 353, 358, 364–366
Griffin, G. L., 448
Griffin, J. P., 167, 168, 178, 457, 461
Griffin, L. A., 188–189
Grigorieff, N., 435
Grigoryev, S. A., 30–31, 42, 225
Grimshaw, P., 184, 185, 189–190
Gron, H., 225
Grootenhuis, P. D., 65, 81, 90
Grootenhuis, P. D. J., 52, 122–124
Grosse, D., 53, 65, 81, 297
Grosse-Kunstleve, R. W., 53–54, 56, 97
Gross, S. S., 100
Grubmuller, H., 308
Gruebele, M., 302
Grunert, T., 9–10
Grütter, M. G., 50, 57
Guan, J. Q., 330
Guan, M., 15
Gubb, D. C., 364, 405–406
Gudewicz, P. W., 276–277
Guerrero, J. A., 121–122
Guerrini, M., 116–117
Guglieri, S., 116–117
Guidolin, A., 110, 230
Guimaraes, J. A., 122–123, 300
Guimond, S. E., 117
Guinto, E. R., 225
Gumbart, J., 309
Gunsalus, G. L., 90–91
Guo, H., 309–311
Gustafson, T. A., 90–91
Guthrie, L. A., 448–449
Gutierrez-Gallego, R., 121–122
Guymer, G., 184, 185, 189–190
Guy, R. K., 166

H

Haak, J. R., 315
Haas, E., 353, 356
Haas, J., 148–149
Hack, C. E., 436–439
Hadzic, N., 382, 383–391, 423, 436, 437, 439–440, 441, 443
Haenni, D., 356
Hagaman, J. R., 96–97
Hagglof, P., 179–180, 182–183, 392–393, 411
Hägglöf, P., 353, 355–356, 358–359, 364–367, 373, 445
Hajduk, P. J., 166
Hajmohammadi, S., 128
Hale, P., 166, 167, 168, 451–452
Halgren, T. A., 151, 152, 153
Halloran, D., 168, 452–453
Halperin, I., 153
Hammond, G. L., 65, 78, 90–94, 182, 296
Hamsten, A., 178
Hannan, N. R., 167, 168, 457, 460, 461
Hänsch, G. M., 448
Hansen, G., 18–19, 26–27, 353, 404–405
Hansen, J. C., 30
Hansen, M., 178
Hansen, S. K., 160–161
Hansson, L., 183
Hansson, T., 299
Hao, G., 100
Haq, I., 139–175, 421–467
Haraguchi, M., 178
Haran, G., 353, 356
Harauz, G., 434–435
Harding, H. P., 445, 446
Hardy, D. J., 301
Hari, S. B., 156
Harley, M. J., 90–91
Harmat, V., 51, 56, 65, 81–82, 298
Harris, J. L., 260–262
Harris, L. E., 65, 81–82, 385, 393–394, 408–410, 413, 414–415
Harrison, D. G., 98–99
Harrop, S. J., 65
Hartl, F. U., 356, 359–362

Hartshorn, M. J., 152
Ha, S., 309–311
Hashimoto, C., 215–217
Haslett, C., 448–449
Hassanali, M., 151–152
Hassine, M., 121
Hassinen, T., 309
Hastie, N. D., 210
Ha, T., 353, 354, 355–356, 357–358, 369–370
Hata, S., 385, 392–393, 454
Hattori, H., 9–10
Hautamaki, R. D., 455–456
Havel, T. F., 342
Hawkins, H. K., 451–452
Hawley, A., 178
Hayashi, T., 130–131
Hayashi, Y., 95
Hayer-Hartl, M., 356, 359–362
Hazes, B., 54, 65, 79, 80, 142, 298, 344–345, 385, 393–394, 408–410
Heck, A. J., 156
Hedayat, H., 291
Hedstrom, L., 109, 353, 354, 355–356, 358, 368, 370–372
Hegedus, D. D., 210–211
Heger, A., 9–10
Heikkilä, V., 309
Heinz, T. N., 309, 316
Heiple, J. M., 451
Hejgaard, J., 262
Hekman, C. M., 9–10, 187
He, L., 128–130
Helin, C., 99–100
Heller, M., 165
Hellman, L., 246, 247
Helmerson, K., 373
Hemsley, P., 194
Henderson, K. N., 408
Henderson, R., 434–435
Hendlich, M., 149, 153
Hendrickx, M. L. V., 121, 189–190
Hendrix, J., 367–368
Hendrix, M. J., 296
Henion, J., 159–160
Henley, D., 95
Hennan, J. K., 178, 184, 185, 189–190
Henrick, K., 50, 311
Henry, B. L., 109
Henson, P. M., 448–449
Herault, J., 128
Herbert, J. M., 65, 76–77, 118–120, 123–124, 128
Hertzog, D. E., 356
Herve, M., 296
Hess, B., 301, 309, 315
Heymann, B., 308
Hibino, T., 14–15
Hidvegi, T., 166, 167, 168, 451–452
Higgins, W. J., 110, 225
Hileman, R. E., 116–117, 122–123
Hilger, F., 356
Hill, M., 382, 383–391, 423, 436, 437, 439–440, 441, 443
Hill, R. E., 210
Hilpert, M., 130
Hinsen, K., 303, 304
Hirschfeld, V., 370
Hitchen, C. R., 14–15, 296
Hjelm, R., 109, 110, 115
Hockney, R. C., 14–15
Hodgin, J. B., 96–97
Hoffman, H., 356
Hoffmann, A., 356
Hofmann, H., 370
Hogan, B. L. M., 296
Hohng, S., 354, 358
Hollfelder, F., 373
Hollup, S. M., 304
Holm, J., 121–122
Holohan, P. D., 358–359, 423, 428, 444
Hol, W. G. J., 52, 65, 81, 90
Homans, S. W., 165
Honig, B., 148, 151
Hook, V. Y., 16
Hopkins, F. G., 81
Hopkins, P. C. R., 14–15, 18, 20–21, 210, 297, 336, 385, 393–394, 410, 412–414
Hoppensteadt, D., 116
Horn, I. R., 181–182
Horowitz, J. C., 276–277
Horrevoets, A. J., 187, 188
Horsley, D. A., 356
Horvath, A. J., 14–15, 65, 210, 225, 226, 228, 230, 233, 296
Ho, S. O., 354
Hostetter, D. R., 256–257
Hou, T., 299
Hou, X., 210–211
Howlett, G. J., 344–345, 383–391, 392–393, 395–398, 397, 402
Hoylaerts, M., 187–188
Hreha, A. L., 178
Hricovini, M., 117
Hsieh, Y. L., 159–160
Hsu, H. J., 256–257
Hsu, J. L., 283–285
Hsu, W. L., 256–257
Huang, C. C., 310
Huang-Doran, I., 167, 168, 457, 461
Huang, S. Y., 283–285
Huang, X., 65, 77–78, 81–82
Huang, Y., 178, 210–211
Huber, R., 53, 54, 57–58, 64, 65, 81, 89–90, 91, 117, 121, 142, 223–224, 291, 296–298, 381, 398, 402–403, 404–405, 408
Huber, T., 306
Huebner, A., 373

Humphrey, W., 310, 316
Hünenberger, P. H., 299, 309, 310, 316
Huntington, J. A., 14–15, 16, 50, 52, 54, 55, 64–78, 80, 81–82, 89–90, 109, 110, 113, 118–121, 122–124, 125–127, 141, 142, 179–180, 181–183, 210–211, 224–225, 238–241, 267–268, 291, 296–298, 326, 344–345, 368, 370–372, 385, 391, 392–394, 404–410, 411, 412–415, 424
Hunt, L. T., 296
Hutchison, G., 309
Huuskonen, J., 309
Hu, X., 303
Hwang, S. R., 16

I

Ibrahimi, O. A., 122–123
Ido, M., 130–131
Ikeda, Y., 128–129
Ikonomou, T., 121–122
Im, H., 65, 338–340, 394–395, 403–404, 405–406, 424
Inghardt, T., 50, 65, 81–82, 183, 184, 185, 189–190, 381, 410
Inoue, I., 99–100
Irving, J. A., 16–17, 31, 34–35, 51, 56, 64, 65, 77, 78–79, 148, 210, 225, 226, 228, 233, 238, 296, 298, 367, 408, 424, 436, 438, 439, 443, 444
Irwin, J. J., 151–152, 154–155
Ishiguro, K., 127–128
Isralewitz, B., 308
Istomina, N. E., 42
Itkin, A., 353, 356
Ito, M., 127–128
Iwase, T., 128–129
Izrailev, S., 308

J

Jachimoviciute, S., 156
Jachno, J., 156
Jacobsen, J. S., 178
Jacobson, M. P., 153, 155
Jacobs, P., 14–15
Jaeger, E., 91, 142, 398, 404–405, 408
Jahnke, W., 165, 166
Jairajpuri, M. A., 109
James, E. L., 306, 338–340, 391, 392, 395–396, 398, 400, 401
James, T. L., 152, 165, 166
Janciauskiene, S., 385, 392–393, 448, 454, 455–456
Jandrot-Perrus, M., 121
Jankova, L., 65
Janse, M. J., 178
Janssen, O. E., 93, 95, 182
Jarvis, D. L., 187
Jarvis, E. E., 210–211
Jauniaux, E., 98–99
Jean, F., 215–217, 225
Jelen, F., 179, 296–298
Jenkins, M. C., 16
Jennette, J. C., 96–97
Jensen, J. K., 65, 121, 178, 181–182, 189–190, 309
Jensen, S., 184, 185
Jerabeck, I., 130
Jernigan, R. L., 298–299, 303, 304
Jeunemaitre, X., 98, 99–100
Jezierski, G., 300
Jha, S., 15, 16
Ji, A., 178
Jiang, H., 210–211
Jiang, W., 434–435
Jin, L., 52, 54, 55, 65, 77, 90, 118–120, 123–124, 141, 142, 182–183, 296–298
Jirasakuldech, B., 91–93
Johansen, H., 14–15
Johansson, J., 364, 405–406
Johansson, L. B. Å., 182–183
Johnson, A. W., 335–336
Johnson, D. J. D., 65, 76–77, 81–82, 90, 118–120, 123–124, 142, 182–183, 240–241, 296–298, 326, 344–345, 385, 391, 392–393, 409, 411, 412–414, 424
Johnson, J. L., 14–15
Johnston, R. B. Jr., 448–449
Jones, D. T., 268–269
Jones, E. A., 276–277
Jones, J. C., 276–277
Jorgensen, W. L., 312
Joseph-McCarthy, D., 309–311
Josephson, A., 91–93
Josic, D., 9–10
Juhan-Vague, I., 178
Jung, C. H., 394–395, 403–404, 405–406
Jungreis, R., 445
Junker, H. D., 147–148
Juraschek, R., 156–157
Justesen, J., 184

K

Kadel, J., 210–211
Kadomatsu, K., 127–128
Kadoya, K., 14–15
Kafatos, F. C., 210–211
Kaiserman, D., 10, 15, 210–211, 238, 240, 241–242, 245, 257, 260–262, 297, 385, 395
Kajihara, Y., 100
Kalamajski, S., 129–130
Kale, L., 309
Kalinin, S., 357–358, 368–370, 373
Kalsheker, N., 382, 383–391, 423, 436, 437, 439–440, 441, 443
Kalyanaraman, C., 153
Kamachi, Y., 15, 225

Kamada, H., 130–131
Kam, C. M., 242
Kaminski, G. A., 312
Kamp, P. B., 217
Kanagawa, Y., 121–122
Kane, A., 356
Kang, U. B., 424
Kanost, M. R., 65, 76–77, 210–211, 240
Kan, W. T., 14–15, 225, 296, 408
Kapanidis, A. N., 358
Karas, M., 156–157
Karlson, U., 246, 247
Karlsson, G., 9–10
Karlsson-Li, S., 179–180, 353, 355–356, 358–359, 364–367, 373, 445
Karolin, J., 182–183
Karplus, M., 299, 307
Kaslik, G., 76, 296
Kass, I., 300
Kastenholz, M. A., 299, 309
Katagiri, C., 14–15
Katagiri, K., 9–10
Katoh, K., 213
Kaufman, R. J., 443, 444, 445
Kaushal, S., 168, 452–453
Kawano, H., 128–129
Kay, L. E., 164–165
Keays, C. A., 215–217
Keeler, J., 179
Keetch, C. A., 158
Keeton, M. R., 178
Ke, H., 303
Keijer, J., 188, 240–241
Keil-Dlouha, V., 276–277
Keller, A., 287
Kelley, D. G., 448, 454
Kelly, J. W., 158
Kemp, C., 168
Kennedy, S. A., 439–440
Kenney, J. M., 186, 385, 391, 412–414
Kerr, F. K., 257
Ke, S. H., 187–188
Keshishian, H., 215–217
Keskin, O., 298–299, 303, 304
Kessler, B. M., 291
Kessler, H., 165
Khalfan, H. A., 194
Khan, K. M. F., 276–277
Kiick, K. L., 114–115, 116–117
Kim, D., 348, 394–395
Kim, H. S., 96–97
Kim, J., 385, 395–396, 402–404, 424
Kim, K. E., 338–340
Kim, M. J., 403–404, 405–406
Kim, S., 65, 451–452
Kim, S. H., 114–115, 116–117
Kim, S.-J., 53, 55
Kim, T. S., 423–424
Kim, Y. S., 215–217, 354
Kinghorn, K. J., 364, 404–406
Kinter, J., 358–359, 423, 428, 444
Kirksey, Y., 178
Kishore, R. B., 373
Kisiel, W., 225
Kissinger, C. R., 53–54
Kjelgaard, S., 178
Kjellberg, M., 16, 65, 110, 120–122
Kjems, J., 121, 189–190
Kjoller, L., 181–182, 184
Klebe, G., 153
Klemke, J. W., 353, 355–356, 358–363, 364, 366, 373
Klenerman, D., 353, 355–356, 358–359, 364–368, 373
Kleywegt, G. J., 57–58
Klicic, J. J., 152, 153
Klieber, M. A., 65, 78, 91–94, 182
Kluckman, K. D., 96–97
Klug, A., 429–430
Knaupp, A. S., 10, 15, 385, 395
Knof, S., 91, 142, 398, 404–405, 408
Knoll, E. H., 152, 153
Knox, C., 151–152
Knudsen, B. S., 197
Kobayashi, D. K., 455–456
Kobori, H., 98–99
Kocher, J. P., 333, 338–340
Kocisko, D. A., 195
Kohrle, J., 182
Kojima, T., 127–128
Kokkinidis, M., 53–54
Kollman, P. A., 311, 313–314
Komiyama, T., 225
Kondo, H., 130–131
Kondrashov, D. A., 304
Kongsgaard, M., 129–130
Kong, X., 358
Koopman, M. M. W., 168
Kopaniak, M. M., 448–449
Kopito, R. R., 423
Koppel, J. L., 260–262
Korlann, Y., 354
Kortemme, T., 303
Korzus, E., 223–224
Koster, J. G., 130–131
Köster, K., 210–211, 215–217
Kosztin, D., 308
Kovacs, H., 164–165
Kowalski, P. S., 110, 225
Kozlov, S. V., 50, 57
Krahenbuhl, O., 242
Krajewski, S., 291
Kramer, B., 153
Krasnewich, D., 358–359, 423, 428, 444
Krautler, V., 309
Krege, J. H., 96–97

Kress, F., 269
Krichevsky, O., 358, 359–361, 362
Krishnamurthy, G., 184, 185, 189–190
Krishnan, R., 370
Krishnasamy, C., 109, 113
Kristensen, S. R., 121–122
Kristensen, T., 178, 181–182
Kroeger, H., 167
Kröger, H., 443, 444, 445
Krönke, G., 130
Kropf, P., 448
Krowarsch, D., 179
Krüger, O., 210–211, 215–217
Kruger, P., 309, 310
Krutchinsky, I. V., 157–158
Kubitscheck, U., 355–356
Kuchnir, L., 309–311
Kufareva, I., 154
Kuhlman, B., 303
Kuhnen, A., 215–217
Kuhn, R. W., 296
Kuizon, S., 195
Kulkarni, M. M., 276–277
Kumar, A., 14–15
Kummer, J. A., 436–439
Kuntz, I. D., 148, 152, 153
Kunz, G., 128
Kurdowska, A., 448
Kuri, B., 345–346, 347
Kurkinen, M., 296
Kusugami, K., 127–128
Kuttner, Y. Y., 353, 356
Kutzner, C., 301, 309, 316
Kvassman, J. O., 179–180, 186–187, 188–189, 198, 224–225, 368
Kwast, L., 130–131
Kwon, K. S., 20–21, 335, 336, 341, 385, 402–403

L

Lacapcre, J. J., 303
Lacbawan, F., 358–359, 423, 428, 444
Ladbury, J. E., 158
Ladewig, J., 210–211, 215–217
Ladjadj, M., 434–435
Laemmli, U. K., 425
Lafitte-Laplace, A. P., 139–175
Lagunoff, D., 109–110
Laio, A., 306
Lakowicz, J. R., 368–369
Lalouel, J. M., 99–100
Lamas, S., 100
Lamba, D., 187–188
Lamb, D. C., 356, 358, 359–362
Lambert Vidmar, S., 276–277
Lambris, J. D., 76, 296
Lamoureux, G., 312
Lamzin, V., 179
Lander, A. D., 114
Landorf, E. V., 354
Lane, D. A., 121–122, 128, 296–298
Lang, D., 408
Langdown, J., 55, 65, 76–77, 81–82, 118–120
Langendorf, C. G., 14–15, 238, 257–258, 296
Lang, P. T., 152, 166
Larkin, M. A., 253, 258
Larsen, J. V., 181–182
Larsson, G., 16
Laskowski, R. A., 149, 153, 154
Latham, C., 184, 185
Laudanna, C., 449
Laurell, C. B., 50, 53, 65, 81, 91, 142, 297, 398, 404–405, 408
Laurence, T. A., 358
Laurie, G. W., 276–277
Lavender, S., 353, 355–356, 356, 358–363, 360, 364, 366, 373
Lavigne, P., 210
Lavinder, J. J., 156
Lawless, M. W., 424–425, 448, 451–452, 453–454
Lawrence, D. A., 50, 51, 55–56, 65, 81–82, 178, 179–180, 181–183, 184, 185, 186–187, 189–190, 192–194, 198, 199, 224–225, 358–359, 368, 423–424, 428, 444
Law, R. H. P., 14–15, 51, 56, 65, 77, 78–79, 225, 226, 228, 230, 233, 238, 257–258, 296, 298, 408
Lawson, M., 450
Le, A., 423
Leach, A. R., 153, 299
Leal, M., 178
Le Bonniec, B. F., 65, 80, 142, 225, 241, 298, 344–345, 393–394, 408–410
Lechaire, I., 178
Lech, M., 128
Leduc, R., 210
Lee, B., 333, 338–340
Lee, C. J., 300–301, 333, 334, 338–340, 343, 404, 424
Lee, C. Y., 307
Lee, J., 152
Lee, K. N., 65, 395–396, 403–404
Lee, M. K., 114
Lee, M. Y., 334, 340, 404
Lee, N. K., 358
Lee, S., 20–21, 152, 451–452
Lee, T. W., 439–440
Lehrer, S. S., 36
Leigh Brown, S. C., 149, 151, 155, 160–161, 167–168
Leik, C. E., 178
Leiserson, W. M., 215–217
Leith, A., 434–435
Lemke, E. A., 370
Lengauer, T., 153
Lengefeld, J., 356

Leonard, E. J., 450–451
Leone, P. A., 184, 185, 189–190
Leontiadou, H., 299
Lesjak, M., 53, 65, 81, 297
Lesk, A. M., 55, 64–76, 96–97, 118–120, 181–183, 238, 336, 385, 393–394, 404–405, 410, 412–414
Leslie, A. G. W., 50, 51, 52, 54–56, 64, 65, 81, 89–90
Letchworth, G. J., 4
Letzel, M. C., 215–217
Levi, M., 178
Levina, V., 10, 15, 385, 395
LeVine, W. F., 276–277
Levitt, D. G., 149
Levitt, M., 303
Levitus, M., 367–368, 373
Levy, R. M., 148, 309, 310
Levy, Y., 335, 341, 342
Lewis, G. M., 65, 81–82, 385, 393–394, 408–410, 413, 414–415
Lewis, J. H., 240
Ley, K., 449
Ley, S. V., 90–93, 95
Leytus, S. P., 194
Liang, A., 109
Liao, Y., 451–452
Liaw, P. C. Y., 110
Libby, P., 225
Liboska, R., 309
Lifton, R. P., 99–100
Li, G., 367–368, 373
Lightman, S., 95
Li, H., 187–188, 353, 358, 364–366
Li, J., 65, 448, 454
Lijnen, H. R., 178, 187–188, 291
Lillehoj, H., 16
Lim, H. K., 159–160
Lindahl, E., 301, 309, 315, 316
Lindahl, U., 296–298
Lindberg, M., 183
Lindblad, D., 451–453
Linders, M., 240–241
Lindorff-Larsen, K., 163–164
Lindquist, S., 370
Linhardt, R. J., 116–118, 122–123
Link, C. A., 22–23
Lin, L., 166
Linli, Y., 291
Lin, P., 109, 300–301
Lin, S. C., 147, 167, 405–408
Linschoten, M., 50, 65, 81–82, 183, 381, 410
Lipinski, C. A., 161
Lipman, E. A., 356
Lippincott-Schwartz, J., 353, 373
Li, S. H., 182–183, 184, 185, 186, 189–190, 192–194
Liu, F., 302
Liu, L., 109, 113, 331, 353, 355–356, 358, 368, 370–372
Liu, X., 189
Liu, Y., 424
Liu, Z., 291
Li, W., 16, 65, 76–77, 81, 82, 110, 118–121, 123–124, 142, 182–183, 240–241, 291, 296–298, 326, 344–345, 385, 391, 392–393, 409, 411, 412–414, 424
Li, X. J., 187–188, 287
Li, Y., 434–435
Li, Z., 455–456
Llop, E., 121–122
Lloyd, G. J., 65, 77, 78–79, 296, 298
Lobermann, H., 65
Loebermann, H., 53, 54, 57–58, 64, 65, 81, 89–90, 381, 402–403
Loethen, S., 451–452
Lohman, T. M., 370
Loh, S. N., 353, 356
Loiseau, F., 90–93, 95
Lokot, T., 217
Löllmann, M., 356
Lomas, D. A., 9–10, 18, 54, 65, 79–80, 142, 144, 145, 148, 149, 151, 154, 155, 160–161, 164, 167–168, 179–180, 210, 238, 296, 297, 298, 299–301, 314–315, 326, 332–333, 334, 341, 344–345, 346–347, 353, 355–356, 358–359, 364–367, 373, 381, 382, 383–391, 392–398, 399–403, 404–410, 411, 412–415, 423–425, 426, 427, 428, 429, 436–440, 441, 443, 444, 445, 448, 453–454, 455
Lombardo, F., 161
Londos, E., 439–440
Longden, I., 213
Longman, E., 194
Lopez, R., 253, 258
Lörincz, Z., 51, 56, 65, 81–82, 298
Lormeau, J. C., 108
Loskutoff, D. J., 9–10, 178, 181–182, 187
Lottspeich, F., 276–277
Loukeris, T. G., 210–211
Lowary, P. T., 40–41
Lu, A., 109
Lu, B. C. G., 228, 230
Lubos, R., 130
Luchinat, C., 165
Luck, L. A., 22–23
Ludtke, S. J., 434–435
Luger, K., 30, 41
Lu, H., 308
Luis, S. A., 118–120
Lukacs, C. M., 65, 76, 296
Luke, C. J., 210–211, 238, 240
Lu, L., 160–161
Lummer, M., 210–211
Lunven, C., 178

Luo, R., 310
Lüthi, A. U., 260–262
Lutter, L. C., 186, 187, 188
Lu, Y., 15

M

Maaroufi, R. M., 121
Ma, B., 153
Maccarana, M., 129–130
MacKerell, A. D. Jr., 309–312
MacKerell, J. A. D., 312
Mackie, G. A., 14–15
MacLeod, I., 167, 443, 444, 445
MacNee, W., 455–456
Madden, T. L., 215
Maddux, J. D., 210–211
Madison, E. L., 187–189
Madrid, P. B., 166
Madsen, J. B., 121, 178, 181–182, 189–190
Maeda, N., 96–97
Maeng, J. S., 333, 338–340
Magde, D., 357–358, 359–361, 363
Magliery, T. J., 156
Magnusson, S. P., 90–91, 129–130
Mahadeva, R., 18, 142, 144, 145–147, 154, 164, 167, 344–345, 366, 385, 395–396, 398, 399–402, 405–408, 414–415, 423–424, 436–439, 448, 453–454, 455–456
Mahmood, K., 241, 256–257
Mahrus, S., 256–257
Maimone, M. M., 125–127
Mainot, S., 460
Mainz, D. T., 152, 153
Ma, J., 148–149, 299
Majdoub, H., 121
Makris, M., 95, 298
Mallya, M., 149, 151, 155, 160–161, 167–168, 334, 341, 385, 392–393, 404–405, 454
Malmstrom, A., 129–130
Ma, N., 215
Mandell, D. J., 302–303
Mangel, W. F., 194
Mankovich, J. A., 225
Manley, S., 353, 373
Mann, D. S., 434–435
Mann, M., 156–157
Mansour, M. B., 121
Ma, Q., 116
Marciniak, S. J., 167, 168, 367, 424, 436, 438, 439, 443, 444, 445, 457, 461
Marcus, N. Y., 451–452
Mares-Guia, M., 307
Margeat, E., 358
Mark, A. E., 299, 300, 309, 310, 312, 315
Markham, B. E., 90–91
Markley, J. L., 311
Marques, O., 304
Marrink, S. J., 299, 302, 312
Martensen, P. M., 184
Martin, D. M. A., 258
Martinet, C., 460
Martinez, A., 460
Martinez, C., 121–122
Martinez-Martinez, I., 121–122, 393
Martin, F., 178
Martin, G. R., 276–277
Martin, S. J., 260–262
Martin, Y. C., 153
Martone, R. L., 178
Maset, F., 110
Mashiba, S., 455–456
Massa, H., 15
Mast, A. E., 291, 298, 383, 384, 385, 393–394, 398, 404–405, 408–410
Mathiasen, L., 178, 189–190
Mathur, S., 159
Matoba, Y., 65
Matsumoto, T., 121–122
Matsumura, T., 100
Matsushita, T., 127–128
Matthews, A. Y., 237–275
Matthews, D. J., 199, 253–254, 298
Matuliene, J., 156
Matulis, D., 156
Maurer-Stroh, S., 260–262
Maurice, N., 168
Mayer, M., 166
Mayne, L., 326–327
Mayr, L. M., 156–157, 189
McCaffrey, T. A., 276–277
McCammon, J. A., 152–153, 299, 307
McCammon, M. G., 156, 158
McClelland, D. B. L., 450
McCoy, A. J., 50, 51, 52, 53–54, 56, 57, 65, 97
McCullagh, P., 455–456
Mcdonald, L., 225
McElvaney, N. G., 424–425, 448, 451–452, 453–454
McFarlane, G., 184, 185, 189–190
McGettigan, P. A., 253, 258
McGowan, S., 29–49, 65, 225, 238, 257–258, 408
McGuffin, L. J., 268–269
McGwire, B. S., 276–277
McKay, A. R., 139–175
McKay, E. J., 50
McKee, C. M., 291
McKeever, B. M., 260–262
McLarney, S., 14–15
McLaughlin, P. J., 64, 65, 81, 89–90
McLean, C., 451–452
McLeod, I., 423–424, 436, 439–440, 443, 444
McMahon, G. A., 178, 181–182
McMaster, W. R., 276–277
McWilliam, H., 253, 258
Meagher, J. L., 108, 109, 334–335

Mee, R. P., 153
Meersseman, G., 41
Meier, T. R., 178
Meijers, J. C. M., 130–131
Melhado, L. L., 194
Mellet, P., 354, 368, 372
Mély, Y., 354, 368, 372
Meng, E. C., 152, 310
Merz, K. M. Jr., 310, 311, 313–314
Metropolis, N., 152–153, 302
Meuvis, J., 367–368
Meyer, B., 166
Meyer, H. A., 182
Michailoviene, V., 156
Michel, C. H., 367, 424, 436, 438, 439, 443, 444
Michl, J., 91–93
Mikami, B., 385, 392, 394–396, 412–414, 424
Mikus, P., 406–408
Millan, E., 460
Miller, S. D., 451–452
Millet, L., 128
Milliat, F., 178
Mills, K., 335–336
Mills, P. B., 335–336
Millward- Sadler, H., 455–456
Milne, J. S., 326–327
Milroy, L., 90–93, 95
Minano, A., 121–122, 393
Mindell, J. A., 435
Minhas, J., 215–217
Minthon, L., 439–440
Miqueo, C., 393
Miranda, E., 65, 144, 149, 167, 168, 334, 341, 344–345, 346–347, 367, 382, 383–391, 385, 392–393, 404, 412–414, 423–424, 436, 437, 438, 439–440, 441, 443, 444, 445, 457, 461
Mirnics, K., 451–452
Miska, K. B., 16
Mistry, H. D., 99–100
Mita, K., 210–211
Mitchell, J. B. O., 153
Mitchell, L., 109
Miteval, M. A., 148, 151
Moerner, W. E., 355–356
Moestrup, S. K., 181–182
Mohammadi, M., 122–123
Molnar, F., 308
Monard, D., 110, 230
Monien, B. H., 109, 113
Moni, R. W., 184, 185, 189–190
Moons, L., 128
Moore, E. G., 181–182
Moore, W. T., 76, 296
Moras, D., 53, 65, 81
Morgan, G. A., 178
Morgenstern, B., 215
Morini, M., 276–277
Mori, Y., 95
Morkin, E., 90–91
Morla, A., 276–277
Moroi, M., 230
Morton, C. J., 65, 404–405
Mosier, P. D., 123–124, 125–127
Moskau, D., 164–165
Mottonen, J., 51, 54, 55–56, 65, 78, 81–82, 90, 179–180, 198, 298, 381
Mourão, P. A. S., 125–127
Mourey, L., 53, 65, 81
Mourier, P., 116–117
Moyer, R. W., 148, 210, 238
Muckenshnabel, I., 156–157
Muegge, I., 153
Muenke, M., 358–359, 423, 428, 444
Mugford, C., 178
Mugford, C. P., 184, 185, 189–190
Muglia, L. J., 451–452
Muhammad, S., 179–180, 181–182, 224–225
Mukherjee, A., 168
Mukherjee, S., 152
Mukhopadhyay, S., 370
Mulders, J., 52, 65, 81, 90
Mulgrew, A. T., 424–425, 448, 451–452, 453–454
Müller, B. K., 356, 358, 359–362
Müller, I., 448
Muller, J. D., 194
Müller-Späth, S., 370
Muller, Y. A., 65, 78, 91–94, 182
Mulloy, B., 122–123, 125–127
Mulnix, A. B., 210–211
Munder, M., 448
Murai, J. T., 296
Muramatsu, T., 127–128
Murata, Y., 95
Murphy, M. P., 52, 56–57, 58, 96–98, 100, 370
Murphy, R. B., 152, 153
Murray, C. W., 152, 153, 161
Murray, D., 194
Muschel, R. J., 291
Mushero, N., 109, 353, 355–356, 358, 368, 370–372
Musil, D., 50, 65, 81–82, 381, 410
Musto, N. A., 90–91
Myatt, L., 98–99
Myers, D., 178
Myers, R. M., 90–93, 95
Myong, S., 355–356
Mysinger, M. M., 153

N

Nachajko, W. R., 188–189
Nachman, R. L., 197
Nagahara, N., 100
Nagase, H., 291
Nagata, K., 445
Naidoo, N., 14–15

Nakagawa, M., 459–460
Nakamura, H., 50, 311
Nakanishi, J., 14–15
Nakayama, Y., 127–128
Nambi, P., 178
Nangaku, M., 98–99
Nangalia, J., 65, 76–77, 120–121, 240
Narasimhan, L., 242
Nar, H., 65, 408
Nathan, C. F., 451
Nath, S., 367–368
Navar, L. G., 98–99
Navarro-Fernandez, J., 121–122
Navaza, J., 53–54, 179
Nayal, M., 148, 151
Na, Y. R., 394–395
Nerelius, C., 364, 405–406
Nerme, G., 178
Nesheim, M. E., 187, 188
Nesi, L., 276–277
Nettels, D., 356, 370
Neumann, T., 147–148
Neuweiler, H., 356
Neve, J., 184, 185, 189–190
Neveu, M., 296
Ng, N., 257
Ng, N. M., 257, 260–262
Nguyen, G., 96–97
Nguyen, K., 51, 56
Nicholas, H. B. Jr., 213
Nicholas, K. B., 213
Nichols, R. J., 117
Nichols, S. W., 451–452
Nickbarg, E. B., 14–15, 156–157
Nielsen, A. F., 121, 189–190
Nielsen, H. M., 439–440
Nielsen, J. E., 110, 225
Nielsen, R. W., 183
Nienhaus, G. U., 356
Niepmann, M., 427
Nilsson, L., 309
Nishimura, M., 127–128
Nishioka, J., 130–131
Nishiyama, A., 98–99
Nisker, J. A., 296
Nobeli, I., 149, 167, 334, 341, 404
Noble, N. A., 178
Noonan, D., 276–277
Nordenman, B., 107, 108
North, P. R., 184, 185, 189–190, 192–194
Northrup, S. H., 307
Notario, V., 14–15
Nourshargh, S., 449
Novoa, I., 445, 446
Nukiwa, T., 382, 423, 428
Nussinov, R., 153
Nyon, M. P., 164–165
Ny, T., 179–180, 182–183, 224, 406–408

O

Oakley, D. J., 395–396, 398, 399–402
O'Brodovich, H., 109
O'Connor, S. E., 178
Odake, S., 242
Oganesian, A., 178
Ohlsson, P. I., 224
Okada, K., 9–10
Okamoto, R., 100
Okamoto, Y., 306
O'Keeffe, D., 113
Okita, K., 459–460
Oldberg, A., 129–130
Olds, R. J., 121–122, 128
Oleksy, A., 296–298
Oley, M., 215–217
Ollivier, V., 121
Olson, S. T., 65, 77–78, 81–82, 108, 109, 110, 111–113, 115, 121, 142, 179–180, 181–183, 186, 187, 188–189, 188, 198, 224–225, 226, 230, 241, 269, 296–298, 334–335, 368
Oltersdorf, T., 166
Olwin, J. H., 260–262
O'Malley, K., 76, 296
Onda, M., 65, 385, 391, 392–393, 394–396, 404–405, 411, 412–414, 424–425, 426, 428, 436–439
O'Neill, S. J., 424–425, 448, 451–452, 453–454
Ong, P. C., 34–35, 225, 408
Onuchic, J. N., 306
Onufriev, A., 310
Oono, Y., 308
Oostenbrink, C., 299, 309, 312, 315
Op den Camp, H. J., 64
Ordoñez, A., 121–122, 393
Orlova, E. V., 434–435
Orte, A., 353, 355–356, 358–359, 364–368, 373
Osterwalder, T., 215–217
Ostrovsky, L. L., 439–440
Ota, N., 307
Otlewski, J., 179, 296–298
Overall, C. M., 276–277, 284, 291
Overington, J., 151–152
Owen, C. A., 455
Owen, M. C., 90–91, 240, 296
Oyadomari, S., 445
Oyola, R., 353, 355–356, 358–363, 364, 366, 373

P

Paczkowski, N. J., 184, 185, 189–190
Pahl, H. L., 445
Pak, S. C., 14, 210–211, 238, 240
Palaniappan, S., 181–182, 186, 198, 368
Palmer, I., 14–15
Pande, V., 309, 310
Pannekoek, H., 178, 181–182, 187, 188, 240–241

Pannu, N. S., 50, 51, 54, 55–56, 65, 77, 78, 80, 81–82, 141, 142, 181–182, 186, 199, 298, 385, 393–394, 408–410
Paolini, G. V., 153
Papadopoulos, J., 215
Parafati, M., 110
Paramo, J. A., 178
Parfrey, H., 154, 164, 404–406
Parker, M. W., 18–19, 26–27, 65, 80, 142, 298, 344–345, 353, 393–394, 404–405, 408–410
Park, H., 152
Park, S. D., 403–404
Park, S. H., 334, 340, 404
Parmar, J., 448, 454
Parmar, J. S., 448, 453–454
Parrinello, M., 306
Parthasarathy, N., 114
Pasenkiewicz-Gierula, M., 300
Paterson, N. A., 91–93, 182
Patston, P. A., 91–93, 121
Patterson, G., 353, 373
Patterson, W. L., 194
Patthy, A., 76, 296
Pavão, M. S. G., 125–127, 129–130
Peake, I. R., 95, 298
Pearce, M. C., 10, 15, 16, 18–19, 26–27, 65, 77, 78–79, 118–120, 225, 296, 298, 353, 385, 395, 396–398, 402, 404–405
Pear, M. R., 307
Pedersen, K. E., 178, 181–182, 183, 186, 385, 391, 412–414
Pedersen, L. G., 300–301
Pedersen, R. A., 459–460
Pedersen, S., 121–122
Peishoff, C. E., 153
Pei, X. Y., 50, 51, 52, 55, 65, 142, 148, 149, 154, 180, 296, 297, 300–301, 314–315, 332–333, 381
Pejler, G., 246, 247
Pellecchia, M., 165
Pellegrini, L., 122–123
Pemberton, P. A., 90–93, 148, 210, 225, 238
Penczek, P., 434–435
Peng, L., 178, 434–435
Pepys, M. B., 90–93
Pérez, J., 144, 167, 344–347, 367, 382, 383–391, 392–393, 412–414, 421–469
Perez-Lara, A., 121–122
Perfumo, S., 455–456
Perlmutter, D. H., 11–12, 166, 167, 404–405, 451–452
Perona, J. J., 372
Perot, S., 148, 151
Perron, M. J., 22–23, 186, 187, 188
Perry, D., 121–122
Perry, J. K., 152, 153
Persson, U., 50
Pertzkall, I., 130
Peter, C., 309
Peterman, E. J., 355–356
Peters, E. A., 247–248
Petersen, H. H., 181–182
Petersen, N. O., 357–358
Petersen, T. E., 90–91
Peterson, F. C., 21–22, 164–165, 372
Petitclerc, E., 178
Petitou, M., 52, 53, 54, 65, 76–77, 81, 90, 108, 109, 111–113, 118–120, 123–124, 128, 141, 182–183, 296–298
Petrassi, H. M., 158
Petrella, R. J., 309
Pettersen, E. F., 152, 310
Pevarello, P., 166
Pfeil, S. H., 356
Phillips, G. N. Jr., 304
Phillips, J. C., 309
Phillips, J. E., 120–121
Phillips, R. L., 149, 151, 155, 160–161, 167–168, 334, 341, 404
Piccardo, P., 423–424
Picheng, Z., 210–211
Pieper, U., 256–257
Piepkorn, M. W., 109–110
Pierce, J., 50
Pierens, G., 184, 185, 189–190
Pietropaolo, C., 110
Piitulainen, E., 455–456
Pike, R. N., 16–17, 52, 54, 55, 64, 65, 77, 80, 90, 95, 98, 118–120, 123–124, 141, 142, 144, 182–183, 225, 226, 228, 233, 238, 241, 242, 243, 254, 256–258, 260–262, 296–298, 344–345, 393–394, 408–410, 424–425, 427
Pinkus, S. N., 81
Pinto, I. G., 166
Pitman, R., 448, 454
Pizzo, S. V., 291
Planchenault, T., 276–277
Plasman, K., 260–262
Plevin, R. J., 450
Plotnick, M. I., 189, 224–225, 372
Plotnikov, A. N., 122–123
Pluquet, O., 167
Poe, M., 242
Poger, D., 300
Popova, E. Y., 225, 408
Portmann, B., 382, 383–391, 423, 436, 437, 439–440, 441, 443
Postma, J. P. M., 315
Post, M. J., 128
Postnova, T. I., 291
Potempa, J., 65, 81, 223–224
Potter, E. E., 120–121
Potter, J. M., 90–93
Povey, S., 455–456
Powers, G. A., 18–19, 26–27, 65, 353
Powers, J. C., 242

Pozzi, A., 276–277
Pozzi, L., 455–456
Pozzi, N., 110
Preissner, K. T., 240–241
Presta, M., 276–277
Preston, F. E., 95, 298
Price, R. G., 194
Princivalle, M., 128
Procter, J. B., 258
Przygodzka, P., 16
Purkayastha, P., 353, 355–356, 358–363, 364, 366, 373, 385
Purkey, H. E., 158
Purkiss, A., 149, 167, 334, 341, 404
Puscau, M. M., 184, 185, 189–190, 192–194

Q

Qian, B., 57
Qian, H. X., 53, 65, 81
Qi, X., 89–106
Quaranta, V., 276–277
Quigley, J. P., 276–277
Quinet, E. M., 178
Quinsey, N. S., 225, 226, 228, 233, 257, 260–262
Qu, J., 450
Quyen-Quyen, A. L., 423

R

Radermacher, M., 434–435
Rafidi, K., 296
Raftopoulos, C., 423–424
Ragan, E. J., 210–211
Ragg, H., 210–211, 215–217, 240
Raghavendra, M. P., 368
Raghuraman, A., 123–124, 125–127
Rahman, I., 455–456
Rahmann, S., 214
Rajagopalan, S., 158
Raja, S., 113
Raman, S., 57
Rampioni, A., 302
Ramsay, M. M., 99–100
Ramstedt, B., 16
Rånby, M., 187–188
Rand-Weaver, M., 194
Ranganathan, S., 256–257
Ranish, J. A., 287
Rarey, M., 153
Rashid, S. T., 167, 168, 457, 461
Rasnik, I., 370
Ravenhill, N., 154, 164
Rayburn, H., 128
Raymoure, W. J., 296
Read, R., 142
Read, R. J., 49–76, 65, 77, 78, 79, 80, 81–82, 89–93, 94–95, 96–98, 100, 141, 179–180, 181–182, 186, 199, 224–225, 238–240, 291, 296, 297, 298, 344–345, 368, 370–372, 385, 393–394, 408–410
Reboul, C. F., 237–273, 295–323
Reckless, J., 450–451
Reddy, V. B., 188–189
Reed, B. J., 448, 453–454
Rees, D. C., 122–123, 161
Rees, I., 434–435
Reetz, A., 182
Refetoff, S., 93, 95
Rege, T. A., 276–277
Reichardt, G., 356
Reichhart, J. M., 210–211, 238, 240
Reid, J. L., 96–97
Rein, C. M., 105–137
Reiner, J. E., 373
Remold-O'Donnell, E., 210, 238, 258
Renatus, M., 65
Rennard, S. I., 276–277
Repasky, M. P., 152, 153
Reuter, N., 304
Reventos, J., 90–91
Reymond, L., 370
Rezaie, A. R., 187, 188, 189, 296–298
Rhoades, E., 353, 356
Ricagno, S., 65
Rice, A. G., 448
Rice, P., 213
Richard, B., 108, 109, 110, 112–113, 115
Richer, M. J., 215–217
Richmond, T. J., 30
Riemersma, R. A., 450
Rifkin, D. B., 276–277
Rijken, D. C., 187–188
Rijneveld, A. W., 130–131
Rippmann, F., 149
Risselada, H. J., 302, 312
Rivera, J., 121–122
Rizzo, R. C., 152
Robbie, L. A., 178
Roberts, C. R., 276–277
Roberts, T. H., 262
Robey, P. G., 276–277
Robinson, C. V., 144, 156, 158, 344–345, 346–347, 385, 391, 392–393, 412–414, 424, 437, 440
Robinson, D., 194
Rodenburg, K. W., 181–182, 184, 185, 210–211
Rogers, B. B., 451–452
Rogers, D. S., 276–277
Rognan, D., 153
Rohde, H., 276–277
Rohrwasser, A., 99–100
Römisch, K., 423–424, 436, 439–440, 443, 444
Ron, D., 423, 445, 446
Rosado, C. J., 238, 257–258
Roseman, A. M., 434–435
Rosenberg, M., 14–15

Rosenberg, R. D., 108, 128
Rosenbluth, A. W., 302
Rosenbluth, M. N., 302
Rossjohn, J., 16–17, 51, 56, 65, 77, 78–79, 80, 142, 225, 226, 228, 233, 296, 298, 344–345, 393–394, 408–410
Rostagno, A. A., 276–277
Rostom, A. A., 158
Rothwell, P. J., 357–358, 368–370, 373
Rotonda, J., 260–262
Roudesli, M. S., 121
Rousseau, F., 260–262
Roussel, B. D., 367, 424, 436, 438, 439, 443, 444, 445
Roux, B., 309, 312
Rovelli, G., 110, 230
Rowley, C., 309
Roy, R., 354, 358
Rozanov, D. V., 291
Rubenstein, R., 195
Rubin, H., 14–15, 65, 76, 189, 224–225, 296, 354, 368, 372
Rudd, T. R., 117
Rudisser, S., 166
Rudnick, D. A., 168, 451–453
Rudy, G. B., 242, 243, 254, 256–257
Rüegger, S., 370
Rugg-Gunn, P., 460
Ruggles, S. W., 260–262
Ruoslahti, E., 276–277
Ruotolo, B. T., 158
Rupin, A., 178
Rupp, B., 50
Ruska, E., 429–430
Rutter, W. J., 372
Ruzicka, T., 225
Ruzyla, K., 51, 56, 65, 77, 78–79, 296, 298
Ryu, S. E., 53, 55, 65
Rzepiela, A. J., 302

S

Saba, T. M., 276–277
Sabino, A. A., 121
Sabourin, J. C., 178
Saffarian, S., 353, 356
Sager, R., 296
Saito, H., 127–128
Saldanha, S. A., 151, 155, 160–161, 167–168, 404
Salensminde, G., 304
Sali, A., 57, 256–257
Salvesen, G. S., 225, 291, 298, 383, 384, 385, 393–394, 398, 404–405, 408–410
Samakur, M., 189
Samama, J. P., 53, 65, 81
Sambrook, J. F., 188–189
San Antonio, J. D., 114
Sancho, E., 14–15
Sander, C., 303
Sandercock, A. M., 158
Sanders, K. L., 184, 185, 189–190, 192–194
Sandstrom, C. S., 455–456
Sanejouand, Y. H., 303, 304
Sang, L., 121, 189–190
Sansom, M. S., 299
Sanyal, I., 15, 16
Sarkar, A., 325–350
Sasaki, T., 211
Sata, M., 128–129
Sathe, G., 14–15
Sauer, M., 354, 356
Saven, J. G., 353, 355–356, 358–363, 373
Savinov, A. Y., 291
Savio, D. A., 178
Savory, W. J., 55, 118–120
Sawdey, M. S., 178
Scapini, P., 276–277
Scaravilli, F., 423–424
Scarsdale, J. N., 64–76
Schack, L., 186, 385, 391, 412–414
Schaeffer, P., 128
Schafer, L. V., 302
Schalch, T., 30
Schapira, M., 446
Schatz, M., 434–435
Schaub, R. G., 178
Schechter, I., 210, 258
Schechter, N. M., 189, 224–225, 372
Schedin-Weiss, S., 109, 110, 113, 115
Scheek, R. M., 148–149, 308
Schick, C., 15, 225
Schimdt, B. Z., 166, 167
Schiraldi, O., 276–277
Schlag, E. W., 158
Schlessinger, J., 122–123
Schmer, G., 109–110
Schmid, G., 98
Schmidt, B. Z., 166, 451–452
Schmidt, K., 147–148
Schmidt, R., 434–435
Schnebli, H. P., 179
Schneiderman, J., 178
Schoenberger, O. L., 14–15
Schöfer, C., 130
Schoichet, B. K., 152, 154–155
Schonbeck, U., 225
Schreiber, G., 94–95
Schreuder, H. A., 52, 65, 81, 90
Schuler, B., 353, 356, 370, 373
Schulten, K., 301–302, 308–310, 316
Schultz, J., 214
Schultz, P. G., 353, 357–358, 369–370
Schulze, A. J., 91, 117, 121, 142, 398, 404–405, 408
Schulz, P., 9–10
Schussler, G. C., 91–93

Schuster-Böckler, B., 214
Schwartz, B. S., 182–183, 186
Schwartz, J. J., 128
Schwimmer, R., 276–277
Schymkowitz, J., 260–262
Scott, D. J., 158
Scott, M., 159
Scott, W. R. P., 309, 310
Sebesteyen, M., 145–147
Seckler, R., 356
Seeliger, D., 148–149
Seftor, E., 296
Seidel, C. A., 357–358, 368–372, 373
Sekul, R., 147–148
Selwood, T., 189
Selzle, H. L., 158
Semple, R., 457, 461
Sendall, T. J., 65, 81–82, 385, 393–394, 408–410, 413, 414–415
Sengupta, D., 302
Sengupta, T., 160–161, 334–335, 340, 345–346, 347
Senior, R. M., 291, 448, 455–456
Seo, E. J., 53, 55, 338–340
Seong, K. M., 424
Serekaite, J., 156
Serpell, L. C., 385, 401–402
Seyama, K., 382, 385, 392–393, 423, 428, 454
Shah, S. A., 307
Shapiro, S. D., 291, 448, 454, 455–456
Sharma, S., 356, 359–362
Sharp, A. M., 50, 51, 55–56, 65, 81–82, 186, 199
Sharp, L. K., 364, 385, 391, 394–395, 404–406, 424–425, 426, 428, 436–439
Shatzman, A., 14–15
Shaw, C. M., 358–359
Shaw, G. L., 179
Shaw, I., 184, 185
Sheffer, R., 113, 115, 241
Shelley, M., 152, 153
Sheng, S., 296
Shen, H., 241, 256–257
Shental-Bechor, D., 335, 341, 342
Sherman, E., 353, 356
Sherman, W., 155
Sheu, S. Y., 158
Shigekiyo, T., 121–122
Shi, G. P., 225
Shiloach, J., 14–15
Shindyalov, I. N.
Shin, H. S., 385, 402–403
Shin, J.-S., 368
Shipley, J. M., 291
Shirk, R. A., 114
Shirvani, A. M., 151, 155, 160–161, 167–168, 404
Shivakumar, D. M., 153
Shi, X., 187
Shoemaker, A. R., 166
Shoichet, B. K., 151–152, 153
Shore, J. D., 22–23, 111–112, 113, 179–180, 181–182, 186–187, 188–189, 198, 224–225, 226, 230, 241, 368
Shortle, D., 340, 341
Shrimpton, A. E., 358–359, 423–424, 428, 444
Shrivastava, S., 151–152
Shukla, A., 144, 145–147, 405–406
Shushanov, S. S., 225
Sibbald, W. J., 91–93, 182
Sidhar, S. K., 344–345, 402–403, 429
Sifers, R. N., 151, 155, 160–161, 167–168, 404, 423, 424, 451–452
Sigler, R., 178
Sigmund, C. D., 96–97
Siiteri, P. K., 296
Silver, G. M., 210–211
Silverman, G. A., 15, 31, 148, 210–211, 225, 238, 240, 257–258
Simmerling, C., 310
Simonovic, M., 65, 297
Simon, R. H., 276–277
Sinha, U., 109
Sisamakis, E., 357–358, 368–370, 373
Sisson, T. H., 276–277
Sivasothy, P., 142, 326, 344–345, 385, 395–396, 406–410, 423–424
Sjolin, L., 65, 81–82, 381, 410
Sjölin, L., 50
Skeel, R. D., 309
Skeldal, S., 178, 181–182
Skepper, J., 167, 168, 457, 461
Skidmore, M. A., 117
Skinner, R., 51, 52, 54, 55, 65, 77, 90, 96–97, 118–120, 123–124, 141, 142, 181–183, 238, 296–298, 404–405
Sklar, J. G., 194–195
Smilde, A., 187, 188
Smith, A. E., 116–117
Smith, A. I., 65, 77, 78–79, 225, 296, 298, 408
Smith, C. L., 90–93, 182
Smith, D. L., 337
Smith, E., 178
Smithers, L. E., 460
Smith, G. P., 243–244
Smithies, O., 96–97
Smith, J. E., 210–211
Smith, J. M., 434–435
Smith, L. H., 178
Smith, P. E., 316
Smolenaars, M. M. W., 210–211
Sofian, T., 14–15, 65, 228, 230, 296
Sommer, J., 110, 230
Sonderegger, P., 50, 57, 358–359, 423–424, 428, 444
Song, H. K., 65
Song, J., 241, 256–257
Soranno, A., 370

Sorge, J. A., 451–452
Sosa, H., 355–356
Sottrup-Jensen, L., 90–91, 184
Speil, J., 355–356
Spellmeyer, D. C., 311, 313–314
Sperandio, O., 148, 151
Sperb, R., 316
Spicer, V. L., 157–158
Spraul, M., 164–165
Sprecher, C. A., 225
Springhetti, E. M., 30–31, 40–41
Sraer, J. D., 96–97
Srinivasan, K. R., 111–112
Srisa-Art, M., 373
Srividya, N., 354, 368, 372
Stack, M. S., 276–277
Stafford, A., 109
Stafford, A. R., 110
Stamler, J. S., 100
Standing, K. G., 157–158
Stanke, M., 215
Stanley, P. L., 51, 52, 56–57, 58, 90–93, 94–95, 96–98, 100, 182
Stassen, J. M., 408
Stavridi, E. S., 76, 296
Steele, F., 14–15
Steenbakkers, P. J., 64
Stefansson, S., 178, 181–182
Steinberg, I. Z., 368–369
Steinkamp, R., 215
Stein, P. E., 50, 51, 52, 53, 54, 56–57, 58, 64, 65, 77, 81–82, 89–93, 94–95, 96–98, 100, 118–120, 181–182, 186, 199, 238–240, 296–298, 385, 392–393, 405–410, 411
Stenflo, J., 16, 110, 120–122
Stepaniants, S., 308
Sternby, N. H., 455–456
Stern, P. S., 303
Stetler-Stevenson, W. G., 276–277
Stewart, S., 448, 454
Stoka, V., 16
Stoller, J. K., 11–12
Stone, J. E., 301–302
Stone, S. R., 14–15, 18, 20–21, 110, 144, 225, 230, 297, 299–300, 383–391, 396–398, 399, 401–402, 436–439
Storoni, L. C., 53–54, 56, 97
Stothard, P., 151–152
Stout, T. J., 65, 199, 298
Strand, A., 51, 54, 55–56, 65, 78, 81–82, 90, 179–180, 298, 381
Strandberg, K., 121–122
Strandberg, L., 179–180, 182–183
Streatfeild-James, R. M., 98
Streich, D., 356
Strelkov, S. V., 65
Streusand, V. J., 108, 109
Strickland, D. K., 181–182
Strick-Marchand, H., 460
Strik, M. C., 436–439
Stringham, J. R., 178
Stromqvist, M., 183
Strongin, A. Y., 291
Stroud, R., 297
Stroupe, C., 307
Stryer, L., 368–369
Suda, S. A., 91–93
Su, E. J., 178, 184, 185, 189–190, 192–194
Sugita, Y., 306
Suhre, K., 304
Sukhova, G. K., 225
Sullivan, B. J., 156
Sullivan, E., 194
Sumitomo, Y., 128–129
Summers, C., 149, 167, 334, 341, 404, 448, 449
Sundstrom, M., 166
Sun, J. R., 225, 241–242, 245, 260–262
Sun, Y. K., 256–257
Sutherland, A. P., 65
Sutiphong, J., 14–15
Sutton, V. R., 225, 241–242
Suzek, T. O., 151–152
Suziki, M., 127–128
Suzuki, K., 130–131, 291
Sveger, T., 423
Swanson, R., 108, 109, 112–113, 187, 188–189, 224–225, 368
Swanson, S. C., 113
Sweet, R. M., 51, 54, 55–56, 65, 78, 81–82, 90, 179–180, 298, 381
Swendsen, R. H., 306
Swift, D., 194
Swillo, R. E., 178
Symersky, J., 51, 54, 55–56, 65, 78, 81–82, 90, 179–180, 298, 381
Symonds, M. E., 99–100
Szabo, A. G., 22–23, 369–372

T

Tachias, K., 187–188
Taggart, C. C., 424–425, 448, 451–452, 453–454
Taggert, C. C., 451–452
Tajkhorshid, E., 309
Takagi, A., 127–128
Takahashi, K., 459–460
Takahashi, N., 385, 392, 394–396, 412–414, 424
Takamori, N., 128–129
Takamoto, K., 114, 330
Takao, M., 423–424
Takeda, K., 93, 95
Takeda, N., 127–128
Takehara, S., 65, 385, 392, 394–396, 412–414, 424
Talanian, R. V., 225
Tama, F., 304
Tame, J. R. H., 158

Tang, G., 434–435
Tanghetti, E., 276–277
Tang, J., 353, 355–356, 358–363, 373
Tang, Y. C., 356, 359–362
Tan, H., 241, 256–257
Tans, G., 187, 188
Tan, T. W., 256–257
Tato, M., 166
Tavan, P., 308
Taylor, A., 296
Taylor, G., 156–157
Taylor, R. D., 152
Teckman, J. H., 144, 166, 168, 344–345, 346–347, 382, 383–391, 392–393, 412–414, 423, 424, 436, 437, 439–440, 441, 443, 451–453
Teller, A. H., 302
Teller, E., 302
ten Cate, J. W., 178
Tengel, T., 16
Teplyakov, A., 53–54
Termine, D. J., 151, 155, 160–161, 167–168, 404
Teruel, R., 121–122
Terwilliger, T. C., 148–149
Tew, D. J., 296, 368, 399–401, 404–405
Tewkesbury, D. A., 90–91
Tewksbury, D., 98
Thalassinos, K., 139–175
Thannickal, V. J., 276–277
Thein, S. L., 121–122, 128
Theunissen, H. J. M., 52, 65, 81, 90
Thomas, D. Y., 167
Thomas, G., 225
Thomas, S. E., 421–467
Thomas, V., 152
Thompson, J. D., 253, 258
Thompson, J. H., 22–23, 186, 187, 188
Thompson, N. L., 356, 359–361, 362
Thompson, P. E., 241–242, 245, 257, 260–262
Thomson, B. A., 157–158
Thor, A., 296
Thornberry, N. A., 260–262
Thornton, J. M., 153
Thorsen, S., 226
Thulin, E., 50
Thunberg, L., 296–298
Tiekstra, M. J., 130–131
Tieleman, D. P., 299, 302, 312
Tildesley, D. J., 299
Timmerman, E., 260–262
Timpl, R., 276–277
Tirado-Rives, J., 312
Tirion, M. M., 295–326
Tironi, I. G., 309, 310, 316
Toh, H., 213
Toida, T., 116–117
Tokuoka, R., 53, 54, 57–58, 64, 65, 81, 89–90, 381, 402–403
Tollefsen, D. M., 125–127, 128–130
Tolmachoff, T., 451–452
Tonge, D. W., 14–15
Tong, Y., 65, 76–77, 240
Torda, A. E., 306
Torresan, J., 156
Torrie, G. M., 307
Torri, G., 116–117
Totrov, M., 148, 151, 152, 154
Touboul, T., 459–460
Trapani, J. A., 225, 241–242, 245, 260–262
Trask, B. J., 15
Travis, J., 65, 81, 223–224, 225, 448
Trew, S., 184, 185
Tripodi, M., 455–456
Trotter, M., 460
Trzesniak, D., 309
Tschesche, H., 276–277
Tschopp, J., 242
Tsutsui, Y., 113, 160–161, 331, 334–335, 337–340, 341, 345–346, 347, 385
Tucker, H. M., 65
Turk, B., 225
Turk, V., 16
Turnbull, J. E., 117
Turnell, W. G., 64, 65, 81, 89–90
Twining, S. S., 291

U

Uchida, K., 455–456
Ueda, M., 455–456
Ufimtsev, I. S., 301–302
Uhrin, P., 130
Ulam, S., 152–153
Ulshöfer, T., 210
Underhill, C., 65, 78, 91–94, 182
Underhill, D. A., 90–91
Unger, J., 276–277
Ung, K., 241–242, 245, 260–262
Unitt, J., 194

V

Vagin, A., 53–54
Valadie, H., 303
Valentin, F., 253, 258
Valeri, A., 357–358, 368–370, 373
Vales, A., 130
Valleau, J. P., 307
Vallez, M. O., 178
Vallier, L., 459–460
van Aalten, D. M. F., 148–149, 308
Van Boeckel., 122–124
Van Damme, P., 260–262
Vandekerckhove, J., 260–262
Vandenabeele, P., 256–257
van den Berg, B. M., 181–182
van den Hurk, C. M., 439–440

van der Poll, T., 130–131
van der Spoel, D., 301, 309, 315, 316
van Diepen, A. C., 439–440
Van Durme, J., 260–262
van Giezen, J. J., 178
van Gunsteren, W. F., 299, 306, 309, 310, 312, 315
van Heel, M., 434–435
Van Hoef, B., 291
Van Meijer, M., 187, 188
van Rijckevorsel, K., 423–424
van Vuuren, A. J. H., 130–131
Van Wynsberghe, A. W., 304
van Zonneveld, A. J., 181–182
Varki, A., 335
Vaughan, D. E., 178
Vaughan, L., 90–91
Velazquez, L., 121–122
Velez, R., 423
Vendruscolo, M., 163–164
Venisse, L., 121
Verbeuren, T. J., 178
Verdonk, M. L., 152
Verhamme, I., 186–187, 188–189, 198, 368
Verhamme, I. M., 179–180, 224
Verli, H., 122–123, 300
Veronesi, M., 166
Verspurten, J., 256–257
Vicente, C. P., 128–130
Vicente, V., 121–122, 393
Vikkula, M., 423–424
Villa, A., 312, 315
Villa, E., 309
Villoutreix, B. O., 148, 151
Vincent, P., 276–277
Visanji, M., 179
Viskov, C., 116–117
Vlahova, P. I., 116–117
Vlasuk, G. P., 178
Vleugels, N., 186–187
Volz, K., 50–51, 297
von Delft, F., 122–123
von Mikecz, A., 225
Vriend, G., 148–149, 308
Vulpetti, A., 166

W

Waack, S., 215
Wade, R. C., 122–124
Wagenaar, G. T. M., 130–131
Wagner, W. D., 114
Wahlund, G., 178
Wakefield, T., 178
Wales, T. E., 327
Wallace, I. M., 253, 258
Wallays, G., 128
Wallqvist, A., 309, 310
Wang, B., 310
Wang, H., 302–303
Wang, J. S., 151–152, 306, 311, 455–456
Wang, R., 153
Wang, S., 153
Wang, W., 309, 353, 355–356, 358–363, 373
Wang, Y., 151–152, 210–211
Wang, Z. M., 14–15, 65, 189, 198, 404–405
Wardell, M. R., 9–10, 52, 65, 77, 81, 90, 118–120, 142, 144, 182–183, 296–298, 336, 385, 393–394, 404–405, 410, 412–414
Ward, K., 99–100
Warnock, M., 184, 185, 189–190, 192–194
Warr, W. A., 151–152
Washington, K., 178
Waterhouse, A. M., 258
Waterhouse, J., 215–217
Watkins, S., 168
Waugh, S. M., 260–262
Webb, G. I., 241, 256–257
Webb, W. W., 353, 356, 357–358, 359–361, 363
Weber, A., 457, 459–460, 461
Wee, L. J., 256–257
Wei, A., 65
Wei, C. M., 188–189
Weipoltshammer, K., 130
Weisberg, A. D., 178
Weiss, S., 353, 354, 357–358, 369–370
Weitz, J., 109
Weitz, J. I., 110, 128
Wei, Z., 50–51, 52, 56–57, 58, 65, 77–78, 90–93, 94–95, 96–98, 100, 182
Wek, R., 446
Wells, J. A., 253–254, 256–257
Welss, T., 225
Wendt, S., 158
Weng, H., 210–211
Werb, Z., 291
Weston, K. D., 354
Westrick, R. J., 128–129, 178
Whisstock, J. C., 14–15, 16–17, 55, 63–89, 65, 96–97, 142, 181–183, 210–211, 225, 226, 228, 233, 238, 240, 241–242, 243, 245, 254, 256–258, 260–262, 296, 298, 304, 336, 338–340, 341–345, 385, 391, 392, 393–394, 408–410, 412–414, 429
Whitehouse, D. B., 335–336
Whitehurst, C. E., 156–157
Whyte, G., 373
Widom, J., 40–41, 367–368, 373
Wight, D. G., 395–396, 398, 399–402
Wijeyewickrema, L. C., 257, 260–262
Wilczynska, M., 16, 182–183, 224
Williamson, D., 98
Willis, L. K., 166, 404–405
Willoughby, C. A., 260–262
Wilm, A., 253, 258

Wilm, M., 156–157
Wilson, K. S., 179
Wilson, V., 99–100
Wiman, B., 178, 187–188
Winchester, B. G., 335–336
Wind, T., 178, 181–182, 186, 189–190, 385, 391, 412–414
Winge, S., 9–10
Winn, M. D., 53–54, 56, 97
Wintrode, P. L., 113, 114, 160–161, 300–301, 331, 333, 334–335, 337–340, 341, 345–346, 347
Wishart, D. S., 151–152
Wisnewski, N., 210–211
Wolf, R. M., 311
Wolfson, H., 153
Wong, M. K., 178
Wong, W., 238, 257–258
Won, Y., 309
Woodcock, C. L., 30–31, 42
Woods, R. J., 310
Woo, J.-R., 53, 55, 65
Woolsey, J., 151–152
Woo, M. S., 65
Woo, S. L., 451–452
Worrall, D. M., 14–15, 51, 56, 110, 225
Wrigers, W., 308
Wright, H. T., 53, 64–76, 65, 81
Wrobleski, S., 178
Wu, J., 178
Wu, S., 4
Wu, Y., 116, 151, 155, 160–161, 167–168, 402–403, 404

X

Xiao, J., 151–152
Xu, D., 291
Xue, Y., 50, 65, 81–82, 381, 410
Xu, G. H., 330
Xu, J., 178

Y

Yagi, S., 128–129
Yamada, T., 128–129
Yamamoto, K., 127–128
Yamanaka, S., 459–460
Yamasaki, M., 65, 81–82, 142, 298, 326, 344–345, 385, 391, 392–394, 408–410, 411, 412–415, 424
Yanada, M., 127–128
Yang, A. S., 256–257
Yang, C. T., 210–211
Yang, D. Y., 158
Yang, E. W., 256–257
Yang, L. W., 304
Yang, M., 373
Yang, W. S., 338–340, 343
Yang, X., 156–157, 385, 392, 394–396, 412–414, 424
Yang, Z. R., 256–257
Yano, M., 9–10
Yan, Y., 52, 56–57, 58, 65, 77–78, 90–93, 95, 96–98, 100
Yap, M. G., 91–93
Yates, E. A., 117
Yayon, A., 122–123
Yee, D. P., 342
Yefimov, S., 302, 312
Yeh, B. K., 122–123
Yepes, M., 178, 181–182, 423–424
Yerby, M. S., 358–359
Ye, S., 65, 76–77, 240
Yi, M., 276–277
Ying, L., 353, 358, 364–366
Ylinenjärvi, K., 142
Young, J. L., 225
Yu, H., 312
Yu, L., 178
Yu, M. H., 20–21, 53, 55, 333, 334, 335, 336, 338–340, 341, 343, 348, 368, 385, 394–396, 402–404, 424
Yun, C. Y., 445
Yunus, S., 186, 187, 188
Yu, R., 299

Z

Zabaleta, A., 181–182
Zakrzewska, M., 296–298
Zaujec, J., 130
Závodszky, P., 51, 56, 65, 81–82, 298
Zaychikov, E., 358
Zechmeister-Machhart, M., 130
Zeng, H., 446
Zenobi, R., 158, 159
Zettlmeissl, G., 109
Zhang, F., 116
Zhang, H., 287
Zhang, J., 151–152, 178, 385, 392, 394–396, 412–414, 424
Zhang, N., 287
Zhang, Q., 65, 77, 78–79, 238, 257–258, 296, 298, 408
Zhang, T., 15
Zhang, W., 368
Zhang, X., 65
Zhang, Y., 445, 446
Zhang, Z., 276–277
Zheng, B., 65
Zheng, J., 427
Zheng, X. J., 113, 114, 160–161, 300–301, 333

Zhong, J. Q., 65
Zhou, A., 14–15, 35, 49–63, 65, 77–78, 89–106, 141, 154, 164, 181–182, 224–225, 267–268, 298, 372, 385, 392–393, 398, 405–410, 407, 411, 454
Zhou, H., 178
Zhou, Q., 65
Zhou, R. H., 306
Zhou, X., 291
Zhou, Z., 148
Zhu, J., 434–435
Ziebell, M. R., 156–157
Zou, Z., 210–211, 296
Zurini, M., 166

Subject Index

A

ACT. *See* Antichymotrypsin
AMBER. *See* Assisted model building with energy refinement
Amino-terminal oriented mass spectrometry of substrates (ATOMS)
  bioinformatics analysis, 277–279, 280
  bona fide proteolytic sites, 277–279
  cleavage sites, liquid chromatography-tandem mass spectrometry, 277
  contaminations, 279–281
  defined, neoproteins, 276–277
  in vitro, proteolytic procession, 282–283
  isotopic labeling and tryptic digestion, 283–285
  mass spectrometry data analysis
    control experiment and ratio data, 287–289
    identification, control experiment, 289–290
    proteolytic cleavage sites identification, 290–291
    TPP software, 287
  N-terminal sequencing, 276
  peptides identification, 286
  protease inhibitors, proteolytic fragments, 291
  ratio cutoff determination and identification
    defined, 281–282
    degradation products, 281–282
  workflow, 278
Antichymotrypsin (ACT), 14–15
Antithrombin (AT)
  crystal structures, 405–406
  heat-generated polymers, 412–414
  incubation, 410
  plasma, 393
  polymerization, 393–394
Antitrypsin (AT)
  preparation
    inclusion bodies, 23–24
    soluble, 24–27
  recombinant production, 14–15
  serpins, 15
α_1-Antitrypsin (α_1AT)
  cleavage, 393–394, 408–410
  crystallographic structures, 393–394
  incubation, 391, 401–402
  plasma, 383
  polymerization and depolymerization monitoring, FCS, 393
  recombinant, 391, 402
  serpinopathies
    cell cultures and DNA transfections, 441–442
    CHO-K1 Tet-On cell line expression, 442
    plasmids, construction, 441
    sandwich ELISA, 443
    screening, survival clones, 442
  soluble polymers, 380–381
  trypsin conformational distributions-α_1AT complexes, 399–401
  Z mutation, 383, 395
Assisted model building with energy refinement (AMBER)
  force fields, 310
  processing units, 313
AT. *See* Antithrombin. Antitrypsin
α_1AT and depolymerization monitoring, FCS
  diffusion time, 359–361
  heat-induced, 362–363
  molecular brightness, 361–362
  molecules diffusion, 359–361
  oligomerization detection, 359, 360
  protocol, 363
  single fluorescent species, calculated correlations, 359–361
ATOMS. *See* Amino-terminal oriented mass spectrometry of substrates

B

Bioinformatic approaches, serpin genes identification
  adaptive evolution, 219–220
  antiproteolytic repertoire, 210
  defense networks, 219–220
  gene programs, 211
  innate immunity, 210–211
  procedure, mRSL cassette exons
    exon-intron structures, 215
    genome, model organism, 215–217
    genomic sequences processing, 213
    HMMs construction, 213
    identification, 213–215
    performance, HMM, 218–219
    software programs and databases, 211, 214
    step-by-step protocol, 217–218
  reactive site loop (RSL), 210
  signature motifs, 211

C

CBG. *See* Corticosteroid-binding globulin
Chemistry at HARvard Molecular Mechanics (CHARMM)
  conjugation, 311
  force field, 309
  software package, 309
  torsion angles, 312
Chromatin immunoprecipitation (ChIP)
  analysis, 43–46
  identification, native MENT-binding sites, 42
Cleaved polymers, polymerization, 408–410
Computational approaches, targeting serpins
  docking
    binding site prediction, 151–152
    in silico docking models, 152
    internal coordinate mechanics (ICM) algorithm, 152–153
    protein side-chain flexibility, 152
  fit evaluation
    efforts, scoring functions, 153
    molecular mechanics functions, 153
  in silico approaches
    binding-site energy, 155
    ICM-based, 154
    initial screening, 155
    pharmacophore binding, 154
    potential energy grid, 154–155
    surface accessible cavity, 154
  site mapping
    binding site prediction, 151
    hydrophobic pocket targeting, 149, 150
  starting model
    cleaved/latent conformers, 149
    highest resolution structure, 148–149
    X-ray crystallographic data, 148
Corticosteroid-binding globulin (CBG), 51, 78, 90–95
Crystallographic approaches, targeting serpins
  crystaliization conditions, 163
  nanocrystalization, macromolecules, 161
  non crystallized protein
    high-throughput approach, 162
    vapor–diffusion experiment, 162
  soaking, 162–163
  stoichiometry, 161

D

Deoxynucleoprotein (DNP), 41–42
Disulfide trapping, polymerization, 412
Dithiothreitol (DTT), 4
Domain swap/β-hairpin, polymerization, 411

E

Electron microscopy (EM), serpinopathies
  biological samples, 429–430
  description, 428
  3D structural data, 428
  image collection, particle picking and data processing
    contrast transfer function (CTF), 435
    photographic film, 434
    programs, automated, 434–435
  image processing, serpin polymer
    BOXER, particle picking, 435
    schematic references, 435
    structural and aymmetric features, 435
  negative staining, 433–434
  TUG-gel electrophoresis, 429
  uranyl acetate staining, 433
  visualization, protein samples, 430–433
Electrophoretic mobility shift assays (EMSAs)
  dsDNA, 36–37
  reconstituted mononucleosomes, 41–42
EM. *See* Electron microscopy
EMSAs. *See* Electrophoretic mobility shift assays
Escherichia coli *(E coli). See* Recombinant serpins production, *E. coli*
Extracellular matrix (ECM) protein
  cell culture isolation, 281–282
  proteolytic procession, 276–277

F

FCS. *See* Fluorescence correlation spectroscopy
Fluorescence-based assays
  8-anilino-1-naphthalene sulfonate (ANS), 156
  intensity, 156
  thermofluor assay, 156
Fluorescence correlation spectroscopy (FCS)
  α_1AT polymerization and depolymerization monitoring
    diffusion time, 359–361
    molecular brightness, 361–362
    molecules diffusion, 359–361
    oligomerization detection, 359, 360
    single fluorescent species, calculated correlations, 359–361
  heat-induced, 361–362
  protocol, 363
  and TCCD utility, polymerization, 367–368
Fluorescence spectroscopy, serpin–glycosaminoglycan interactions
  affinity measurement, 108
  emission wavelength, 106–107
  energy transfer, 106–107
  kinetic studies
    AT–heparin, 113
    conformational activation pathway, 112–113
    stopped-flow fluorometer, 111–112
  light scattering, 107
  PCI, 110
  reactive center loop (RCL), 109–110
  squamous cell carcinoma antigens (SCCAs), 110
  structures, DS hexasaccharide, 108

AT thermodynamic/kinetic studies, 109
titration, heparin pentasaccharide, 106–107

G

GROMACS. *See* GROningen MAchine for Chemical Stimulations
GROningen MAchine for Chemical Stimulations (GROMACS)
chemical simulations, 309
numerical analysis, 316
polar and apolar solvents, 311–312
software package, 309

H

High-throughput and structure-based drug design, targeting serpins
cellular pathways, 141
computational approaches
choosing starting model, 148–149
docking, 151–153
drug design, 148
evaluation, fit, 153
identification, binding sites, 148
in silico approaches, 154–155
site mapping, 149–151
in vitro screening
crystallographic approaches, 161–163
fluorescence-based assays, 156
mass spectrometry, drug design, 156–161
NMR spectroscopy, 163–166
PAGE, 155
scalability, 155
ligand selection and optimization, 140–141
mammalian cell models
α1-antitrypsin deficiency, 166–167
drug effects, evaluation, 168
functional activity restoration, 167
hepa1 cell line, 167–168
neuroserpin polymerization, 166–167
serpinopathies, 168
metastable structural scaffold, 140
pathogenic variant, α1-antitrysin
benchmark peptide, 142, 143
binding affinity and specificity, 142
biomolecular interactions analysis, 147–148
combinatorial approach, 145–147
formation, pathogenic polymers, 142
library synthesis, 144
nondenaturing and urea-native PAGE, 144–145
pharmaceutical industry development, 168–169
physiological systems, 141–142
rational design, 141
High-throughput screening (HTS)
automated instruments, 185
choice, protease
inhibitory rate constants, 187, 188
Michaelis complex conversion, 188–189
plasminogen, 187–188
protease active site, 189
compound libraries
development, 189
pharmacokinetics, 189–190
splenocyte immortalization, 190
dependence, compound concentration, 190–192
enzymatic activity, 184–185
ionic strength and pH, 185
molar ratio
effect, 192, 193
hit compounds, 192–194
protease inhibition, 192
PAI-1 selection
bacterial expression systems, 187
inhibitory monoclonal antibodies, 186
reversible equilibrium, 186–187
reporter substrates
amido-4-methylcoumarin (AMC), 194
emission spectra, Rhod 110, 194
fluorescent signal recovery, 194–195
p-nitroanalide (pNA), 194
small-scale screens, 195, 196
Hormone carriers, serpins
and allosteric modulation
antithrombin, heparin pentasaccharide, 90
predicted modulatory mechanisms, 89–90, 91
angiotensinogen and interaction, renin
disulphide bond, 98
key mediator, RAS system, 96–97
peptide cleavage site, 97–98
preeclampsia, 99–100
rat and mouse, 97
redox switch, 98–99
structure, 96–97
tissue interaction, 100
hormone carriage-TBG and CBG
antithrombin, heparin bound and unbound structures, 93–94
flip-flop modulatory mechanism, 93
ligand affinity, 93
modulation, release, 94–95
protease inhibitors, 90–91
reactive loops proteolytic cleavage, 91–93
thyroxine release, binding pocket, 91–93
protein thermocouple, 95
HTS. *See* High-throughput screening
HXMS. *See* Hydrogen/deuterium exchange by mass spectrometry
Hydrogen/deuterium exchange by mass spectrometry (HXMS)
advantages and disadvantages, 331

Hydrogen/deuterium exchange by mass spectrometry (HXMS) (*cont.*)
cavity-filling mutation, α_1-antitrypsin
allosteric rigidification, 334
crystal structure, G117F, 334–335
inhibitory function *vs.* thermodynamic stability, 334
conformational mobility, 326
denaturants, 328
equilibrium unfolding, α_1-AT
CD spectroscopy, 341–342
ellipticity, 342
glycolysation, defined, 341
mass spectra, peptides, 342
EX1 exchange, 328
"functional unfolding", native → cleaved transition
β-strand 5A and β-strand 2C, 344
correlation, inhibitory efficiency and ability, 343
deutrium uptake, pattern, 344
D_2O, α_1-antitrypsin, 343
G117F mutant, equilibrium unfolding
destabilization, molten globule form, 341
inhibitory conformational change, 341
overlapping denaturation curves, 340
stabilization and inhibitory activity, 340
glycosylation, human α_1-antitrypsin
Asn residues and bi- and triantennary complex glycans, 335–336
metastable → stable transition, 336
peptides, 335–336
posttranslational modification, 335
interpretation, 327
intrinsic exchange rate, 326–327
ion mobility MS, α_1-AT
deuterium uptake, 347
β-hairpin polymerization, 344–345
IM-MS, 346–347
incubation, 345–346
monomer and polymers, 346
serpinopathies, defined, 344–345
serpin polymers, 347
NMR, 326
•OH protein footprinting, 330
plotting, 329–330
principles and procedures, 326
protein fluctations, 327–328
and radiolytic footprinting
flexibility, α_1-AT, 332–333
HPLC column, 332
mobility, 331
protocols, 331
rapid deuterium uptake, 333
thermodynamic stability
denaturants, 337
protein flexibility, 336–337
WT α_1-AT, pulse-labeling technique
CD spectroscopy, 338–340
denaturation curves, peptide fragment, 339
GuHCl concentration, 337–338
isotope and peptide curve, 338
protease inhibition, 338–340

I

Immunochemical methods, serpin–glycosaminoglycan interactions
antithrombin
blood coagulation, 128
prenatal development, 127–128
subendothelial basement membrane, 128
heparin cofactor II
expected Mendelian frequency, 128–129
thrombotic imbalance, 129–130
in silico methods
combinatorial virtual library screening (CVLS), 123–124
genetic algorithm-based docking, 123
GOLD, 124
library screening, 124–125
modeled geometries, 122–123
putative binding geometry, 125–127
mouse models, 127–131
PCI, 130–131
specificity, antibodies, 121–122

L

LED. *See* Local elevation dynamics
Local elevation dynamics (LED)
algorithms and protocols, 310
description, 306–307
Loop-sheet mechanism, polymerization, 410

M

MARTINI, 312
Mass spectrometry, drug design
automation, 160
electrospray ionization (ESI) technique, 156–157
hydroxyl-mediated protein footprinting, 114
ligand binding characterization
gas-phase complexes, 158
nano-ESI mass spectra, 158
mass-to-charge (m/z) ratio, 156
noncovalent assemblies
gel filtration columns, 157
kinetic energy, 157–158
peptide fragments, 160–161
solution based titration protocol
data analysis, 159
gas-phase methods, 160
optimization, instrumental parameters, 159
sample preparation, 159
titration, infusion and ionization, 159

structural and dynamic properties, 113
Monte Carlo simulations, 302–303
Myeloid and erythroid nuclear termination (MENT). *See* Nuclear serpin MENT isolation and characterization

N

Neuroserpin polymerization measurement, TCCD
application, 366
association quotient, 364–366
protocol, 366–367
Neuroserpin, serpinopathies
cell lines *vs.* polymerogenic serpin mutants, 444
COS-7 cells, transient transfection, 443–444
NMA. *See* Normal mode analysis
NMR spectroscopy
ligand binding
compound screening, 165
paramagnetic relaxation enhancement (PRE), 166
saturation transfer difference (STD), 166
structure–activity relationships (SAR), 166
preparation, isotopically labeled α1-antitrypsin, 164
protein structure and dynamics, 163–164
site directed mutagenesis, 164
spectra, serpins, 164–165
Normal mode analysis (NMA)
computational tools, 309
description, 303
theoretical limitations, 303
Not (just) Another Molecular Dynamics (NAMD) program
force fields, 310
parallelization, 310
software package, 309
Nuclear serpin MENT isolation and characterization
analysis, native chromatin *in situ*
avian erythrocytes, lymphocytes and granulocytes, 42
ChIP, nuclear DNA, 43–46
and fractionation, chicken blood cells, 42–43
chromatin-associated cysteine proteinase inhibitor, 31
chromatin association assays
and fractionation, chicken erythrocyte nuclei, 37
oligomeric suprastructures, 40
soluble preparation, chicken erythrocytes, 38–39
sucrose gradient purification, soluble, 39–40
deoxynucleoprotein electrophoresis, reconstituted nucleosomes
DNA positioning and chicken erythrocyte core histones, 40–41
DNase I protection experiments, trimers, 42
DNP, monomers, 41–42
monomers, 213-bp-long DNA templates, 41
trimers preparation, 639-bp-long DNA templates, 41
EMSAs, dsDNA
agarose gels, 36
preparation, 36
reactions, 37
eukaryotes, 30
"M-loop", 31
protease inhibition/serpin activity
intrinsic tryptophan fluorescence, conformational change assess, 35–36
kinetic parameters determination, 34–35
native acid PAGE, 35
purification, protein
cell nuclei, white blood cells, 32
chicken blood collection and processing, 31–32
and expression, recombinant, 33–34
white blood cell nuclei, 33
tertiary chromatin structures, 30–31

O

Oligomerization detection, 359, 360
OPR detection
Ca^{2+} flux measurements, 447
luciferase assay, NFkB ATF6 signaling, 447–448
Optimized potentials for liquid simulations (OPLS), 312

P

PAI-1. *See* Plasminogen activator inhibitor-1
PAI-1 inhibitors mechanism
aminoterminal sequencing, 199
antiproteolytic activity, 198
biological and biochemical properties, 199
potential drugs, 198
surface plasmon resonance, 198–199
X-ray crystallography, 199
Pathogenic variant, α1-antitrysin
combinatorial approach
molecular diversity, 145
split-and-mix method, 145–147
nondenaturing and urea-native PAGE
conformation-sensitive assay, 145
readout, polymerization, 144
surface plasmon resonance
chip-based SPR, 147–148
library screening, 147
Phage display methods, serpin–protease interactions
application
analysis, recombinant sequences, 244

Phage display methods, serpin–protease interactions (*cont.*)
biopanning cycle, 244, 245
T7 bacteriophage, 243–244
data quality, biopanning
ClustalW, 253
hexahistidine tag, 252–253
PCR amplification, 252
execution steps, 242–243
library generation
cloning kit, 246
codons, nonamer sequence, 246
sense and antisense oligo, 247
screening, substrate phage library
biopanning procedure, 249–250
recombinant proteins, 248
sequence analysis, 251–252
sublibrary enrichment, 250–251
substrate design
hexahistidine affinity tag, 244
recombinant phage, 244–245
synonymous codons, 254
Plasminogen activator inhibitor-1 (PAI-1)
development
efforts before screens, 184
HTS, 184–197
validation, 197
work-up of mechanism, action, 198–199
drug targets
biochemistry, canonical inhibitors, 179
biologic functions, 180–182
physiologic ligand-binding, 182–183
uniqueness, serpin fold, 179–180
effects, pathological processes, 200
modulating fibrinolysis, 178
physiologic processes, 178
serpin fold, 78
structure, 77, 81–82
Prediction of protease specificity (PoPS) program
artificial anchoring, 255
human granzyme B, 256
specificity model, 254
Protein C inhibitor (PCI)
differential organ expression profiles, 130
disseminated intravascular coagulopathy, 130–131
neomycin resistance gene, 130
Protein Z (PZ), 77–78
PZ. *See* Protein Z

R

RCLs. *See* Reactive center loops
Reactive center loops (RCLs)
cleavage, N-terminal portion, 408–410
cleave serpin, 2, 6–7, 9
insertion, 76, 79, 80
and loop-sheet model, 412
MENT, 31
mutations, 414–415
nicking, 393–394
proteolysis, 9
RCL-cleaved serpins, 402
sequence, 40–41
serpin, 64–77
Recombinant serpins intracellular production, yeast
assessing activity and removing inactive forms
protease, 9
susceptibility advantage, RCL, 9
eukaryotes, 2
growth, 3
large-scale growth and induction
aeration, 5
batch-based culturing, scalability, 6
lysis, 6–7
polymerogenic serpins production
serpinopathies, 10
Z AAT disease, 10, 11–12
protein, 1–2
purification
elution points, proteins, 7–9
N-terminal hexahistidine tags, 7
screening transformants, 4–5
strain and expression plasmid selection, 2–3
transformation, 4
Recombinant serpins production, *E. coli*
expression
prokaryotic and eukaryotic, 14
proteins, 14
host strain, expression, 16
improvement, soluble expression, 16–17
insoluble expression, AT, 20–21
lysis, 18
nonnatural amino acid incorporation, 22–23
AT preparation
inclusion bodies, 23–24
soluble, 24–27
recovering overexpressed protein, soluble *vs.* insoluble purification
ACT elution profile, phenyl sepharose column, 19, 20
cell pellet, 18
AT elution profile, Q sepharose column, 19
serpin structure, 18–19
strategies, refolding proteins, 18
soluble expression, AT, 21
typical approach, purification, 17
uniform isotopic labeling, 21–22
vector choice and gene construct
mRNA, 14–15
N-terminal hexahistidine tag, 15
soluble *vs.* insoluble expression systems, AT, 15
REMD. *See* Replica exchange molecular dynamics

Renin–angiotensin system (RAS), 96–97
Replica exchange molecular dynamics (REMD)
advantages, 306
defined, 306
scale energy barriers, 306

S

SDS-PAGE. *See* Sodium dodecylsulfate-polyacrylamide gel electrophoresis
Sequence analysis methods, serpin–protease interactions
bioinformatics approaches
empirical scoring function, 256–257
in silico discovery, 256
machine learning-based, 257
construction, granzyme B
procedures, PoPS model, 262–264
substrate specificity models, 262
inhibitory likelihood
cleavage score, PoPS, 268, 269
human granzyme substrates B, 267
molecules function, 269
P7-P6 sequences, 263, 267–268
proteome wide search
amino acid sequence alignment, 258
human serpins, 258, 259
PoPS program, 262
RCL identification, 257–258
Serpin conformational change and structural plasticity
case study, 314–316
cost ratio, CPU and disk storage, 316
directed simulations
random fluctuations, 307
slow progress forward, 307–308
SMD and TMD, 308
dynamic stimulations, local to global
coarse-grained, 302
MD, 300–302
Monte Carlo, 302–303
NMA, 303–304
secondary structure elements, 299–300
force fields
bonded *vs.* nonbonded atoms, 311
parameters, 311–312
PDB files, 3D structure, 311
SPCs, 312
S-to-R transition, 312
hardware
computational resources, 312–313
multicore architectures, 313
powerPC processors, 313
superposition, initial structures, 315
"video card", defined, 313
inhibitory and noninhibitory proteins, 316
"lock and key" mechanism, 296–298
nondynamical methods, 308
sampling
high-temperature simulations, 305
LED, 306–307
REMD, 306
search problem, defined, 304
serpin native fold, 296
software
AMBER, 310
CHARMM, 309
GROMACS, 309
NAMD, 310
visualization, 310
theoretical methods, 299
theoretical methods and algorithms, 316, 317
Serpin conformational distributions, SMF
encapsulation, 373
FCS, TCCD and spFRET, 353
polymerization (*See* Serpin polymerization)
protease-serpin complexes
protocol, spFRET, 372–373
SpFRET, 368–370
trypsin-α_1AT, 370–372
and proteases, fluorophores
proteins labeling, maleimide, 354–355
solvent-accessible Cys, 354
protein sample, 353
structural remodeling, 353
super-resolution methods, 373
techniques, 355–358
Serpin crystal structures
conformational states, serpin
α_1-antitrypsin polymer, 54
RCL, 54
crystallization
antithrombin, 52
method, 52
proteins, 51–52
electron density model, 56–57
ensembles usages, multiple models
maspin, 56
program *Phaser*, 56
experimental phasing
cleaved α_1-antitrypsin, 53
diffraction, 52
molecular replacement method, 53
homology models
molecular replacement, 57
neuroserpin, 57
Rosetta, 57
modifications, aid crystallization
glycosylation, 51
maspin, 51
PAI-1, 51
orientation and position, model, 53–54
phase improvement, density modification
angiotensinogen, 58
solvent flattening and copies average, 57–58
Protein Data Bank, 53

Serpin crystal structures (*cont.*)
protein production and purification
eukaryotic, 50
His-tags addition, 50–51
refinement and validation, 58
rigid-body movements
cartoon diagram, native α_1-antitrypsin, 54–55
structural database, 55
subdomains, 55–56
technique, molecular replacement, 53
Serpin genes, mRSL cassette exons
databases and software programs, 211, 212
genome, model organism
A. pisum serpin genes, 215, 216
endoplasmic retrieval signal, 215–217
target specificity determination, 217
logos, profile HMM, 213, 214
step-by-step protocol
HMM construction, 217–218
serpin genes identification, 218
translated genomic sequences, 213–215
Serpin–glycosaminoglycan interactions
qualitative methods
affinity chromatography, 121
electromobility shift assays, 121
immunochemical methods, 121–122
quantitative methods
affinity co-electrophoresis, 114
circular dichroism (CD), 117
fluorescence spectroscopy, 106–110
isothermal titration calorimetry, 116–117
kinetic studies, 111–113
mass spectrometry, 113–114
NMR, 117–118
solid-phase binding, 114–115
surface plasmon resonance, 116
X-ray crystallography, 118–121
Serpinopathies
assessment, EM (*See* Electron microscopy (EM), serpinopathies)
characterization, α_1-antitrypsin and neuroserpin, 424
description, 423
in vivo, biophysical techniques
monomers and polymers, 425–427
quantitation, glycosylation and proteolutic digestion, 425
stable, native and polymeric states, 427
iPS, hepatocyte-cell lines generation
derivation and maintenance, human, 459–460
dermal fibroblasts, patient, 457–459
description, 457
ER entrapment, 457
hIPSCs characterization, 460–461
validation, 461
mAbs, α_1-antitrypsin and neuroserpin
ELISA assays, formats, 439–440
immunocytochemistry, 440
nondenaturion, PAGE, 440
polymer detection, 2C1 monoclonal antibody, 437
types, 436–439
OPR detection
Ca^{2+} flux measurements, 447
luciferase assay, NFkB ATF6 signaling, 447–448
relationship, 423–424
serpin polymers *vs.* neutrophils
functional assays, 449–451
preparation, 448–449
transgenic mice, polymerization
hepatic consequence, 451–453
pulmonary consequence, 453–456
UPR detection
lusiferase assay, ATF6 signaling, 446–447
RT-PCR and XBP1 splicing assay, 445–446
western blot, 446
Serpin polymerization *in vitro*
biochemical methods
disulfide trapping, 412
limited proteolysis, 412–414
loop-sheet and β-hairpin mechanism, 411
mutations, RCL, 414–415
folding process, 408
inducing methods
α1AT, 383
chemical modification, 393
denaturants, 391
heat, 383–391
low pH, 392–393
proteolytic cleavage and short peptides, 393–394
refolding, 394–395
selection, techniques, 383, 385
translation, 395
kinetic scheme and features, 395–396
“loop-sheet” hypothesis, 381–383
monomer loss
activity, 396–398
native PAGE, 398
size-exclusion chromatography, 399
mutations effect and drugs
and cavity-filling, 404
destabilizing and stabilizing, 403–404
inhibition, peptides, 405–406
inhibition, small molecules, 404–405
loop-sheet mechanism, 402–403
reversal, peptides, 406–408
Z-α1AT, 402–403
polymerigenic mutation, 416
ribbon diagrams, native and β-sheet A-expanded forms, 380–381
spectroscopic methods

circular dichroism, 401–402
extrinsic fluorescence, 401
intrinsic fluorescence, 399–401
light scattering, 402
stable intermolecular polymer linkage, 381–383
structures and models
cleaved polymers, 408–410
domain swap/β-hairpin, 411
loop-sheet mechanism, 410
Serpin polymers *vs.* neutrophils, serpinopathies
defined, 448
functional assays
chemotaxis, 450–451
myeloperoxidase, 451
shape, change, 450
LPS, 448
preparation, 448–449
Serpin–protease interactions
development, inhibitors, 269–270
phage display methods
application, 243–244
biopanning, 252–253
execution steps, 242–243
PoPS, 254–256
protocols and requirements, 244–248
screening, substrate library, 248–252
specificity, granzyme B, 253–254
sequence analysis methods
bioinformatics approaches, 256–257
construction, granzyme B, 262–266
RCL identification, 257–262
theoretical P1 site, 253–254
structure and function
hemophilia, 240
inhibitory pathway, 238–240
kinetic nature, 241
macromolecular cofactors, 240–241
reversible Michaelis complex, 238–240
substrate specificity
biopanning and analysis, 242
human granzyme B, 242
phage display, 242
potent cytotoxin, 241–242
site-specific proteolysis, 241
Serpins crystallography and complexes
abnormal conformational change, δ-form, 79
conformational change and formation
PAI-1, 78
tengpin structure, 78–79
and crystallization, 81–82
first glimpses, structures, 64
Michaelis complex, 76–77
polymers
alpha-1 disease, 79–80
antitrypsin, 80
domain-swapped dimer, 80, 81
PZ and ZPI, 77–78
serpin-enzyme
conformations, molecules, 64–76
serpin-protease, 76
serpins-antithrombin and heparin, 77
structures, serpins, 64, 65
TBG and CBG, 78
Serpins, drug targets
biologic functions
hormone-binding serpins, 180
PAI-1 function, 181–182
reversible hormone binding, CBG, 182
physiologic ligand-binding
modulation, binding sites, 183
native antithrombin, 182–183
protease inhibition, 179
uniqueness
devastating pathobiological consequences, 179–180
reactive center loop (RCL), 179–180
Simple point charges (SPCs)
description, 312
dielectric permittivity constant, 316
water representation, 315
Single molecule fluorescence (SMF) techniques
applications, biological problems, 355–356
concentration and volume limitations, 356
instrumentation
dichroic mirror, 358
inverted microscope, 357–358
TCCD and spFRET experiments, 358
one-photon microscope design, 355–356, 357
Single pair Förster resonance energy transfer (spFRET)
dipole-dipole interactions, 368–369
energy transfer efficiency, 369–370
protocol, protease-serpin complexes, 372–373
reported R_0 values, donor-acceptor pairs, 369, 370
SMF instrumentation, 357–358
Sodium dodecylsulfate-polyacrylamide gel electrophoresis (SDS-PAGE)
N-terminal sequencing, 276
proteolytic fragments, 282
PVDF membrane, 283
Solid-phase binding
hydroxyl-mediated protein footprinting, 114–115
serpin-ligand interactions, 115
Spectroscopic methods, polymerization
circular dichroism, 401–402
extrinsic fluorescence, 401
intrinsic fluorescence
equation rates, exponential terms, 399–401
partial unfolding and kinetics monitoring, 399–401
light scattering, 402
Steered molecular dynamics (SMD)
algorithms and protocols, 310

Steered molecular dynamics (SMD) (*cont.*)
atomic force microscopy, 308
biasing potential, 308

T

Targeted molecular dynamics (TMD)
algorithms and protocols, 310
description, 308
umbrella sampling, 308
TBG. *See* Thyroxine binding globulin
TCCD. *See* Two-color coincidence detection
Thyroxine binding globulin (TBG)
and CBG, 90–95
S-to-R change, 95
TMD. *See* Targeted molecular dynamics
TPP. *See* Trans-proteomic pipeline
Trans-proteomic pipeline (TPP), 287
Two-color coincidence detection (TCCD)
events, neuroserpin polymerization
amyloid formation, 364
application, 366
association quotient, 364–366
α_1AT, 364
protocol, 366–367
and FCS utility, polymerization, 367–368

X

X-ray crystallography
global conformational changes, 118–120
HCII inhibition, thrombin, 118–120

Y

Yeast. *See* Recombinant serpins intracellular production, yeast

Z

Z α_1-antitrypsin (Z AAT) disease
mechanisms, 11–12
recombinant production, 10
Z-dependent proteinase inhibitor (ZPI), 77–78
ZPI. *See* Z-dependent proteinase inhibitor

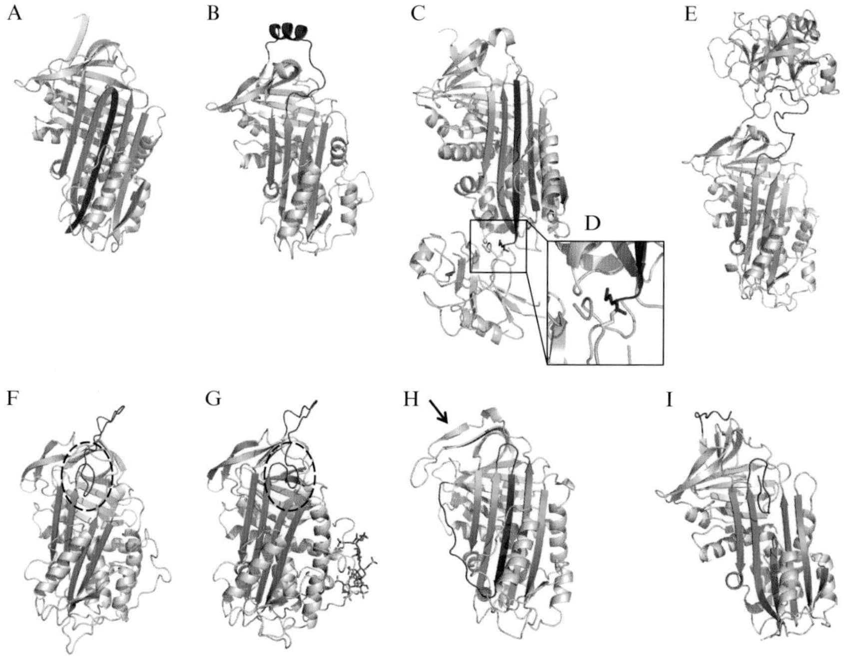

M. A. Dunstone and James C. Whisstock, Figure 5.1 The different conformations that can be adopted by serpin molecules. In all structures the A β-sheet is colored as blue, the reactive center loop (RCL) is colored red, and complexed proteases are colored green. (A) Cleaved and inserted structure of antitrypsin (PDB ID: 7API). (B) The uncleaved structure of ovalbumin (PDB ID: 1OVA). (C) The structure of antitrypsin in complex with trypsin trapped in the acyl-enzyme intermediate state showing distortion of the catalytic triad (PDB ID: 1EZX). (D) Inset of the antitrypsin–trypsin complex showing the bond between the catalytic triad Ser195 side chain (yellow sticks) and the P1-Arginine of the serpin (red sticks). (E) The Michaelis serpin–protease complex where the protease is noncovalently bound to the RCL (PDB ID: 1K9O). (F) The structure of native antithrombin shows partial insertion of the RCL into the A-sheet (dashed circle) (PDB ID: 2ANT). (G) In the presence of bound heparin (purple sticks), the RCL of antithrombin is expelled out of the top of the A-sheet (dashed circle) (PDB ID: 1AZX). (H) The latent conformation of PAI-1 where the RCL is fully inserted into the A-sheet. This is achieved by peeling away the outside β-strand of the C-sheet (C-sheet shown with arrow) (PDB ID: 1C5G). (I) The δ-conformation of antichymotrypsin where there is partial insertion of the RCL into the top of the A-sheet and insertion of the region from the last turn of the F-helix and subsequent loop into the bottom of the A-sheet (colored orange) (PDB ID: 1QMN).

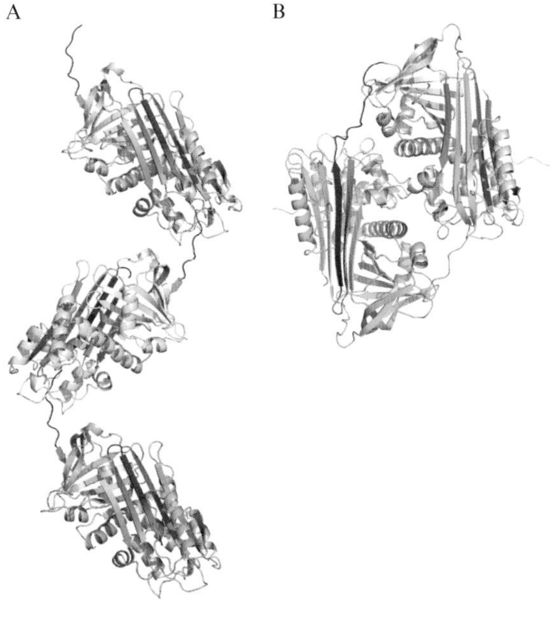

M. A. Dunstone and James C. Whisstock, Figure 5.2 Polymerization of serpins. Colors as for Fig. 5.1. (A) Cleaved polymer: The structure of the cleaved antitrypsin polymer (PDB ID: 1D5S). (B) Domain-swapped dimer: The structure of antithrombin dimer showing swapping of both the RCL (red) and the fifth β-strand (blue) of one antithrombin molecule into the A-sheet of the neighboring antithrombin molecule (green) (PDB ID: 2ZNH).

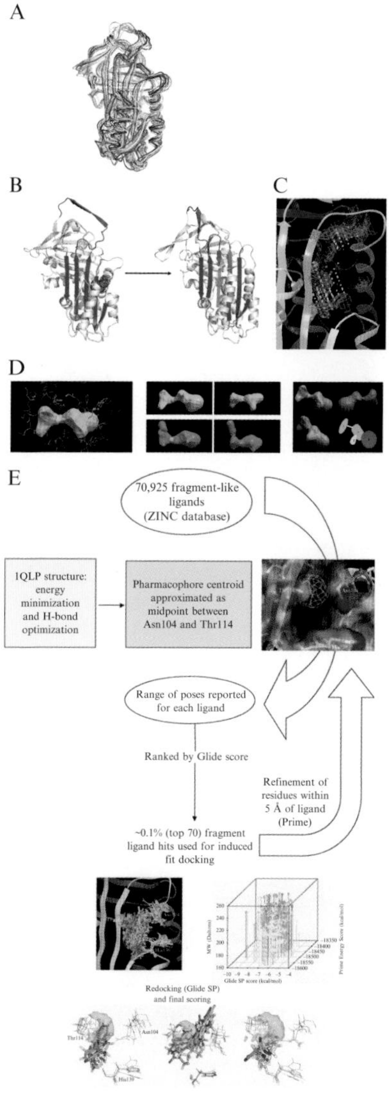

Yi-Pin Chang *et al.*, Figure 8.3 (A) Computationally generated, native-like conformers of α_1-antitrypsin. A hundred conformers were generated using the program CONCOORD. A representative sample of 8 (from 100) conformers is shown. All 100 structures were screened for persistence of binding sites observed in crystallographic structures. (B) Screening targets identified through interpretation of prior crystallographic and biochemical data. The s4A site (see Section 2) believed to be patent in the polymerogenic intermediate of α_1-antitrypsin (right) is highlighted in orange. The hydrophobic pocket flanking β-sheet A is shown on the native conformer (1QLP) as identified by SURFNET (purple mesh). As this cavity is abolished by expansion of β-sheet A, compounds binding here should prevent polymerization.

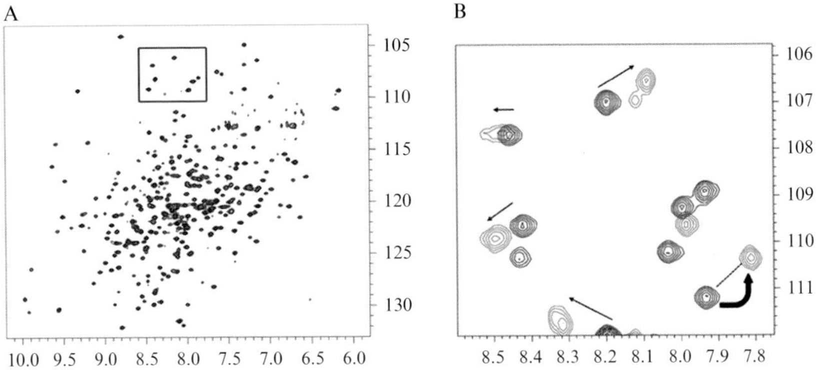

Yi-Pin Chang *et al.*, Figure 8.5 (A) [^{1}H–^{15}N]-TROSY-HSQC spectrum of native α_1-antitrypsin at 25 °C (Nyon *et al.*, manuscript submitted). The ^{1}H chemical shift (ppm) data are plotted on the *x*-axis, ^{15}N chemical shift data (ppm) on the *y*-axis. (B) Changes seen in one part of the spectrum (boxed in A) upon ligand binding. Comparison of [^{1}H–^{15}N]-TROSY-HSQC spectra (at 37 °C) of native α_1-antitrypsin (black) and α_1-antitrypsin when saturated with ligand (red) shows changes in chemical shift (straight arrows). Two possible patterns of chemical shift change are indicated in the lower right corner. Gradual migration of cross-peaks from the apo- to ligand-saturated states (e.g., along the path of the dashed line) is the characteristic pattern of chemical shift change for weak binding with rapid conformational exchange. Discrete jumps in cross-peaks (indicated by curved arrow) indicate tight binding with very slow rates of conformational exchange between the bound and apo-state.

(C) Hydrophobic pocket flanking β-sheet A as identified using SiteMap. Each site is highlighted using white spheres (i.e., site points), around which hydrophobic (yellow grid), hydrogen acceptor (red grid), and hydrogen donor (blue grid) maps are highlighted. (D) The same site shown in (B) and (C), as identified by PocketFinder (following relaxation of Asn104). Changes in pocket dimensions are demonstrated by comparison of volumes pre- (blue, PocketFinder only) and post- (green) ligand-induced changes modeled using the program SCARE. Division of this cavity into subvolumes defined chemically and computationally is shown (right). (E) Upper panel; overview of process in which a cavity subvolume (shown as blue mesh within the pocket flanking β-sheet A) was identified by comparison of the high-resolution crystal structures of Thr114Phe (PDB code: 3DRM) and wildtype (1QLP) α_1-antitrypsin. This was then used as the target for a computational screen of >70,000 fragment-like compounds. Asn104, Thr114, and His139 were used as coordinating residues. Lower panel; The best hits (ensemble top left) were classified (top right) by two docking scores, molecular weight and proximity to target centroid (heatmap coloring $\geq$5.0 Å, yellow; 2.5–5.0 Å, orange; $\leq$2.5 Å, red). Three ensembles of five fragments that scored highly by different criteria are shown (lower row, target in cyan, coordinating α_1-antitrypsin residues shown as per induced-fit, color coded with appropriate ligand).

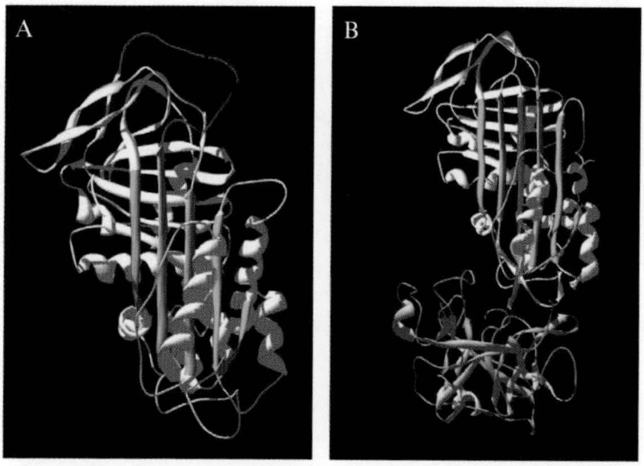

Shih-Hon Li and Daniel A. Lawrence, Figure 9.1 Native and protease-bound serpin structures. An archetypal inhibitory serpin shown in its (A) active conformation (PDB accession 1QLP) (Elliott *et al.*, 2000) and (B) in complex with a cognate protease (PDB accession 2D26) (Dementiev *et al.*, 2006). The RCL (red) serves as a substrate loop and inserts into the central β-sheet (green) upon acylation by a protease (light blue), stabilizing the covalent protease–serpin complex.

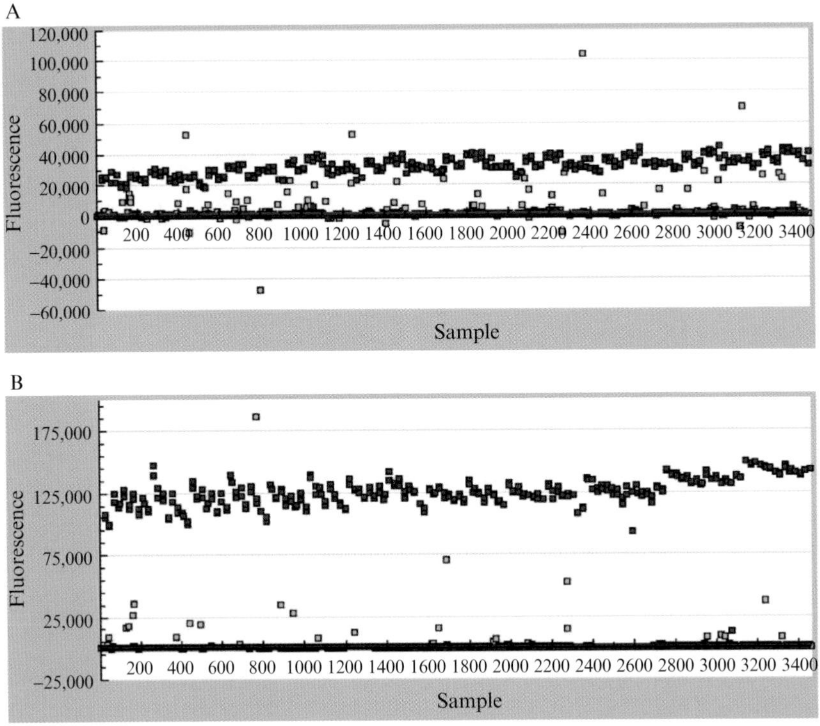

Shih-Hon Li and Daniel A. Lawrence, Figure 9.5 Two small-scale screens. The MicroSource SPECTRUM and the BioFocus NIH Clinical collections were screened under different conditions. (A) PAI-1 at 10 n*M* was mixed with 10 μ*M* of each compound, before the addition of 5 n*M* uPA followed by 50 μ*M* zGly-Gly-Arg-AMC. (B) Alternatively, 10 n*M* PAI-1 was mixed with 10 μ*M* of each compound before the addition of 1 n*M* trypsin followed by 2.5 μ*M* (CBZ-Ala-Arg)$_2$-Rhod110. In each case, HTS was performed in 40 m*M* HEPES, pH 7.8, 100 m*M* NaCl, and 0.005% Tween-20 at room temperature. Reactions containing PAI-1 alone are in *blue* while those containing uPA and substrate are in *red*. Mixtures of PAI-1 with compound and uPA are shown in *green*.

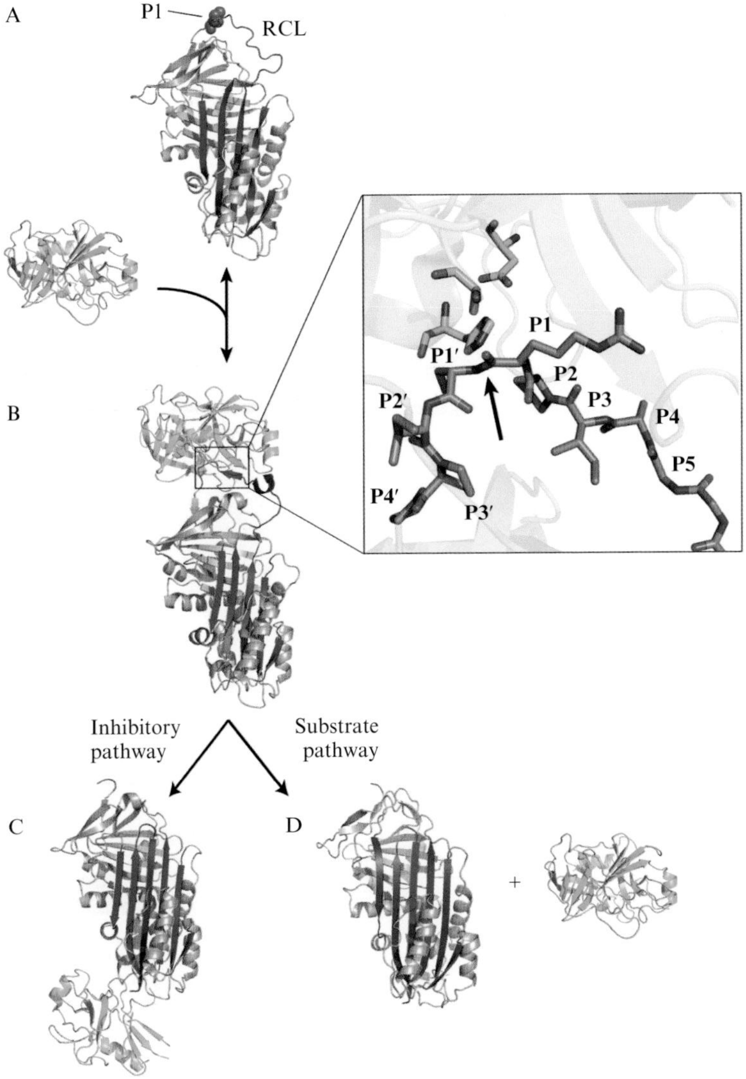

Jiangning Song *et al.*, Figure 12.1 Serpin structural forms and inhibitory mechanism. The serpin β-sheet A is in red, the RCL in blue. The remaining of the serpin body is in gray; the target protease is in green. (A) Crystal structure of the serpin α-1-antitrypsin in the native state. Position of the P1 residue is indicated. (B) Reversible Michaelis complex. The boxed insert illustrates RCL/protease interactions prior to cleavage of the P1–P1′ bond (indicated by an arrow) and formation of the acyl-enzyme intermediate. Residues of the RCL are colored blue and numbered according to the nomenclature of Schechter and Berger (1967). Trypsin catalytic triad is displayed in green. (C) Inhibitory complex formed following the inhibitory pathway. (D) Serpin cleaved state and release of the protease, following the substrate pathway. PDB (Berman *et al.*, 2000) accession codes: (a) α-1-antitrypsin: 1ATU, trypsin: 1DP0; (b) cartoon and boxed representations: 1OPH; (c) 1EZX; (d) 2ACH.

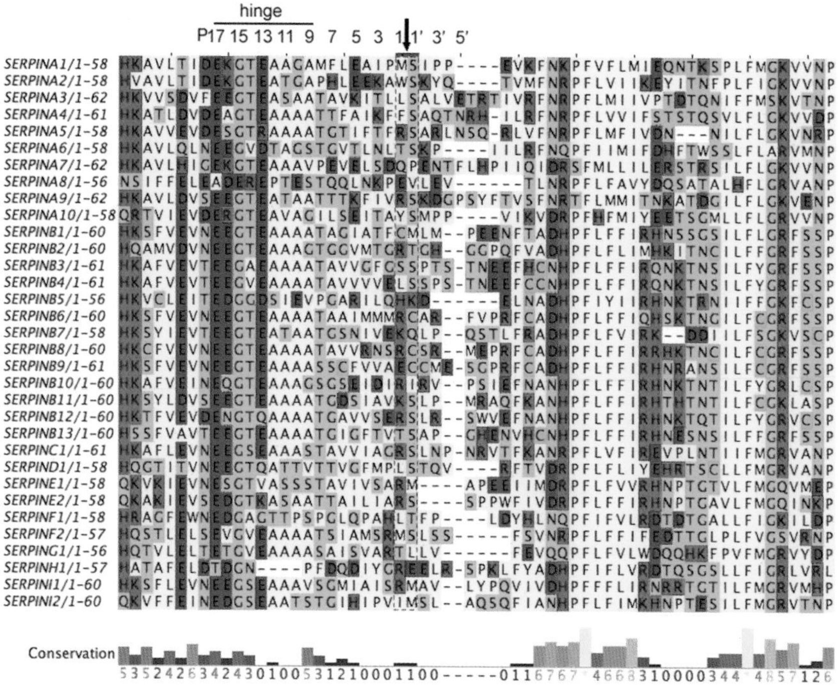

Jiangning Song *et al.*, Figure 12.6 Amino acid sequence alignment of the RCL sequences of the 33 human serpins. Amino acid sequences were aligned using ClustalW 2.0 (Larkin *et al.*, 2007) and displayed using Jalview 2.4 (Waterhouse *et al.*, 2009). Using the nomenclature of Schechter and Berger (1967), the RCL sequences are numbered from P17 to P5′, which are considered to be protease specificity determining. The hinge region in the RCL is underlined. The scissile bond between P1 and P1′ sites is marked by a black arrow. Residues are colored according to their side chain chemical properties: Glu + Asp (negatively charged; red), Asn + Gln + Gly + Ser + Thr (polar uncharged; green), Ala + Leu + Ile + Val + Met + Phe + Pro + Trp + Tyr (nonpolar; yellow), Arg + Lys + His (positively charged; blue), Cys (light blue). The conservation, quality, and consensus scores, as three of the automatically quantitative alignment annotations, are also shown.

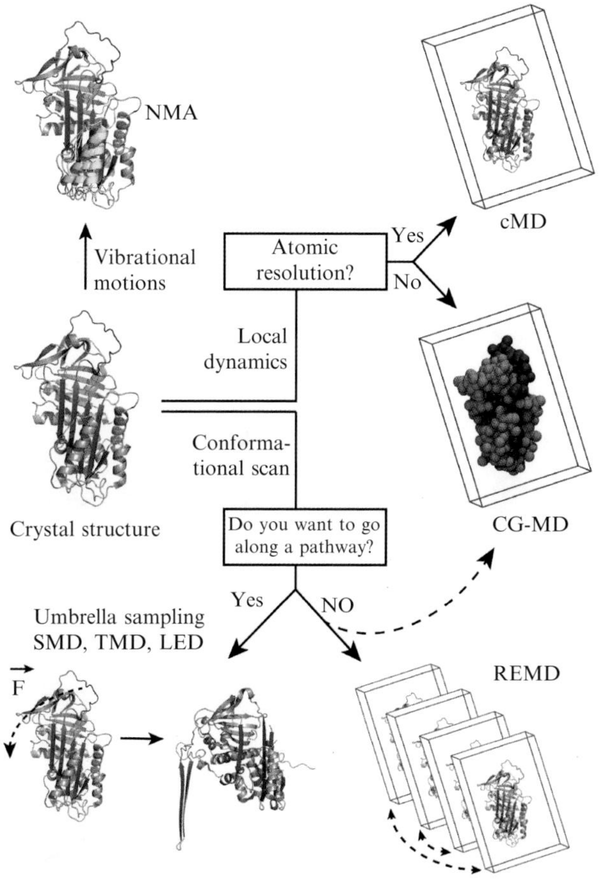

Itamar Kass *et al.*, Figure 14.3 Schematic flowchart of the various methods presented in this chapter. See text for abbreviations.

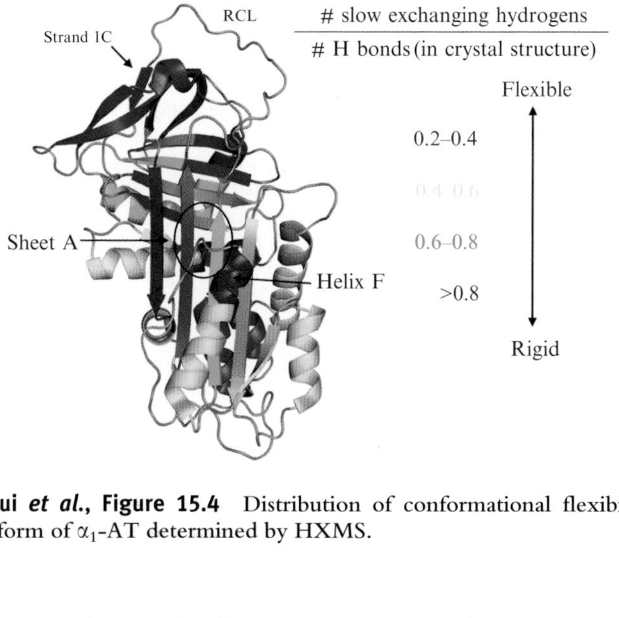

Yuko Tsutsui *et al.*, Figure 15.4 Distribution of conformational flexibility in the metastable form of α_1-AT determined by HXMS.

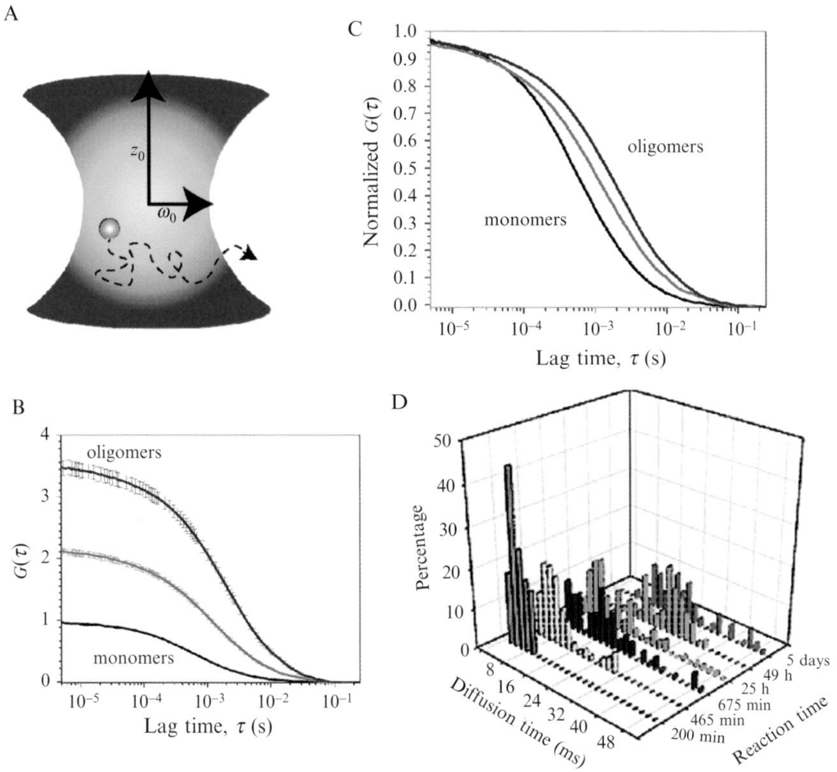

Nicole Mushero and Anne Gershenson, Figure 16.2 Detecting oligomerization with FCS. (A) Schematic of a molecule diffusing through the observation volume, assumed to be a Gaussian volume with *x–y* radius ω_0 and axial extent z_0. (B, C) FCS polymerization data for α_1AT incubated at 45 °C for 0 h (black), 24 h (blue), and 48 h (red). (B) As polymerization proceeds, the number of independent proteins decreases, increasing the amplitude of the correlation curves. (C) Polymerization also increases the average size of the molecular assemblies, increasing the diffusion time and shifting the correlations to longer time as shown for the normalized correlation functions. (D) As the α_1AT polymerization reaction proceeds, the distribution of diffusion times, τ_D, shifts to longer times. These data, adapted with permission from Purkayastha *et al.* (2005) copyright 2005 American Chemical Society, are for fits to Eq. (16.6) rather than Eq. (16.7) so the percentage of large polymers may be overestimated (see Section 4.1.1).

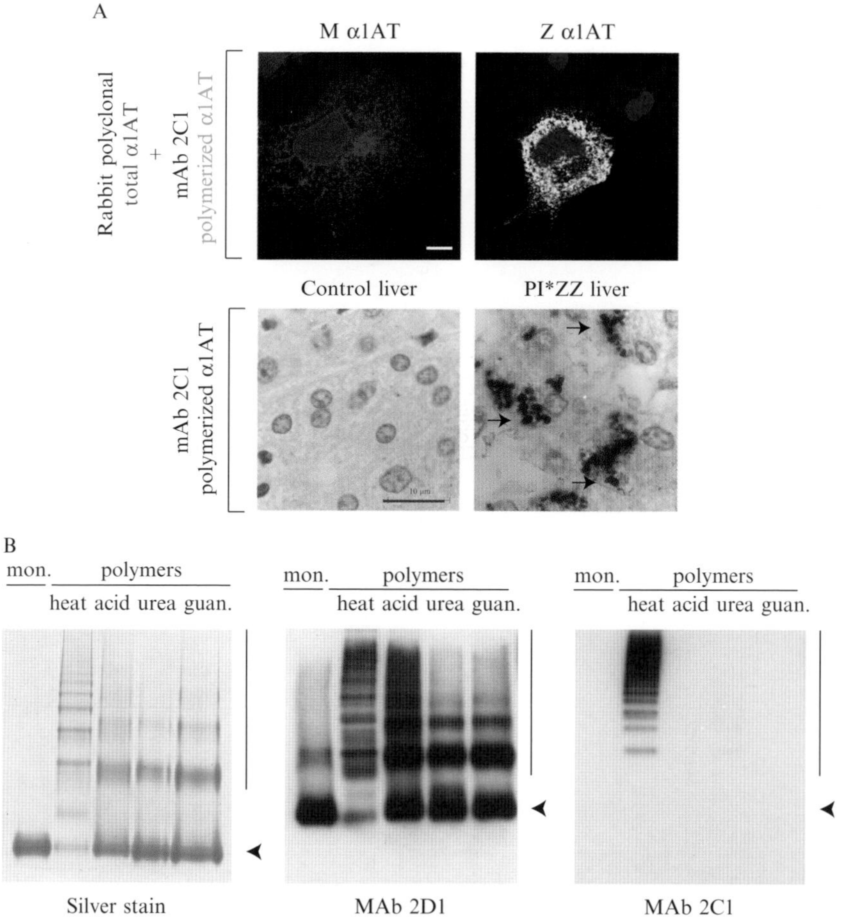

James A. Irving *et al.*, Figure 18.5 Detection of polymers with the 2C1 monoclonal antibody. The mAb 2C1 detects polymers of α_1-antitrypsin in immunocytochemistry. In (A) and (B), COS-7 cells transiently transfected with M or Z α_1-antitrypsin (a1AT) were fixed and immunostained with mAb 2C1 (green) and with a polyclonal antibody that recognizes all forms of α_1-antitrypsin (red). Only the merged panels are shown, with overlapping signals in yellow. The DNA is stained blue (DAPI). The polyclonal antibody detected α_1-antitrypsin in cells expressing the M and Z protein (both panels), while mAb 2C1 only reacted with Z α_1-antitrypsin contained in perinuclear inclusions (green and yellow staining in the right panel). In the lower panels, immunostaining of paraffin embedded liver sections with mAb 2C1. The mAb-detected polymers of α_1-antitrypsin within hepatocytes of a PI*ZZ individual (right panel, black arrows) but gave no signal in a control liver (left panel). Scale bars: 10 μm. (B) MAb 2C1 recognizes polymers of M α_1-antitrypsin prepared by heating but not by other denaturing conditions. Equal amounts of total protein were loaded in nondenaturing PAGE and analyzed either by silver stain (left panel) or by Western blot, either with mAb 2D1 (middle panel) or with mAb 2C1 (right panel) in the same membrane. The 2C1 mAb only recognized polymers prepared by heating of M α_1-antitrypsin. It gave no signal in polymers prepared by treating M α_1-antitrypsin at low pH (4.5) or 3 *M* guanidine (guan.), and only a low signal was detected with longer exposures of polymers of M α_1-antitrypsin prepared by incubation with 4 *M* urea (results not shown). Vertical line: polymers; black arrowhead: monomers. Reproduced from Miranda *et al.* (2010) and Ekeowa *et al.* (2010).